Joachim Gottsmann · Jürgen Neuberg
Bettina Scheu
Editors

Volcanic Unrest

From Science to Society

Editors
Joachim Gottsmann
School of Earth Sciences
University of Bristol
Bristol, Bath and North East Somerset
UK

Bettina Scheu
Ludwig Maximilian University of Munich
Munich, Bayern
Germany

Jürgen Neuberg
Faculty of Environment
University of Leeds
Leeds, UK

https://doi.org/10.1007/978-3-319-58412-6

Preface

Volcanic unrest is a manifestation of complex subsurface processes which might or might not herald an imminent volcanic eruption. Short-term (hours to few weeks) unrest activity before an eruption may in hindsight be interpreted as "pre-eruptive". Protracted periods or waning of unrest without a clear relationship between unrest and future eruptive activity may be deemed "non-eruptive".

Periods of unrest challenge both scientists and decision-makers to interact effectively and efficiently to minimise societal risk during the unrest and in anticipation of a possible imminent eruption.

The scientific challenge is to identify whether unrest phenomena are pre-eruptive or non-eruptive early on in a developing unrest crisis. The challenge for decision-makers is to respond with the "right" decision during an emerging crisis.

Expectations from many different stakeholders including the media and the general public need to be managed amid epistemic (the unknown subsurface processes hidden from the observer) and aleatoric uncertainty (their stochastic variability).

Numerous examples in the recent past have shown that periods of unrest are hazardous and costly even without leading to imminent eruption (e.g. at La Soufriere, Guadeloupe, in 1976; at Campi Flegrei, Italy, in 1982–1984; at Long Valley Caldera/Mammoth Mountain, USA, in the 1980s and 1990s).

Tensions between different stakeholders can arise particularly during protracted periods (years to decades) of unrest at volcanoes with significant and growing populations, when trust in scientific knowledge is challenged amid the need for economic and societal development and growth.

Volcanic Unrest: From Science to Society—highlights the complexities of volcanic unrest from both scientific and societal perspectives and provides a summary of findings from the project entitled "Volcanic Unrest in Europe and Latin America: Phenomenology, eruption precursors, hazard forecast, and risk mitigation (VUELCO)". The multi-disciplinary and cross-boundary research project was funded by the European Commission's 7th framework programme for research, technological development and demonstration under grant agreement no. 282759 and brought together an international team of academics from the social and natural sciences, as well as personnel from volcano observatory and civil protection agencies.

The project was designed in response to a call by the EC for a "collaborative project (small- or medium-scale focused research project) for specific cooperation actions (SICA) dedicated for international cooperation partner countries with focus on Latin America".

The work was conducted at ten partner institutions across Europe and Latin America during 2011–2015. Most of the findings have been published in dedicated scientific journals over the past few years, and others will be published in the years to come.

The purpose of this open-access book is to summarise and publicise the key findings of the project for a broad audience. Seventeen and one appendix chapters focus on four broad topics relevant to the understanding of volcanic unrest in the context of scientific and societal challenges:

1. The significance of volcanic unrest at the natural hazard and risk interface,
2. Geophysical and geochemical fingerprints of unrest and precursory activity,
3. Subsurface dynamics leading to unrest phenomena,
4. Stakeholder interaction and volcanic risk governance.

This book aims to make our research accessible to both scientific and non-scientific audiences with interest in the different aspects of volcanic unrest, its impact and consequences. The chapters have been written with the intention to make the findings accessible to a broad audience. This entails a balancing act between keeping the scientific jargon at bay, whilst also satisfying the curiosity of scientific readers from different disciplines. In addition, most chapters are accompanied by Spanish-language abstracts.

As a consequence, the style of the chapters is different in several ways from the general peer-reviewed scientific literature. Most chapters have a summary/review character of findings from the project, which were originally published in dedicated journals. The chapters hence provide syntheses and articulations of concepts rather than comprehensive compilation of data and the available literature. However, all chapters provide the reader with references to original publications that will permit a wider reading and study of the findings behind the chapters.

The book is a joint effort between editors, authors, reviewers and publishers.

All chapters have undergone peer review, and we are indebted to all reviewers from the VUELCO community as well as the following external reviewers who provided their expert opinions and comments:

S. de Angelis, F. Arzilli, O. Bachmann, F. Costa, N. Deligne, J. Gardner, H. Gonnermann, A. Hicks, S. Hurwitz, S. Jenkins, P. Lesage, C. Newhall, C. Pritchard, G. Woo.

We thank J. Schwarz at Springer for her patience and expert handling of all things related to publishing this book.

Munich, Germany — Bettina Scheu
Bristol, UK — Joachim Gottsmann
Leeds, UK — Jürgen Neuberg

Contents

Volcanic Unrest and Pre-eruptive Processes: A Hazard and Risk Perspective

J. Gottsmann, J.-C. Komorowski and J. Barclay

Abstract

Volcanic unrest is complex and capable of producing multiple hazards that can be triggered by a number of different subsurface processes. Scientific interpretations of unrest data aim to better understand (i) the processes behind unrest and their associated surface signals, (ii) their future spatio-temporal evolution and (iii) their significance as precursors for future eruptive phenomena. In a societal context, additional preparatory or contingency actions might be needed because relationships between and among individuals and social groups will be perturbed and even changed in the presence of significant uncertainty. Here we analyse some key examples from three international and multidisciplinary projects (VUELCO, CASAVA and STREVA) where issues around the limits of volcanic knowledge impact on volcanic risk governance. We provide an overview of the regional and global context of volcanic unrest and highlight scientific and societal challenges with a geographical emphasis on the Caribbean and Latin America. We investigate why the forecasting of volcanic unrest evolution and the exploitability of unrest signals to forecast future eruptive behaviour and framing of response protocols is

J. Gottsmann (✉)
School of Earth Sciences, University of Bristol, Bristol, UK
e-mail: j.gottsmann@bristol.ac.uk

J. Gottsmann
The Cabot Institute, University of Bristol, Bristol, UK

J.-C. Komorowski
Institut de Physique du Globe de Paris et Université Paris Diderot, Université Sorbonne Paris Cité, CNRS UMR 7154, Paris, France

J. Barclay
School of Environmental Sciences, University of East Anglia, Norwich, UK

Advs in Volcanology (2019) 1–21
DOI 10.1007/11157_2017_19

Published Online: 17 August 2017

challenging, especially during protracted unrest. We explore limitations of current approaches to decision-making and provide suggestions for how future improvements can be made in the framework of holistic volcanic unrest risk governance. We investigate potential benefits arising from improved communication, and framing of warnings around decision-making timescales and hazard levels.

Resumen

La agitación volcánica es compleja y capaz de generar múltiples peligros que pueden ser desencadenados por un número diferente de procesos subsuperficiales. Las interpretaciones científicas sobre datos de agitación volcánica tienen como objetivo el mejor entendimiento de (i) los procesos detrás de la agitación volcánica y sus señales superficiales asociadas, (ii) su evolución espacial-temporal y (iii) su significado como precursores de fenómenos eruptivos a futuro. Dentro de un contexto social, acciones adicionales preparatorias o de contingencia podrían ser requeridas debido a que las relaciones entre individuos y dentro de grupos sociales serán perturbadas e inclusive modificadas ante la presencia de incertidumbre significativa. Aquí nosotros analizamos algunos ejemplos clave a partir de tres proyectos internacionales y multidisciplinarios (VUELCO, CASAVA y STREVA) en los cuales las cuestiones alrededor de los límites del conocimiento volcánico tienen impacto en la gestión pública del riesgo volcánico. Proveemos una perspectiva general del contexto regional y global de la agitación volcánica y sobresaltamos retos científicos y sociales con énfasis geográfico en el Caribe y América Latina. Investigamos porqué el pronóstico de la evolución en la agitación volcánica y el aprovechamiento de señales de agitación volcánica para el pronóstico de comportamiento eruptivo a futuro y el enmarque de protocolos de respuesta es un reto, especialmente durante periodos de agitación prolongada (años a décadas) en los que algunos retos surgen desde la utilización de señales de agitación para pronosticar la evolución de agitación a largo plazo y sus eventuales consecuencias. Exploramos las limitantes de actuales enfoques para la toma de decisiones y proveemos sugerencias acerca de cómo pueden hacerse reformas a futuro dentro del marco holístico de gobernabilidad ante el riesgo de agitación volcánica. Investigamos los potenciales beneficios que surgen por comunicación mejorada, y delimitando alertas alrededor de escalas de tiempo para la toma de decisiones y los niveles de alerta. Proponemos la necesidad de la cooperación a través de las fronteras científicas tradicionales, una valoración más amplia del riesgo natural y una mayor interacción de los sectores interesados.

1 Introduction

Volcanic unrest is a complex multi-hazard phenomenon of volcanism. Although it is fair to assume that probably all volcanic eruptions are preceded by some form of unrest, the cause and effect relationship between subsurface processes and resulting unrest signals (geophysical or geochemical data recorded at the ground surface, phenomenological observations) is unclear and surrounded by uncertainty (e.g., Wright and Pierson 1992). Unrest may, or may not lead to eruption in the short-term (days to months). If an eruption were to ensue it may involve the eruption of magma or may be non-magmatic and mainly driven by expanding steam and hot water (hydrothermal fluids) (Table 1). These conundrums contribute significant uncertainty to short-term hazard assessment and forecasting of volcanic activity and have profound impact on the management of unrest crises (e.g., Marzocchi and Woo 2007).

While institutional and individual decision-making in response to this unrest should promote the efficient and effective mitigation or management of risk, informed decision-making

Table 1 Summary of processes contributing to unrest signals in space and time, possible outcomes and hazards/impact of unrest

Nature of processes	Processes	Signals	Hazards/Impact	Unrest Outcome
Magmatic	Magma and/or melt and/or volatile migration (input, loss or ascent from reservoir), chemical differentiation, thermal convection, thermal perturbation (heating or cooling), pore fluid migration reservoir rejuvenation, crystallization and other phase changes	Seismicity, ground deformation, changes in potential fields, changes in gas and/or ground water chemistry, changes in heat flux, changes in volatile flux	Ground deformation, shaking and rupture and associated infrastructure damage; water table level changes; toxic gas emissions, contamination of ground water, atmosphere and crops; edifice destabilization; toxic gas emissions	Waning and return to background activity; eruptive activity (magmatic and/or phreatic)
Tectonic/gravitational	Faulting, changes in local/regional stress fields, edifice gravitational spreading, crustal loading, pore fluid migration			Waning and return to background activity; eruptive activity (magmatic and/or phreatic)
Hydrothermal	Fluid migration, phase changes, changes in temperature and/or pressure, chemical changes, pore pressure variations, porosity and permeability changes (sealing), host-rock alteration			Waning and return to background activity; phreatic eruptive activity

Processes can act individually, in unison or in any combination is fundamentally dependent on the early and reliable identification of changes in the subsurface dynamics of a volcano and their "correct" assessment as precursors to an impending eruption. However, uncertainties in identifying the causative processes of unrest impact significantly on the ability to "correctly" forecast the short-term evolution of unrest.

When a volcano evolves from dormancy through a phase of unrest, scientific interpretations of data generated by this unrest relate to (i) the processes behind unrest and their associated surface signals, (ii) their potential future spatio-temporal evolution (i.e., hydrothermal vs. phreatic vs. magmatic processes and their intensity) and (iii) their significance as precursors for future eruptive phenomena. Scientific interpretations framed towards the governance of and social responses to the risk implicit in the potential onset of an eruption focus on: (i) understanding the epistemic (relating to the limits of existing knowledge) and aleatoric (relating to the intrinsic variability of natural processes) uncertainties surrounding these data and their impact on decision making and emergency management, (ii) the communication of these uncertainties to emergency managers and the citizens at risk, and (iii) understanding how best to manage evolving crises through the use of forecasted scenarios.

2 Motivation

The analysis presented in this chapter synthesises wider results and experiences gained in three major research consortia with focus on volcanic hazards and risks: (1) The VUELCO project, (2) the CASAVA project, and (3) the STREVA project.

The European Commission funded VUELCO project (2011–2015; "Volcanic unrest in Europe and Latin America: Phenomenology, eruption precursors, hazard forecast, and risk mitigation; www.vuelco.net) focused on multi-disciplinary research on the origin, nature and significance of volcanic unrest and pre-eruptive processes from the scientific contributions generated by collaboration of ten partners in Europe and Latin America. Dissecting the science of monitoring data from unrest periods at six target volcanoes in Italy (Campi Flegrei caldera), Spain (Tenerife), the West Indies (Montserrat), Mexico (Popocatepetl) and Ecuador (Cotopaxi) the consortium created strategies for (1) enhanced monitoring capacity and value, (2) mechanistic data interpretation and (3) identification of eruption precursors and (4) crises stakeholder interaction during unrest.

The CASAVA project (2010–2014; Agence nationale de la recherche, France; *Understanding and assessing volcanic hazards, scenarios, and risks in the Lesser Antilles—implications for decision-making, crisis management, and pragmatic development;* https://sites.google.com/site/casavaanr/, last accessed 11-10-2016) implemented an original strategy of multi-disciplinary fundamental research on the quantitative assessment of volcanic risk for the Lesser Antilles region with emphasis on Guadeloupe and Martinique. The aim of the project was to improve the capacity to anticipate and manage volcanic risks in order to reduce reactive 'repairing' post-crisis solutions and promote the emergence of a society of proactive volcanic risk prevention in case of a future eruption. Part of this was achieved via a forensic analysis of past crises, described here.

The STREVA Project (2012–2018 funded by the UK Natural Environment and Economic and Social Research Councils; www.streva.ac.uk) was designed as a large interdisciplinary project to develop new means to understand how volcanic risk should be assessed and framed. It uses the 'forensic' interdisciplinary analysis of past volcanic eruptions in four settings to understand the key drivers of volcanic risk. The aim is to use this analysis to generate future plans that will reduce the negative consequences of future eruptions on populations and their assets. STREVA works closely with partners in the Caribbean, Ecuador and Colombia, focussing the forensic analysis on long-lived eruptions of Soufrière Hills Volcano (Montserrat) and Tungurahua (Ecuador) and shorter duration eruptions of La Soufrière (St. Vincent) and Nevado del Ruiz (Colombia). The focus of the 'forensic analysis' process in the STREVA project has been to understand the key drivers of risk and resilience during long-lived volcanic crises. Nonetheless the analysis of the initial phases of activity from these eruptions provide some insights into the acute uncertainties of unrest and the social, political and scientific consequences of that uncertainty.

3 Volcanic Unrest: Scientific and Social Context

Volcanic unrest can be defined in a scientific context: "The deviation from the background or baseline behaviour of a volcano towards a behaviour or state which is a cause for concern in the short-term (hours to few months) because it might prelude an eruption" (Phillipson et al. 2013). The term "eruption" in the context of a possible unrest outcome could either relate to a magmatic or non-magmatic (phreatic or hydrothermal) origin including the possible evolution from phreatic to magmatic activity or an alternation or mix between the two (e.g., Rouwet et al. 2014). In a social context, these concerns might necessitate additional preparatory or contingency actions in response to the unrest phenomena or the preparation for an eruption given that the organisation and preparedness of communities and those who manage them will be perturbed and even changed in the context of significant uncertainty (Barclay et al. 2008 and next section).

4 Challenges and Key Questions Relating to Volcanic Unrest

4.1 Wider Perspective

Whether or not unrest results in eruption, either of magmatic or non-magmatic origin, and whether (in hindsight) "correct" or "false" forecasts are issued to suggest there could be an imminent eruption are among the central questions that need answering as soon as unrest is detected.

The cost of scientific uncertainty regarding the causes and outcome of volcanic unrest may be substantial not only in terms of direct or indirect financial implications such as explored in Sect. 5, but also regarding knock-on (secondary) effects such as public trust in the accuracy or inaccuracy of scientific knowledge, public perception of the relationships between signals of unrest and volcanic risk and future public compliance with orders to evacuate or improve preparedness in the medium to long term.

A multitude of subsurface processes may contribute to unrest signals and some are summarised in Table 1. Not all processes are pre-eruptive and the challenge lies in deciphering the causes of unrest with a view to establish early on in a developing crises whether a volcanic system develops towards a state where an eruption may ensue. Whether or not unrest leads to eruption depends on many parameters. In general the main concern during volcanic unrest lies with the potential for a magmatic eruption. For this to occur magma must rise from depth and break through the surface. The dilemma for scientists is that magma movement does not create uniquely attributable unrest signals and does not necessarily lead to eruption (Table 1). For example, seismicity and ground uplift, both common indicator of unrest, may be induced by the replenishment of a magma reservoir, the ascent of magma towards the surface or the redistribution of aqueous fluids and fluid phase changes (see Salvage et al. 2017; Hickey et al. 2017; Mothes et al. 2017 for examples from VUELCO volcanoes). Similarly, an increase in the gas and heat flux (Christopher et al. 2015) at the surface may be induced by magmatic or hydrothermal processes and even tectonic stress changes have also been shown to trigger such behaviour (e.g., Hill et al. 1995). In fact, non-magmatic eruptions are associated with significant hazards and have or could have caused fatalities in the past such as for example Bandai in 1888 (Sekiya and Kikuchi

1890), Te Maari Tongariro in 2012 (e.g., Jolly et al. 2014) and recently at Ontake in 2014 (e.g., Maeno et al. 2016). Many unrest processes contribute to non-eruptive secondary hazards such as flank instability and collapse (e.g. Reid 2004).

4.2 Uncertain Causes and Uncertain Effects

Substantial uncertainties surround both the interpretation of the drivers of unrest and the assessment of the potential evolution and outcome of unrest. Critical questions include: Will an eruption ensue? If so, will it occur in the short-term (days to months) or long-term (years to decades)? What will be the nature and intensity of the eruption (magmatic vs. phreatic)?

In the case of magmatic unrest, magma ascent towards the surface can lead to a magmatic eruption with potential for the formation of lava flows, pyroclastic flows, lahars, ash-fall and ballistics. These processes impact the proximal (few tens to hundreds of meters), medial (kilometers) and distal (tens of kilometres or more) areas around the volcano. Conversely unrest driven by sub-surface hydrothermal activity may peak in a phreatic eruption and while impacted areas are rather proximal to the volcano, associated ballistics and dilute pyroclastic density currents triggered by laterally-directed explosions and emplacement of a debris avalanche from a partial edifice collapse can lead to an anomalously high loss of lives as recently shown by the September 27, 2014 Mount Ontake eruption, the deadliest eruption in more than 100 years in Japan (e.g. Maeno et al. 2016).

The challenge, however, is to identify and discriminate signals that are indicative of reactivation leading towards a major expulsion of magmatic material from those associated with a slight deviation from background levels and potential waning of unrest phenomena (Table 1).

The fundamental limitation for volcanologists is that it is not possible to directly observe causative processes at depth. Thus interpretations of these drivers rely on the secondary interpretation of observable signals associated with those processes (Salvage et al. 2017) or the reproduction of interpreted processes via laboratory experiments (Wadsworth et al. 2016). In addition, many volcanic processes are intrinsically non-linear and characterized by a chain-link reaction such that minor variations of some uncertain parameters might have ultimately significant consequences on the eruptive outcome. Such non-linear processes coupled with epistemic and aleatoric uncertainties are complex to understand and model. This chapter analyses some key examples across the three aforementioned projects where issues around the limits of volcanic knowledge exacerbated risk and makes suggestion for how future improvements can be made.

4.3 The Hazard and Risk Interface

Scientific Challenges

In the light of the above, from a scientific point of view the early identification of the cause of unrest and its likely outcome and evolution is pivotal for effective and efficient risk assessment, risk management and the design of mitigation efforts. In order to address the key scientific question of whether unrest is a prelude to imminent eruption or whether it will wane after some time without eruption several questions require answering first (note, that the list is not exhaustive):

- Is the anomalous behaviour unambiguously indicative for a change in the volcano's behaviour and for a deviation from its background state?
- How reliable is the assessment of unrest as a prelude to eruption, particularly in the absence of data on past events?

- What are the mechanistic processes at depth leading to observed unrest signals?
- Are monitoring signals indicative of magmatic, hydrothermal or tectonic unrest?
- Can the unrest be caused by perturbations and changes in the host-rock properties (e.g. porosity, permeability, mechanical properties) rather than by distinct endogenic processes of hydrothermal or magmatic origin?
- What are the uncertainties surrounding monitoring signals and inferred sub-surface processes (see Hickey et al. 2017 and Salvage et al. 2017)?
- Do secondary processes (e.g. hydrothermal system perturbation, meteorological forcing) modify primary signals from deeper-seated magmatic processes?
- What are the consequences of signal modification for the assessment of the process-to-signal-to-outcome causal link?
- Does one follow a deterministic or probabilistic approach for observations and forecasting (e.g., Hincks et al. 2014; Aspinall and Woo 2014; Rouwet et al. 2017)?
- What is the likelihood of a specific eruptive or non-eruptive scenario to manifest (e.g., Bartolini et al. 2017)?
- Which types of eruptions did the volcano produce in the past?
- If an eruption is to occur, what is its likely nature: magmatic, or phreatic or a mix?
- How much lead-time before eruption is there based on previous experience; how much lead-time is there in the absence of previous experience?
- Which eruptive or non-eruptive unrest episodes at analogue volcanoes can provide clues for the interpretation of signals and forecasting of unrest evolution and outcome (e.g., Sheldrake et al. 2016)?
- What is the likely size of the eruption and the associated hazards and risks and impacted area?
- What is the temporal evolution of eruptive intensity once the eruption has started? i.e., what is the likelihood that the eruption (a) will have its paroxysmal phase in the first 24 h of eruption (42% of eruptions do, according to Siebert et al. (2015)); or (b) will have a more progressive escalation over several months that will culminate in a paroxysm; or (c) will be characterised by peaks in activity separated by more or less long-lasting pauses or strong decline of activity preceding another rapid increase and peak of activity?

Societal Challenges

At the same time, the political, sociological, cultural and economic (grouped here under the term 'societal') implications from unrest need addressing in order to respond appropriately to the emerging natural hazard (Wynne 1992) Here we provide a (non-exhaustive) list of questions for risk managers and/or politicians in the context of risk governance during volcanic unrest:

- What is the best-practice to provide maximum response time, while minimizing vulnerability and optimizing the cost/benefit ratio (see Fig. 1) of mitigation actions in a developing unrest crises?
- What is the best practice to issue or raise an alert?
- When and how to decide to raise an alert and to take action?
- What are the potential (legal) consequences of a false positive or false negative (see Table 2 and Bretton et al. 2015)?
- What are the consequences of a true positive (Table 2)?
- What is the basis for raising an alarm: the outcomes of unrest (e.g., instability of buildings due to ground deformation or seismicity; toxic degassing and environmental contamination) or the potential for eruption?
- How to best disseminate what information on unrest and its potential consequences, when, and via which communication vehicle(s) to the public?

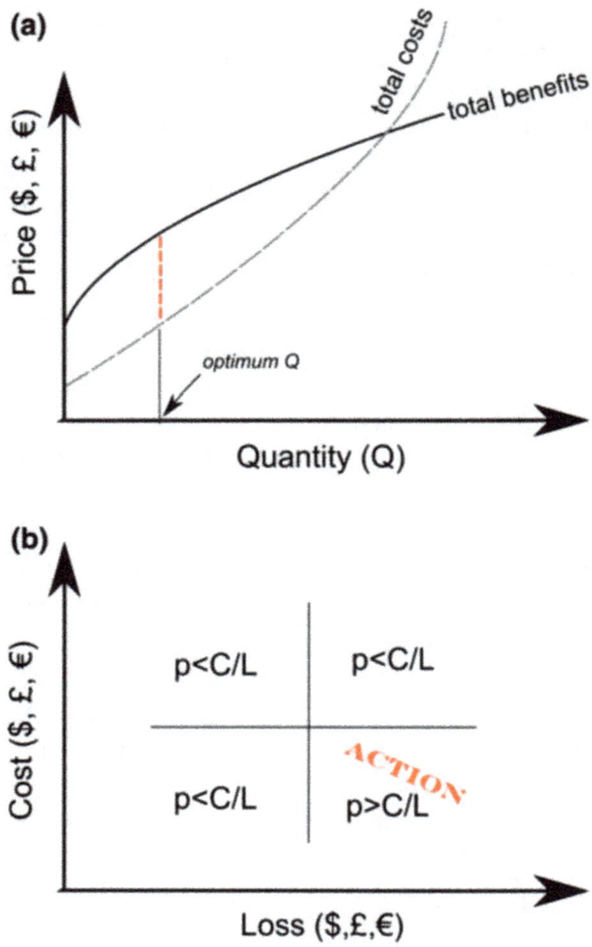

Fig. 1 Cost-benefit relationship as a tool for decision-making. **a** In the context of volcanic unrest risk management, actions of given quantity Q (for example, number of shelters or evacuees) are associated with costs in relation to their expected benefits (expressed by a financial value). An optimal relationship between costs of mitigation efforts and resultant benefits can be achieved when the difference between investment and benefit is greatest (shown by *stippled red line*). The example is based on concepts of capital management theory presented in Brealey et al. (2011). **b** Cost (C) versus loss (L) model for volcanic risk management (after Marzocchi and Woo 2007). If, in this decision-making framework, the expected expense (cost) for mitigation action is to be minimised, then action is required if the probability (*p*) of an adverse event to occur exceeds the ratio between the cost of the action and the expected loss (L/C). See discussion for a wider appraisal of the challenges arsing from such an analysis

Table 2 Concept of successful and unsuccessful forecasting

	Event forecast	Event not forecast
Event occurs	True positive	False negative (Type II error)
Event does not occur	False positive (Type I error)	True negative

- How to account for uncertainty and the diversity of expert opinions in deciding the alert level?
- In what context does this occur such as political pressures, concurrent natural or other hazards (pandemic, famine, cyclone, etc.)

4.4 Cost-Benefit Analysis (CBA)

In the previous paragraphs, several questions related to how to get both the scientific analysis ('what is going on?') and the societal response ('how to respond?') 'right'. One measure employed to quantify the economic consequences of action or no-action under imminent threat and a tool for informed institutional decision-making is the cost-benefit analysis (Marzocchi and Woo 2007) whereby one aims to find a good answer to the question: "Given an assessment of costs and benefits related to risk mitigation efforts, which actions should be recommended?" Figure 1 shows the concept of evaluating the optimum ratio between the cost and benefit of mitigation efforts and provides a cost-benefit matrix for the design of action plans in response to a future [short-term in context of this chapter) adverse event of given probability (*p*) (Brealey et al. 2011; Marzocchi and Woo 2007)]. A critical issue in CBA is the 'minimum value of a human life', which we will not discuss further here. The interested reader is referred to, for example, Woo (2015) for further details on this quantification. Another interesting point relates to what might be regarded as a 'cost' and a 'benefit' in a response to an unfolding unrest crisis with an uncertain outcome (see also Sect. 6).

5 Global and Regional Context of Volcanic Unrest

5.1 Unrest Durations and Characteristics

Phillipson et al. (2013) reviewed global unrest reports of the Smithsonian Institution Global Volcanism Program (GVP) between January 2000 and July 2011 to establish the nature and length of unrest activity, to test whether there are common temporal patterns in unrest indicators and to test whether there is a link between the length of inter-eruptive periods and unrest duration across different volcano types.

Using available formation on unrest at 228 volcanoes they defined unrest timelines to demonstrate how unrest evolved over time and highlight different classes of unrest including reawakening, pulsatory, prolonged, sporadic and intra-eruptive unrest (see Fig. 2 for an example from Cotopaxi volcano). Statistical analyses of the data indicate that pre-eruptive unrest (where there is a causal link between unrest and an eruption within the observation period) duration was different across different volcano types with 50% of stratovolcanoes erupting within one month of reported unrest. The median average

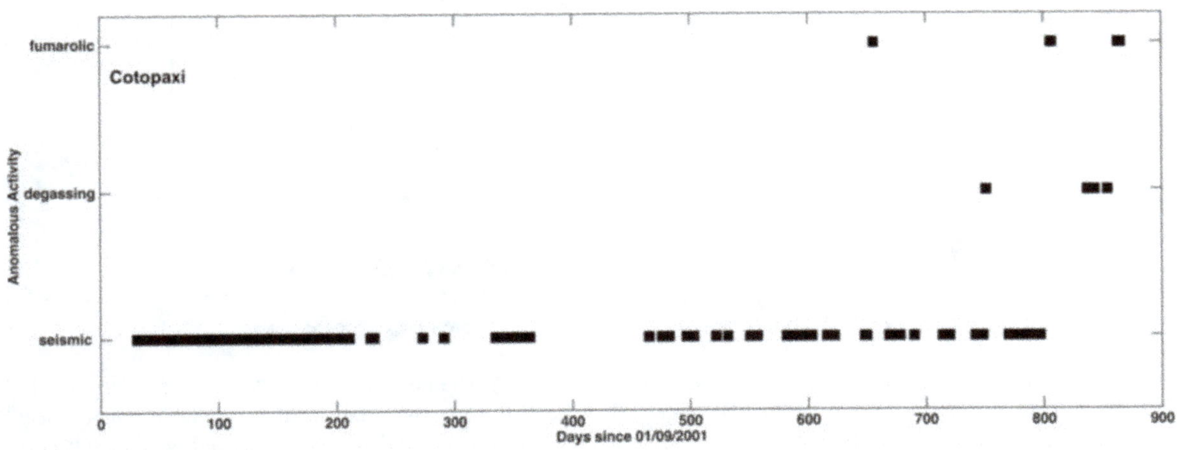

Fig. 2 Timeline of reported anomalous activity at Cotopaxi volcano (Ecuador) in 2001/2002. This period of pulsatory unrest lasted for more than 3 years with a heightened level of activity in 2001 and 2002. The unrest did not lead to an eruption in the short-term (weeks to months), but Cotopaxi entered an eruptive phase in August 2015 after a short-period of renewed unrest activity starting in April 2015 (see Mothes et al. 2017 for details). The data shown in the graph are from Phillipson et al. (2013)

duration of pre-eruptive unrest at large calderas was about two months, while at shield volcanoes a median average five months of unrest was reported before eruptive activity. The shortest median average duration is reported for complex volcanoes where eruptive unrest was short at only two days. Overall there appears to be only a very weak correlation between the length of the inter-eruptive period and pre-eruptive unrest duration. This may indicate that volcanoes with long periods of quiescence between eruptions will not necessarily undergo prolonged periods of unrest before their next eruption (Fig. 3). Phillipson et al. (2013) found statistically relevant information only from reports of anomalous seismic behaviour, most other monitoring signals are either not recorded or not reported as unrest criteria. The authors reported a noteworthy lack of geodetic data/information and in particular satellite remote sensing data in the available reports. Recently Biggs et al. (2014), addressed the latter and systematically analysed 198 volcanoes with more than 18 years of satellite remote sensing deformation data for their deformation behaviour. 54 volcanoes that showed deformation also erupted during the observation period. Their analysis does not imply any causal link, or even a temporal relationship between any specific eruptions and episodes of deformation and is hence not directly comparable to the causal and predictive analysis by Phillipson et al. (2013). However, given that 46% of deforming volcanoes erupted while 94% of non-deforming volcanoes did not erupt provides "strong evidential worth of using deformation data as a trait association with eruption" (Biggs et al. 2014).

It is important to note that exploitable records on volcanic unrest are limited and the available

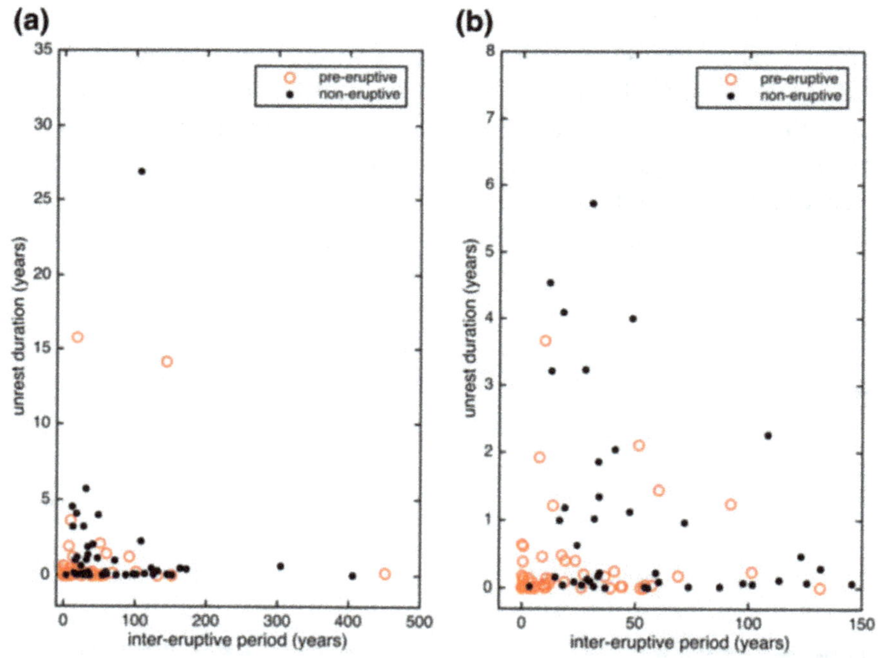

Fig. 3 Comparison between the inter-eruptive period (IEP) and unrest duration (UD) from the data set presented in Phillipson et al. (2013). **a** shows entire data set (n = 118) **b** shows a subset of the data for clarity of inter-eruptive periods <150 years. The *p*-values of the Pearson's correlation test are $p = 0.93$ for the entire data set, $p = 0.60$ for the subset of non-eruptive unrest (n = 58) and $p = 0.20$ for the subset of eruptive unrest (n = 60). The null hypothesis ("the UD is independent of the IEP") is hence statistically acceptable when considering the entire data set. Considering the subset of pre-eruptive unrest, however, the statistical tests do not provide enough evidence to fully accept the null hypothesis since the associated *p*-value of 0.20 might hint towards some weak correlation between the two variables

data sets are far from complete. Key issues are the lack of or poor instrumentation at most volcanoes, the lack of reporting by observers particularly if an unrest turns out to be minor and without immediate consequences, and the lack of integrating unrest data from satellite remote sensing. The GVP generally lacks the post-facto integration of unrest indicators from satellite-remote sensing data (e.g., Fournier et al. (2010) and Biggs et al. (2014) for deformation and Carn et al. (2011) for degassing). In this respect, it is vitally important to recognise and support initiatives to collate and exploit worldwide volcano monitoring data such as for example the WOVOdat project (Venezky and Newhall 2007). Only by significantly increasing the knowledge-base on the spatial and temporal evolution of the unrest-eruption relationship can we embark on statistically sound exploitations of the data with a potential to improve forecasting capabilities early on in developing unrest crises.

5.2 Socio-Economic Contexts

The Wider Perspective

Nowadays, about 800 million people live on or in direct vicinity of active volcanoes (Brown et al. 2015). The overwhelming majority of this population lives in low and middle income countries (countries with an annual gross national income per capita of less than US$12,700) including the focus area of the VUELCO, STREVA and CASAVA projects: the wider Latin American (LA) region extending from Mexico, through Central America and the Caribbean to South America. This region hosts about 330 Holocene volcanic centres compared to 84 in Europe and one quarter of the reported global fatalities attributed to volcanic events occurred there (Global Volcanism Program 2013).

Volcanic disasters are among the least audited of all natural disasters and therefore our knowledge on the impact of volcanic activity beyond claiming lives is largely incomplete (Benson 2006; Auker et al. 2013). Huge uncertainty surround estimates for indirect losses from for example disease or starvation as a result of volcanic activity. Beyond increased human vulnerability, the direct and indirect financial impacts from volcanic activity can be immense as demonstrated by the relatively small-scale eruption of Iceland's Eyjafjallajökull volcano in April 2010 and the associated air travel disruption. This eruption demonstrated the vulnerability of modern infrastructure to volcanic hazards on an unprecedented scale with losses to the aviation industry alone at a minimum of US$2.5 Billion (European Commission 2010).

Equally there are social, political and financial implications for "false positives" related to volcanic unrest. In these instances actions are taken in response to an imminent threat, which then did not manifest. In the case of volcanic unrest the imminent threat is generally defined as a volcanic eruption, although the multi-hazard nature of volcanic unrest (e.g., ground shaking, ground uplift or subsidence, ground rupture, ground instability, toxic gas emissions, contaminated water supplies) and possibly ensuing eruptive activity (magmatic vs. phreatomagmatic vs. phreatic) makes the definition of 'imminent threat' rather complex.

Although there is little systematic gathering and synthesis of data relating to financial or social losses associated with these episodes there are some well-documented analyses. Examples include:

(1) On Guadeloupe in the French West Indies a major evacuation over a period of 4 months in excess of 70,000 individuals was initiated in 1976, as a result of abnormal levels of volcanic seismicity and degassing (see also next section). The estimated cost of the unrest was about US$340 Million at the 1976 exchange rate (data compiled using Lepointe 1999; Tazieff 1980; Blérald 1986; Baunay 1998; Kokelaar 2002; Annen and Wagner 2003), which translates to more than US$ 1.2 Billion at the time of this writing (July 2016). At the time the cost equaled to ca. 60% of the Gross National Product of the Guadeloupe economy (Blérald 1986). 90% of these costs were incurred by the costs of the evacuation,

and the costs associated with the rehabilitation and salvage of the economy in Guadeloupe after the evacuation.

(2) Unrest at Rabaul volcano in Papua New Guinea (an LDC) between 1983 and 1985, had significant adverse implications for both the private and public sectors. Considerable economic costs were incurred, estimated at over US$22.2 Million at the 1984 rate of exchange although an eruption did not occur until 10 years later (Benson 2006).

(3) Evacuation and rehousing of 40,000 inhabitants of the Pozzuoli area in the Campi Flegrei volcanic area of Italy resulted as a response to intense seismicity and ground uplift in the early 1980s. Although decision-makers did not release notice that this was in part due to the threat from an imminent eruption (see also Sect. 4.3.2), it is true that the re-location of these inhabitants moved them from the area of highest threat in the event of an eruption. At the time there was no agreement amongst scientists as to the cause of the unrest (Barberi et al. 1984) and the scientific discussion as to the cause of these events is still ongoing more than 30 years after the crisis.

The following paragraphs focus on two examples of short-term and long-term volcanic unrest crises response and provide more detailed insights into the volcanic risk governance in two different jurisdictions.

Short-Term Crisis Example: The 1976–1977 La Soufrière of Guadeloupe Unrest

The unrest on Guadeloupe culminated in a series of explosive eruptions of hot gas, mud and rock (termed phreatic eruption) without the direct eruption of magma before waning in 1977 (Feuillard et al. 1983; Komorowski et al. 2005; Hincks et al. 2014). Fortunately no fatalities were caused by the activity. Had the unrest on Guadeloupe led to a magmatic eruption, then the cost of the unrest would have likely been negligible. Although the precautionary evacuation caused a substantial economic loss with severe social consequences, it is acknowledged that the "proportion of evacuees who would have owed their lives to the evacuation, had there been a major eruption, was substantial" (Woo 2008). The CASAVA project undertook an exhaustive hindsight analysis of the process of scientific decision-making for the unrest and eruptive crisis of 1976–1977 at La Soufrière de Guadeloupe. The crisis caused significant hardships and loss of livelihood for the evacuated population and the whole society in Guadeloupe as a result of controversial crisis management associated with a forecast of a major magmatic eruption that did not occur (false positive) (Feuillard et al. 1983; Fiske 1984; Komorowski et al. 2005; Hincks et al. 2014). Given the evidence of continued escalating pressurisation and the uncertain transition to a devastating magmatic eruption, authorities declared a 4-month evacuation of ca. 70,000 people on August 15, 1976 that provoked severe socio-economical consequences for months to years thereafter. This evacuation is still perceived as unnecessary and reflecting an exaggerated use of the "principle of precaution" on behalf of the government.

However, some level of risk governance (i.e. evacuation of the most exposed area) was justified in hindsight given the persistent ashfalls and environmental contamination from acid degassing as well as the hazards from a series of non-magmatic eruptions (e.g., pyroclastic flows from laterally directed explosions, partial edifice collapse, mudflows) (Komorowski et al. 2005; Hincks et al. 2014).

The (in hindsight) erroneous identification of the presence of '*fresh glass*' in the ejecta and its interpretation as evidence of the magmatic origin of the unrest and thus of its possible outcome, led to a major controversy amongst scientists that was widely echoed in the media. Lack of a comprehensive monitoring network prior to the crisis, limited knowledge of the eruptive history, and living memory of past devastating eruptions in the Lesser Antilles contributed to a high degree of scientific uncertainty and a publically-expressed lack of consensus and trust in available expertise. Consequently analysis, forecast, and crisis response were highly challenging for scientists and authorities in the context of escalating and fluctuating activity and societal pressure. The high uncertainty about a so-called "unequivocal" impending disaster fostered a binary zero-sum strongly opinionated approach in the scientific discourse. The public debate thus became polarized on issues of opposing "truths" served by contrasted scientific expertise rather than on how science could help constrain epistemic and aleatoric uncertainty and foster improved decision-making in the context of uncertainty (Komorowski et al. 2017). This situation acted as an ideal crucible to fuel a media-hyped controversy on the crisis and its management. A recent retrospective Bayesian Belief Network analysis of this crisis (Hincks et al. 2014) demonstrates that a formal evidential case could have been made to support the authorities' concerns about public safety and decision to evacuate in 1976.

As part of the CASAVA project we conducted focus group interviews, issued questionnaires, and ran role playing games with the population currently living in areas potentially threatened by renewed unrest and eruptive activity from La Soufrière, (be it magmatic or non-magmatic). We found that the current population's risk perception increases to a level of preparing to evacuate chiefly on the basis of the timing and nature of scientific information issued publically by the volcano observatory. This implies that the population is prone to self-evacuate ahead of any official evacuation order given by the authorities in charge of civil protection and crisis response.

Long-Term Crises Examples: Soufrière Hills (Montserrat) and Tungurahua (Ecuador)

The forensic analyses of the STREVA project have focussed on the integration of new social-science based understandings of population response and recovery with the scientific insights prompted by these long-lived eruptions. This has similarities with the 'FORIN' approach advocated by the International Program on Integrated Risk for Disaster Reduction (Burton 2010). In this description we focus particularly on the initial stages of the eruptions.

The long-lived volcanic crisis of the Soufrière Hills Volcano is probably one of the most written about volcanic eruptions, encompassing a wide variety of perspectives, scientific, social-scientific and personal, in that writing. As a consequence of the activity on the island of Montserrat a population of over 10,500 was reduced to just 2850 (the population has since risen to 4922 [2011 census], Hicks and Few 2015). At the onset of eruption (1995) an assessment of risk existed (Wadge and Isaacs 1988) but was not acted on or acknowledged by the authorities, and so preparedness was low, exacerbated by the recent passage of Hurricane Hugo (1989) which had caused 11 fatalities and rendered 3000 homeless. Governance on Montserrat was reforming in the wake of the economic and social crisis induced by the hurricane (Wilkinson 2015). The protracted uncertainty in the early stages of the eruption coupled with a lack of coherence in governance between the UK and local governments lead to the protracted evacuation of 1300 people in temporary public shelters, which suffered from overcrowding, lack of privacy, poor sanitation and lack of access to good nutrition. Ultimately, this led to a partial disregard for evacuation advice and a strong pulse of outwards migration. In the longer term, the long-lived volcanic eruption has acted to exaggerate pre-existing vulnerabilities in the local population (Hicks and Few 2015).

The early stages of the current Tungurahua (Ecuador) eruptive episode that started in 1999 typify a further challenge for the management of unrest prior to or between surface activity at the early stages of a volcanic crisis. Initially the local population were evacuated by a compulsory evacuation order but when the initial phases proceeded more slowly than had been expected by local authorities and communities, civil unrest and disturbance happened with the re-occupation by force and ultimately abandonment of the evacuation order. These arose from the acute economic and social pressures visited on the population by the evacuation (Mothes et al. 2015). Subsequently, the response of the monitoring organisation to these pressures represents a new archetype for collaborative monitoring and management of restive volcanoes (Mothes et al. 2015; Stone et al. 2014). The growth of trust, and attempts to maximise resilience in the face of repeated unrest episodes provides strong evidence for collaborative approaches to risk management (Few et al. 2017). Nonetheless tensions still exist, largely arising from our current incapacity to predict the intensity or magnitude of eruptions from signals relating to new unrest. There can be problems in this risk system implicit in anticipating the 'maximum expected' outcome from unrest.

6 Discussion

6.1 The Caveats of Volcanic Unrest Response

Managing volcanic unrest episodes is extremely complex and challenging due to the multi-hazard nature of unrest. The risks to be assessed and mitigated include both those associated with the unrest itself as well as those from the potential future eruptive activity. Whilst ground deformation, seismicity, thermal flux or anomalous degassing are indicators of possible future activity these phenomena also pose significant immediate threats to population, infrastructure and other assets in affected areas during the unrest.

From a scientific point of view, hazard assessment relating to eruptive activity has made considerable progress in recent years partly through the deployment of increasingly powerful computational models and simulation capabilities (e.g., Esposti Ongaro et al. 2007; Manville et al. 2013) as well as through advances in the development of probabilistic eruption forecasting tools (e.g., Marzocchi et al. 2008; Aspinall 2006; Aspinall and Woo 2014) and improvements to fundamental understandings of the root drivers of changing activity (e.g., Cashman and Sparks 2013).

Despite these crucial advances for short-term eruption forecasting, the knowledge-base on volcanic unrest, its significance as an eruption precursor, its exploitability regarding forecasting of potential eruptive behaviours and framing of response protocols (e.g., CBA) remains weak for a number of reasons:

(1) The scientific interpretation of volcanic unrest is surrounded by substantial uncertainty, ambiguity and ignorance (Stirling 2010) regarding causes and eventual outcome. Since the contributing subsurface processes cannot be directly observed, volcanic unrest is likely among the least understood phenomena in volcanology for a variety of reasons:

(i) Incomplete knowledge of the mechanistic processes and their dynamic behaviour over time within a magma reservoir and its surroundings (host-rock, hydrothermal system, meteoric recharge, local and regional structural context) that trigger the geophysical, geochemical and geodetic signals recorded at the surface during unrest periods (Table 1).

(ii) Consequently, the interpretation, of the departure of monitoring signals from a long-term baseline level or in the absence of baseline data a crescendo or decrescendo of signals collected during periods of unrest are often ambiguous or non-unique. While this can in practice be addressed in models through epistemic and aleatoric uncertainties, ambiguities in the interpretation will remain.

(2) Ambiguity, uncertainty and ignorance (Stirling 2010) have impact on probabilistic forecasting of duration, spatio-temporal evolution, causal relationship between sequential events and outcomes of unrest episodes (see Sandri et al. 2017) and on remedial actions to mitigate current and future adverse effects. Uncertainties in the decision-making process may give rise to "false alerts" (i.e., false positives; see Table 2) and actions by civil protection with adverse impacts on the compliance of affected communities in future unrest events.

(3) Lack of globally accepted and standardised approach for the terminology, methodology,

criteria, protocols and best practice employed to evaluate and respond to volcanic unrest by different stakeholders such as academia, volcano observatories and the Civil Protection Agencies. This absence of commonly recognised standards can result in the critical issue of managerial risk vulnerability (i.e., standard equivocality' after Bretton et al. 2015). It also often impacts negatively on the effectiveness of communication between stakeholders, hinders or delays effective and efficient decision-making processes and hampers the dialogue among members of scientific, governmental and civil communities (De la Cruz-Reyna and Tilling 2008). However, it is important to note that internationally-defined standards should not be rigidly imposed irrespective of local, cultural, political and social practices (e.g., Bretton et al. 2015; IAVCEI 2016).

(4) Globally, there is no commonly accepted and standardised denominator between those that provide and those that receive scientific advice regarding the level of appropriate scientific complexity to be considered. This may hamper a wider discourse on scientific and technological advances in the quantification of unrest phenomena and resultant uncertainties with other stakeholders. From the scientist's perspective this may generate the notion that the public, administrators, mass media and governmental entities do not appreciate the "excellence of the science" behind unrest characterisation and use the inherent uncertainty as a rationale to go into denial over the hazards posed during unrest. From a sociological point of view, however, decision-making apparently prompted solely by the present or likely volcanic hazards, does not account for local context and can result in a lack of trust in either scientific expertise or government representatives, (Johnson 1987; Haynes et al. 2008; Christie et al. 2015; Komorowski et al. 2017).

6.2 Some Ways Forward

The issues identified above can contribute less optimal unrest response and risk mitigation actions. Although there are other strong contributors to societal vulnerability, we have shown that scientific uncertainty combined with a lack of social awareness and preparedness does act to increase the vulnerability of a society to hazardous unrest phenomena with possibly adverse outcomes. Here, we propose future avenues which can form part of a Risk Governance Framework (IGRC 2017; Fig. 4) including research that could gather critical evidence for some of the key drivers of decisions that result in adverse outcomes for affected populations in the face of an unrest crisis. Such research could also contribute to the analysis of and identification of key targets for future research in volcanology and the social sciences.

(a) Cost-Benefit Analysis

CBA, where the economic impacts of different decisions are quantified, can be difficult at the unrest hazard and risk interface. The case studies presented here demonstrate that intangible assets such as social and cultural cohesion and capital as well as trust (in the context of CBA analysis this would be intentional trust in the sense of Dasgupta (1988); i.e., the subjective probability assigned to compassionate action by an individual or a group of individuals) between different stakeholders can have a strong impact on individual and institutional vulnerabilities during crises.

Analyses that include a wider range of definitions and types of 'costs' and 'benefits' of mitigation efforts (e.g. loss of empowerment, a loss of cultural identity or cultural references), informed by past experiences would facilitate a discourse between different stakeholders. This would entail the need to attribute a financial value to, for example, mental well-being, social networks and cohesions and would necessarily trigger a wider discourse of the impacts of decision-making beyond the avoidance of 'cost

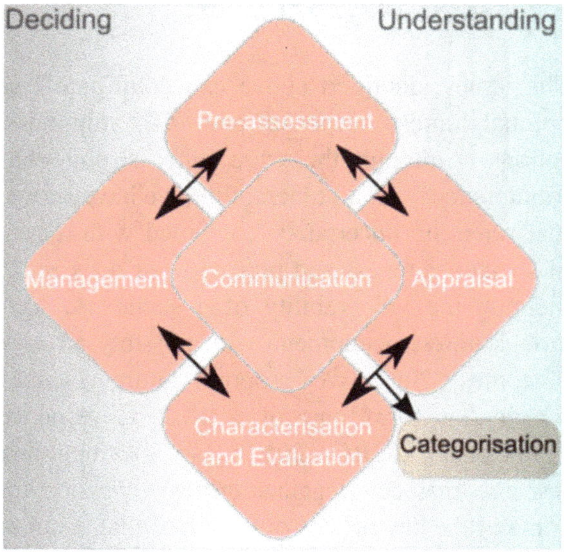

Fig. 4 The *International Risk Governance Council* (IRGC 2017) Risk Governance Framework adapted for the specific case of volcanic unrest. *Hazard and Risk Pre-assessment*—"peacetime framing" the hazard and risk in order to provide a structured definition of the baseline behaviour of the volcano and its consequences, of how the hazard and risk are framed by different stakeholders, of how the risk may best be handled, and of the thresholds to be met or exceeded to declare a state of unrest. *Hazard and Risk Appraisal*—combining a scientific risk assessment of the current unrest hazards (using for example a rating scheme of unrest intensity, e.g., Potter et al. 2015) and its probability with a systematic concern assessment (of public concerns and perceptions) to provide the knowledge base for subsequent decisions in an emerging unrest crises. *Risk Characterisation and Evaluation*—in which the scientific data and a thorough understanding of societal values affected by the risk are used to evaluate the risk as acceptable, tolerable (requiring mitigation), or intolerable (unacceptable). *Risk Management*—the actions and remedies needed to avoid, reduce transfer or retain the unrest risk and risks from probable unrest outcomes. *Risk Communication*—how stakeholders and civil society understand the unrest risk and participate in the risk governance process. *Risk Categorisation and Evaluation*—categorising the knowledge about the cause-effect relationships as either simple, complex, uncertain or ambiguous. In the context of volcanic unrest this may include the categorisation of the outcome of unrest and probable future eruptive activity

to lives'. Ideally this discourse would be framed during 'peace-time' (i.e. not in response to an unfolding unrest crisis) and involve participation from a wide spectrum of scientific and societal stakeholders. The necessity to move beyond circumscribed appraisal methods such as the CBA is also evident from the response to protracted (several years or decades) unrest requiring an above back-ground level of long-term vigilance (e.g. yellow/vigilance level for La Soufrière of Guadeloupe since 1999, Komorowski et al. 2005; OVSG-IPGP 1999–2016 or at the Campi Flegrei caldera since 1969; Ricci et al. 2013). In such cases there are obvious long-term strategies that could be developed to improve social well-being and economic development (e.g. developing resilient critical infrastructures such as roads, bridges, public electrical water and sewer systems and communications networks) that would significantly enhance the quality of life for years of "peace time" from the volcano while ensuring a more efficient crisis response and recovery should the volcano erupt and impact the society.

(b) Improved communication

Open and multi-directional communication processes are of paramount importance in fostering the development of a shared representation and understanding among all stakeholders of the nature, magnitude, dynamics, and societal and environmental consequences of unrest and its potential eruptive outcome on multiple spatio-temporal scales (e.g. Barclay et al. 2008, 2015; Komorowski et al. 2017). While communicating this information in a timely and comprehensible format is challenging, the evidence presented here suggests that a continuous discourse is needed between different stakeholders ideally both before, during and after an unrest situation. The case studies presented here demonstrate that part of this discourse should involve a discussion about the appropriate scientific complexity in the communication between scientific and non-scientific stakeholders is essential, so that the information exchange is 'useful, usable and used' (Aitsi-Selmi et al. 2016) and fit for the decision-making purpose to which it is intended (Fischhoff 2013). Wider discourse could for example include regular information bulletins from monitoring agents to the civil society and authorities, the development of scenario-based approaches in simulation exercises involving the civil society and the wider appraisal of less tangible 'assets' (i.e., live stock or cultural capital) in risk governance efforts.

Dialogues between those responsible for monitoring hazards and those responsible for managing risk, as well as the communities at risk cannot only help to understand the most important aspects of scientific information to convey but could also lead to an improved understanding of the context into which emergency response actions must be made (e.g. Christie et al. 2015), and encourage citizens at risk to act on advice. In particular more systematic studies that analyse the effectiveness of different techniques and strategies in achieving these goals would be very useful (see Fearnley et al. 2017 for a recent compilation). These efforts should help address reluctance by the public to follow emergency-response advice in an emerging unrest crises.

(c) Wider natural risk appraisal

In a similar vein, the implementation of advice on volcanic risk could be more effective if it is considered in the context of other natural risks and social challenges (e.g., Wilkinson et al. 2016). By definition the onset of a volcanic eruption involves the anticipation of impacts from multiple hazards but the risk associated with volcanic hazards are often considered in isolation, and as a low probability, high consequence hazard, ignored in advance of an unrest crisis. This lack of dialogue and preparation has been identified above as a strong contributor to tensions during unrest crises. Volcanic regions only very rarely suffer solely from the impacts of a single natural hazard (e.g. volcanic small-island developing states discussed in Wilkinson et al. 2016; Komorowski et al. 2017). Therefore methods that consider the multi-hazard context more clearly may ultimately help communities at risk cope with uncertainty in face of volcanic hazards. This may be particularly the case, if they are able to identify 'co-benefits' during volcano "peace time" where preparedness or mitigation measures yield benefits for more than one hazard scenario (Wilkinson et al. 2016). This improvement of social well-being is likely to allow the society to take better decisions when times of impeding adversity arise.

(d) Framing of warnings around decision-making timescales and hazard level

Typically changes in alert levels are strongly tied to pre-determined changes in geophysical and geochemical signals or phenomenological observations and have carefully worked out associated actions. In our case studies, difficulties have arisen when the time-scale over which mitigating actions can be taken is much shorter than needed to implement mitigating actions such as evacuation or much longer than the timescale over which unrest or new eruptive activity impacts on the population at risk. In the case of the former, lives or assets may be put at risk and in the case of the latter, possessions and livelihoods can be negatively impacted with repercussions on trust and political stability. Managing decade or longer periods of protracted moderate-level unrest amid significant epistemic and aleatoric uncertainty on its outcome constitutes major challenges for scientists, authorities, the population, and the media.

The development of novel probabilistic formalism for decision-making could help reduce scientific uncertainty and better assist public officials in making urgent evacuation decisions and policy choices should the current and ongoing unrest lead to renewed eruptive activity. To improve decision-making around changing alert or hazard levels, improved modelling efforts of the time-scales and pathways of population mobilisation or actions (both as forward modelling and as analysis of past events) and better understanding of the consequences of protracted unrest or eruptive activity on the vulnerabilities of affected populations (e.g. Few et al. 2017) could improve choices to be made in responding to changing or escalating activity as well as chain-link scenarios.

Further, focussing on the time-scales associated with the *responses* to unrest (from the time taken to mobilise populations in an acute emergency, to the time-limits of tolerability of evacuation processes and finally the time-scales over which services and livelihoods deteriorate in response to protracted unrest) could provide important indicators for the time-scales over which alert levels (and attendant actions) need attention. In turn this perspective could inform scientific targets for improved forecasting, with strong effort expended to reduce uncertainty over time intervals that match those most critical to effective societal action.

7 Conclusions

We have identified a number of scientific and sociological problems surrounding volcanic unrest and have highlighted key aspects of risk governance at the interface between scientists, emergency managers and wider societal stakeholders. We have in particular focussed on the issue of scientific uncertainty and its impact on preparatory or contingency actions that might be needed because relationships between and among individuals and social groups will be perturbed or even changed. Especially, during periods of protracted unrest (years to decades) challenges arise from the exploitability of unrest signals to forecast long-term unrest evolution and its eventual outcome. This impacts directly on establishing the probability for and the timing and type of future eruptive behaviour as well as the definition of appropriate response protocols. To improve communication and trust between stakeholders as well as the framing of warnings around decision-making timescales and hazard levels of unrest, we propose that bridging across traditional scientific boundaries, wider natural risk appraisal and broader stakeholder interaction is needed.

Acknowledgements The VUELCO project received funding by the EC-FP7 program under grant agreement number 282759, the CASAVA project (ANR-09-RISK-002) was funded by the Agence Nationale de la Recherche (France) and the STREVA Project (NE/J020052/1) was funded by the UK NERC and ESRC. We acknowledge the useful insights and comments from two anonymous reviewers.

References

Aitsi-Selmi A, Blanchard K, Murray V (2016) Ensuring science is useful, usable and used in global disaster risk reduction and sustainable development: a view through the Sendai Framework lens. Palgrave Commun (2) doi:10.1057/palcomms.2016.16

Annen C, Wagner J-J (2003) The impact of volcanic eruptions during the 1990s. Nat Hazards Rev 4:169–175

Aspinall WP (2006) Structured elicitation of expert judgment for probabilistic hazard and risk assessment in volcanic eruptions. In: Mader HM, Coles SG, Connor CB, Connor LJ (eds) Statistics in volcanology. The geological society for IAVCEI, 15–30

Aspinall WP, Woo G (2014) Santorini unrest 2011–2012: an immediate Bayesian belief network analysis of eruption scenario probabilities for urgent decision support under uncertainty. J Appl Volcanol 3:12

Auker MR, Sparks RSJ, Siebert L, Crossweller S, Ewert J (2013) A statistical analysis of the global historical volcanic fatalities record. J Appl Volcanol 2:2

Barclay J, Haynes K, Mitchell T, Solana C, Teeuw R, Darnell A, Crosweller HS, Cole P, Pyle DM, Lowe C, Fearnley C, Kelman I (2008) Framing volcanic risk within disaster risk reduction: finding ways for the social and physical sciences to work together. In: Liverman DGE, Pereria CPG, Marker B (eds) Communicating environmental geoscience. Geological Society Special Publication, 305

Barclay J, Haynes K, Houghton BF, Johnston DM (2015) Social processes in volcanic risk reduction. In: Sigurdsson H et al (eds) Encyclopaedia of volcanology, 2nd edn. Elsevier, Amsterdam.

Barberi F, Corrado G, Innocenti F, Luongo G (1984) Phlegraean Fields 1982–1984: brief chronicle of a volcano emergency in a densely populated area. Bull Volcanol 47:175–185

Bartolini S, Marti J, Sobradelo R, Becerril L (2017) Probabilistic e-tools for hazard assessment and risk management. Adv Volcanol 1–15

Baunay Y (1998) Gestion des risques et des crises: l'exemple de deux crises d'origines volcaniques dans les Caraïbes: Montserrat 1995–1998 et la Guadeloupe 1976–1977. Mémoire de Maîtrise de Géographie, Université de Paris 1, Panthéon-Sorbonne, Paris, unpublished, 164 pp.

Biggs J, Ebmeier SK, Aspinall WP, Lu Z, Pritchard ME, Sparks RSJ, Mather TA (2014) Global link between deformation and volcanic eruption quantified by satellite imagery: Nat Commun 5

Blérald APh (1986) Histoire éruptive de la Guadeloupe et de la Martinique du XVIIème siècle à nos jours, Editions Karthala, Paris, 336 p

Benson C (2006) Volcanoes and the economy. In: Marti J, Ernst GGJ (eds) Volcanoes and the environment. Cambridge University Press, pp 440–467.

Brealey R, Myers S, Allen F (2011) Principles of corporate finance. McGraw, 872 p

Bretton R, Gottsmann J, Aspinall WP, Christie R (2015) Implications of legal scrutiny processes (including the L'Aquila trial and other recent court cases) for future volcanic risk governance. J Appl Volcanol 4:18

Brown SK, Auker MR, Sparks RSJ (2015) Populations around Holocene volcanoes and development of a population exposure index. In: Loughlin SC et al (eds) Global volcanic hazards and risk 'populations around volcanoes and the development of a population exposure index. Cambridge

Burton I (2010) Forensic disaster investigations in depth: a new case study model. Environ Mag 52(5):36–41

Cashman KV, Sparks RSJ (2013) How volcanoes work: a 25 year perspective. Geol Soc Am Bull. doi:10.1130/b30720.1

Carn SA, Froyd KD, Anderson BE, Wennberg P, Crounse J, Spencer K, Dibb JE, Krotkov NA, Browell EV, Hair JW, Diskin G, Sachse G, Vay SA (2011) In situ measurements of tropospheric volcanic plumes in Ecuador and Colombia during TC4. J Geophys Res Atmos, 116

Christie R, Cooke O, Gottsmann J (2015) Fearing the knock on the door: critical security studies insights into limited cooperation with disaster management regimes. J Appl Volcanol 4:19

Christopher TE, Blundy J, Cashman K, Cole P, Edmonds M, Smith PJ, Sparks RSJ, Stinton A (2015) Crustal-scale degassing due to magma system destabilization and magma-gas decoupling at Soufrière Hills Volcano, Montserrat: Geochemistry. Geophys Geosyst 16(9):2797–2811

Dasgupta P (1988) Trust as a commodity. In: Gambetta D (ed) Trust: making and breaking of cooperative relations. Blackwell, Oxford, 49–72

De la Cruz-Reyna S, Tilling RI (2008) Scientific and public responses to the ongoing volcanic crisis at Popocatépetl Volcano, Mexico: importance of an effective hazards-warning system. J Volcanol Geotherm Res 170:121–134

Esposti Ongaro T, Cavazzoni C, Erbacci G, Neri A, Salvetti MV (2007) A parallel multiphase flow code for the 3D simulation of volcanic explosive eruptions. Parallel Comput 33:541–560

European Commission (2010) Press Release 27. April, 2010). http://ec.europa.eu/commission_2010-2014/kallas/headlines/news/2010/04/doc/information_note_volcano_crisis.pdf

Fearnley C, Bird D, Haynes K, Jolly G, McGuire B (eds) (2017) Observing the volcano world: volcano crisis communication. Advances in volcanology, IAVCEI. Springer, Berlin, 350 p

Feuillard M, Allegre CJ, Brandeis G, Gaulon R, Le Mouel JL, Mercier JC, Pozzi JP, Semet MP (1983) The 1975–1977 crisis of La Soufrière de Guadeloupe (FWI): a still-born magmatic eruption. J Volcanol Geotherm Res 16:317–334

Few R, Armijos T, Barclay J (2017) Living with Volcan Tungurahua: the dynamics of vulnerability during prolonged volcanic activity. Geoforum 80:72–81

Fischhoff B (2013) The sciences of science communication. In: Proceedings of the National Academy of Sciences of the United States of America, 110 (Supplement 3), 14033–14039. doi:10.1073/pnas.1213273110

Fiske R (1984) Volcanologists, journalists, and the concerned local public: a tale of two crises in the eastern Caribbean. In: Boyd F (ed) Explosive volcanism: inception, evolution and hazards. Studies in geophysics. National Academy Press, Washington DC, pp 170–176

Fournier TJ, Pritchard ME, Riddick SN (2010) Duration, magnitude, and frequency of subaerial volcano deformation events: new results from Latin America using InSAR and a global synthesis. Geochem Geophys Geosyst 11:Q01003. doi:10.01029/02009GC002558

Global Volcanism Program (2013) Volcanoes of the World, v. 4.5.5. In: Venzke E (ed) Smithsonian Institution. Downloaded 04 May 2017. http://dx.doi.org/10.5479/si.GVP.VOTW4-2013

Haynes K, Barclay J, Pidgeon NF (2008) The issue of trust and its influence on risk communication during a volcanic crisis. Bull Volcanol 70:605–621

Hicks A, Few R (2015) Trajectories of social vulnerabiltiy during the Soufriere Hills volcanic crisis. J Appl Volcanol 4:10

Hill DP, Johnston MJS, Langbein JO, Bilham R (1995) Response of Long Valley caldera to the Mw = 7.3 Landers, California, Earthquake. J Geophys Res 100 (B7):12985–13005

Hickey J, Gottsmann J, Mothes P, Odbert HM, Prutkin I, Vajda P (2017) The ups and downs of volcanic unrest: insights from integrated geodesy and numerical modelling. Adv Volcanol 1–17

Hincks T, Komorowski J-C, Sparks RSJ, Aspinall WP (2014) Retrospective analysis of uncertain eruption precursors at La Soufrière volcano, Guadeloupe, 1975–77: volcanic hazard assessment using a Bayesian Belief Network approach. J Appl Volcanol 3:3

IAVCEI task force on crises protocols (2016) Toward IAVCEI guidelines on the roles and responsibilities of scientists involved in volcanic hazard evaluation, risk mitigation, and crisis response. Bull Volcanol 8(31). doi:10.1007/s00445-00016-01021-00448

IRGC (2017) https://www.irgc.org/risk-governance/irgc-risk-governance-framework/. Last accessed 05/02/2017.

Johnson BB (1987) Accounting for the social context of risk communication. Sci Technol Stud, 103–111

Jolly AD, Jousset P, Lyons JJ, Carniel R, Fournier N, Fry B, Miller C (2014) Seismo-acoustic evidence for an avalanche driven phreatic eruption through a beheaded hydrothermal system: an example from the 2012 Tongariro eruption. J Volcanol Geotherm Res 286:331–347

Kokelaar BP (2002) Setting, chronology and consequences of the eruption of Soufrière Hills Volcano, Montserrat (1995–1999). In: Druitt TH, Kokelaar BP (eds) The eruption of Soufrière Hills Volcano, Montserrat, from 1995 to 1999, vol 21. Geological Society, London, Memoirs, pp 1–44

Komorowski J-C, Boudon G, Semet M, Beauducel F, Anténor-Habazac C, Bazin S, Hammouya G (2005) Guadeloupe. In: Lindsay JM, Robertson REA, Shepherd JB, Ali S (eds) Volcanic Atlas of the Lesser Antilles, Seismic Research Unit, The University of the West Indies, Trinidad and Tobago, WI, 65-102

Komorowski J-C, Morin J, Jenkins S, Kelman I (2017) Challenges of volcanic crises on small islands states. In: Fearnley C, Bird D, Haynes K, Jolly G, McGuire B (eds) Observing the volcano world: volcano crisis communication', Advances in Volcanology, IAVCEI, Springer, Berlin, pp 1–18. doi:10.1007/11157_2015_15

Lepointe E (1999) Le réveil du volcan de la Soufrière en 1976: la population guadeloupéenne à l'épreuve du danger. In: Yacou A (ed) Les catastrophes naturelles aux Antilles – D'une Soufrière à l'autre. CERC Université Antilles et de la Guyane, Editions Karthala, Paris, pp 15–71

Maeno F, Nakada S, Oikawa T, Yoshimoto M, Komori J, Ishizuka Y (2016) Reconstruction of a phreatic eruption on 27 September 2014 at Ontake volcano, central Japan, based on proximal pyroclastic density current and fallout deposits. Earth, Planets and Space 68:82

Manville V, Major JJ, Fagents SA (2013) Modeling lahar behavior and hazards. In: Fagents S, Gregg T, Lopes R (eds) Modeling volcanic processes: the physics and mathematics of volcanism. Cambridge University Press, pp 300–330

Marzocchi W, Woo G (2007) Probabilistic eruption forecasting and the call for an evacuation. Geophys Res. Lett 34:L22310. doi:10.21029/22007GL031922

Marzocchi W, Sandri L, Selva J (2008) BET_EF: a probabilistic tool for long- and short-term eruption forecasting. Bull Volc 70:623–632

Mothes P, Yepes HA, Hall ML, Ramon PA, Steele AL, Ruiz MC (2015) The scientific-community interface over the 15 year eruptive episode of the Tungurahua Volcano, Ecuador. J Appl Volcanol 4:9

Mothes P, Ruiz MC, Viracucha EG, Ramón PA, Hernández S, Hidalgo S, Bernard B, Gaunt EH, Jarrín P, Yépez MA, Espín PA (2017) Geophysical footprints of Cotopaxi´s unrest and minor eruptions in 2015: an opportunity to test scientific and community preparedness. Adv Volcanol 1–30

OVSG-IPGP (1999–2013) Bulletin mensuel de l'activité volcanique et sismique de Guadeloupe. Monthly public report, Observatoire Volcanologique et Sismologique de Guadeloupe - Institut de Physique du Globe de Paris, Gourbeyre (ISSN 1622-4523). http://www.ipgp.fr/fr/ovsg/bulletins-mensuels-de-lovsg. Last accessed 11-10-2016

Phillipson G, Sobradelo R, Gottsmann J (2013) Global volcanic unrest in the 21st century: an analysis of the first decade. J Volcanol Geotherm Res 264:183–196

Potter S, Scott B, Jolly G, Neall V, Johnston D (2015) Introducing the volcanic unrest index (VUI): a tool to quantify and communicate the intensity of volcanic unrest. Bull Volcanol 77:1–15

Reid ME (2004) Massive collapse of volcano edifices triggered by hydrothermal pressurization. Geology 32:373–376

Ricci T, Barberi F, Davis MS, Isaia R, Nave R (2013) Volcanic risk perception in the Campi Flegrei area. J Volcanol Geoth Res 254:118–130

Rouwet D, Sandri L, Marzocchi W, Gottsmann J, Selva J, Tonini R, Papale P (2014) Recognizing and tracking volcanic hazards related to non-magmatic unrest: a review. J Appl Volcanol 3:17

Rouwet D, Constantinescu R, Sandri L (2017) Deterministic versus probabilistic volcano monitoring: not "or" but "and". Adv Volcanol. doi:10.1007/11157_2017_8

Sandri L, Tonini R, Rouwet D, Constantinescu R, Mendoza-Rosas AT, Andrade D, Bernard B (2017) The need to quantify hazard related to non-magmatic unrest: from BET_EF to BET_UNREST. Adv Volcanol. doi:10.1007/11157_12017_11159

Salvage RO, Karl S, Neuberg J (2017) Volcano seismology: detecting unrest in wiggly lines. Adv Volcanol 1–17

Sekiya S, Kikuchi Y (1890) The eruption of Bandai-san. Tokyo Imp Univ Coll Sci J 3:91–172

Sheldrake TE, Sparks RSJ, Cashman KV, Wadge G, Aspinall WP (2016) Similarities and differences in the historical records of lava dome-building volcanoes: Implications for understanding magmatic processes and eruption forecasting. Earth Sci Rev 160:240–263

Siebert L, Cottrell E, Venzke E, Andrews B (2015) Earth's volcanoes and their eruptions: an overview. In: Sigurdsson H et al (eds) The encyclopedia of volcanoes, 2nd edn. Elsevier, Amsterdam. http://dx.doi.org/10.1016/B978-0-12-385938-9.00012-2

Stone J, Barclay J, Simmons P, Cole PD, Loughlin SC, Ramon P, Mothes P (2014) Risk reduction through community-based monitoring: the vigías of Tungurahua. Ecuador. J. Appl. Volcanol. 3:11

Stirling A (2010) Keep it complex. Nature 468:1029–1031

Tazieff H (1980) About the Soufrière of Guadeloupe. J Volcanol Geotherm Res 8:3–6

Venezky D, Newhall C (2007) "WOVOdat Design Document; The Schema, Table Descriptions, and Create Table Statements for the Database of Worldwide Volcanic Unrest (WOVOdat Version 1.0)." U.S. Geological Survey Open-File Report, 184.

Woo G (2008) Probabilistic criteria for volcano evacuation decisions. Nat Hazards 45(1):87–97

Woo G (2015) Cost-Benefit analysis is volcanic risk. In: Papale P (ed) Volcanic Hazards, Risks and Disasters. Springer, pp 289–300. https://www.elsevier.com/books/volcanic-hazards-risks-and-disasters/papale/978-0-12-396453-3

Wadge G, Isaacs MC (1988) Mapping the volcanic hazards from Soufriere Hills Volcano, Montserrat, West Indies using an image processor. J Geol Soc 145(4):541–551

Wadsworth FB, Vasseur J, Scheu B, Kendrick JE, Lavallée Y, Dingwell DB (2016) Universal scaling of fluid permeability during volcanic welding and sediment diagenesis. Geology 44(3):219–222

Wilkinson E (2015) Beyond the volcanic crisis: co-governance of risk in Montserrat. J Appl Volcanol 4:3

Wilkinson E, Lovell E, Carby B, Barclay J, Robertson REA (2016) The dilemmas of risk sensitive development on a small volcanic island. Resources 5 (2):21. doi:10.3390/resources5020021

Wright TL, Pierson T (1992) Living with volcanoes. No. 1073. US Geologic Survey

Wynne B (1992) Uncertainty and environmental learning: reconceiving science and policy in the preventive paradigm. Glob Environ Change 2:111–127

The Role of Laws Within the Governance of Volcanic Risks

R. J. Bretton, J. Gottsmann and R. Christie

Abstract

The governance of volcanic risks does not take place in a vacuum. In many cultures, volcanic risks are perceived to be susceptible to governance with the objective of achieving their effective mitigation, and have become the responsibility of the institutions and stakeholders of relevant social communities. An array of international, national and local laws dictate governance infrastructures, the roles of duty holders and beneficiaries and the relationships between them (the stakeholders), duties and rights (the stakes) and acceptable standards of safety and wellbeing (the ultimate rewards). Many regional, national and local stakeholders (individuals and entities) have a range of different, yet complementary, roles, duties, rights and powers. Much of this chapter, which has two main sections, represents a summary of a longer paper (Bretton et al. 2015) that addresses legal aspects of the future governance of volcanic risks. After a general introduction to relevant terminology in the first section, the second section describes the significant threat posed by periods of volcanic unrest.

R. J. Bretton (✉) · J. Gottsmann
School of Earth Sciences, University of Bristol, Wills Memorial Building, Queens Road, BS8 1RJ Bristol, UK

R. J. Bretton · J. Gottsmann · R. Christie
The Cabot Institute, University of Bristol, Wills Memorial Building, Queens Road, BS8 1RJ Bristol, UK

R. Christie
School of Sociology, Politics and International Studies, University of Bristol, 4 Priory Road, BS8 1TU Bristol, UK

Advs in Volcanology (2019) 23–34
DOI 10.1007/11157_2017_29

Published Online: 28 November 2017

The third section contains a general introduction to the critical concept of risk which lies at the heart of governance and provides a more detailed description of the many roles that national laws play. Reference is also made to international law which has an increasingly important role in the absence of relevant national laws, or when national laws are inadequate, ineffective or unenforced.

Keywords
Hazard · Risk · Risk governance · Legal duties

1 Introduction

This chapter describes the ways in which laws create the administrative and functional infrastructures that facilitate the effective mitigation of volcanic risks.

For the sake of brevity and clarity, we draw upon the existing rich discourse on relevant terminology (e.g. Fournier d'Albe 1979; Luhmann 1998, 1992; Power 2007, 2009; UN/ISDR 2009; Smith and Petley 2009; MIAVITA 2012) and adopt a number of brief working definitions.

A '*hazard*' is an event (defined by risk-related temporal, spatial and other parameters) that may cause adverse effects. It is a complex function being the "probability of any particular area being affected by a destructive volcanic manifestation within a given period of time" (Fournier d'Albe 1979, 321). A '*volcanic hazard*' is a volcanic scenario (defined by risk-related temporal, spatial and physical parameters) that may cause adverse consequences to people and/or valued assets.

In the early 1970s, definitions of risk identified the product of three separate and distinct elements—'vulnerability' on 'exposure' to a defined 'hazard' (UNESCO 1972). Bankoff et al. (2004) refers to volcanic hazards being one of three variables (hazard, exposure and vulnerability) that are convolved to produce volcanic risks. For the purposes of governance, it may be helpful to quantify each identified exposure in units of both number (i.e. the number of people exposed) and time (given that, by way of example, non-resident workers may be less exposed than full-time residents and 'vulnerability' may be related to length of exposure).[1]

'*Governance*' is a complex concept attracting a multitude of definitions (Walker et al. 2010). Although it "encompasses an array of organisations, practices and ideas" (Rothstein et al. 2012) that change over time, for sake of clarity, we embrace following definition.

> Governance is the sum of the many ways individuals and institutions, public and private, manage their common affairs. It is a continuing process through which conflicting or diverse interests may be accommodated and co-operative action may be taken. It includes formal institutions and regimes empowered to enforce compliance, as well as informal arrangements that people and institutions either have agreed to or perceive to be in their interest (Commission on Global Governance 1995, 4).

'Risk governance' includes all attempts to manage the three constituent variables of risk including steps to mitigate volcanic hazards (there are very few successful examples of this), reduce the exposure of people, assets etc. and reduce their vulnerability when exposed. This expression is adopted not only as an analytic term to describe stakeholders undertaking mitigation activities but also for 'normative' (i.e. evaluative standard) purposes. Risk governance

[1]The 18th report of the Scientific Advisory Committee on Montserrat Volcanic Activity contains a good example of a risk assessment which adopts this approach. It differentiates between the risks faced by residents in Zone A and those of workers involved in the shipment of sand from Plymouth Jetty.

has a set of definable 'good' qualities that provide for the effective integration of the key components of how risks are handled by risk stakeholders (Walker et al. 2010; IRGC 2009).

2 Geological Background

Active volcanism can involve complex multi-hazard phenomena. Precursory unrest provides, by means of its signals, the monitoring data upon which evidence-based short-term hazard analysis is grounded. However, periods of mild unrest, even if they may not lead to an eruption, can themselves present a range of hazards including earthquakes, ground deformation, hydrothermal changes/eruptions and gas/water chemistry changes. These precursory hazards can create societal risks that can escalate unnecessarily and therefore require very careful management. Unrest periods create, not only uncertainty about what is happening and resulting public alarm, anxiety and speculation, but also demands for information and advice (Johnston et al. 2002).

The evolution of an unrest period will depend upon its underlying causative processes, which can lead to different outcomes in different locations and with different spatial and physical properties (Rouwet et al. 2014; Sobradelo and Marti 2015).

Volcanic hazard communications and risk mitigation decisions rely upon the suitable and sufficient collection, and the correct analysis and interpretation of monitoring data, and the geological record (Newhall and Hoblitt 2002; Sparks et al. 2012; Rouwet et al. 2014). The analysis of monitoring data, which will often be limited in both quantity and quality, is challenging and there are many uncertainties in identifying causes and thereafter anticipating the evolution of unrest and imminent eruption (Sparks et al. 2012; Phillipson et al. 2013; Sobradelo and Marti 2015).

Hazard analysis is difficult and the risk governance stakes are high. Poorly handled unrest periods cause social, economic and political problems, even without an eruption. "Adverse response may take the form of the release of inappropriate advice, media speculation, unwarranted emergency declarations and premature cessation of economic activity and community services" (Johnston et al. 2002, 228).

3 Risk Governance and Roles of Law

The concept of risk as something that can be managed through human intervention is a relatively new one and important because it has become an increasingly pervasive concept in many societies. Risk is also associated with notions of choice, responsibility and blame (OECD 2015).

Risk evolved from its modest origins in the seventeenth century and became in the nineteenth century a principle for the objectification of possible experience—not only of the hazards of personal life and private venture, but also of the common venture of society (Gordon 1991).

By the late nineteenth century, risk had "become central to the rhetoric of regulation". State regulation of risk emerged as the means by which the state controlled economic activities in Western societies. The traditional objects of state regulation were manufactured risks, most particularly those resulting from scientific and technological innovation within manufacturing processes. The usual style of state regulation was "command and control" by imposing formal, structured and active risk management duties. The state exercised control through the promulgation of primary (i.e. enabling) and secondary (i.e. detailed implementing) laws and policing through specialist inspectorates.

In the twenty first century, regulation is no longer confined to non-natural, human-made risks. Many risks are, in whole or in part, recurring social manifestations (i.e. human-made phenomena) with negative consequences (Lauta 2014). In many cultures, particularly western cultures, they are no longer perceived as the consequences of external forces occurring independently of society and insusceptible to mitigation by society. Accordingly, they are now positioned within, and have become the responsibility of, the institutions and stakeholders of relevant social communities (Lauta 2014). These human-made risks are perceived to be susceptible to regulation with the objective of achieving their effective mitigation. By way of illustration, the population of Naples has greatly increased since 1944 and many would argue that the resulting increase in volcanic risk exposure is human-made and capable of regulation.

Low probability-high impact risks pose a particular challenge for legislators. In fact there are three related challenges, namely scientific uncertainty, a low likelihood of occurrence, and significant societal consequences. Whilst the elevated consequences of these risks call for some level of regulation, the intrinsic uncertainty and low probability of their occurrence make it difficult to review the evidentiary scientific justification, to assess costs and benefits, and to identify means by which chosen regulatory goals can be pursued (Simoncini 2013).

In the absence of a tragedy, it is difficult to measure the performance of law-backed societal risk governance by the usual measures of: (1) economy (e.g. value for money) for input and process; (2) efficiency (e.g. quality delivered on time) for process and output; and (3) effectiveness for output and outcome. The indicators of outcome (the intended and unintended results) of the integrated governance system will be related to the impacts on, and the consequences for, public good, safety, security, health and welfare but it will be a challenge for any related targets (e.g. benchmarks and performance standards) to be SMART—Specific, Measurable, Achievable, Relevant and Timed (OECD 2002).

By contrast, in a fact-finding process of scrutiny after a tragedy, the use of SMART targets may become more practicable. It may be possible to measure hazard characterisation outputs against planned targets for timely delivery, user-friendliness, outcome-focussed, and temporal/spatial/intensity forecast accuracy. Based upon findings of fact, it may be feasible to quantify the resulting risk-mitigation impact measured in lives and assets saved.

Notwithstanding these challenges, many jurisdictions have national laws that attempt to regulate the management of risks arising from natural hazards. Many reflect the shift in paradigm, at both international and national levels, from focussing on ex-post, reactive response (the phases of emergency response and post-disaster longer term recovery) to ex-ante, pro-active risk management and mitigation (the phase of planning and preparedness) (UN SC-DRR 2009).

As illustrated in Fig. 1, national laws create governance infrastructures, duties of care and duty holders, rights and rights holders, enabling powers, regulators, enforcement powers, and lastly scrutiny venues. Each will now to considered in turn.

3.1 The Creation of National Risk Governance Infrastructures

National laws tend to identify, authorise and fund risk governance bodies (e.g. government departments and agencies, and public corporations) and public officials (e.g. individuals such governors, mayors, prefects and village heads) within a coherent legal and administrative framework, in other words, a risk governance infrastructure. These laws often use and build upon existing entities within existing administrative frameworks that have multi-level national, regional, district, municipal etc. political divisions and subdivisions.

In some jurisdictions, formal legal infrastructures anticipate and rely upon less formal structures and relationships at local levels nearer at-risk communities. For example, in Ecuador, the risk governance infrastructure relies upon the engagement and commitment of local representatives (e.g. chiefs and elders) and volunteers, such as hazard wardens/monitors, for both hazard data gathering and risk mitigation.

In some jurisdictions, such as Italy, the laws favour the imposition of duties upon individuals, rather than impersonal legal entities such as government departments/agencies and public companies. Legal duties may be founded upon an

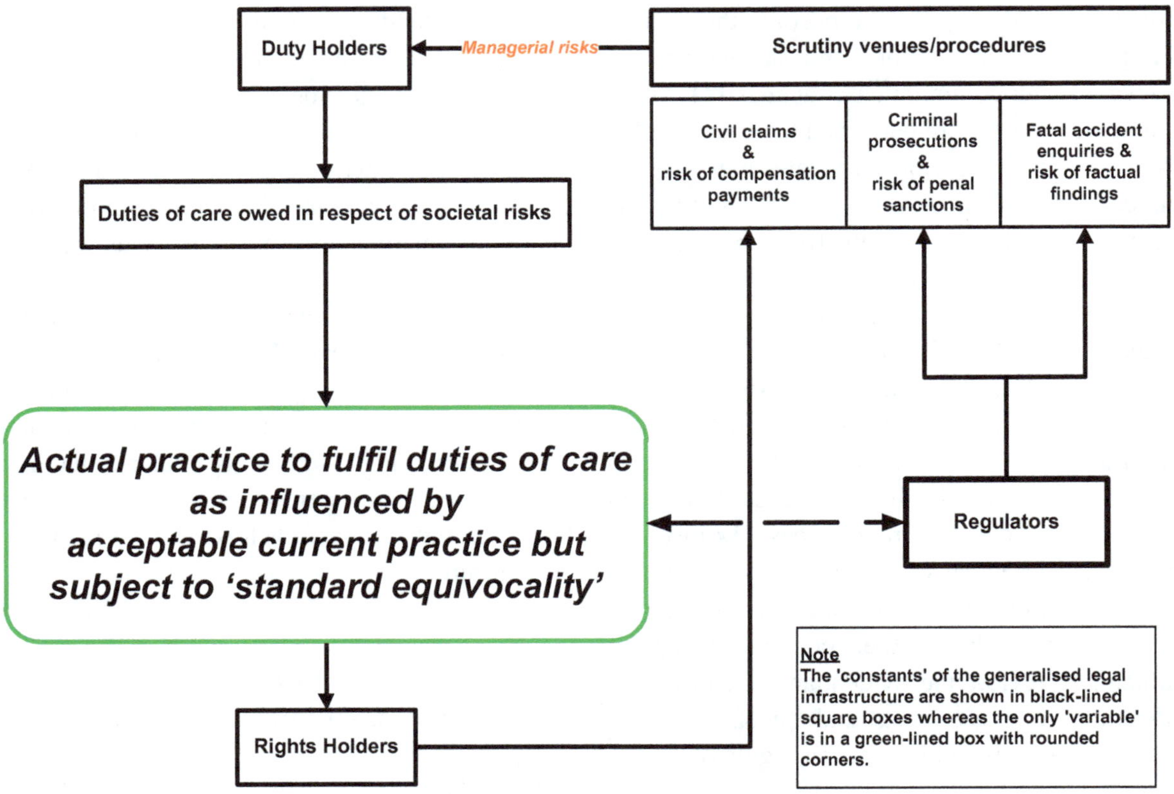

Fig. 1 The many roles of law in the governance of natural hazards set out in a generalised legal framework

individual having effective decision-making powers and control over financial resources rather than upon an individual holding a particular title or occupying a particular post (Bergman et al. 2007).

These infrastructures can be complex, confusing, fragmented and multi-level. They are often the creations of multiple sets of national primary (enabling) and secondary (detailed implementing) legislation supplemented as necessary by further provisions at ministerial, inter-ministerial, regional, provincial and local levels of government.

Occasionally additional specialised bodies are established (e.g. emergency management agencies, research/monitoring institutes and volcano observatories) with the creation of statutory roles to be filled by appointed individuals.

In a few known jurisdictions (e.g. USA, Canada and the Philippines) laws also regulate to varying degrees the qualification, licensing and registration of geologists and the practice of geology per se.

3.2 The Creation of Duty and Rights

National laws allocate to bodies and individuals (duty holders such as volcano observatories and civil protection authorities) high level management functions with responsibilities (duties of care), which are owed to the particular classes of people for whose benefit the duties were created (rights holders).

Since modest beginnings in the 1840's near Vesuvius Italy, the role of over 100 volcano observatories around the world has evolved. Observatories have at least two overlapping roles which involve a synergy of observation and theory. They have been described as 'critical in the volcanic risk reduction cycle'[2] (Jolly 2015, 302), and employ and/or engage scientists who practice at the hazard-risk interface. The World Organisation of Volcano Observatories (WOVO), a Commission of the International

[2]This cycle includes periods before, during and after periods of volcanic unrest that may or may not lead to an eruption (Jolly 2015, 302).

Table 1.1 Volcano observatories—three contrasting regimes

Montserrat—One individual with no powers of delegation
Section 8 of the Montserrat Volcano Observatory Act 2002 states that the Director of the Observatory shall be responsible for "reporting on the status of the volcanic activity in a regular and timely manner to the appropriate authorities" and "assisting in the dissemination to the public of information concerning the status of volcanic activity"
Alaska, USA—A group of institutions
"The Alaska Volcano observatory is a joint programme of the United States Geological Survey (USGS), the Geophysical Institute of the University of Alaska Fairbanks (UAFGI) and the State of Alaska Division of Geological and Geophysical Surveys (ADGGS)" (Jolly 2015, 299)
New Zealand—One institution under powers of delegation given to one individual
The Institute of Geological and Nuclear Sciences Limited (or GNS Science) is a Crown research institute created in 1992. The GNS has "sole responsibility for providing volcanic activity warnings and hence provides the function of a volcano observatory" (Jolly 2015, 299) under wide powers of delegation given to the Director of Civil Defence Emergency Management in the Civil Defence Emergency Management Act 2002

Association of Volcanology and Chemistry of the Earth's Interior (IAVCEI), is an organization of and for volcano observatories of the world. WOVO's website notes that its "members are institutions that are engaged in volcano surveillance and, in most cases, are responsible for warning authorities and the public about hazardous volcanic unrest".

Examples of three contrasting regimes are given in Table 1.1.

In some jurisdictions, general disaster management obligations are also imposed on 'the community' (i.e. members of the public) and non-government business entities. At-risk individuals and communities, businesses (such as aeroplane makers and operators within the aviation industry) and insurers have active and critical roles to play in the governance of volcanic risks, however, their roles are not the principal focus of either this chapter or Bretton et al. (2015). The needs and sentiments of all duty and rights holders, who depend upon and use geoscientific knowledge of volcanic hazards, must be identified and reflected carefully and clearly in the roles and interface practices of volcanologists.

During an emerging period of volcanic unrest, the relevant duty holders may change as the defined duties are transferred from one duty holder to another—sometimes as a result of changing hazard or risk characterisations. These duties of care can be framed in a wide variety of ways. They may relate to general health and safety, not specifying any risk creator (a particular hazard, natural or otherwise) or they may be more specific, identifying a particular hazard (ground uplift, earthquakes etc.).

Rights holders may be given the right to a safe and healthy environment, and to be represented, consulted or engaged in risk decisions and/or given information. Additional rights may be given to certain categories of persons due to special vulnerabilities and/or the influence of social structures and practices. These categories may include women, the very poor, older persons, children and people with disabilities (IFRC 2015).

There are two main types of duty of care, which are called here, respectively, 'functional' and 'goal-setting'. Functional duties dictate the fulfilment of a particular role (e.g. a duty to undertake monitoring, to prepare plans and programmes for emergency preparedness or to provide emergency preparedness communications and warnings). Goal-setting duties require the achievement of an outcome (e.g. a duty to ensure the safety and wellbeing of identified rights holders). Not even within the highly regulated field of occupational health are these safety goals absolute (i.e. unqualified). The imposition of an unrealistic absolute duty would give rights holders a theoretical guarantee of health and safety within a risk-free environment.

As a general rule, 'qualified' duties of care are therefore laid down. These duties represent democratic statements or mandates of 'acceptable' or 'optimal' risk after mitigation, and express a rational trade-off between safety and risk (Hood and Rothstein 2001; Rothstein 2014). Rothstein (2014) notes "After all, what is an acceptable risk other than a euphemistic boundary between an *acceptable* adverse outcome and an *unacceptable* failure".

Compliance with qualified duties inevitably requires duties holders to perform a risk-focussed cost/benefit analysis and a test of proportionality. A duty holder wishing to establish that societal risks have been reduced 'as low as reasonably practicable' has to show that the costs (the sacrifice) of further feasible safety measures would be grossly disproportionate to the additional safety benefits of those measures (based upon UK Office for Nuclear Regulation 2013).

Laws rarely, if ever, attempt to dictate, in either general or more detailed terms, the societal risk management arrangements that will be required to either fulfil a functional role or achieve a stated safety outcome. In practice, an assessment of societal vulnerability has two main stages. Firstly, the nature and scope of duties of care must be identified and delineated. This is essentially a matter of law and involves the legal interpretation of primary and secondary legislation and, if relevant, case law. Secondly, it is necessary to identify the actions that the duty holder should take to fulfil those duties. This is far more difficult. Competent lawyers can describe the safety function or outcome required in law—in football terms the dimensions and position of the goal. However, they can offer very limited guidance regarding the practical measures that the competent societal risk manager will need to take to achieve legal compliance (i.e. how to actually get the ball over the goal line).

In the case of food standards and occupational health and safety standards, it is common for national laws to set up government agencies to carry out research, to set performance standards and offer approved codes of practice or authoritative guidance. By contrast, in respect of volcanic hazards, at both the international and national levels, there appear to be neither: (1) law-based performance standards offering guidance to societal risk managers; or (2) law endorsed self-regulatory regimes such as those that frequented the early stages of food regulation.

Within the 'goal-setting' legislative approach, referred to above, it is implicit that there is an obligation on duties holders to establish the nature and suitability (i.e. the legal adequacy) of their societal risk management arrangements. This difficult justification will usually be done post-facto, in other words, only after the risk outcome (perhaps a disaster properly so-called) is known and legal consequences are already being considered (Simoncini 2011). The justification will cover, but not be limited to, the arrangements that were necessary for the planning, organisation, control, monitoring and review of societal risk mitigation measures. To complete the authors' footballing analogy, post-facto legal processes are analogous to slow motion TV replays, in full view of partisan onlookers and experts with hindsight. They determine what has happened, whether or not a goal has been scored and, if not, why not and what effect any missed goal had on the final score (i.e. whether legal compliance has actually been achieved and, if not, why not and what the consequences should have been if compliance had been achieved).

'Standard equivocality', which is the absence of commonly recognised standards (norms), is likely to exist in the absence of clear 'legal' requirements, approved codes of practice or guidance. The resulting challenges faced by duty holders are: (1) to find or design authoritative standards or benchmarks to steer their societal risk management arrangements; and thereby (2) to increase their chances of fulfilling their societal risk duties of care and achieving legal compliance; and thereby (3) to minimise their vulnerability to managerial risks.

Rothstein (2002) and Hood (1986) have noted that, in the absence of commonly agreed and practical principles or methodologies by which compliance can be measured ('standard-unequivocality'), process compliance is difficult to monitor and enforce. Other obstacles to monitoring, surveillance and enforcement include inherent scientific uncertainty, a dynamic state of scientific knowledge, a lack of expertise within regulatory agencies, and often complex and fragmented multi-level infrastructures. Donovan and Oppenheimer (2014) note complexities in governmental structures presented major challenges to managing volcanic eruptions in Montserrat. Recent crises including the 2010 Icelandic Eyjafjallajökull eruption have highlighted the difficulty of co-ordinating and synthesising scientific input from many different disciplines and institutions and translating these into useful policy advice at very short notice (OECD 2015).

3.3 The Creation of Powers

National laws have traditionally granted defined 'authorities' to government duty holders backed by administration, protection and intervention powers (ordinary, extraordinary and emergency). Governance, with an emphasis more on control than protection, has often been achieved by the exercise of authority using linear "coercion and enforcement" (Walker et al. 2010).

3.4 The Creation of Regulators, Enforcement Powers and Scrutiny Venues

Laws often establish, resource and empower regulators to monitor the performance of duty holders and to take enforcement actions, including prosecutions, against them if necessary. Examples of regulators include the Labour Standards Agency in Japan, the Department of Labour Health and Safety Service in New Zealand and the Occupational Safety and Health Agency in the United States of America. These regulators often have very wide powers similar to, and sometimes exceeding, those of police forces. They include the power to enter premises, to investigate and inspect, to acquire and preserve evidence and to serve notices.

Laws provide the formal scrutiny venues (courts, tribunals etc.) and related procedures for: (1) the ex-ante pro-active enforcement of duties of care by regulators, generally health and safety agencies; and (2) the ex-post facto reactive scrutiny of events, the identification of duty holders, the assessment of what happened and what should have happened and, if appropriate, the imposition of sanctions and/or the granting of remedies. The latter procedures are required at a national level to comply with the international law ex-post facto obligations which are now considered.

3.5 The Role of International Law

In the absence of relevant national laws, or when national laws are inadequate, ineffective or unenforced, there is room for the intervention of international law. The European Court of Human Rights (ECHR) has taken the lead and it is suggested here that in time the Inter-American Court of Human Rights will follow. The European Convention of Human Rights (EConHR) lays down a positive obligation on States to take appropriate steps to safeguard the lives of citizens within their jurisdiction. Article 2(1) EConHR provides that "Everyone's right to life shall be protected by law. No one shall be deprived of his life intentionally save in the execution of a sentence of a court following his conviction of a crime for which this penalty is provided by law."

In the context of the management of natural hazards, the most important case involving Article 2 arose in 2008. Budayeva and others v Russia (2008) ECHR 15339/02 concerned a mudslide. In this case, the EHCR considered principles that had been applied in Oneryildiz v Turkey (2004) to a human-made hazard—an industrial risk or dangerous activity such as the operation of a waste site. They were subsequently adopted in Kolyadenko and others v Russia (2012) in respect of natural flash floods.

Budayeva and others v Russia

The town of T was situated in a mountain district. Two tributaries passed through it and were known to have associated mudslides. A mud collector and a dam were constructed in order to protect T. The dam was seriously damaged by a mud and debris

> flow in August 1999, so funds were requested to construct observation posts to warn of mudslides until it could be repaired, and to carry out certain emergency works to the dam.
>
> Those measures were never implemented. A number of mudslides occurred in July 2000, killing eight residents, including the first applicant's husband, and destroying the applicants' homes.
>
> It was decided to dispense with a criminal investigation into the circumstances of the death of the first applicant's husband, and claims of compensation by the first applicant and others were refused on the basis that a mudslide of such exceptional force could neither have been predicted nor stopped. However, the applicants were granted substitute housing and a lump-sum emergency allowance.
>
> The applicants complained to the ECHR, inter alia, that the authorities had violated the substantive limb of Article 2 of the EConHR. The first applicant asserted that the authorities were responsible for the death of her husband and she and the other applicants asserted that the authorities had failed to take appropriate measures to mitigate the risks to their lives posed by natural hazards.
>
> The Court concluded that the relevant authorities were aware of the mudslides (the hazards) and their capacity to cause devastating consequences (the risks). There was no ambiguity about the scope and timing of the work that needed to be performed (the risk mitigation actions). After 1999, risk mitigation was not given proper consideration by the decision makers and budgetary bodies (the duty holders) and there was no functioning early warning system. State responsibility for the deaths had never been investigated. Each applicant was awarded compensation.

The EHCR determined that the obligation in Article 2 entails, above all, a primary duty on the State to put in place a clear legislative and administrative framework designed to provide effective deterrence against threats to the right to life. This applies in the context of any activity, whether public or not, in which the right to life may be at stake and extends not only to industrial risks and dangerous activities but also to actions and omissions to control natural hazards.

In the cases of Oneryildiz v Turkey (2004), Budayeva v Russia (2008), and Kolyadenko v Russia (2012), the EHCR determined that there is a positive obligation: (1) ex-ante to take substantive regulatory measures to manage risks; and (2) ex-post facto to ensure that any risk eventuated fatalities are followed by a public investigation. In relation to the latter, procedures must exist for identifying not only shortcomings in the ex-ante regulatory measures but also any errors committed by those responsible (i.e. duty holders). If there are any shortcomings and the infringement of the right to life was not intentional, it is not necessary for criminal proceedings to be brought in every case. It may satisfactory to make available to the victims civil law remedies (either alone or in conjunction with a criminal law remedy), enabling any responsibility of the parties concerned to be established and any appropriate civil redress, such as an order for the payment of damages, to be obtained.

The positive obligations of EConHR State duty holders under the ECHR are summarised in Fig. 2.

3.6 The Role of International Institutions and Agencies

In March 2015, the International Federation of Red Cross and Red Crescent Societies (IFRC) and the United Nations (UN) Development Programme issued the pilot version of "The checklist on law and disaster risk reduction". It encourages accountability mechanisms within legislation to address failures to fulfil risk governance responsibilities. In particular it advocates laws: (1) to establish public reporting or parliamentary oversight mechanisms and transparency requirements for government entities tasked with risk governance responsibilities; (2) to give a mandated role to the judiciary in enhancing accountability; (3) to provide enforceable incentives for compliance and disincentives for non-compliance; and (4) to establish legal and/or administrative sanctions (as appropriate) for public officials individuals and businesses for a gross ("particularly egregious") failure to fulfil their duties (IFRC 2015, 16).

The prioritisation of mitigation before response and recovery was recognised within the Sendai Framework for Disaster Risk Reduction 2015–2020 (the Sendai Framework) which

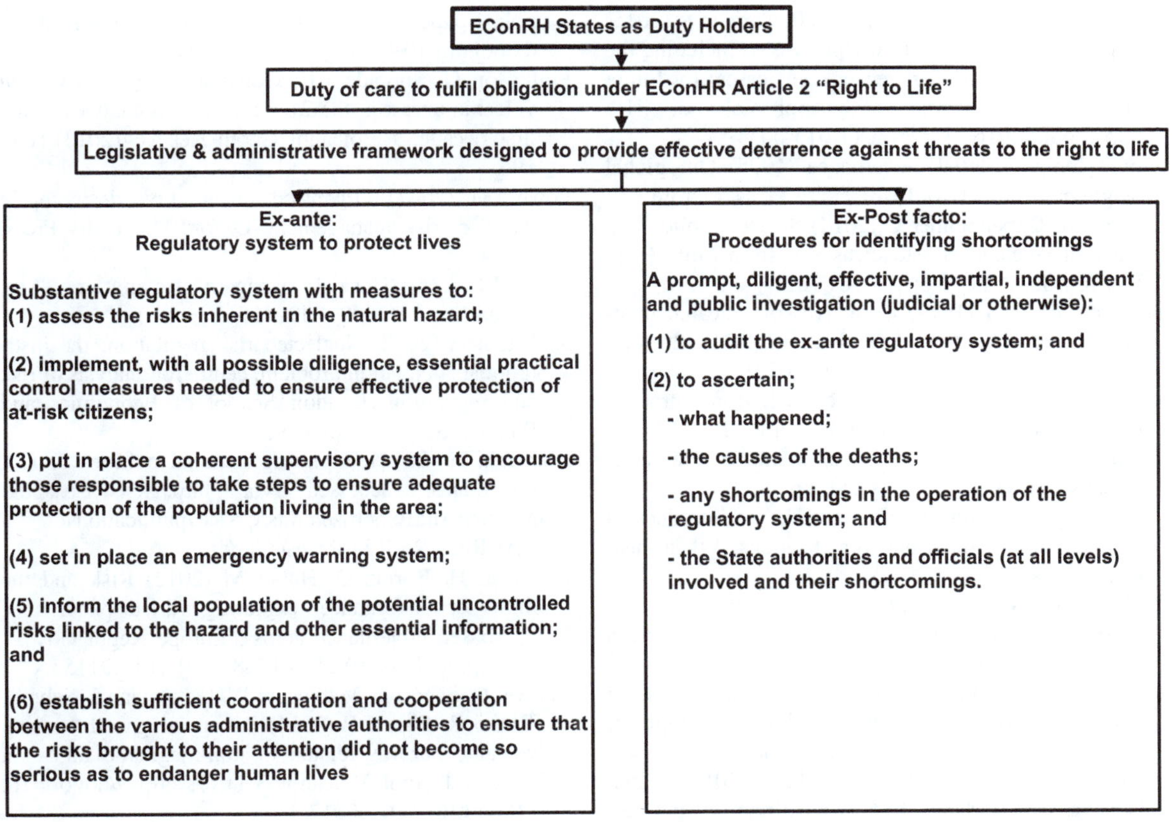

Fig. 2 The obligations of EConHR 'States' to manage natural hazards

emerged from the United Nations 3rd World Conference on Disaster Risk Reduction (UN/ISDR 2015). It is suggested here that one inevitable effect of the Sendai Framework will be to enhance the importance of not only the collection and interpretation of monitoring data but also the better characterisation of unrest periods. For the reasons stated in Sect. 2, periods of volcanic unrest, even if they do not lead to an eruption, present multiple hazards and risks which require very careful assessment and mitigation.

4 Conclusions

In many cultures, volcanic risks are perceived to be susceptible to governance and have become the responsibility of the institutions and stakeholders of relevant social communities.

Laws create the stakeholders, the stakes and the standards of risk governance. Without national laws, supplemented by international laws and initiatives, few countries would have the complex administrative infrastructures necessary for the mitigation of volcanic risks.

Although emergency response may still dominate thinking and funding in some jurisdictions, national laws are unlikely to diminish in number and/or reach in the light of the emerging international law governance norms, the IFRC/UN law checklist and the Sendai Framework.

References

Bankoff G, Frerks G, Hilhorst D (2004) Mapping vulnerability – disasters, development & people. Earthscan, London and New York

Bergman D, Davis C, Rigby B (2007) International comparison of health and safety responsibilities of company directors, HSE research report RR 53, 207. http://www.hse.gov.uk/leadership/international.pdf

Bretton RJ, Gottsmann J, Aspinall WP, Christie R (2015) Implications of legal scrutiny processes (including the L'Aquila trial and other recent court cases) for future volcanic risk governance. J Appl Volcanol 4:18. https://doi.org/10.1186/s13617-015-0034-x

Commission on Global Governance (1995) Our global neighbourhood. Oxford University Press, Oxford

Donovan A, Oppenheimer C (2014) Science, policy and place in volcanic disasters: insights from Montserrat. Environ Sci Policy 38:150–161 (May)

Fournier d'lbe EM (1979) Objectives of volcanic monitoring and prediction. J Geol Soc136:321–326. doi: https://doi.org/10:1144/gsjgs.136.3.0321

Gordon C (1991) Governmental rationality: an introduction. In Burchell G, Gordon C, Miller P (eds) The foucault effect: studies in Governmentality. Harvester Wheatsheaf, Hemel Hempstead, pp 1–52

Hood C (1986) Administrative analysis: an introduction to rules, enforcement and organisations. Wheatsheaf Books, Sussex

Hood C, Rothstein H (2001) Risk regulation under pressure: problem solving or blame shifting? Adm Soc 33(1):21–53

International Federation of Red Cross, Red Crescent Societies (IFRC) and United Nations Development Programme (2015) The checklist on law and disaster risk reduction, Pilot version, March 2015. IFRC, Geneva, Switzerland. http://www.ifrc.org/PageFiles/115542/The-checklist-on-law-and-drr.pdf

International Risk Governance Council (IRGC) (2009) What is risk governance? www.irgc.org

Jolly G (2015) The role of volcano observatories in risk reduction. In: Loughlin SC, Sparks RSJ, Brown SK, Jenkins SF, Vye-Brown C (eds) Global volcanic hazards and risk. Cambridge University Press, Cambridge

Johnston D, Scott B, Houghton B, Paton D, Dowrick D, Villamor P, Savage J (2002) Social and economic consequences of historic caldera unrest at the Taupo volcano, New Zealand and the management of future episodes of unrest. Bull New Zealand Soc Earthquake Eng 35(4):215–229

Lauta KC (2014) Disaster law. Routledge, Oxford and New York

Luhmann N (1998) Observations on modernity. Stanford University Press, Stanford, CA

Luhmann N (1992) Risk: a sociology theory. de Gruyter, Berlin

MIAVITA (2012) Handbook for volcanic risk management: prevention, crisis management and resilience. MIAVITA team, Orleans

Newhall CG, Hoblitt RP (2002) Constructing event trees for volcanic crises. Bull Volcanol 64:3–20

OECD (Organisation for Economic Co-operation and Development) (2002) Outcome focussed management in the United Kingdom by Ellis K, Mitchell S. OECD Publications, Paris, France

OECD (2015) "Scientific advice for policy making: the role and responsibility of expert bodies and individual scientists", OECD science, technology and industry policy papers, no 21, OECD Publishing, Paris. http://dx.doi.org/10.1787/5js33l1jcpwb-en

Phillipson G, Sobradelo R, Gottsmann J (2013) Global volcanic unrest in the 21st century: an analysis of the first decade. J Volcanol Geoth Res 264(2013):183–196

Power M (2007) Organised uncertainty: designing a world of risk management. Oxford University Press, Oxford

Power M (2009) The risk management of nothing. Acc, Organ Soc 34(2009):849–855

Rothstein H (2002) Neglected risk regulation: the institutional attenuation problem, centre for analysis of risk and regulation. London School of Economics and Political Science, London

Rothstein H (2014) Exploring national cultures of risk governance accessed via http://www.lse.ac.uk/researchAndExpertise/units/CARR/publications/CARRmagR&R25-Rothstein.pdf

Rothstein H, Borraz O, Huber M (2012) Risk and the limits of governance: exploring varied patterns of risk-based governance across. Europe Regul Gov 1–21 http://doi.org/10.111/j.1748-5991.2012.01153.x

Rouwet D, Sandri L, Marzocchi W, Gottsmann J, Selva J, Tonini R, Papale P (2014) Recognizing and tracking volcanic hazards related to non-magmatic unrest: a review. J Appl Volcanol 3(1):17. https://doi.org/10.1186/s13617-014-0017-3

Simoncini M (2011) Regulating catastrophic risks by standards. EJRR (1):37–50

Simoncini M (2013) Governing air traffic management in the single European sky: the search for possible solutions to safety issues. Eur Law Rev Issue 2 (2013):209–228

Smith K, Petley DN (2009) Environmental hazards: assessing risk and reducing disaster. Routledge, London and New York

Sobradelo R, Martí J (2015) Short-term volcanic hazard assessment through Bayesian inference: retrospective application to the Pinatubo 1991 volcanic unrest. J Volcanol Geoth Res 290(2015):1–11

Sparks RSJ, Aspinall WP, Auker M, Crosweller S, Hincks T, Mahony S, Nadim F, Polley J, Syre E (2012) Mapping and characterising volcanic risk, magmatic rifting and active volcanism conference, Addis Ababa, 12 Jan 2012

United Kingdom Office for Nuclear Regulation (2013) Guidance for the demonstration of ALARP (As Low As Reasonably Practicable) NS-TAST-GD-005 Revision 6

United Nations Educational, Scientific and Cultural Organization (UNESCO) (1972) Report of consultative meeting of experts on the statistical study of natural hazards and their consequences Document SC/WS/500 11pp, Paris, France

United Nations Development Programme Safer Communities through Disaster Risk Reduction in Development Programme SC-DRR (2009) Lessons learned: disaster management legal reform—The Indonesian experience UND-, Jakarta, Indonesia

UN/ISDR (2009) Terminology on disaster risk reduction. Switzerland, Geneva

UN/ISDR (2015) Sendai framework for disaster risk reduction 2015-30

Walker G, Whittle R, Medd W, Watson N (2010) Risk governance and natural hazards, Cap Haz-Net WP2 Report, Lancaster Environment Centre, Lancaster University, Lancaster. http://caphaz-net.org/outcomess-results

Legal Authorities and Case Law

Budayeva, others v Russia (2008) Applications 15339/02, 21166/02, 20058/02, 11673 & 15343/02, judgment of 20 March 2008, ECHR 15339/02

Kolyadenko, others v Russia (2012) Applications 17423/05, 20534/05, 20678/05, 23263/05, 24283/05, 35673/05, ECHR 17423/05

Oneryildiz v Turkey (2004) Application 48939/99 18 BHRC 145, 41 EHRR 20

Deterministic Versus Probabilistic Volcano Monitoring: Not "or" But "and"

D. Rouwet, R. Constantinescu and L. Sandri

Abstract

Volcanic eruption forecasting and hazard assessment are multi-disciplinary processes with scientific and social implications. Our limited knowledge and the randomness of the processes behind a volcanic eruption yield the need to quantify uncertainties on volcano dynamics. With deterministic and probabilistic methods for volcanic hazard assessment not always being in agreement, we propose a combined approach that bridges the two schools of thoughts in order to improve future volcano monitoring. Expert elicitation has proven to be an effective way to bind deterministic research within a probabilistic framework aiming to reduce the uncertainties related to any hazard forecast; yet, numerous exercises based on expert elicitation have revealed that the attempt to reduce uncertainties led to the creation of new ones, often unquantifiable, created by human nature and reasoning during stressful situations. Such reasoning ignores the complexity of volcanic processes and the fact that every scenario has a probability to occur. The recent probabilistic methods and tools marry probabilistic and deterministic approaches and lead to unprecedented models. Nevertheless, probabilistic hazard assessment is often misunderstood as not all of the researchers involved have backgrounds in such matters. A probabilistic method cannot stand-alone

D. Rouwet (✉) · L. Sandri
Istituto Nazionale di Geofisica e Vulcanologia, Sezioni di Bologna, Bologna, Italy
e-mail: dmitrirouwet@gmail.com; dmitri.rouwet@ingv.it

R. Constantinescu
Seismic Research Centre, University of West Indies, Saint Augustine, Trinidad and Tobago

Present Address:
R. Constantinescu
School of Geosciences, University of South Florida, 4202 E. Fowler Avenue, NES 107, Tampa, FL 33620-5550, USA

Advs in Volcanology (2019) 35–46
DOI 10.1007/11157_2017_8

Published Online: 21 June 2017

as it depends on data input obtained by deterministic approaches. We propose that, given the symbiotic relationship between the two methods, a probabilistic framework can play a role of moderator between various deterministic disciplines, thus creating a coherent environment for discussion and debate among seismologists, geodesists, geochemists. This can be achieved by training all scientists involved in hazard assessment, probability theory and data interpretation, while at least one group member objectively uses the information provided to produce the probabilities. Hence, numerical outcomes can be interpreted transparently as they represent the quantification of experts' knowledge and related uncertainties. A probabilistic method that incorporates the joint opinions of a group of multi-disciplinary researchers facilitates a more straightforward way of communicating scientific information to decision-makers.

Resumen extendid

La previsión de erupciones volcánicas y la evaluación del peligro son procesos multidisciplinarios, con implicaciones tanto científicas como sociales. Nuestro conocimiento limitado de los procesos detrás de una erupción volcánica y su aleatoriedad genera la necesidad de cuantificar las incertidumbres sobre las dinámicas del volcán y de mejorar la política de la toma de decisiones durante una crisis volcánica. Sabiendo que existe un desacuerdo sobre el uso de métodos determinísticos o probabilísticos durante la evaluación de la peligrosidad volcánica, revisamos ambos métodos y proponemos un enfoque que sirve como puente entre las dos escuelas de pensamiento y que pueda mejorar las capacidades de monitoreo volcánico en el futuro, hacia el reconocimiento en tiempo de manifestaciones volcánicas y amenazas relacionadas. La elicitación de expertos resulta ser una manera efectiva para relacionar la investigación determinística con el marco probabilístico para poder reducir la incertidumbre relacionada a cualquier intento de previsión de erupción; sin embargo, numerosos ejercicios basados en elicitaciones de expertos revelaron el hecho que este intento de reducir la incertidumbre creó nuevas incertidumbres, a menudo imposible de cuantificar, siendo generada por la naturaleza del pensamiento humano durante situaciones de estrés. El proceso general es sujeto a la personalidad de un/a investigador/a o un grupo de investigadores y sus ideas basadas en su experiencia. Esta manera de pensar interfiere con la complejidad intrínsica de los procesos volcánicos y con el hecho que cada escenario tiene una probabilidad de ocurrencia. Los métodos e instrumentos probabilísticos recientes juntaron los investigadores probabilísticos y determinísticos lo que resultó en modelos e interpretaciones de información sobre volcanes sin precedentes. Sin embargo, la novedad de la evaluación probabilística de peligrosidad es, a menudo, incomprendida debido al hecho que no todos los investigadores involucrados tienen una formación en estas materias teóricas. El método probabilístico no puede existir autónomamente ya que depende de datos de entrada obtenido a través de los estudios determinísticos. Proponemos que, dada la relación simbiótica entre ambos métodos, un marco probabilístico puede jugar el papel como moderador entre las varias disciplinas deterministas, creando un ambiente coherente para discusiones y debates entre científicos (e.g., sismólogos, geodetas, geoquímicos). Este se puede obtener por medio de entrenamiento de todos los científicos involucrados en el monitoreo volcánico en la teoría probabilística y la interpretación de datos, mientras que al menos un miembro del grupo utiliza de manera objetiva la información disponible para producir las probabilidades numéricas. Así, los resultados numéricos pueden ser interpretados sin duda alguna, ya que representan la cuantificación del conocimiento de los expertos y las incertidumbres relacionadas. Un método probabilístico que incorpora las opiniones conjuntas de un grupo de investigadores multidisciplinarios facilitará una manera más transparente de comunicación de la información científica hacia las autoridades civiles, así mejorando (1) el proceso de toma de decisiones durante una crisis y la mitigación a largo plazo, y (2) el estado de medidas de preparación que incorpora los varios aspectos sociales. En el futuro, los reportes de previsiones emitidos por los científicos deberían incluir los resultados numéricos de los modelos probabilísticos; la arquitectura de monitoreo se debería expander más allá del arreglo clásico "sismo-deformación-gas" hacia un arreglo "sismo-deformación-gas-probabilidad". Este capítulo de opinión pretende proponer una ideología posible, con el máximo respeto para el volcán, la sociedad, los científicos individuales o los grupos de científicos, y para las autoridades que toman las decisiones, con un objetivo común: mejorar la previsión de amenazas relacionadas a los volcanes para proteger la sociedad.

Keywords

Volcano monitoring · Probabilistic hazard assessment · Deterministic research · Bridging and symbiosis · Best practice scheme

Palabras clave

Monitoreo volcánico · Evaluación probabilística de amenaza · Investigación determinística · Puente y simbiosis · Esquema de mejor práctica

1 Introduction

Volcanoes are intrinsically complex and unpredictable systems manifesting non-linear behaviour in space and time, on the long- and short-term. Understanding how volcanoes evolve with time through the various stages of activity—from quiescence through unrest to eruption—is highly challenging. Awareness of these facts is a basic requisite when working with/on volcanoes. A major goal in volcanology is quantifying uncertainties on volcano dynamics, and learning how to translate these to decision makers and, occasionally, to the population. The combination of social implications and forecasting future behaviour of complex natural systems makes volcanology a rather unique but "tricky business". Besides the need to quantify uncertainties and better understand our limited knowledge on volcanoes it is necessary to legally protect volcanologists when exporting their knowledge outside their protected professional community (Bretton et al. 2015).

During the past 40–50 years, volcano monitoring and eruption forecasting during volcanic crises has been largely dominated by the deterministic approach (Sparks 2003, i.e. most likely scenarios). Experts with different research backgrounds track changes in "their" parameters related to volcanic activity, afterwards discussed in "protected" round tables, generally behind locked doors, to eventually come up with a single voice. This single voice does not and should not reflect possible internal conflicts or disagreements behind the closed door. Such disagreement is an often-unstated expression of the uncertainty due to our lack of knowledge on volcanic processes, and the possible unavailability of data (i.e. epistemic uncertainty) and due to the intrinsic randomness of the volcanic process studied (i.e. aleatory uncertainty). The power of the single voice from the group of experts often misleads the receivers of the message (decision makers, authorities or lay public), believing the experts are "sure" on the evolution of volcanic activity. This is one of the reasons why volcanologists are often, correctly or incorrectly, highly trusted professionals by the public (Haynes et al. 2008; Donovan et al. 2011).

During the last decade, this "untouchable aura" around volcano monitoring based on deterministic research has vanished with the introduction of probabilistic hazard and eruption forecasting (e.g., Newhall and Hoblitt 2002; Sparks 2003; Marzocchi et al. 2004, 2008; Sparks and Aspinall 2004; Marzocchi and Bebbington 2012; Sobradelo et al. 2014; Sobradelo and Marti 2015). A recent chapter by Newhall and Pallister (2015) starts from the same false dichotomy, aiming to spouse the deterministic and probabilistic points of view in the highly applicable method of "Multiple Data Sets".

Marzocchi and Woo (2007) propose a rational on decision-making based on the hazard/risk separation principal, using a cost-benefit analyses as the guiding tool.

Among the methods for probabilistic hazard assessment and eruption forecasting, many are based on a Bayesian approach (e.g., Marzocchi et al. 2004, 2008; Sobradelo et al. 2014; Sobradelo and Marti 2015), that allows describing the probability of interest not as a single numerical value, but as a probability distribution. In this view, the probability of an eruption occurring, or of a given hazardous event hitting a target area, is described both by a best-evaluation value (for example the mean or the median of the probability distribution), and by a dispersion around such value (represented by standard deviations or by a confidence interval). These two quantities can be directly related to two different sources of uncertainty: the aleatory one and the epistemic one. In this way, Bayesian approaches allows quantifying also to what extent our probabilistic assessments are constrained by data and knowledge. In other words, now we know, as a group of volcano-experts, to which degree we can be wrong in our forecasts behind the closed doors. This black-on-white awareness brought to light by probability density functions has led to some key questions, from the in- and outside worlds: (1) are we, as volcano-experts, replaceable by a numerical approach?, and (2) we thought you, volcanologists, knew what was happening, but it seems you don't know.

This chapter critically reviews both "philosophies" of eruption forecasting and tracking of volcanic unrest and related hazards, in search of a combined approach that could become a guideline for future volcanic surveillance architectures. But we still need bridges between two schools (deterministic and probabilistic) apparently speaking a different language. Remember that both methodologies aim for the same goal: the timely recognition when volcanoes become hazardous in their various ways of expressions. This is our common professional and social responsibility as volcanologists.

2 Forecasts based on Deterministic Research

The goal of volcano monitoring based on deterministic research is to link temporal variations of physical-chemical parameters with variations in the state of unrest of the physical object volcano (i.e. unrest, magmatic unrest, non-magmatic unrest, eruption, hazard; Rouwet et al. 2014). Every volcanic eruption is intrinsically preceded by magma rise towards the surface. The major aim in eruption forecasting is the quick recognition of such magma rise by changes in the physical parameters (deformation, seismicity) and chemical parameters. The most direct way to do so is to determine how, where, when and why the physical object "volcano" responds to magma rise.

Despite the straight-to-goal approach, large uncertainties exist: (1) a volcano remains a complex system (aleatory uncertainty), and (2) our knowledge on the volcano remains limited (epistemic uncertainty). How do we know if we have detected all the signals the volcano eventually releases? Which of these signals are we considering in our forecast framework, and why? Sometimes we may dismiss some signals as being not pertinent, or simply because we are unable to correlate them either with our 'understanding', or with the rest of the signals. Some 'signals' are considered as stand-alone, at the moment they occur, others are considered within a time evolution.

The quality of the forecast largely depends on the interpretation of the signals and the hypothesis/model of future activity developed as a result of this assumed scientific stringency. The latter is related to the experience and expertise of the deterministic researcher, or better, on how the experience and expertise is perceived by individual researchers or groups, decision makers and the researcher her/himself. It is known that the most informative and valid opinion may not always be that of the most respected or distinguished professional (Selva et al. 2012).

A big advantage in volcano monitoring based on deterministic research is the fact that, if independent monitoring approaches (e.g., geochemistry vs. geophysics) point toward a similar hypothesis on future hazard in time and space, the future scenario will become more likely. Finding a larger number of arguments in favour of certain scenarios is surely an efficient way to decipher volcanic unrest.

The timescale of the forecast is highly ambiguous and based on the limited knowledge on how the volcano (or analogue volcanoes) behaved in the past within the desired time-scale. Sometimes all the 'signals' converge towards an obvious conclusion, yet, there are numerous cases in which activity stopped or pulled back, sometimes for years before an actual eruption. So, the deterministic approach, which is based mostly on a recurrence interval and 'experience' of the volcanologist, is limited in providing a sound time scale for the evolution of the 'activity', and hence, for a forecast. In deterministic monitoring the "time" concept is not unambiguously defined, can be case dependent or even change within the evolution of an unrest phase. Fortunately, with converging signals through time, the monitoring time window often becomes narrower when building up towards increased unrest or eruption, although the exact time window cannot be rigorously chosen.

Besides the instrumental accuracies (detection limits, analytical errors, data quality), the uncertainty of the forecast cannot be quantified before an eruption. The only way to decrease this "unquantifiable" uncertainty is by increasing our knowledge on volcanoes, be it the specific volcano in unrest or any volcano that has shown similar behaviour in the past. The current development of methods to increase the quality (e.g., novel approaches, numerical modelling) and quantity of data (increase frequency, e.g., by remote sensing and real-time transmission) will undoubtedly help to achieve better insights into volcanic systems.

3 Probabilistic Forecasts

Probabilistic methods and tools for both short- and long-term time windows are more and more in the spotlight (Marzocchi et al. 2008; Sandri et al. 2009, 2012, 2014; Lindsay et al. 2010; Selva et al. 2010, 2011, 2012; Sobradelo et al. 2014; Sobradelo and Marti 2015; Bartolini et al. 2013; Becerril et al. 2014). A key review on probabilistic volcano monitoring can be found in Marzocchi and Bebbington (2012).

Within this opinion chapter, we highlight some critical aspects of the probabilistic forecasting method, without entering in the technical and operational details (see Tonini et al. 2016 and Sandri et al. 2017 for further reading).

A probabilistic forecast can provide a global but clear, numerical view of the opinion of, generally, a group of people, based on the volcanic history and knowledge of the volcano. Lately, probabilistic forecasts are more and more applied in real crisis situations. Thus far, the efficiency or accuracy have hardly been evaluated, probably due to the fact that only recently we are reaching statistically relevant numbers of cases to test this critical issue (Newhall and Pallister 2015). Once high numbers of applications are reached, the numerical outcomes of probabilistic methods can even be considered to support long-term hazard analyses, by becoming input information itself.

In practice, probabilistic hazard assessment and eruption forecasting frameworks constructed on the Event-Tree methodology (Newhall and Hoblitt 2002; Newhall and Pallister 2015) rely on a dataset of information about the past activity of a volcano (i.e. past data), theoretical/mathematical models (i.e. prior data) and a series of monitoring signals (i.e. parameters) that allow us to track the changes in the system with time (Marzocchi et al. 2008; Sobradelo et al. 2014). This information allows us to compute the probabilities of a specific hazardous outcome. As any such application reveals, the quality of the numerical output depends on the quality and quantity of the input. The risk exists that using a dataset for long-term probabilistic assessment will introduce an uncertainty, since the operators are often biased by the hypothesis or model coming forth of the dataset. Data should hence be considered "just" facts. In both deterministic and probabilistic hazard assessment, volcanologists rely on information about past eruptions: eruptive behaviour, eruption frequency, and eruption style. Such catalogues of information are inevitably incomplete. For instance, traces of smaller scale events could have been literally eroded away from the geological record, buried or masked by larger events and, hence, relics of precursory activity cannot be deduced. Consequently, one should limit the part of the catalogue used, for the period and specific kind of event you desire to forecast, for which it is reasonably complete. As such, the foundation of a probabilistic framework represents a source of intrinsic uncertainty, that can somehow be overcome by quite robust methods for estimating completeness of sections of catalogues (Moran et al. 2011). The most intuitive solution, simply choosing a smaller dataset, usually representing the most recent years/centuries of a volcano's activity reduces this uncertainty. But this choice alone will be reflected in the quality of the probabilistic assessment. The unknowns of the data catalogue represent the uncertainty in a probabilistic framework, thus forcing volcanologists to 'select' how far to track back in time, and which information to use. In other words, we select e.g., only the last 300 years of activity of a volcano just simply because we believe to be more certain on what happened, instead of choosing the last 2000 years. What is the real scientific control of these choices? We cannot say with an acceptable certainty that a volcano will behave like it did in the last 300 or 2000 years. Indeed, with the recent probabilistic methods we are able to quantify the uncertainty of such choices but we are yet to find a sound scientific mechanism that allows us to make objective decisions regarding the data set. After all, the end goal of eruption forecasting is to give a prediction by analysing signals from an extremely complex system governed by a large degree of freedom.

Another aspect to tackle is the use of monitoring information, especially for short-term forecasts. Asking for numerical thresholds for monitoring parameters at the various nodes of event tree structures is intrinsically wrong, as a numerical threshold is an expression of certainty on something we cannot be certain about. For this, volcanologists use monitoring parameters in order to detect anomalies with respect to the volcano's background activity to be able to track their evolution with time. Moreover, from the beginning, we rely on a subjective choice when we define the unrest, unrest being commonly agreed upon as a state of elevated activity above background that causes concern (Phillipson et al. 2013). This cause of concern, expressed in numerical thresholds is a subjective choice: experts involved in volcano monitoring usually decide thresholds above/below which the volcano is considered in unrest. But is a volcano really in unrest simply because we observe one day an anomaly in one of the parameters? And if so, how is the choice of a threshold scientifically sound, since most of the time it is based on the "expert's experience"? Of course, expert's experience should not be dismissed and never replaced by computer codes, but indeed, such a choice is associated with a large uncertainty that is extremely difficult to quantify. Even the act of reducing the uncertainty of such a choice relies on corroboration with information from other sources (e.g., analogue volcanoes) that is again subjective itself.

The "fuzzy-threshold approach" (upper and lower thresholds) somehow resolves this problem, as it tracks the degree of anomaly, emphasising from a state in which the volcano is 'not causing concern' to one 'causing a degree of concern'. Thresholds can be case-dependent and are therefore in most cases themselves biased. Especially in the case of poorly monitored volcanoes, or of volcanoes without a monitored stage of unrest (e.g., towards higher nodes in the event tree), Boolean (Y/N) parameters are highly preferable.

During an unrest crisis it might be tempting to adapt the numerical values of the thresholds, when e.g., the previous threshold is exceeded while the volcano does not "react on this parameter" as we thought it would have. Nevertheless, once thresholds for parameters are fixed, they should not be modified, in order to track the time evolution of probabilities (and related uncertainties).

4 Recommendations: Not "or" But "and"

4.1 Expert Elicitation: A Solution?

In general, any choice made by an expert panel regarding when a volcano enters a phase of unrest, what information is pertinent for hazard analyses and how to interpret the precursory signals is done by a discussion-based elicitation process. Each expert in a specific volcanological sub-domain will exercise their opinions, based on the experience they have, on every parameter within a monitoring setup, with the goal to reach consensus about the most likely scenario/threshold. All data and interpretations should be heard and evaluated. However, this process is still unduly influenced by the "stronger voice" of the group. We may be certain on something until someone else makes us doubt it. At the moment we start doubting our opinions we will be easily influenced by other, stronger opinions. On the other hand, volcanologists are often forced to make such decisions and be liable for their choices (in court of law, Bretton et al. 2015). The pressure of a volcanic 'crisis' and the feeling of liability increases scientists' reservations when faced with such choices.

A suggestion to improve the expert elicitation process is to introduce a person to act as a "Devil's advocate". This will mean that one of the experts is supposed to do exactly the opposite to the group's decision. If most of the experts agree on a scenario, it is the duty of the latter to completely disagree and evaluate the opposite scenario. This can be a good way to "account" for surprise scenarios. Other ways are to weight final results according to anonymous calibration tests (Cooke method; Cooke 1991; Aspinall et al. 2003; Aspinall 2006, 2010) and/or anonymous estimation of the most reliable members of the group, not necessarily the loudest.

The introduction of probabilistic hazard assessment methods in the multi-disciplinary volcanological community has first led to a discredit of the purely deterministic approach. After the usefulness of the probabilistic approach has been demonstrated, and confusion on the different philosophies has disappeared, or at least decreased, the awareness on framing the various "niche" research branches in a bigger picture resulted into constructive discussions among the various research groups and individuals involved. This results in coherent group thinking and a more collaborative atmosphere among volcanologists. Expert elicitation on its turn has obliged researchers with various backgrounds to absorb new data and ideas from one another. This is definitely a positive side-effect of the probabilistic approaches and expert elicitations.

4.2 How to Interpret Uncertainties?

Probabilistic volcanic hazard assessment and eruption forecasting is a relatively new concept in modern volcanology, and often reserved for those with a background in statistics. But, as reality showed (Constantinescu et al. 2016), most of the volcanologists are not fully aware of the probability theory and how its results should be interpreted. A panel of volcano experts usually comprises seismologists, geochemists, geodesists, geologists, petrologists, and not all of them necessarily have a background in probabilistic approaches, especially when such approach is based on the integration of opinions of all members of such a group. One idea to cope with this problem is to train the group members in how the probabilistic methodology works and how results should be interpreted, while another member of the group (the so-called "PROB-runner") objectively uses the information provided by the expert panel to produce the probabilistic forecast. In this way, the members of the group can interpret the numerical outcomes accounting for the associated uncertainty without questioning and second-guessing the output because they are already aware of the process (Newhall and Pallister 2015; Constantinescu et al. 2016).

The whole idea of the elicitation approach is to allow your mind to explore each possibility without influencing one of the possible outcomes just because one expert believes more in one outcome than the other. It is some sort of letting go. People don't like to admit they might be wrong, so the Event Tree and Cooke elicitation approaches allow them to anonymously change their views upon elicitation. If one is capable of admitting fallibility and look at the big picture with an open mind, allowing all possibilities to unfold, then full discussion can occur and resulting estimates of probabilities will eventually have lower uncertainties. In the end, the idea of probabilities is that all scenarios are possible to happen, some with a larger probability others with a lower one; all have probabilities (and related uncertainties) and nothing should be dismissed simply because 'I strongly believe it can't be, so I don't agree'.

4.3 Trust in Scientists?

Even the best scientists can make mistakes. If volcanic unrest or activity is badly forecasted, initial trust in scientists may dissolve in legal proceeding. Ideally, scientists who act to the best of their knowledge should be protected rather than being blamed if they make a bad forecast (Bretton et al. 2015). Trust in scientists depends on how these scientists are portrayed to society and actually how scientific aspects of volcanology are presented to the public. This should be a system with two-way feedback (Christie et al. 2015). Many countries lack volcano-education among communities, but people know that there are some scientists that 'know what they are doing'. People feel protected, but this is a false feeling of safety, propped up by ignorance of the real situation. When disaster strikes, scientists are often the first to blame. If the community will be involved fully in the mitigation and preparedness process (e.g., Gregg et al. 2004; Rouwet et al. 2013; Dohaney et al. 2015), they may be guided by 'compassion' and will understand that volcanologists cannot stop an eruption and protect people, and eventually the blame or trust too often falls on the elected authorities. Elected authorities should be the liaison between science and the general public. Trust is inherent when you are aware of the problem and the person dealing with it. Trust in scientists may grow because they successfully predict an eruption, but sustained growth in trust is due to multi-yearly exposure of the scientific staff to the public (Christie et al. 2015). This involves long years of planning, investing and engagement in educational campaigns. Trust is something that comes in time and involves a feedback loop between people and the scientists.

Within the current scope of this opinion chapter, the probabilistic method often seems to serve as a more transparent way of bridging between the scientists and the elected authorities (decision makers). First, the use of probabilities inherently implies some uncertainty, which in scientifically literate society is essential for public trust. Second, elicitation tools reflect a joint-opinion of a group of experts rather than of one. There is a need for training and a full understanding of probabilistic results by the officials. The role of probabilistic tools should, in our opinion, not intervene in communication protocols with the lay public, but should be rather "restricted" to transmit information to decision-makers, representing part of the voice of the group of volcanologists. The efforts in communicating towards the lay public, in order to build "trust" amongst the population, should be decoupled from the background of the involved scientist (deterministic or probabilistic). Communication protocols are independent of the applied scientific method, and researchers should become more skilled to transmit their information openly towards the public, with the awareness of the uncertainty their information contains (Hicks and Few 2015).

4.4 Towards Collaborative Volcano Monitoring

A major accomplishment of the probabilistic method is to have increased harmony among the various deterministic research environments. This new dynamic favours the refining of previous conceptual models that originate from deterministic research, as the reference frame has become more complete.

Nevertheless, the probabilistic research approach is not yet fully accepted by the deterministic community due to criticism and anxiety to be "replaced" by the probabilistic method (VUELCO simulations Colima, Campi Flegrei, Cotopaxi and Dominica). This concern is unnecessary, since the first requisite for the probabilistic method to function is the availability of data, information, a priori believes and models, originating from deterministic research. More input information for the probabilistic method means significant decreases in the epistemic uncertainty of probabilistic outcomes.

Moreover, during volcanic crisis situations, deterministic researchers still stick to the "round table" approach and the lack of time inhibits to efficiently interact with the researchers that run the probabilistic models (e.g., VUELCO simulation exercises). The latter need more detailed feedback and input information for single parameters at the various nodes of the event-tree. Despite refining the probabilistic model during pre-crisis by expert elicitations, in the heat of the moment of the crisis, a wrong interpretation of a numerical value provided by determinists will sometimes lead to disastrous numerical outcomes in the probabilistic models. The PROB-runner cannot be blamed for not being an expert in all fields in volcanology, incorporated through parameters in e.g., BET or HASSET.

Three solutions to this crucial issue are proposed: (1) probabilistic model runners should actively take part in the "closed-door" discussions by the experts from the various fields, before incorporating numerical values in probabilistic models, (2) a separate team of experts that profoundly know the needs and functioning of probabilistic models should flank the PROB-runners during crisis. Both realities are not yet accomplished, and/or (3) an event tree structure, put forward by a facilitator between the deterministic and probabilistic research teams should serve as the base to guide scientific discussion and get fast to the nucleus of the crisis (Newhall and Pallister 2015). Future simulation exercises on volcanic crisis situations should focus on this training approach, in order to be prepared when real the crisis strikes.

Moreover, the outcomes of probabilistic models have to be included in the final reports transmitted to authorities, and be respected as one of the many monitoring tools, without decreasing or increasing their weight and value within the still deterministic-dominated general opinion. Since a probabilistic framework offers a measure of the uncertainty, any interpretation should not be taken for granted, neither decision makers should make decisions based solely on a probability. Probabilities should be considered as an addition to the information upon which decisions are made, and not as a decisive factor.

5 Take Home Ideas

Probabilistic forecasting has become an inherent part within a multi-facet view of research and volcano monitoring; neither deterministic, nor probabilistic methods, can or should stand alone. More than being a means to transmit information between volcanologists and decision making authorities, probabilistic models should also be based on, and promote, deterministic research that can be written up after "round table" discussions (Fig. 1).

Incorporation of probabilistic models in volcano monitoring has many advantages: (1) protecting against oversimplified, over-confident forecasts. Even though decision makers may initially have difficulties to understand uncertainties and prefer black-on-white numbers, Y/N forecasts, they will soon come to appreciate probabilities if they are represented in an understandable way; (2) creating harmony amongst the volcanological community because probabilities will reflect the general view of the monitoring team, and (3) legally protecting the entire monitoring team by probabilities and their uncertainties as forecasts are perfectly traceable and reproducible, if disaster strikes after "erroneous forecasts". However, to date, probabilistic forecasts have not been rigorously evaluated to know whether they are an improvement over traditional, non-probabilistic forecast methods. For sure, they do better than traditional methods at estimating uncertainty. We still need tests on whether they are more accurate, and more useful for decision makers than older methods (Newhall and Pallister 2015).

In the future, reports should include the forecast of probabilistic models; a monitoring architecture should expand beyond the classical "seismo-deformation-gas" setup and become "seismo-deformation-gas-probability" setup (in random order of importance) (Fig. 1). Probabilistic models cannot stand alone, as they need the input and feedback from deterministic research. "Probabilists" should not communicate their numerical outcomes directly to the decision-making authorities: it is better to convey the opinion of an entire group. Probabilistic methods can serve as "moderator" among the various disciplines, while expert elicitations are the "glue" between the deterministic and probabilistic approaches (Fig. 1). Probabilistic methods should knock down walls and stimulate discussion and coherence amongst the various research branches (seismologists, geodesists,

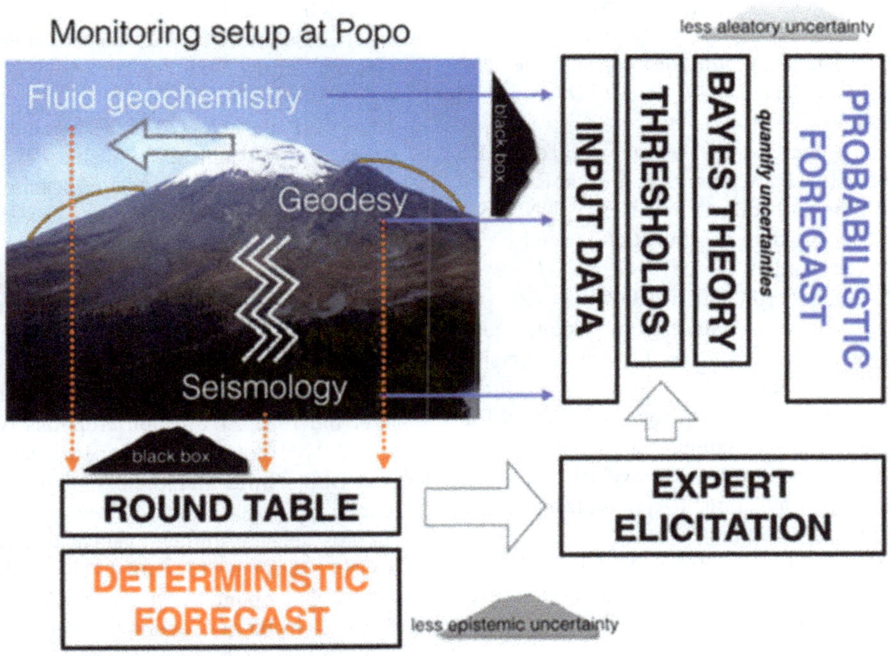

Fig. 1 From dichotomic monitoring setups (deterministic vs. probabilistic forecast, "or" setups) to an "and" strategy. The picture shows VUELCO target volcano Popocatépetl, Mexico (November 2011, D.R.). Before interpretation of the data (monitoring data, eruptive history or any a priori model) the volcano is considered a "*black box*"

geochemists, petrologists, geologists). This requires time, effort and an open-mind by all involved parties/volcanologists in volcano monitoring.

References

Aspinall WP (2006) Structured elicitation of expert judgement for probabilistic hazard and risk assessment in volcanic eruptions. In: Mader HM, Coles SG, Connor CB, Connor LJ (eds). Statistics in volcanology. IAVCEI Publications. ISBN: 978-1-86239-208-3. pp 15–30

Aspinall WP (2010) A route to more tractable expert advice. Nature 463(21):294–295

Aspinall WP, Woo G, Voight B, Baxter PJ (2003) Evidence-based volcanology: application to vol canic crises. J Volcanol Geotherm Res 128:273–285

Bartolini S, Cappello A, Martí J, Del Negro C (2013) QVAST: a new quantum GIS plugin for estimating volcanic susceptibility. Nat Hazards Earth Syst Sci 13:3031–3042. doi:10.5194/nhess-13-3031-2013

Becerril L, Bartolini S, Sobradelo R, Marti J, Morales JM, Galiendo I (2014) Long-term volcanic hazard assessment on El Hierro (Canary Islands). Nat Hazards Earth Syst Sci 14:1853–1870. doi:10.5194/nhess-14-1853-2014

Bretton RJ, Gottsmann J, Aspinall WP, Christie R (2015) Implications of legal scrutiny processes (including the L'Aquila trial and other recent court cases) for future volcanic risk governance. J Appl Volcanol 4:18. doi:10.1186/s13617-015-0034-x

Christie R, Cooke O, Gottsmann J (2015) Fearing the knock on the door: critical security studies insights into limited cooperation with disaster management regimes. J Appl Volcanol 4:19. doi:10.1186/s13617-13015-10037-13617

Constantinescu R, Robertson R, Lindsay JM, Tonini R, Sandri L, Rouwet D, Patrick Smith P, Stewart R (2016) Application of the probabilistic model BET_UNREST during a volcanic unrest simulation exercise in Dominica, Lesser Antilles. Geochem Geophys Geosyst 17:4438–4456. doi:10.1002/2016GC006485

Cooke RM (1991) Experts in uncertainty: opinion and subjective probability in science. Oxford University Press. ISBN: 9780195064650

Dohaney J, Brogt E, Kennedy B, Wilson TM, Lindsay JM (2015) Training in crisis communication and volcanic eruption forecasting: design and evaluation of an authentic role-play simulation. J Appl Volcanol 4:12. doi:10.1186/s13617-015-0030-1

Donovan A, Oppenheimer C, Bravo M (2011) Social studies of volcanology: knowledge generation and expert advice on active volcanoes. Bull Volcanol. doi:10.1007/s00445-011-0547-z

Gregg CE, Houghton BF, Johnston DM, Paton D, Swanson DA (2004) The perception of volcanic risk in Kona communities from Mauna Loa and Hualālai volcanoes, Hawai'i. J Volcanol Geotherm Res 130:179–196

Haynes K, Barclay J, Pidgeon N (2008) The issue of trust and its influence on risk communication during a volcanic crisis. Bull Volcanol 70:605–621. doi:10.1007/s00445-007-0156-z

Hicks A, Few R (2015) Trajectories of social vulnerability during the Soufrière Hills volcanic crisis. J Appl Volcanol 4:10. doi:10.1186/s13617-015-0029-7

Lindsay J, Marzocchi W, Jolly G, Constantinescu R, Selva J, Sandri L (2010) Towards real-time eruption forecasting in the Auckland Volcanic Field: application of BET_EF during the New Zealand National Disaster Exercise 'Ruaumoko'. Bull Volcanol 72:185–204. doi:10.1007/s00445-009-0311-9

Marzocchi W, Woo G (2007) Probabilistic eruption forecasting and the call for an evacuation. Geophys Res Lett 34:L22310. doi:10.21029/22007GL031922

Marzocchi W, Bebbington M (2012) Probabilistic eruption forecasting at short and long time scales. Bull Volcanol. doi:10.1007/s00445-012-0633-x

Marzocchi W, Sandri L, Gasparini P, Newhall C, Boschi E (2004) Quantifying probabilities of volcanic events: The example of volcanic hazard at Mount Vesuvius. J Geophys Res 109:B11201. doi:10.1029/2004JB003155

Marzocchi W, Sandri L, Selva J (2008) BET_EF: a probabilistic tool for long- and short-term erup tion forecasting. Bull Volcanol 70:623–632. doi:10.1007/s00445-007-0157-y

Moran SC, Newhall C, Roman DC (2011) Failed magmatic eruptions: late-stage cessation of mag ma ascent. Bull Volcanol 73:115–122. doi:10.1007/s00445-010-0444-x

Newhall C, Hoblitt RP (2002) Constructing event trees for volcanic crises. Bull Volcanol 64:3–20. doi:10.1007/s004450100173

Newhall C, Pallister J (2015) Using multi data sets to populate probabilistic volcanic event trees. In: Schroder JF, Papale P (eds) Volcanic hazards, risks, and disasters. Hazards and disasters series. Elsevier, Amsterdam, pp 203–232

Phillipson G, Sobradelo R, Gottsmann J (2013) Global volcanic unrest in the 21st century: an ana lysis of the first decade. J Volcanol Geotherm Res 264:183–196

Rouwet D, Iorio M, Polgovsky D (2013) A science & arts sensitization program in Chapultenango, 25 years after the 1982 El Chichón eruptions (Chiapas, Mexico). J Appl Volcanol 2:6. doi:10.1186/2191-5040-2-6

Rouwet D, Sandri L, Marzocchi W, Gottsmann J, Selva J, Tonini R, Papale P (2014) Recognizing and tracking hazards related to non-magmatic unrest: a review. J Appl Volcanol 3:17. doi:10.1186/s13617-014-0017-3

Sandri L, Guidoboni E, Marzocchi W, Selva J (2009) Bayesian event tree for eruption forecasting (BET_EF) at Vesusvius, Italy: a retrospective forward application to the 1631 eruption. Bull Volcanol 71:729–745. doi:10.1007/s00445-008-0261-7

Sandri L, Jolly G, Lindsay J, Howe T, Marzocchi W (2012) Combining long-term and short-term probabilistic volcanic hazard assessment with cost-benefit analysis to support decision making in a volcanic crisis from the Auckland Volcanic Field, New Zealand. Bull Volcanol 74:705–723. doi:10.1007/s00445-011-0556-y

Sandri L, Thouret J-C, Constantinescu R, Biass S, Tonini R (2014) Long-term multi-hazard assessment for El Misti volcano (Peru). Bull Volcanol 76:771. doi:10.1007/s00445-013-0771-9

Sandri L, Tonini R, Rouwet D, Constantinescu R, Mendoza-Rosas AT, Andrade D, Bernard B (2017) The need to quantify hazard related to non-magmatic unrest: from BET_EF to BET_UNREST. In: Gottsmann J, Neuberg, J, Scheu B (eds) Volcanic unrest: from science to society—IAVCEI advances in volcanology. Springer

Selva J, Costa A, Marzocchi W, Sandri L (2010) BET_VH: exploring the influence of natural uncertainties on long-term hazard from tephra fallout at Campi Flegrei (Italy). Bull Volcanol 72:717–733. doi:10.1007/s00445-010-0358-7

Selva J, Marzocchi W, Papale P, Sandri L (2012) Operational eruption forecasting at high-risk vol canoes: the case of Campi Flegrei, Naples. J Appl Volcanol 1:5. http://www.appliedvolc.com/content/1/1/5

Selva J, Orsi G, Di Vito MA, Marzocchi W, Sandri L (2011) Probability hazard map for future vent opening at Campi Felgrei caldera. Bull Volcanol, Italy. doi:10.1007/s00445-011-0528-2

Sobradelo R, Bartolini S, Marti J (2014) HASSET: a probability event tree tool to evaluate future volcanic scenarios using Bayesian inference. Bull Volcanol 76:770. doi:10.1007/s00445-013-0770-x

Sobradelo R, Marti J (2015) Short-term volcanic hazard assessment through Bayesian inference: retrospective application to the Pinatubo 1991 volcanic crisis. J Volcanol Geotherm Res 290:1–11. doi:10.1016/j.jvolgeores.2014.11.011

Sparks RSJ (2003) Forecasting volcanic eruptions. Earth Planet Sci Lett 210:1–15. doi:10.1016/S0012-821X(03)00124-9

Sparks RSJ, Aspinall WP (2004) Volcanic activity: frontiers and challenges in forecasting, pre diction and risk assessment. The State of the Planet: Frontiers and Challenges in Geo physics. Geophysical Monograph 150, IUGG Volume 19. doi:10.1029/150GM28

Tonini R, Sandri L, Rouwet D, Caudron C, Marzocchi W, Suparjan P (2016) A new Bayesian Event Tree tool to track and quantify unrest and its application to Kawah Ijen volcano. Geochem Geophys Geosyst 17:2539–2555. doi:10.1002/2016GC006327

Probabilistic E-tools for Hazard Assessment and Risk Management

Stefania Bartolini, Joan Martí, Rosa Sobradelo and Laura Becerril

Abstract

The impact of a natural event can significantly affect human life and the environment. Although fascinating, a volcanic eruption creates similar or even greater problems than more frequent natural events due to its multi-hazard nature and the intensity and extent of its potential impact. It is possible to live near a volcanic area and take advantage of the benefits that volcanoes offer, but it is also important to be aware of the existing threats and to know how to minimise risks. In this chapter, we present an integrated approach using e-tools for assessing volcanic hazard and risk management. These tools have been especially designed to assess and manage volcanic risk, to evaluate long- and short-term volcanic hazards, to conduct vulnerability analysis, and to assist decision-makers during the management of a volcanic crisis. The methodology proposed here can be implemented before an emergency in order to identify optimum mitigating actions and how these may have to be adapted as new information is obtained. These tools also allow us identifying the most appropriate probabilistic and statistical techniques for volcanological data analysis and treatment in the context of quantitative hazard and risk assessments. Understanding volcanic unrest, forecasting volcanic eruptions, and predicting the most probable scenarios, all imply a high degree of inherent uncertainty, which needs to be quantified and clearly explained when transmitting scientific information to decision-makers.

Resumen

El impacto de un evento natural puede afectar significativamente la vida humana y el medio ambiente. Aunque fascinante, una erupción volcánica

S. Bartolini (✉) · J. Martí · L. Becerril
Group of Volcanology, SIMGEO (UB-CSIC), Institute of Earth Sciences Jaume Almera, ICTJA-CSIC, Lluís Solé I Sabarís S/N, Barcelona 08028, Spain
e-mail: sbartolini.1984@gmail.com

J. Martí
e-mail: joan.marti@ictja.csic.es

L. Becerril
e-mail: lbecerril@ictja.csic.es

R. Sobradelo
WRN Earth Hub Leader, Analytics Technology & Willis Research Network, London, UK
e-mail: rosa.sobradelo@willistowerswatson.com

Advs in Volcanology (2019) 47–61
DOI 10.1007/11157_2017_14

Published Online: 15 July 2017

puede generar problemas parecidos o incluso mayores que otros eventos naturales más frecuentes, ya que se trata de un fenómeno multipeligro y con un impacto potencial de gran intensidad y magnitud. Vivir cerca de una zona volcánica es posible considerando sus innumerables beneficios, pero sin embargo debemos ser conscientes de la amenaza existente y tomar las medidas oportunas para minimizar el riesgo. En este capítulo, se presenta un enfoque integrado que utiliza herramientas para la evaluación de la peligrosidad y del riesgo volcánico, el análisis de vulnerabilidad, y para ayudar en la gestión de una crisis volcánica. La metodología aquí propuesta puede ser implementada antes de una emergencia con el fin de determinar las medidas óptimas de mitigación y como pueden adaptarse a medida que se obtiene nueva información. Además, estas herramientas se basan en las técnicas probabilísticas y estadísticas más adecuadas para el análisis de datos vulcanológicos y su implementación en el contexto de las evaluaciones cuantitativas del peligro y riesgo. La interpretación del *unrest*, la predicción de las erupciones volcánicas, y la identificación de los escenarios más probables, implican un alto grado de incertidumbre, que debe ser cuantificado y claramente explicado para transmitir correctamente la información científica a los gestores de las crisis volcánicas.

Keywords
Volcanic risk · Volcanic hazard · Vulnerability · Risk management · Decision-making · E-tools

Introduction

One of the most important tasks in modern volcanology is to manage volcanic risk and, consequently, to minimise it. Forecasting volcanic eruptions and predicting the most probable scenarios are tasks that are subject to high degrees of uncertainty but which need to be quantified and clearly explained when transmitting scientific information to decision-makers. Assessing eruption risk scenarios in probabilistic ways has become one of the main challenges tackled by modern volcanology (Newhall and Hoblitt 2002; Marzocchi et al. 2004; Aspinall 2006; Neri et al. 2008; Sobradelo et al. 2014).

The volcanic management cycle consists of four phases (Sobradelo et al. 2015): (i) the pre-unrest phase that includes long-term assessment, hazard and risk mapping, and estimation regarding expected scenarios, volcano monitoring, and emergency planning; (ii) the unrest phase, which includes short-term assessment, alert, communication, and information procedures, the implementation of emergency measures, and the interpretation of eruption precursors; (iii) the volcanic event itself, represented by a major change in the state of the volcano; decisions can be revised as new information is obtained and the evolution of the volcanic event is updated; (iv) the post-event phase characterized by rescue and recovery.

Previous studies, focusing attention on the first phase of volcanic crisis management, have developed different methodologies for evaluating hazards. Most of these studies are based on the use of simulation models and Geographic Information Systems (GIS) that enable volcanic hazards such as lava flows, pyroclastic density currents (PDCs), and ash fallout to be modeled and visualised (Felpeto et al. 2007; Toyos et al. 2007; Cappello et al. 2012; Martí et al. 2012; Becerril et al. 2014; Bartolini et al. 2014a, b). The use of spatio-temporal data and GIS has now become an essential part of integrated approaches to disaster risk management in light of the development of new analysis/modelling techniques. These spatial information systems are used for storage, situation analysis, modelling, and visualisation (Twigg 2004). Normally, these studies present a systematic approach based on the estimation of spatial probability, that is, a susceptibility analysis (Martí and Felpeto 2010), a temporal analysis based on Bayesian inference (Marzocchi et al. 2004; Sobradelo et al. 2014), or the evaluation of hazards (Felpeto et al. 2007; Bartolini et al. 2014a, b; Becerril et al. 2014) and vulnerability (Marti et al. 2008; Scaini et al. 2014). These studies have been applied in a number of volcanic areas such as Etna, Sicily (Cappello et al. 2013), Tenerife, Spain (Martí et al. 2012), Peru (Sandri et al. 2014), the island of El Hierro, Spain (Becerril et al. 2014), and Deception Island, Antarctica (Bartolini et al. 2014a). Other procedures have been employed to assess volcanic hazards in Campi Flegrei, Italy (Lirer et al. 2001), Furnas (São Miguel, Azores), Vesuvius, Italy (Chester et al. 2002), and Auckland, New Zealand (Sandri et al. 2012).

These studies underline the fact that scientists are aware of the relationship between volcanic hazards and socio-economic impacts and are in the process of developing new approaches and models to assess its importance. One positive aspect is that, as a result, there is now a choice of freely available models; on the other hand, these models are not integrated into a single platform and have been developed in a variety of different programming languages.

Despite the fact that these tools have been created for application in real situations and have been successfully tested in different volcanic areas and retrospectively for a number of volcanic crisis (Martí et al. 2012; Sobradelo et al. 2014; Bartolini et al. 2014a, b; Becerril et al. 2014; Scaini et al. 2014), to date they only exist as academic exercises. To convert them into practical tools ready to be used by Civil Protection managers and decision-makers, they must be checked and tested, and then adapted to the real needs of end users.

Probabilities are still the best outcome of scientific forecasting. However, they are not easily understood. Understanding the potential evolution of a volcanic crisis is crucial for designing effective mitigation strategies. One of the main issues when managing a volcanic crisis is how to make scientific information understandable for decision-makers and Civil Protection managers. Thus, we need quantitative risk-based methods for decision-making under conditions of uncertainty that can be developed and applied to volcanology. In order to resolve this problem and to take a step forward in minimising risks, we have defined an integrated approach using user-friendly e-tools, which can be run on personal computers. They are specifically useful for long- and short-term hazard assessment, vulnerability analysis, decision-making, and volcanic risk management. In this chapter, we describe the e-tools designed to manage and to minimise volcanic risk.

Volcanic Risk: Hazard, Vulnerability, and Value

In general, risk is defined as the probability or likely magnitude of a loss (Blong 2000). In volcanic risk assessment, risk depends on the adverse effects of volcanic hazards and can be defined as the product of three main factors: volcanic hazard, vulnerability to those hazards, and the value of what is at risk.

Volcanic hazard is defined as the probability of any particular area being affected by a destructive volcanic event within a given period of time (Blong 2000). The quantification and evaluation of the volcanic hazard allow us determining which areas will be affected by a given volcanic event, and to design appropriate emergency plans and territorial planning.

A vulnerability assessment uses indicators to quantify the physical vulnerability of the elements of each sub-system (buildings, transportation system, urban services, and population) and is a measurement of the proportion of the value likely to be lost as a result of a given event (Blong 2000). In fact, the values of elements-at-risk that can be directly or indirectly damaged by a given hazard will vary. Each hazardous phenomenon affects elements and infrastructures in different ways in terms of their specific physical vulnerability; this in turn will be depend on their number (of buildings, people, etc.), monetary value, surface area, and the importance of the elements-at-risk. Indicators are used to determine the specific physical vulnerability of each hazardous phenomenon and sub-system (Scaini et al. 2014).

The value is the number of human lives at stake, together with the capital value (land, buildings, etc.) and productive capacity (factories, power plants, highways, etc.) exposed to the destructive events (Blong 2000).

Risk management is a complex process (Fig. 1) since different steps are necessary for evaluating and minimising risk. It can be thought of as the sum of risk assessment—which includes risk analysis and risk evaluation—and risk control. Risk analysis aims to improve prevention tools through the collection and acquisition of data on hazards and risks, and then to disseminate it in the form of maps and scenarios (Thierry et al. 2015). This phase is characterised by five different steps: hazard identification, hazard assessment, elements-at-risk/exposure analysis, vulnerability assessment, and risk estimation (Van Westen 2013). In particular, it is important to distinguish between long- and short-term hazard assessments, which will vary according to the expected period of time over which the process will display significant variations.

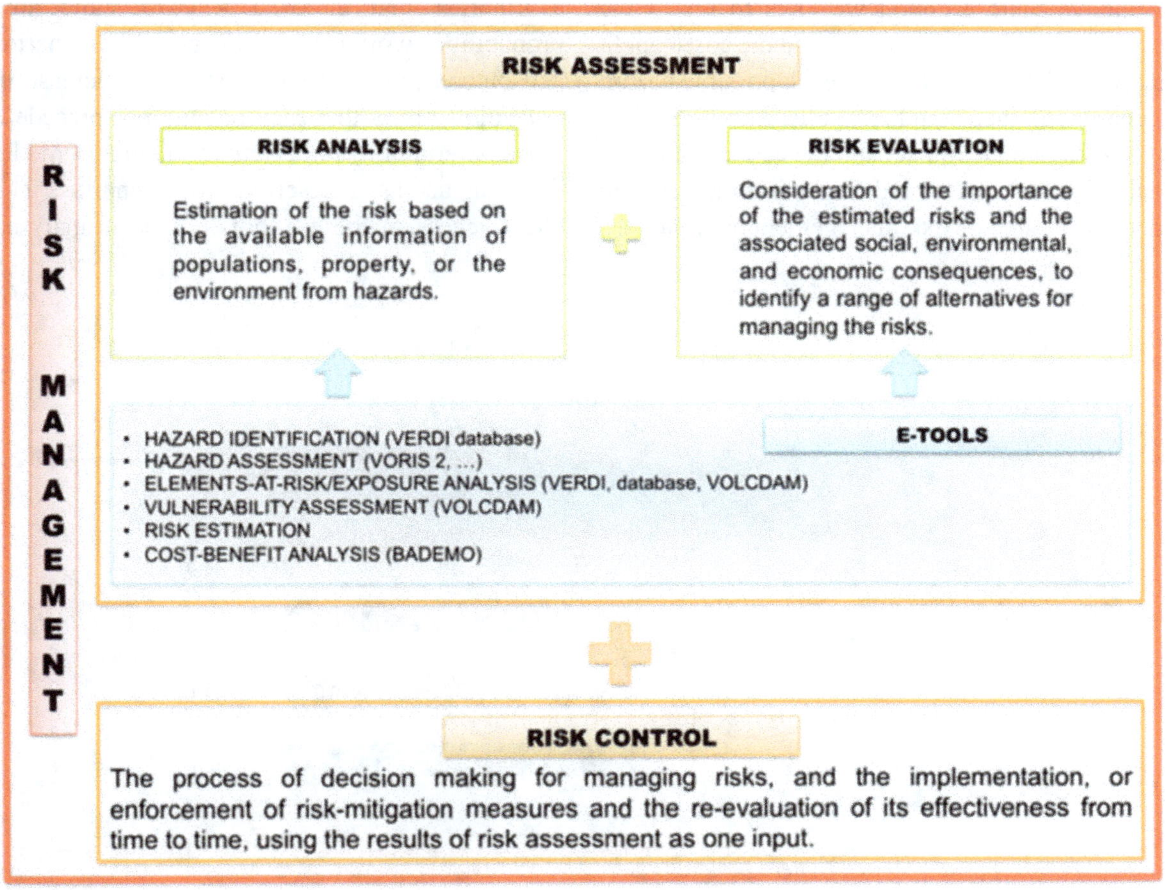

Fig. 1 Risk management schematic: steps for evaluating and minimising risk

Long-term assessment is based on historical and geological data, as well as theoretical models, and refers to the time window available—during which time the volcanic system shows no signs of unrest—before an unrest episode occurs. On the other hand, short-term assessment refers to the unrest phase, and complementary information derived from the combination of a long-term analysis with real-time monitoring data is needed to update the status of the volcanic hazard (Blong 2000). Short-term evaluation helps forecast where and when the eruption may take place and the most likely eruptive scenarios.

Once the long-term risk analysis is computed, it is then possible to adopt mitigation measures such as land-use planning and emergency preparations to reduce the risk. In addition, this long-term analysis will help manage the volcanic crisis as it will constitute the basis for the short-term analysis and, combined with a cost-benefit analysis, will assist in correct decision-making (Sobradelo et al. 2015). To evaluate the total risk related to a particular volcanic eruption we have to repeat the evaluation of the vulnerability and the cost-benefit analysis (risk evaluation) for each possible hazard scenario and then sum the results. This will allow us to estimate the impact and the economic losses that will affect society and the environment, and to identify a range of risk management alternatives.

Finally, the second part of risk management is risk control, which consists of the decision-making process involved in managing risks whose aim is to improve crisis management capabilities and implement risk-mitigation measures using the results of risk assessment as an input (Western 2013). During this phase measures must be adopted for reducing vulnerability (people and infrastructure) and developing recovery and resilience capacities after an event has taken place.

E-tools for Volcanic Hazard and Risk Management

In this section, we present different e-tools that have been specifically designed to assess and manage volcanic risk (Fig. 2). The objective is to combine freely available models to produce a new approach for minimising and managing volcanic risk. These e-tools are based on the assumption that the best way to show how probabilities work is to use the possible scenarios and outcomes of volcanic unrest (an increase in volcanic activity that may or may not precede a volcanic eruption) to design an integrated model that can act as a descriptor of scenarios. The effectiveness of these e-tools has been analysed

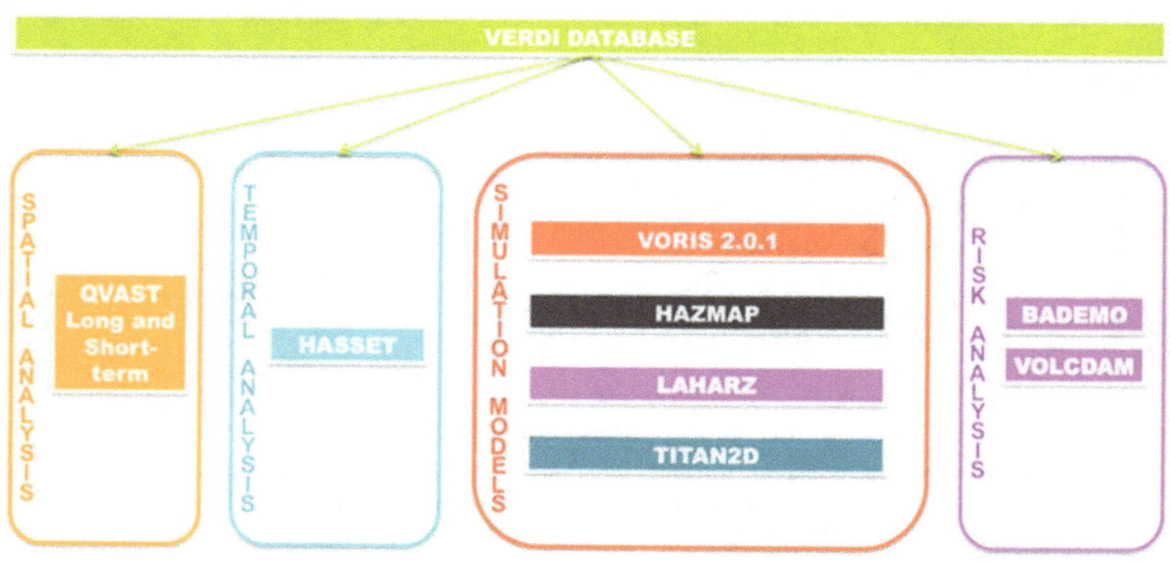

Fig. 2 E-tools for assessing and managing volcanic risk that allow to evaluate the possible hazards that could affect a volcanic area and develop appropriate hazard and risk maps

in different volcanic areas showing pros and contras of these approaches, such as in the case of El Hierro Island in Canary Islands (Becerril et al. 2014; Sobradelo et al. 2015), Tenerife in Canary Islands (Scaini et al. 2014), Deception Island in Antarctica (Bartolini et al. 2014a), and La Garrotxa Volcanic Field in Spain (Bartolini et al. 2014b). User manuals are available at https://volcanbox.wordpress.com/ and http://www.vetools.eu.

These e-tools are designed to be implemented before an emergency in order to identify (i) the optimum mitigating actions and (ii) the most appropriate probabilistic and statistical techniques for volcanological data analysis and treatment in the context of quantitative hazard and risk assessment, and (iii) how appropriate responses may change as new information is obtained. They constitute different steps, as shown in Fig. 3, in hazard assessment and risk management.

Storing Data

The results of our analyses are as important as the input parameters that are used. Given the nature, variety, and availability of the data we need to handle, one of the most important aspects of risk assessment is the management and exchange of information (De la Cruz-Reyna 1996). The quality of the data will determine the evaluation of the volcanic risk, which is an essential part of risk-based decision-making in land-use planning and emergency management. Some of the most relevant issues include how and where to store the data, in which format should they be made available, and how to facilitate its use and exchange. Thus, it is essential to have an appropriate database that is specifically adapted to the task of evaluating and managing volcanic risk. For that reason, we designed VERDI (Volcanic managEment Risk Database desIgn) (Bartolini et al. 2014c), a

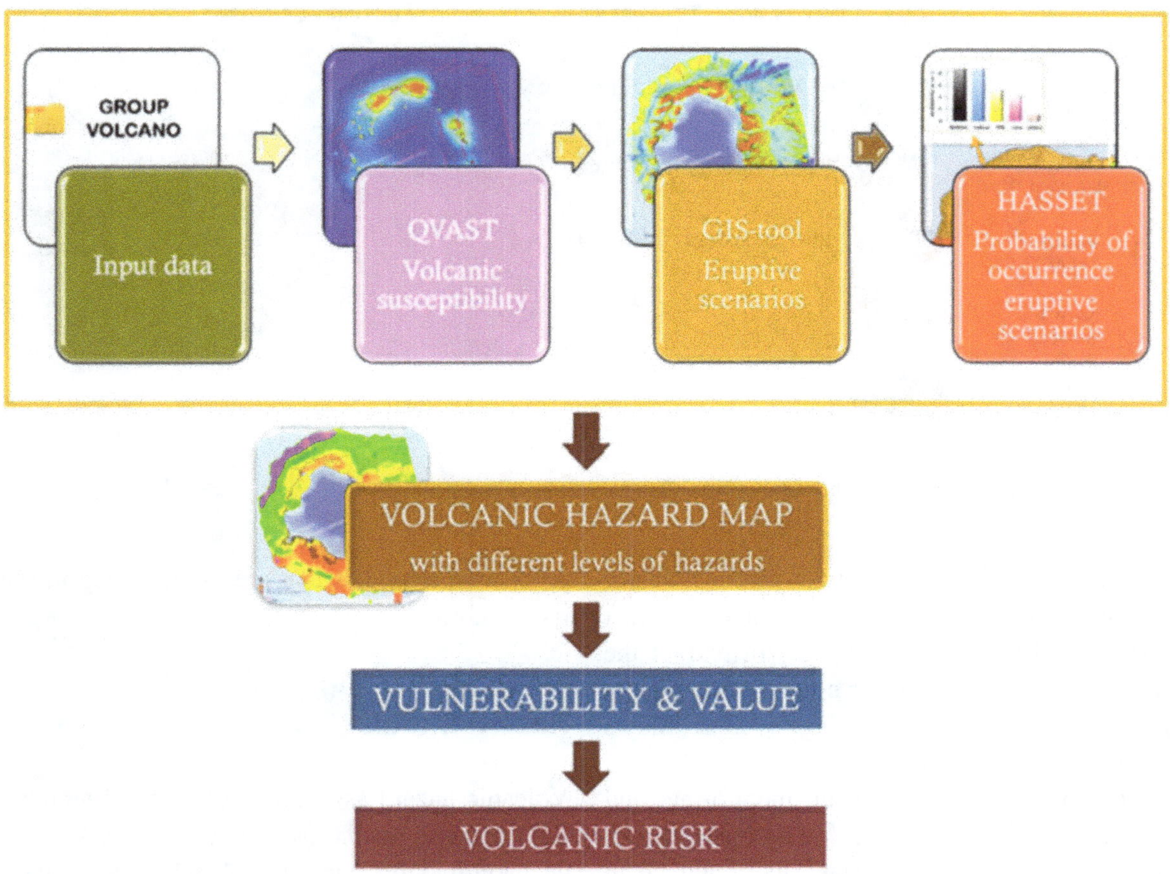

Fig. 3 Steps in volcanic risk assessment applying e-tools to be implemented before an emergency

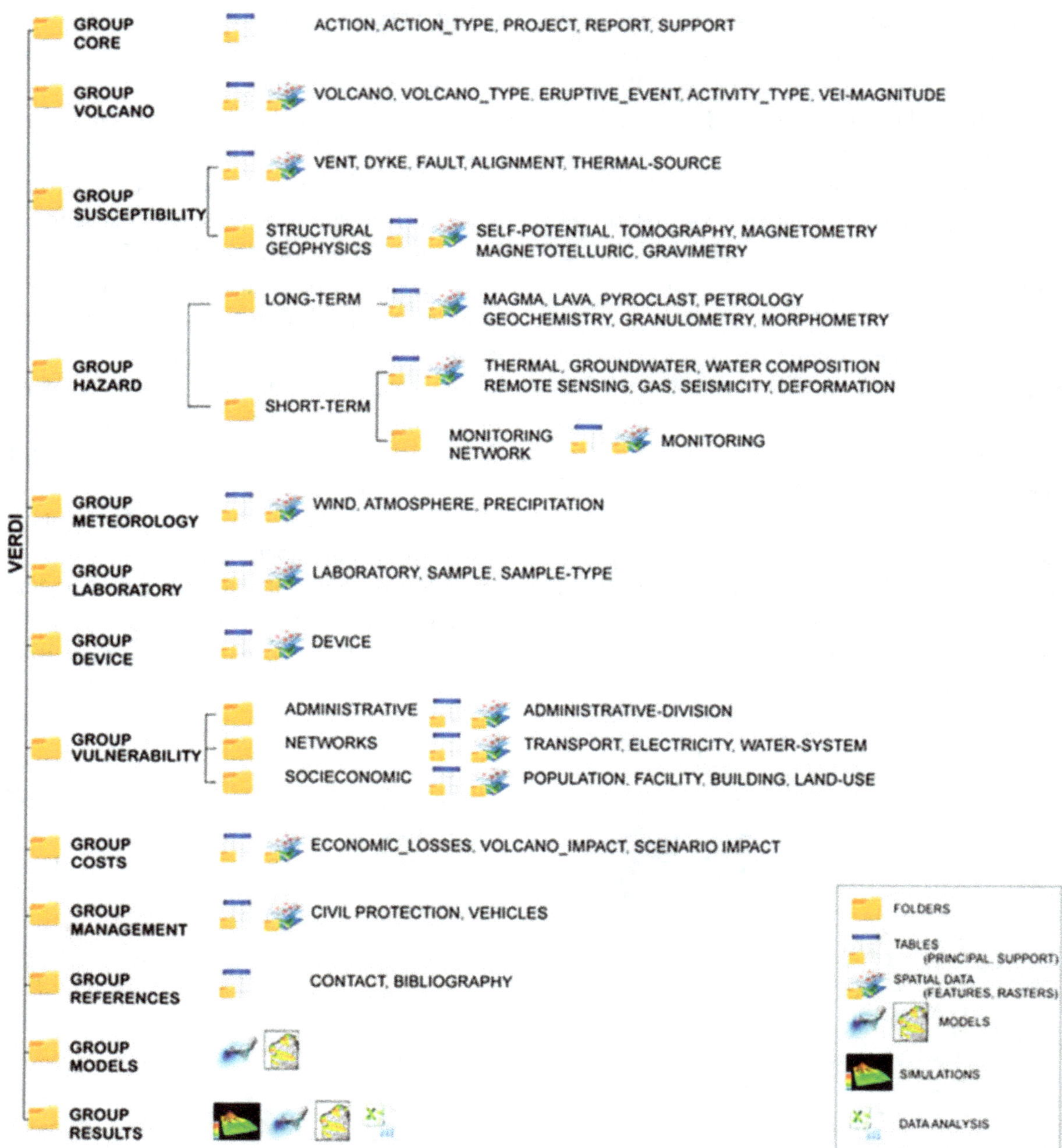

Fig. 4 The design of the VERDI database (after Bartolini et al. 2014c)

geodatabase with an appropriate architecture for volcanic risk assessment and management that stores the data from which we will extract the input parameters to run our e-tools. VERDI is a spatial database structure (Fig. 4) that allows different types of data, including geological, volcanological, meteorological, monitoring, and socio-economic information, to be manipulated, organised, and managed. The data contained in this database are the basis for applying the probabilistic models that are the first step in our risk analysis.

Hazard E-tools

Volcanic hazard assessment consists of simulating eruptive scenarios for use in risk-based decision-making, land-use planning, and emergency management. They must necessarily be based on a good knowledge of the past eruptive history of the volcano or volcanic area, and will reveal how volcanoes erupted in the past, thereby providing clues to *how* they will erupt in the future. The first step in the quantitative assessment of volcanic hazards is the spatial probability of occurrence of a hazard, i.e. *where* the next eruption may take place and its extent. This analysis is based on the development of susceptibility maps (Martí and Felpeto 2010), that is, the spatial probability of a future vent opening given the past eruptive activity of a volcano and the simulation of possible eruptive scenarios. Another important task is to investigate the temporal probability, in other words, *when* the next eruption will occur in the future and the type of scenarios that are most likely to be involved. Thus, an evaluation of volcanic hazard enables us to infer where and when the next eruption may take place and its magnitude.

Spatial Analysis

Susceptibility analysis is the evaluation of the spatial distribution of future vent openings (Fig. 5a). This challenging issue is generally tackled using probabilistic methods that use the calculation of a kernel function at each data location to estimate probability density functions (PDFs). Commonly, a Gaussian kernel, describing a normal distribution, is used to estimate local event densities in volcanic fields, which will give the intensity of a new vent opening. This method is based on the distance from nearby volcanic structures and a smoothing parameter, also known as bandwidth. This factor is the most important parameter in the kernel function and represents the degree of randomness in the distribution of past events.

QVAST (QGIS for VolcAnic SuscepTibility), developed by Bartolini et al. (2013), is a new tool designed to generate user-friendly quantitative assessments of volcanic susceptibility (e.g. the probability of hosting a new eruptive vent). QVAST allows an appropriate method for evaluating the bandwidth for the kernel function to be selected on the basis of input parameters and the shapefile geometry, and can also evaluate the PDF with the Gaussian kernel. When different input data sets are available for the area, the total susceptibility map is obtained by assigning different weights to each of the PDFs, which are then combined via a weighted sum and modeled in a non-homogeneous Poisson process. This e-tool has been used to evaluate susceptibility on the island of El Hierro (Canary Islands) (Becerril et al. 2014) and on Deception Island (Antarctica) (Bartolini et al. 2014a).

When monitoring data generated during an unrest phase are available, the QVAST e-tool can also be used to update the susceptibility map. In fact, seismicity and surface deformation are good indicators of magma movement and during volcanic unrest variations in shallow volcano-tectonic and long-period seismicity, as well as ground deformation, are observed as the magma migrates within the volcanic system (Martí et al. 2013). Thus, QVAST is a highly useful tool that can be applied to both long- and short-term evaluations.

Temporal Analysis

HASSET (Hazard Assessment Event Tree), developed by Sobradelo et al. (2014), is a probability tool built on an event tree structure that uses Bayesian inference to estimate the probability of occurrence of a future volcanic scenario. It also evaluates the most relevant sources of uncertainty in the corresponding volcanic system. Event tree structures (Newhall and Hoblitt 2002) constitute one of the most useful and necessary tools in modern volcanology for assessing the volcanic hazard of future eruptive scenarios. An event tree is a graphic representation of events in the form of nodes and branches. It evaluates the most relevant sources of uncertainty when estimating the probability of occurrence of a future volcanic event.

The objective of this e-tool is to outline all relevant possible outcomes of volcanic unrest at progressively greater detail and to assess the hazard of each scenario by estimating its probability of occurrence within a future time interval. Each node of the event tree represents a step and contains a set of possible branches (the outcomes for that particular category). The nodes are

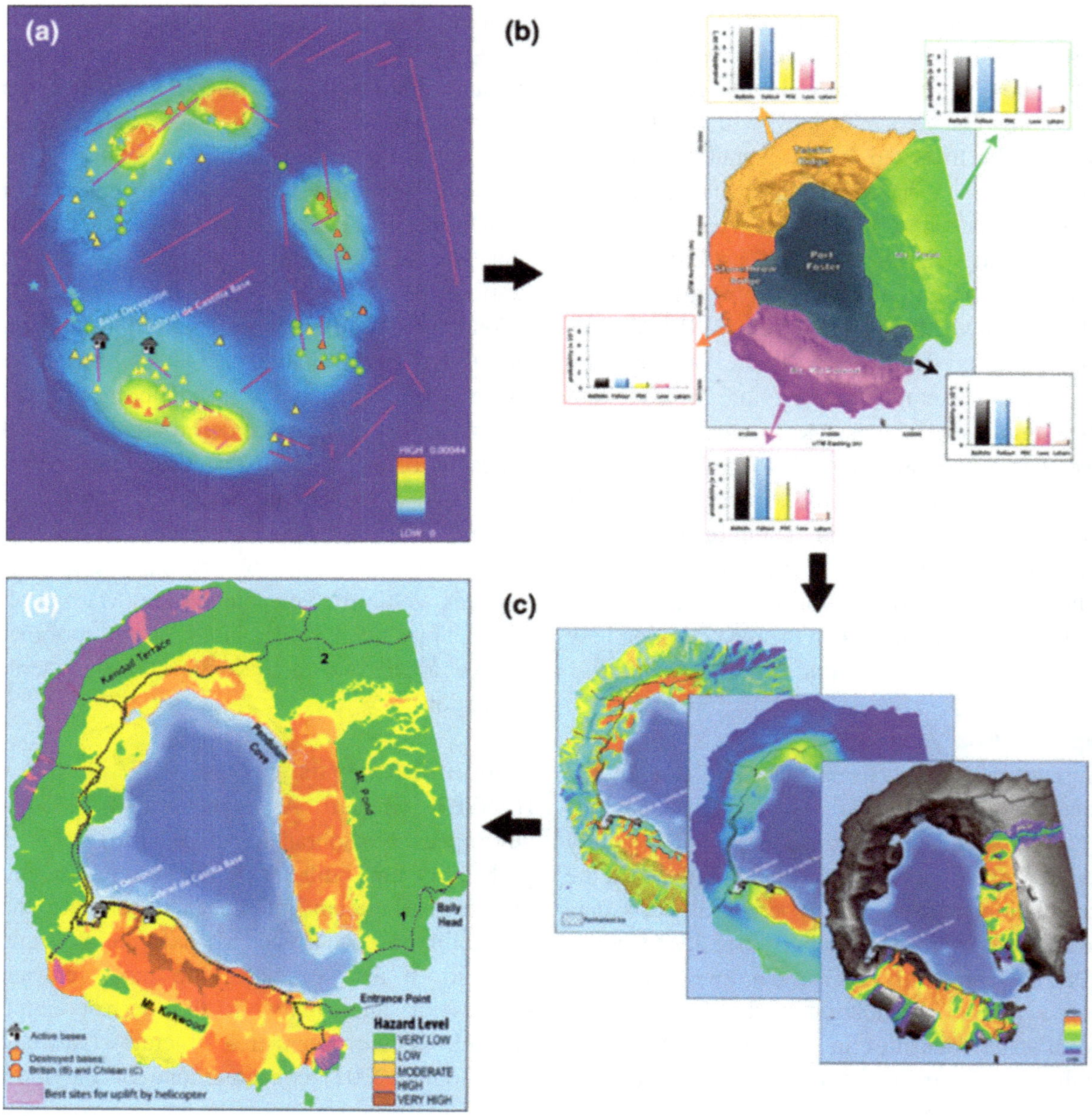

Fig. 5 Methodological approach for obtaining a qualitative hazard map: its application to Deception Island (after Bartolini et al. 2014a)

alternative steps from a general prior event, state, or condition that move towards increasingly specific subsequent events and a final outcome. HASSET (Fig. 5b) uses this event tree structure to make estimations of the probabilities for each possibility (branches and nodes) using a statistical methodology known as Bayesian Inference (Newhall and Hoblitt 2002; Marzocchi et al. 2004; Sobradelo et al. 2014). In particular, and based on comparisons with previous event trees for volcanic eruptions, HASSET accounts for the possibility of (i) flank eruptions (as opposed to only central eruptions), (ii) geothermal or tectonic unrest (as opposed to only magmatic unrest), and (iii) felsic or mafic lava composition, as well as (iv) certain volcanic hazards as possible outcomes of an eruption, and (v) the distance reached by each hazard.

A user-friendly interface guides the user through all steps and helps

- enter all the data needed for the analysis;
- compute the estimated probability for each branch in the event tree;

- compute the total estimated probability and compare up to five different scenarios.

This tool has been used for determining the eruption probability on El Teide (Tenerife, Spain) (Sobradelo and Martí 2010), El Hierro (Canary Islands) (Becerril et al. 2014), and Deception Island (Antarctica) (Bartolini et al. 2014a).

In both long-term and short-term evaluations HASSET can be useful for determining the occurrence probabilities of eruptive scenarios and can assist decision-makers assess the required mitigation actions associated with each scenario and estimate the corresponding potential risk.

Simulation Models

Simulating eruptive scenarios caused directly (e.g. lava flows, fallout, surges) and indirectly (earthquakes, landslides) by an eruption requires a detailed analysis of the past activity of the volcano or volcanic area, and must take into account all the possible hazards associated with the eruptive activity. Volcanic hazard can be assessed via two different types of approaches: deterministic, which defines a maximum extension area affected by an eruptive episode based on deposits generated by past activity, or probabilistic, based on the probability that a certain area will be affected by an eruptive process. In order to generate hazard maps (Fig. 5d), it is important to understand past eruptive behaviour and to employ physical simulation models that will permit the behaviour of future volcanic activity to be foreseen. In this type of approach, accurate and detailed geographic and cartographic data are required for high-quality analysis with a GIS.

Here, we describe some of the e-tools freely available for download that allow volcanic hazards to be evaluated (Fig. 5c):

- VORIS 2.0.1 is a GIS-based tool, developed by Felpeto et al. (2007) that allows users to simulate lava flows, fallout, and pyroclastic density current scenarios. Lava-flow simulations are based on a probabilistic model that assumes that topography is the most important factor determining the path of a lava flow. The determination of the probability of each point being invaded by lava is performed by computing several random paths with a Monte Carlo algorithm. Fallout simulation models are advection diffusion models that assume that away from the vent the transport of the particles from a Plinian column is controlled by the advective effect of the wind, by diffusion due to atmospheric turbulence, and by the settling velocity of the particles. The model for simulating pyroclastic density currents is the energy cone model proposed by Sheridan and Malin (1983). The input parameters are the topography, the collapse equivalent height (H), and the collapse equivalent angle (θ). The intersection of the energy cone, originating at the eruptive source, with the ground surface defines the distal limits of the flow.
- HAZMAP is a free program for simulating the sedimentation of volcanic particles at discrete point sources that predicts the corresponding ground deposits (deposit mode) (Macedonio et al. 2005). HAZMAP is also able to evaluate the probability of overcoming a given loading threshold in ground deposits by using a set of different wind profiles recorded on different days (probability mode). Using a statistical set of recorded wind profiles (and/or other input parameters), it can also be used to draw hazard maps for ashfall deposits. In HAZMAP, settling velocities can be calculated using several models as a direct function of particle diameters, densities, and shapes. The advantage of HAZMAP is that it is a simple tool able to predict ashfall during hypothetical or real eruptions of a given magnitude and wind profile.
- LAHARZ is a semi-empirical code for creating hazard-zonation maps that depict estimates of the location and extent of areas inundated by lahars (Schilling 1998). The input parameters for this model are the Digital Elevation Model and the lahar volume, which provide an automated method for mapping areas of potential lahar inundation. These hazard zones can be displayed in a GIS with

other types of volcano hazard information such as the proximal hazard zone, infrastructure, hydrology, and population, as well as contours and shaded relief, to produce volcano hazard-zonation maps. Such maps show the proximity and intersection of potential hazard zones to people and infrastructures.

- TITAN2D is a computer program model developed by the University at Buffalo (Patra et al. 2005) that simulates granular flows over digital elevation models, based on a "thin layer model". The input for the computer code includes simulation time, minimum thickness of the final deposit, internal and bed friction angles, starting coordinates, and the initial speed and direction of the flow. Additionally, this program allows users to define the specific starting pile dimensions or a dynamic flow source. The outputs from the program (represented dynamically) are flow depth and momentum, which yield the deposit limit, run-out path, average flow velocity, inferred deposit thickness, and travel time.

Vulnerability E-tool

Once we have obtained hazard maps, the next step consists of adding population, infrastructures, and land-use data to evaluate the vulnerability associated with the impact of a determined hazard. The data required for generating vulnerability maps are very complex and varied, and depend on the observation scale. Vulnerability is directly dependent on the type of phenomena in question and on the socio-economic characteristics of the environment. VOLCANDAM is a new e-tool based on the methodology developed by Scaini et al. (2014) that generates maps estimating the expected damage caused by volcanic eruptions. VOLCANDAM (Fig. 6) consists of three main parts: exposure analysis, vulnerability assessment, and the estimation of expected damages. The exposure analysis identifies the elements exposed to the potential hazard and focuses on the relevant assets of the study area (population distribution, social and economic conditions, and productive activities and their role in the regional economy). The vulnerability analysis defines a physical vulnerability indicator for all exposed elements, as well as a corresponding qualitative vulnerability index. Systemic vulnerability considers the possible relevance of each element in the system and their interdependencies by taking into account all exposed and non-exposed elements (people, buildings, transportation network, urban services, and productive activities). Damage assessment is performed by associating a qualitative damage rating to each combination of hazard and vulnerability, bearing in mind their specific contexts and roles in the system. The way one element can be damaged—and thus lose its functionality—depends in fact on the type of hazardous event and the characteristics of the element. The result is damage maps that can be displayed at different levels of detail, depending on user preferences. This tool aims to facilitate territorial planning and risk management in active volcanic areas.

Decision-Making

The evaluation of the "direct costs" and "factors" (indirect costs) that have an impact on the economic growth of an area affected by a volcanic event needs to take into account a number of elements. A cost-benefit analysis may assist the decision-making process by evaluating the economic impact of the different scenarios. The approach used by Sobradelo et al. (2015) (Fig. 7)

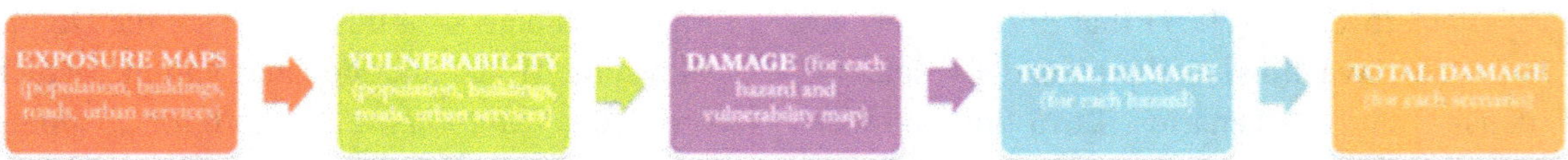

Fig. 6 Steps in the vulnerability analysis in the VOLCANDAM approach (after Scaini et al. 2014). See text for details

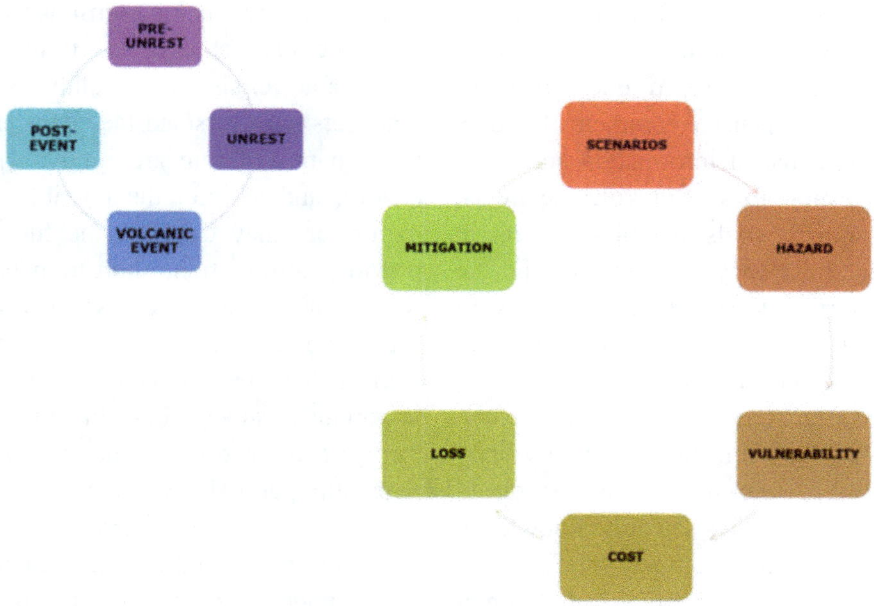

Fig. 7 The volcanic crisis management cycle: stages and phases (after Sobradelo et al. 2015). See text for details

is a Bayesian decision model that applies a general, flexible, probabilistic approach to the management of volcanic crises by combining the hazard and risk factors that decision-makers need for a holistic analysis of a volcanic crisis. These factors include eruption scenarios and their probabilities of occurrence, the vulnerability of populations and their activities, and the costs of false alarms and erroneous forecasts. This model can be implemented before an emergency to (i) pinpoint actions for reducing the vulnerability of a particular area, (ii) identify the optimum mitigating actions during an emergency and how these may change as new information is obtained, (iii) assess after an emergency the effectiveness of a mitigating response and, in light of results, (iv) how to improve strategies before another crisis occurs. BADEMO (BAyesian DEcision MOdel) is part of this integrated approach, and enables the previous analysis of the distribution of local susceptibility and vulnerability to eruptions to be combined with specific costs and potential losses. Indeed, BADEMO should be seen as a tool for improving communication between the monitoring scientists who provide volcanological information and those responsible for deciding which action plans and mitigating strategies should be put into practice.

Discussion and Conclusions

Modern volcanology is a scientific discipline with social and practical applications that derive from its direct involvement in risk reduction. Today, a multidisciplinary approach is required to risk assessment since new methodologies are constantly being developed and explored, and, for example, geo-spatial technologies are becoming extremely helpful in the disaster-risk management. A defined and solid methodology for assessing volcanic hazard and managing risk is fundamental in both long-term evaluations and in unrest phases when monitoring information is available. Furthermore, e-tools can be progressively modified and implemented in light of the outcomes obtained in order to integrate as many models as possible into the assessment and management of volcanic risk.

The possibility of foreseeing an event such as a volcanic eruption is more effective—and guarantees greater economic savings—than acting after a disaster. The potential of simple e-tools such as those presented in this chapter lies in the fact that they are freely available and have been developed on accessible and dynamic graphical user interfaces. The main advantage of using these e-tools is that new data or new model results can be easily incorporated into the procedures for updating the hazard assessment. By contrast, the use of expensive commercial e-tools hampers the exchange of information and complicates their testing in situ in volcanic fields. Furthermore, non-free tools may hinder effective risk assessment since they are often beyond the means of advisory and management groups with limited financial, technological, and manpower resources (Leidig and Teeuw 2015). However, this does not mean that anybody can use the e-tools discussed here for, on the contrary, they require expertise in volcanic hazards and related issues and their use implies scientific knowledge that will avoid incorrect outcomes.

In this chapter we have presented different statistical methodologies and e-tools for interpreting volcanic data and assessing long- and short-term volcanic hazards and vulnerability, and for carrying out cost-benefit analyses. Statistical analysis enables us to extract information about the future behaviour of a volcano by looking at the geological and historical activity of the volcanic system (VERDI database, Bartolini et al. 2014c). The QVAST tool (Bartolini et al. 2013) can be used to analyse past activity and to calculate the possibility that new vents will open (volcanic susceptibility), while the Bayesian event tree statistical method HASSET (Sobradelo et al. 2014) can be applied to calculate eruption recurrence. Using these calculations, we can identify a number of significant scenarios using GIS-based e-tools (i.e. VORIS 2.0.1, HAZMAP, …) and evaluate the potential extent of the main volcanic hazards expected to occur in volcanic areas. The results obtained allow us to generate volcanic hazard maps for different levels of hazards, evaluate vulnerability (VOLCANDAM e-tools, Scaini et al. 2014), conduct cost-benefit analysis (BADEMO e-tools, Sobradelo et al. 2015), and, finally, manage volcanic risk.

During a volcanic crisis, emergency plans must be put into practice and so different government departments need to be prepared in advance. Therefore, it is important to have conducted previous long-term hazard assessment —among many other tasks—to properly manage a volcanic crisis. They allow scientists and managers to understand the characteristics of the volcano or volcanic area and its past eruptive history, and to infer the possible eruptive scenarios that may occur in the future. With this previous information, short-term hazard assessment can be conducted when volcanic unrest starts and hazard maps can be drawn up and alert levels be defined; nevertheless, it is important to always bear in mind that the best form of protection is the evacuation of the population at risk. Volcanic monitoring is an essential part of short-term assessment and so should be performed by experts and based on a good understanding of volcanic processes. Despite the possibility of conducting cost-benefit analysis, which can help maintain the economic order, the security and health of the population should always be the main concern.

One of the purposes in the near future is to create a new software platform (VolcanBox, see VETOOLS European Project—www.vetools.eu) with a user-friendly interface. This platform will contain different e-tools that, via a homogeneous and systematic methodology, will help minimise risk. However, the feasibility and applicability of each tool will have to be analysed by different groups of experts with experience in regions possessing different volcanological and socio-economic scenarios. This evaluation must also bear in mind potential end users (i.e. Civil Protection agencies), not only to test the ability of existing tools but also to understand decision-makers' needs and requirements, and train them in the use of these tools. If we are to reduce volcanic risk we must ensure that scientists, managers, and decision-makers are all fully prepared to confront this phenomenon since the best way to guarantee risk reduction is to possess good knowledge of its causes.

Acknowledgements This work was supported by the European Commission (FP7 Theme: ENV.2011.1.3.3-1; Grant 282759: VUELCO). The English text was corrected by Michael Lockwood.

References

Aspinall WP (2006) Structured elicitation of expert judgment for probabilistic hazard and risk assessment in volcanic eruptions. In: Mader HM et al (eds) Statistics in volcanology. Special Publication of IAVCEI, Geological Society of London

Bartolini S, Cappello A, Martí J, Del Negro C (2013) QVAST: a new Quantum GIS plugin for estimating volcanic susceptibility. Nat Hazards Earth Syst Sci 13 (11):3031–3042

Bartolini S, Geyer A, Martí J, Pedrazzi D, Aguirre-Díaz G (2014a) Volcanic hazard on deception Island (South Shetland Islands. J Volcanol Geotherm Res, Antarctica). doi:10.1016/j.jvolgeores.2014.08.009

Bartolini S, Bolós X, Martí J, Riera E, Planagumà LL (2014b) Hazard assessment at the Quaternary La Garrotxa Volcanic Field (NE Iberia). Nat Hazards. doi:10.1007/s11069-015-1774-y

Bartolini S, Becerril L, Martí J (2014c) VERDI: a new volcanic management risk database design. J Volcanol Geotherm Res 288:132–143

Becerril L, Bartolini S, Sobradelo R, Martí J, Morales JM, Galindo I (2014) Long-term volcanic hazard assessment on El Hierro (Canary Islands). Nat Hazards Earth Syst Sci 14:1853–1870

Blong R (2000) Volcanic hazards and risk management. In: Sigurdsson H et al (eds) Encyclopedia of volcanoes. Academic, San Diego, pp 1215–1227

Cappello A, Neri M, Acocella V, Gallo G, Vicari A, Del Negro C (2012) Spatial vent opening probability map of Etna volcano (Sicily, Italy). Bull Volcanol 74:2083–2094. doi:10.1007/s00445-012-0647-4

Cappello A, Bilotta G, Neri M, Del Negro C (2013) Probabilistic modelling of future volcanic eruptions at Mount Etna. J Geophys Res Solid Earth 118:1925–1935. doi:10.1002/jgrb.50190

Chester DK, Dibben CJL, Duncan AM (2002) Volcanic hazard assessment in Western Europe. J Volcanol Geotherm Res 115:411–435

De La Cruz-Reyna S (1996) Long term probabilistic analysis of future explosive eruptions. In: Scarpa R, Tilling RI (eds) Monitoring and mitigation of volcanic hazards. Springer, Berlin, pp 599–629 (IAVCEI/UNESCO Volume)

Felpeto A, Martí J, Ortiz R (2007) Automatic GIS-based system for volcanic hazard assessment. J Volcanol Geotherm Res 166:106116

Leidig M, Teeuw R (2015) Free software: a review, in the context of disaster management. Inter J Appl Earth Observ Geoinfo 42:49–56

Lirer L, Petrosino P, Alberico I (2001) Volcanic hazard assessment at volcanic fields: the Campi Flegrei case history. J Volcanol Geotherm Res 101(1–4):55–75

Macedonio G, Costa A, Longo A (2005) A computer model for volcanic ash fallout and assessment of susbsequent hazard. Comput Geosci 31:837–845

Martí J, Felpeto A (2010) Methodology for the computation of volcanic susceptibility. An example for mafic and felsic eruptions on Tenerife (Canary Islands). J Volcanol Geotherm Res 195:69–77

Marti J, Spence R, Calogero E, Ordoñez A, Felpeto A, Baxter P (2008) Estimating building exposure and impact to volcanic hazards in Icod de los Vinos, Tenerife (Canary Islands). J Volcanol Geotherm Res 178:553–561

Martí J, Sobradelo R, Felpeto A, García O (2012) Eruptive scenarios of phonolitic volcanism at Teide-Pico Viejo volcanic complex (Tenerife, Canary Islands). Bull Volcanol 74:767–782. doi:10.1007/s00445-011-0569-6

Martí J, Pinel V, López C, Geyer A, Abella R, Tárraga M, Blanco MJ, Castro A, Rodríguez C (2013) Causes and mechanisms of the 2011–2012 El Hierro (Canary Islands) submarine eruption. J Geophys Res Solid Earth 118(3):823–839

Marzocchi W, Sandri L, Gasparini P, Newhall C, Boschi E (2004) Quantifying probabilities of volcanic events: the example of volcanic hazard at Mount Vesuvius. J Geophys Res 109. doi:10.1029/2004JB003155

Neri A, Aspinall W, Cioni R, Bertagnini A, Baxter P, Zuccaro G, Andronico D, Barsotti S, Cole P, Ongaro TE, Hincks T, Macedonio G, Papale P, Rosi M, Santacroce R, Woo G (2008) Developing an event tree for probabilistic hazard and risk assessment at Vesuvius. J Volcanol Geotherm Res 178 (3):397–415

Newhall CG, Hoblitt RP (2002) Constructing event trees for volcanic crisis. Bull Volcanol 64:3–20

Sandri L, Jolly G, Lindsay J, Howe T, Marzocchi W (2012) Combining long-and short-term probabilistic volcanic hazard assessment with cost-benefit analysis to support decision making in a volcanic crisis from the Auckland Volcanic Field, New Zealand. Bull Volcanol 74(3):705–723

Sandri L, Thouret JC, Constantinescu R, Biass S, Tonini R (2014) Long-term multi-hazard assessment for El Misti volcano (Peru). Bull Volcanol 76(2):1–26

Scaini C, Felpeto A, Martí J, Carniel R (2014) A GIS-based methodology for the estimation of potential volcanic damage and its application to Tenerife Island, Spain. J Volcanol Geotherm Res 278–279:40–58

Schilling SP (1998) LAHARZ: GIS program for automated mapping of lahar-inundation hazard zones. US Geological Survey Open-File Report, pp 98–638

Sheridan MF, Malin MC (1983) Application of computer-assisted mapping to volcanic hazard evaluation of surge eruptions: Vulcano, Lipari. J Volcanol Geotherm Res 17:187–202

Sobradelo R, Martí J (2010) Bayesian event tree for long-term volcanic hazard assessment: application to Teide-Pico Viejo stratovolcanoes, Tenerife, Canary islands. J Geophys Res 115. doi:10.1029/2009JB006566

Sobradelo R, Bartolini S, Martí J (2014) HASSET: a probability event tree tool to valuate future volcanic scenarios using Bayesian inference presented as a plugin for QGIS. Bull Volcanol 76:770

Sobradelo R, Martí J, Kilburn C, López C (2015) Probabilistic approach to decision-making under uncertainty during volcanic crises: retrospective application to the El Hierro (Spain) 2011 volcanic crisis. Nat Hazards 76(2):979–998

Thierry P, Neri M, Le Cozannet G, Jousset P, Costa A (2015) Preface: approaches and methods to improve risk management in volcanic areas. Nat Hazards Earth Syst Sci 15:197–201

Toyos GP, Cole PD, Felpeto A, Martí J (2007) A GIS-based methodology for hazard mapping of small pyroclastic density currents. Nat Hazards 41:99–112

Twigg J (2004) Disaster risk reduction: mitigation and preparedness in development and emergency programming. Overseas Development Institute, Humanitarian Practice Network, London, Good Practice Review no. 9

Van Westen CJ (2013) Remote sensing and GIS for natural hazards assessment and disaster risk management. In: Shroder J (editor in Chief), Bishop MP (ed) Treatise on geomorphology. vol 3, Remote Sensing and GIScience in Geomorphology. Academic Press, San Diego, pp 259–298

The Need to Quantify Hazard Related to Non-magmatic Unrest: From BET_EF to BET_UNREST

Laura Sandri, Roberto Tonini, Dmitri Rouwet, Robert Constantinescu, Ana Teresa Mendoza-Rosas, Daniel Andrade and Benjamin Bernard

Abstract

Most volcanic hazard studies focus on magmatic eruptions and their accompanying phenomena. However, hazardous volcanic events can also occur during non-magmatic unrest, defined as a state of volcanic unrest in which no migration of magma is recognised. Examples include tectonic unrest, and hydrothermal unrest that may lead to phreatic eruptions. Recent events (e.g. Ontake eruption, September 2014) have demonstrated that the successful forecasting of phreatic eruptions is still very difficult. It is therefore of paramount importance to identify indicators that define the state of non-magmatic unrest. Often, this type of unrest is driven by fluids-on-the-move, requiring alternative monitoring setups, beyond the classical seismic-geodetic-geochemical architectures. Here we present a new version of the probabilistic model BET (Bayesian Event Tree), called BET_UNREST, specifically developed to include the forecasting of non-magmatic unrest and related hazards. The structure of BET_UNREST differs from the previous BET_EF (BET for Eruption Forecasting) by adding a dedicated branch to detail non-magmatic unrest outcomes. Probabilities are calculated at each node by merging prior models and past

L. Sandri (✉) · D. Rouwet
Sezione di Bologna, Istituto Nazionale di Geofisica e Vulcanologia, Via Donato Creti 12, Bologna, Italy
e-mail: laura.sandri@ingv.it

R. Tonini
Sezione di Roma 1, Istituto Nazionale di Geofisica e Vulcanologia, Via di Vigna Murata 605, Rome, Italy

R. Constantinescu
Seismic Research Centre, University of West Indies, Gordon street, Saint Augustine, Trinidad & Tobago

Present Address:
R. Constantinescu
School of Geosciences, University of South Florida, 4202 E. Fowler Avenue, NES 107, Tampa, FL 33620-5550, USA

A.T. Mendoza-Rosas
Instituto de Geofísica, Universidad Nacional Autónoma de México. C. Universitaria, Coyoacán 04510, México, D.F., Mexico

Present Address:
A.T. Mendoza-Rosas
CONACYT-Centro de Ingeniería y, Desarrollo Industrial, Av. Playa pie de la cuesta 702, Desarrollo San Pablo, Querétaro, Qro. 76125, Mexico, C.P., Mexico

D. Andrade · B. Bernard
Escuela Politécnica Nacional, Instituto Geofísico, Quito, Ecuador

Advs in Volcanology (2019) 63–82
DOI 10.1007/11157_2017_9

Published Online: 12 August 2017

data with new incoming monitoring data, and the results can be updated any time new data has been collected. Monitoring data are weighted through pre-defined thresholds of anomaly, as in BET_EF. The BET_UNREST model is introduced here, together with its software implementation PyBetUnrest, with the aim of creating a user-friendly, open-access, and straightforward tool to support short-term volcanic forecasting (already available on the VHub platform). The BET_UNREST model and PyBetUnrest tool are tested through three case studies in the frame of the EU VUELCO project.

Resumen extendido

La mayoría de los estudios sobre amenazas volcánicas están enfocados en las erupciones magmáticas y fenómenos relacionados. Sin embargo, fenómenos volcánicos peligrosos pueden también ocurrir durante una fase de “unrest” no-magmático, definido por el estado de unrest volcánico en el cual no se reconoce la migración de un magma. Ejemplos de esto son unrest tectónico (capaz de causar preocupación independientemente del resultado posterior) y unrest hidrotermal, que pueden resultar en erupciones freáticas. Eventos recientes (e.g. la erupción de Ontake en septiembre 2014) han demostrado que las erupciones freáticas siguen siendo difícilmente previsibles. Por estas razones, es de extrema importancia identificar señales que permitan definir un estado de unrest no-magmático. Muchas veces, este tipo de unrest es provocado por fluidos en movimiento, y requiere la instalación de un sistema de monitoreo alternativo, más allá de la clásica arquitectura sismo-geodético-química. En este capítulo, presentamos la nueva versión del modelo probabilístico BET (Arbol de Eventos Bayesiano, por sus siglas en inglés), llamado BET_UNREST, específicamente desarrollado para incluir la previsión de unrest no-magmático y sus peligros relacionados. La estructura de BET_UNREST difiere de la versión anterior BET_EF (BET para Previsión de Erupciones, por sus siglas en inglés), añadiendo una rama dedicada para detallar los resultados potenciales de unrest no-magmático. Las probabilidades están calculadas para cada nodo juntando modelos a priori y datos pasados con los datos nuevos, provenientes del monitoreo. Los datos de monitoreo están ponderados mediante umbrales predefinidos de anomalía, como es el caso en BET_EF. Este capítulo ilustra el modelo, y su herramienta, con tres casos de estudio, en el marco del proyecto EU VUELCO:

(i) un análisis retrospectivo para el volcán Popocatépetl, en donde no hay necesidad de la rama hidrotermal, debido al carácter magmático; Popocatépetl ha permanecido en estado de unrest desde diciembre 1994 hasta el presente. Para esta aplicación, BET_UNREST fue corrido usando la base de datos de la UNAM (1997–2012), con una aplicación retrospectiva para prever erupciones mayores (columnas eruptivas >8 km) durante el período abril-junio 2013.

(ii) una aplicación basada en un ejercicio de simulacro en el Cotopaxi; en este caso se probó con BET_UNREST de manera retrospectiva, pero, esta vez, usando datos creados específicamente para el simulacro, junto a datos de entrada basados en la historia real del volcán preparados antes del simulacro. Presentamos la previsión de erupciones magmáticas resultantes del simulacro mismo.

(iii) el simulacro en tiempo casi-real organizado en el marco de VUELCO en Dominica (mayo 2015). El sistema volcánico de Dominica es el prototipo para BET_UNREST debido a su carácter hidrotermal. Actividad freática/freatomagmática ocurrió realmente durante el simulacro, lo cual de hecho era bastante probable según BET_UNREST (la probabilidad media de unrest hidrotermal fue de 0.73, mientras la probabilidad media de una erupción hidrotermal fue de 0.32). También se produjó un mapa de probabilidades para la apertura de ventos eruptivos en caso de erupciones magmáticas y freáticas.

Con estos ejercicios, estamos convencidos de haber llevado BET un paso más cerca hacia una implementación completa en situaciones de crisis. Al final, BET_UNREST funcionó como se esperaba. Sin embargo, es importante ser consiente de algunos puntos críticos que han resultado de estas aplicaciones, incluso realizar más pruebas para mejorar su diseño y comprobar su utilidad en casos reales en el futuro. BET_UNREST se introdujo junto a su implementación digital PyBetUnrest con el objetivo de crear un instrumento de fácil uso, libre y de acceso directo (disponible en el sitio web Vhub) para ayudar en la evaluación de la amenaza volcánica a corto plazo.

Keywords
Volcanic unrest · Forecasting · Hydrothermal · Magmatic · Bayesian inference

Palabras clave
Unrest volcánico · Previsión · Hidrotermal · Magmático · Inferencia Bayesiana

1 Introduction

Monitoring activities represent the main source of information to understand the behaviour of volcanic systems on short time-scales and, possibly, during emergency crises. In this framework, one of the main challenges of volcano monitoring is the identification and characterisation of the phase defined as "unrest", which consists of a relevant physical or chemical change in the volcanic system with respect to its background behaviour, leading to cause for concern. Unrest can be due to several factors and depends on the local characteristics of each volcanic system, making it very difficult to find general features or patterns (Phillipson et al. 2013). Unrest may be followed by volcanic eruptions due to the movement of magma, but can also be associated with other dangerous phenomena: indeed, in addition to magma-related hazards (e.g., tephra fallout, lava flows, ballistics), hydrothermal and tectonic activities, without evidence for "magma-on-the-move", can also lead to dangerous outcomes (i.e., flank collapses, gas emissions, phreatic explosions, lahars). Such hazardous events related to non-magmatic unrest are not easy to track and, in volcanic hazard evaluations, are sometimes underestimated (Rouwet et al. 2014). For instance, many volcanoes pass through a phase of hydrothermal unrest for years, decades or even centuries. Due to this long-term behavioural similarity, it is often difficult to recognise how hydrothermal unrest can lead to related hazards in the short-term. Where the driving agent and the main eruptive product is not magma, but water (liquid or vapour) and occasionally liquid sulphur, or gas, this type of unrest can lead to non-magmatic eruptions. On the other hand, non-eruptive hydrothermal unrest can also promote volcanic hazards after prolonged gas emissions, acidic fluid infiltration into aquifers, soils and the hydrologic network, or deformation induced by a rising fluid front (see Rouwet et al. 2014).

In this light, although most volcanic hazard assessments focus only on magmatic eruptions as potential hazard sources, hazardous events can also occur during non-magmatic unrest, which in this chapter is defined as a state of volcanic unrest in which no migration of magma is recognised. Examples of non-magmatic unrest include the tectonic (which causes concern independently on how it evolves and eventually ends), and hydrothermal unrest types; the latter may eventually lead to phreatic eruptions. Recent occurrences of phreatic eruptions (e.g. Ontake eruption, September 2014, Japan) have demonstrated that they are still very hard to anticipate from classical observations based on seismic-geodetic-geochemical monitoring architectures. For these reasons, it is of paramount importance to identify indicators that define the state of non-magmatic unrest. Often, this type of unrest is driven by "fluids-on-the-move", requiring alternative and innovative monitoring setups, beyond the classical ones.

In the last decade it has become crucial to provide forecasts of the possible outcomes of volcanic unrest, to give quantitative support and scientific advice to decision makers (e.g., Woo 2008; Marzocchi and Woo 2007, 2009). Because of this, event tree schemes have been proposed (e.g., Newhall and Hoblitt 2002; Marzocchi et al. 2004), and a few probabilistic tools based on event trees and Bayesian inference have been developed (e.g., BET_EF, Marzocchi et al. 2008; HASSET, Sobradelo et al. 2013) with the ability to quantify the probability of different possible outcomes related to magmatic unrest. However, the need for recognising and tracking the evolution of *any* type of volcanic unrest, and to quantify the probability linked to non-magmatic unrest as well, have led us, within the VUELCO project, to the development of a new probabilistic model, able to forecast both magmatic and non-magmatic hazardous events related to volcanic unrest: BET_UNREST. The BET_UNREST model is based on an event tree, whose structure is extended with respect to the previous schemes such as BET_EF (see the generalisation from BET_EF to BET_UNREST in Fig. 1, highlighted in red) by adding a specific branch to detail the track and outcome of non-magmatic unrest. Nonetheless, BET_UNREST adopts from BET_EF the Bayesian inferential paradigm and the ability to account both for long-term data (typically from the geological record) and short-term information from monitoring networks.

In this chapter, we briefly present the BET_UNREST model and its implementation in the PyBetUnrest software tool (Tonini et al. 2016), made with the aim of providing a user-friendly, open-access, and straightforward tool to handle probabilistic forecasts and visualise results, and that has already been included on the Vhub platform (https://vhub.org/resources/betunrest). The new event tree and tool are applied here as illustrative examples to the VUELCO target volcanoes Popocatépetl (Mexico), Cotopaxi (Ecuador) and Dominica (West Indies).

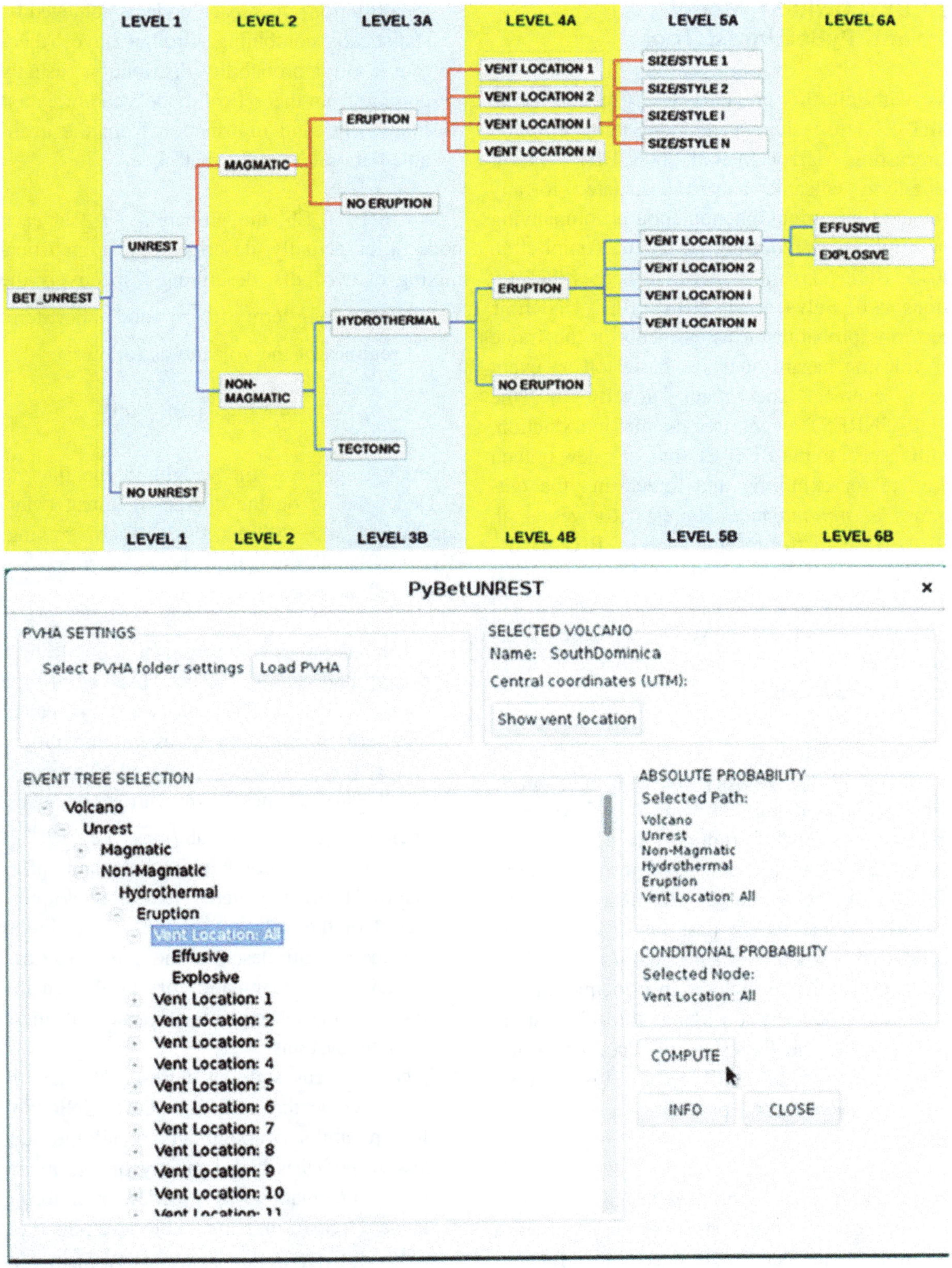

Fig. 1 The new event tree as defined for the BET_UNREST model (*on top*) and its visual implementation in the software PyBetUnrest (*on bottom*). The *red branch* corresponds to the previous BET_EF model

2 BET_UNREST Model and PyBetUnrest Tool

As with all the previous BET models (e.g., BET_EF, for short- and long-term eruption forecasting, Marzocchi et al. 2008; BET_VH, for long-term volcanic hazard associated to any potential hazardous phenomenon accompanying an eruption, Marzocchi et al. 2010; Tonini et al. 2015; BET_VHst, a model that merges the previous two, Selva et al. 2014), BET_UNREST performs probabilistic assessments in the frame of volcanic hazard analysis, based on an event tree scheme. The main novelty in the BET_UNREST event tree is the introduction, with respect to the BET_EF tree, of a new branch (Fig. 1) for exploring and forecasting the outcomes of *non-magmatic* unrest (Rouwet et al. 2014). Due to the resemblance of BET_UNREST to other BET models from a methodological and computational point of view, here we will only give a brief overview. The papers by Marzocchi et al. (2004, 2008) provide a more detailed description.

BET_UNREST probabilities are evaluated by a Bayesian inferential procedure, in order to quantify both the aleatory and epistemic uncertainty characterising the impact of volcanic eruptions in terms of eruption forecasting and/or hazard assessment. Such a procedure allows merging all the available information, such as models, a priori beliefs, past data from volcanic records and, when available, real-time monitoring data in order to include, in principle, all the knowledge about the considered volcanic system.

In general, the Bayesian inference procedure at the basis of BET_UNREST assigns a probability to each node, providing a framework where:

- probabilities are expressed through a probability density function (pdf), and not as a single number, to account for a best-evaluation value (for example the mean of the probability density function, representing a degree of aleatory uncertainty) and for a measure of the epistemic uncertainty (the dispersion of the pdf);
- the posterior pdf, at each node, is achieved by statistically combining, through Bayes' theorem, a prior probability distribution (usually coming from theoretical models and/or expert judgement) and information from the available data relevant for that node.

As in BET_EF, the probability $[\theta_k]$ at each node k is actually described by a statistical mixing of two pdfs, describing respectively the "so-called" long-term $[\theta_k^{\{\bar{M}\}}]$ and short-term $[\theta_k^{\{M\}}]$ regimes of the volcano as follows:

$$[\theta_k] = \gamma_k[\theta_k^{\{M\}}] + (1 - \gamma_k)[\theta_k^{\{\bar{M}\}}]$$

where γ_k represents the weight in the interval [0,1] depending on the degree of unrest (Marzocchi et al. 2008). With such mixing, BET_UNREST switches between the two "regimes". In practice:

- When anomalies with respect to the volcano's background activity are not observed at time $t = t_0$, BET_UNREST relies on the so-called long-term information to assign the probabilities (hereinafter also referred to as background probabilities) at the various branches. Such background probabilities (i.e., $[\theta_k^{\{\bar{M}\}}]$) are based on theoretical models and information from the geological and eruptive record of the volcano studied, or of similar volcanoes, and describe the long-term frequencies of magmatic or non-magmatic unrest, and subsequent outcomes at these volcanic systems.
- When a clear state of unrest of whatever nature is detected at $t = t_0$ by BET_UNREST, the probabilistic assignment at all the successive nodes is based mainly on the monitoring information. In practice, monitoring data are transformed into subjective pdfs (i.e., $[\theta_k^{\{M\}}]$) relative to the occurrence of magmatic or non-magmatic unrest and the following branches. Actually, at some nodes, monitoring data are not considered as relevant (for example, in forecasting the size of an

eruption, magmatic or not), and here BET_UNREST continues to rely on theoretical models and long-term frequencies.

- When, at time $t = t_0$, BET_UNREST observes a "degree of unrest" (of whatever nature) without it being completely clear, the statistical mixing provides a resulting pdf which accounts for both the regimes, giving the short-term regime a weight equal to the degree of unrest, and to the long-term regime its complement.

In this way, during a phase of unrest, the past data have less (null, in the case of complete unrest) importance. The short term hazard/eruption forecasting depends exclusively on the translation of observed anomalies into pdfs describing all the branches of the event tree. This is done, separately at each node, by weighting monitoring data through pre-defined thresholds of anomaly (Marzocchi et al. 2008) and converting the resulting "degree of anomaly" into a best-evaluation probability, to which a degree of variance is associated (Fig. 2). This is a very simple and intuitive procedure, in which the basic assumptions are:

1. the first anomaly detected is the most informative
2. subsequent anomalies contribute less and less to the increase of the degree of anomaly
3. strong non-linear coupling among anomalies are neglected.

At each node, BET_UNREST evaluates the following probabilities (see also Fig. 1) by means of Bayesian inference (we give the acronyms used throughout the chapter to indicate the probability at each node in brackets):

- *Unrest*: probability ($P(U)$) of unrest in the time period [t_0; t_0 + |], given the monitoring observations at time $t = t_0$; the time window | is defined by the user;
- *Magmatic unrest*: probability ($P(MU)$) that the unrest is due to "magma-on-the-move", given the unrest;
- *Magmatic eruption*: probability ($P(MEr)$) of a magmatic eruption, given magmatic unrest; the following sub-branches mirror the BET_EF structure, so we point the reader to Marzocchi et al. (Marzocchi et al. 2008) for them;
- *Non-magmatic unrest*: this is the complementary of the *Magmatic unrest* branch, so by definition is the probability of non-magmatic unrest, given an unrest;
- *Hydrothermal unrest*: probability ($P(HU)$) of hydrothermal unrest, given a non-magmatic unrest;
- *Tectonic unrest*: this is the complementary of the *Hydrothermal unrest* branch, so it describes the probability ($P(TU)$) of a tectonic unrest, given a non-magmatic unrest;
- *Hydrothermal eruption*: probability ($P(HEr)$) of a hydrothermal eruption, given a hydrothermal unrest;
- *Vent of hydrothermal eruption*: here we explore the spatial probability of vent opening in a hydrothermal eruption, given a hydrothermal eruption occurring; this node is an extension with respect to the event tree proposed in Rouwet et al. (2014);
- *Size of hydrothermal eruption*: probability of an explosive hydrothermal eruption, given a hydrothermal eruption occurring from a specific vent; its complementary branch is the effusive hydrothermal eruption.

In order to keep the structure of BET_UNREST as simple as possible, an effort has been made to maintain, where possible, a dichotomic branching into complementary (i.e., exhaustive and mutually exclusive) events. This is why the *Unrest* node does not branch directly into magmatic, hydrothermal and tectonic, but first it branches into magmatic-or-not. This allows a simplification in the evaluation of short-term probabilities. In particular, with this type of ramification, the user defines which monitoring measurements (plus thresholds and weight) affect the pdf of one of the two branches; the pdf of the complementary branch then comes automatically.

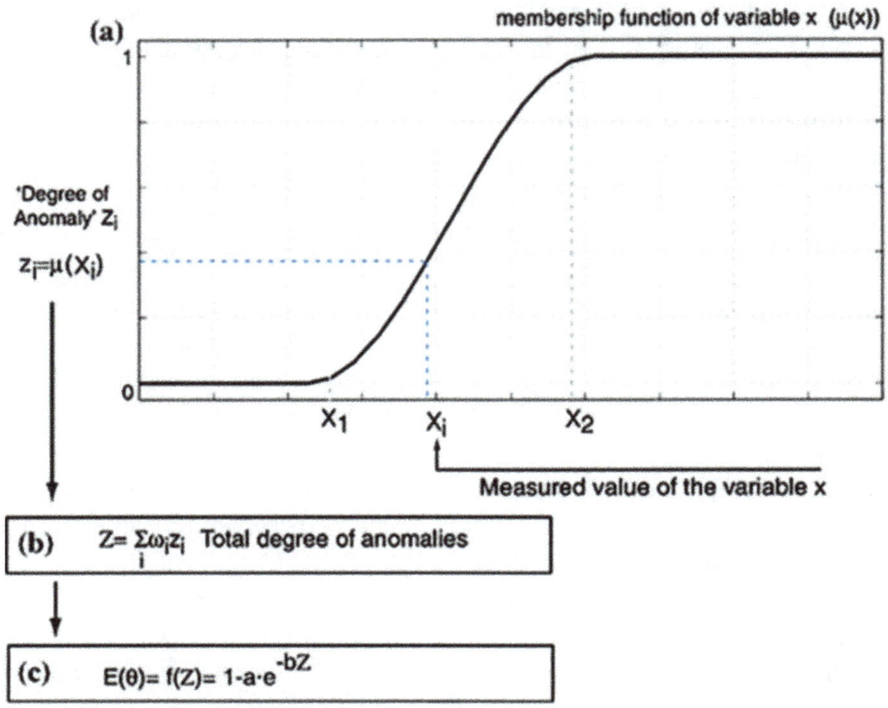

Fig. 2 This figure explains how monitoring measures are transformed into a best-evaluation probability at a given node of the event tree. First, a monitoring measure x_i is translated in a degree of anomaly z_i according to a selected anomaly function $\mu(\cdot)$ (**a**). In the above example, a measure below x_1 is considered background, above x_2 is anomalous, and in between it has a certain degree of anomaly. After collecting the degree of anomaly for all parameters considered at the node, we combine them using a weighted average (ωi is the weight of the i-th parameter) in order to obtain the total degree of anomaly (**b**). Then the total degree of anomaly is transformed into an average probability using a predefined function, in BET_UNREST, we use the function in (**c**). The parameters, weights, and thresholds are selected by the user, possibly through expert opinions' elicitation. Figure modified from (Marzocchi and Bebbington 2012)

The new BET_UNREST model is applied here with its software implementation PyBetUnrest presented in Tonini et al. (2016), which aims to provide an open and usable tool to bridge between the scientific community and decision makers, with a graphical user interface which allows the exploration of the event tree and the visualisation of the results (see Fig. 1). This solution was also implemented in the VHub cyber-infrastructure (http://vhub.org/resources/betunrest). In the present PyBetUnrest tool only one file needs to be adapted when new monitoring information is gathered. This structure makes PyBetUnrest extremely fast and user-friendly during crisis situations. More on the technical background of the BET_UNREST model and PyBetUnrest tool can be found in the VUELCO Deliverable 7.3 (at http://vhub.org) and in Tonini et al. (2016).

So far BET_UNREST and PyBetUnrest have not yet been blindly tested in real-time during an actual volcanic crisis, but only retrospectively (Tonini et al. 2016) at Kawah Ijen (Indonesia), for the time period 2010–2012 (after a learning period based on the observations from 2000 to 2010). The term "blindly" signifies that the rules of BET_UNREST (the long-term pdfs, and the monitoring parameters, thresholds and weight at the different nodes) are set *before* the beginning of the application, on different data (the *learning* dataset), and then the model is applied untouched to new data (the voting dataset), typically covering a different time period (as in the case of Tonini et al. 2016).

In the next section of this chapter, results and performances of the new model and tool will be discussed and validated by analysing the unrest crises for VUELCO target volcanoes Popocatépetl, Cotopaxi and Dominica through blind applications of BET_UNREST. The latter two applications show the results of the VUELCO crisis simulation exercises held in Quito (November 2014) and Dominica (May 2015).

3 BET_UNREST Applications

3.1 Popocatépetl, Mexico: A Retrospective Application Based on the Popo-DataBase

Here we apply the BET_UNREST model to Popocatépetl Volcano (Mexico), based on a catalog of monitored parameters of the 1994-ongoing eruptive period. Popocatépetl volcano awakened in December 1994, after almost 48 years of volcanic quiescence. Since 1994, Popocatépetl volcano has been one of the most active volcanoes in the world, and magmatic activity has been nearly constant. This fact raises the need to first redefine the concept of volcanic unrest for Popocatépetl, as BET_UNREST, at the *Unrest* node, requires indicative parameters to verify if the given volcano is in a state of unrest, or not. In *stricto* sensu, Popocatépetl has remained at least in a state of unrest, or even magmatic or eruptive unrest, since 1994, as its common manifestations are dome growth and vulcanian eruptive phases. The continuous state of unrest is reflected by the decision to never decrease the level of alert from orange to green (traffic light, De la Cruz-Reyna and Tilling 2008). Nevertheless, many of these eruptions are of no cause of concern (so, no unrest in *lato* sensu), neither for volcanologists nor for population. On the other hand, a practical scope of the BET_UNREST application at Popocatépetl is to forecast *major* eruptions, which can be considered a deviation from its current background activity. During the past 23,000 years, nine Plinian eruptions occurred at Popocatépetl (Mendoza-Rosas and De la Cruz-Reyna 2008), while, since 1994, three eruptions with an eruption column >8 km have occurred. No Plinian eruptions have occurred during the 1994-ongoing eruption cycle, and thus none of the past Plinian eruptions have been monitored. For practical purposes, we thus define a *major* eruption for Popocatépetl as an eruption *with an eruption column >8 km*, as they are recorded during the current monitoring period. These eruptions have caused ash fall in the Puebla-Mexico City metropolitan area, thus having an impact on human activity. We aim at finding precursory signals for major eruptions (>8 km, VEI 3) for the period 1997–2012 (the learning period), and test the BET_UNREST retrospectively, using monitoring data of the volcanic activity observed during 2013 (the voting period). The time window, |, is defined as 1 month.

In Table 1 we report the activity carried out 24/7 with regards to monitoring at Popocatépetl, available as short-term information for unrest, origin of unrest and eruption. However, for the time period 1994–2012, the available data (as listed in Mendoza-Rosas, *VUELCO deliverable 5.1*), are restricted mainly to seismicity (VT, tremor, number of events) and visual observations (i.e. number of eruptions, column height). No real-time SO_2 flux is available for our purpose, and deformation data would need further processing. Regarding past data (long-term information for unrest, origin of unrest, and eruption), there have been 13 unrest episodes, and constant unrest since December 1994 (so, a priori probability to be in a state of unrest for the next month is about 85%). Out of the 13 unrest episodes, 6 were due to magma-on-the-move (magmatic unrest), of which 3 lead into a magmatic *major* eruption. The monitoring parameters listed in Table 2, along with respective thresholds and weight, have been identified in the UNAM (Universidad Nacional Autónoma de México) database for the period 1997–2012, and used to set BET_UNREST for Popocatépetl. The volcano is a stratocone with a higher probability of an eruption to occur from the central vent. For the period of observation (1997–2012) all eruptions were magmatic and occurred at the central crater. The a priori spatial distribution of vent opening is assigned as in Table 3. As a prior

Table 1 Activity carried out 24/7 as regards monitoring at Popocatépetl

Observations	4 cameras for visual observations
	5 three-component seismic stations
	5 BB seismic stations
	1 video camera + microwaves
	1 doppler radar
	3 biaxial inclinometers
	Geochemical observations (3 sites)
Routine actions	Automatic alarm for anomalies in seismicity
	Cell phone messages to personnel
	Comité Técnico Cientifico Asesor UNAM/CENAPRED
	Reports by SMS to population

model to define the size/style of magmatic eruptions we take the power law from Simkin and Siebert (1994). As past data we take the Mendoza-Rosas and De la Cruz-Reyna (2008) catalog for the past 23,000 years, and assume it to be complete for VEI >= 2 (Table 3).

We retrospectively applied BET_UNREST for the voting period April–June 2013, in which respectively 10, 11 and 2 eruptions of 2, 3 and 4 km-high columns were observed. No *major* eruption occurred. Observed anomalies include ash eruptions up to 130/day (all with columns <4 km), seismic tremor, incandescence in the crater/dome, and VT events (but no shallow event with depth <5 km). There was no anomalous deformation, no dome growth, and no SO_2 data available. Results of *P*(*MEr*) for the retrospective application period (weekly updated) are presented in Fig. 3. For the whole period, *P* (*MEr*) of a *major* eruption (>8 km eruption column) was <1% per month.

3.2 Cotopaxi, Ecuador: Retrospective Application Inspired by the VUELCO Simulation Exercise in Quito

A volcanic unrest simulation exercise for Cotopaxi volcano (5897 m.a.s.l.) was performed on November 13th, 2014 in Quito, Ecuador. The ice-capped stratovolcano, with an andesitic to

Table 2 Monitoring parameters set for BET_UNREST at Popocatépetl

	Parameters
Unrest	– # exhalations with ash (<4 km) > 20/day – Tremor Y/N – Increase VT Y/N
Magmatic unrest	– Incandescence dome Y/N, weight 2 – Duration tremor >6000 s, weight 1 – SO_2 >2000 t/d, weight 1
Magmatic eruption	– Dome growth Y/N – SO_2 >9000 t/d – Tectonic EQ > M5.5 (along the coast/arc Michoacán-Chiapas) Y/N – Incandescent debris Y/N – Change in # tremor Y/N – VT depth <5 km – Increase # VT >M2 Y/N – Duration tremor >30,000 s (inertia 2 months) Y/N – Increase # ash eruptions >2000–4000 (inertia 2 months) – Deformation Y/N

Table 3 *Left Part:* Spatial probability of vent opening for magmatic eruptions assigned for BET_UNREST at Popocatépetl: best guess a priori values. No past data are used. *Right Part:* Parameters of the magmatic eruption size distribution assigned for BET_UNREST at Popocatépetl: best guess a priori values and past data

Spatial probability of vent opening in magmatic eruptions		Size of magmatic eruption		
Vent location	A priori probability (best guess values; equivalent number of data = 1)	Size	A priori (best guess values; equivalent number of data = 1)	Past data
Central vent	0.99	VEI 1	0.83	975
North flank	0.0025	VEI 2	0.14	13
East flank	0.0025	VEI 3	0.023	3
South flank	0.0025	VEI 4	0.0038	7
West flank	0.0025	VEI $\geq$ 5	0.0008	2

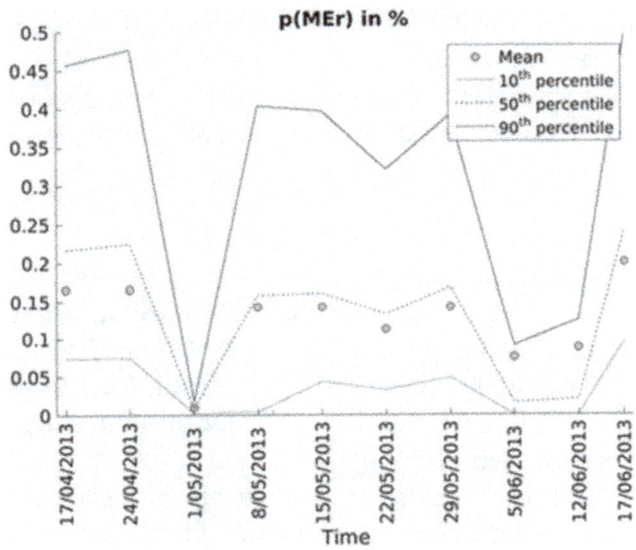

Fig. 3 Time history of probability (expressed in percentage) to have a magmatic eruption in the retrospective analysis at Popocatépetl

rhyolitic composition, is one of the most active and hazardous volcanoes in Ecuador. Historic eruptions at Cotopaxi produced large lithic-rich pyroclastic flows, ash flows, lava flows as well as large lahars (Barberi et al. 1995; Hall and Mothes 2008; Biass and Bonadonna 2011). Some lahars reached the Pacific Ocean at >200 km distance (Aguilera et al. 2004; Pistolesi et al. 2013). Recent unrest periods at Cotopaxi occurred in 1975–1976 and 2001–2002 and were characterised by increased fumarolic activity, elevated seismicity and edifice deformation (Molina et al. 2008). Fumarolic activity is a concern due to the heat transfer that may affect the ice cover resulting in non-eruptive debris flows or lahars.

A still unstable version of PyBetUnrest was set up (along with parameters and thresholds at each node for Cotopaxi volcano derived from monitoring information) before the simulation exercise, based on the available data in the literature up to the beginning of the simulation (the learning period stopped with the beginning of the exercise), in order to preliminarily test its value in decision support by providing near-real time probabilities of (i) the occurrence of unrest, (ii) the origin and nature of unrest and (iii) eruptive activity. However, during the simulation, the reports from the "volcano team" did not reflect the real eruptive and unrest history of Cotopaxi, as the past activity for the simulation was "invented". A different setting of BET_UNREST (and consequently of PyBetUnrest) on site was not possible due to the lack of time and the still premature customisability of the tool. This obliged us to set up and run the old BET_EF tool during the exercise (Constantinescu et al. 2015). Obviously, this prevented us from providing probabilistic assessment of non-magmatic events during the exercise at Cotopaxi: this would have been possible with BET_UNREST, enabling the calculation of probabilities for hydrothermal unrest and hydrothermal eruptions (*P*(*HU*) and *P*(*HEr*)). Nevertheless, the unrest scenario proposed by the "volcano team" (Bulletins 1–5) did not emphasise a significant state of hydrothermal unrest, which, on the one hand, made our output less biased in not providing an evaluation for *P*(*HU*) and *P*(*HEr*); but on the other hand this simulation was probably not the best case to test BET_UNREST.

Here, we will re-run BET_UNREST and PyBetUnrest at Cotopaxi retrospectively for the unrest phases described in the five bulletins provided by the "volcano team" during the simulation exercise and using the BET_UNREST setup prepared prior to the simulation based on the *real* past activity of the volcano (Table 4). The time window | was set to 1 month. In Table 5 we show the probabilities resulting from the run of the code, after each bulletin:

(1) *Phase 0: The background activity of Cotopaxi (NO anomalies)*: results are based on the past activity of Cotopaxi, with all observation within background limits.
(2) *Phase 1 (Bulletin 1)*: the observed anomalies in this phase were limited to an increase in seismic activity compared to background level. Such an increase is indicative, according to pre-set parameters, of magma-on-the-move (*P*(*MU*) = 0.68). The considerable uncertainty is summarised by the 10th to 90th percentiles confidence interval.
(3) *Phase 2 (Bulletin 2)*: the observed anomalies in this phase were: a drastic increase in seismicity, an increase in SO_2 emission (5 times background levels), and a crater thermal anomaly. As a consequence, the mean *P*(*MU*) increases, along with a decrease in the associated uncertainty.
(4) *Phase 3 (Bulletin 3)*: the observed anomalies in this phase were: an increase in VT and LP events, occurrence of tremor, appearance of new fumaroles, an increase in SO_2 emission, and an increase in the crater thermal anomaly. As a consequence, the *P*(*MU*) is similar to Bulletin 2, but the *P*(*HU*) increases slightly, due to the new fumaroles.
(5) *Phases 4 and 5 (Bulletins 4 and 5)*: the observed anomalies in these phases were similar, and included: intense fumarolic activity, occurrence of hybrid seismic events, an increase in SO_2 emission, and an increase in the crater thermal anomaly. As a consequence, *P*(*MEr*) increases from 0.21 (phase 3) to 0.57, combined with a lower uncertainty.

3.3 Dominica, West Indies, Lesser Antilles: VUELCO Simulation Exercise, Dominica, May 2015

Dominica is characterised by hydrothermal activity manifested as thermal springs (up to boiling temperature), boiling-temperature fumarolic emissions (e.g. Valley of Desolation) and a crater lake, known as 'Boiling Lake', with a particular hydrodynamic behaviour (Fournier et al. 2009; Joseph et al. 2011; Rouwet et al. 2017). No high-temperature manifestations occur on the island, so no clear evidence of active magmatic degassing exists at the present time.

The simulation exercise, and consequently the BET_UNREST application, for the VUELCO target island of Dominica mainly focused on an unrest scenario for the southern part of the island. The purpose of the exercise was to test the tracking/assessment of an unrest period, and the decision making process undertaken by the scientific advisory group and local authorities.

Due to the hydrothermal character of Dominica, the application of BET_UNREST is highly suited. *Before* the simulation exercise, the

Table 4 Monitoring parameters set for BET_UNREST at Cotopaxi

Node-parameter#	Parameter and threshold(s) (Y/N indicates a Boolean observation)
Unrest-parameter 1	LP/month (205–335) (Garcia-Aristazabal 2010)
Unrest-parameter 2	VT/month (24–32) (Garcia-Aristazabal 2010)
Unrest-parameter 3	M Tectonic EQ (3–4)
Unrest-parameter 4	SO_2 (Y/N)
Magmatic unrest-parameter 1	EQ depth (>4.5–5.5 km)
Magmatic unrest-parameter 2	Deep VLP (Y/N)
Magmatic unrest-parameter 3	T fumarole (>119 °C)
Magmatic unrest-parameter 4	Appearance of acidic gas (Y/N)
Magmatic unrest-parameter 5	VT/month (>32)
Magmatic unrest-parameter 6	Increased deformation (Y/N)
Magmatic unrest-parameter 7	VLP + LP together (Y/N)
Magmatic unrest-parameter 8	Harmonic LP tremor (Y/N)
Magmatic unrest-parameter 9	SO_2 flux (t/d) (>100–350)
Magmatic eruption-parameter 1	sudden stop (Y/N)
Magmatic eruption-parameter 2	SO_2 flux (t/d) (>2000–2500)
Magmatic eruption-parameter 3	Tornillos (Y/N)
Hydrothermal unrest-parameter 1	New fumarole (Y/N)
Hydrothermal unrest-parameter 2	Anomalous glacier volume decrease (defrosting) (Y/N)
Hydrothermal unrest-parameter 3	LP/month (>205–335) (Garcia-Aristazabal 2010)
Hydrothermal eruption-parameter 1	Increase in T of fumarole (>120–200 °C)
Hydrothermal eruption-parameter 2	Increase in extension of fumarolic field (Y/N)
Hydrothermal eruption-parameter 3	Inflation of fumarolic field (Y/N)
Hydrothermal eruption-parameter 4	Landslides in hydrothermal areas (Y/N)
Hydrothermal eruption-parameter 5	New/extension of alteration areas (Y/N)

PyBetUnrest tool was set for Dominica, based on (1) existing literature of the past volcanic activity; (2) insights on the current hydrothermal activity; (3) discussion-based expert elicitation sessions (4 sessions at SRC and 1 at INGV-Bologna); and (4) exchanges with local experts in order to fine-tune the code with the monitoring parameters. We remark that all of this was done *prior* to the start of the simulation exercise (the learning period stopped at the beginning of the simulation exercise, as for Cotopaxi), and again no hindsight tuning was made. The long-term setup of PyBetUnrest is done by filling up a configuration file that includes the a priori and past data specifically for Dominica, whose main information is summarised in Table 6. The short-term information is listed in Table 7 (parameters and thresholds identified prior to the exercise onset, see above). Further details on the Dominica simulation exercise and on the BET_UNREST application are given in Constantinescu et al. (2016).

During the simulation exercise (May 14–15, 2015) three phases of changes in volcanic activity, each with a duration of six months, were distributed by the "volcano team" to the operators of the unrest crisis. The reports included four types of observations: (1) seismic bulletin, (2) GPS, (3) geothermal monitoring data, and (4) other observations.

The translation of the reported bulletins into the values for the selected parameters in the BET_UNREST for Dominica setup were reported back to the team of experts in real-time

Table 5 Resulting probabilities from retrospective application of BET_UNREST at Cotopaxi

		P(U)	P(MU)	P(MEr)	P(HU)	P(HEr)
Phase 0 (Background)	Mean	0.005	0.002	0.0005	0.001	0.0006
	10th prctile	0.0013	0.0002	0	0	0
	50th prctile	0.004	0.001	0.0002	0.0006	0.0002
	90th prctile	0.009	0.004	0.001	0.003	0.001
Phase 1	Mean	1	0.68	0.18	0.08	0.02
	10th prctile	1	0.07	0	0	0
	50th prctile	1	0.84	0.02	0.001	0
	90th prctile	1	1	0.69	0.30	0.04
Phase 2	Mean	1	0.83	0.22	0.05	0.013
	10th prctile	1	0.27	0	0	0
	50th prctile	1	1	0.04	0	0
	90th prctile	1	1	0.75	0.13	0.008
Phase 3	Mean	1	0.80	0.21	0.13	0.07
	10th prctile	1	0.14	0	0	0
	50th prctile	1	1	0.04	0.002	0.0003
	90th prctile	1	1	0.72	0.54	0.22
Phase 4 and 5	Mean	1	0.81	0.57	0.12	0.07
	10th prctile	1	0.23	0.02	0	0
	50th prctile	1	1	0.65	0.0004	0.0002
	90th prctile	1	1	1	0.49	0.28

Table 6 Set up of BET_UNREST at Dominica in terms of long-term information

	A priori mean (equivalent n data in brackets)	Past data
Unrest	0.5 (1)	Past data (successes) = 14 Past data (total) = 608
Magmatic	0.5 (1)	Past data (successes) = 13 past data (total) = 14
Magmatic eruption	0.58 from Phillipson et al. (2013) (1)	Past data (successes) = 0 Past data (total) = 13
Magmatic vent location	file	file
Hydrothermal vent location	file	file
Size distribution (Magmatic)	Dome extrusion: 0.83 Small explosive: 0.14 Large explosive: 0.03 (1)	Dome extrusion: 0 Small explosive: 5 Large explosive: 2

Some of the data are too many to be listed (this is indicated by the label "file" in the table). They can be provided in the form of files on request

during the simulation. In Table 8 we provide the probabilities resulting from the run of the code after each bulletin. In Fig. 4 we also provide the time evolution of some of the most relevant probability distributions, across all the time periods spanned by the simulation exercise in Dominica. For each bulletin, among the output information from PyBetUnrest, there were two

Table 7 Monitoring parameters set for BET_UNREST at Dominica

Node–parameter#	Parameter and threshold(s) (Y/N indicates a boolean observation)
Unrest-parameter 1	Increased CO_2 flux above background (Y/N)
Unrest-parameter 2	Increase in T of hot springs and/or fumaroles (Y/N)
Unrest-parameter 3	Changes in H_2O/CO_2 (Y/N)
Unrest-parameter 4	Appearance of new fumaroles and/or hot springs (Y/N)
Unrest-parameter 5	Vegetation die back (Y/N)
Unrest-parameter 6	Appearance of LPs and hybrid EQs (Y/N)
Unrest-parameter 7	Large regional tectonic event ($M > 7$) (Y/N)
Unrest-parameter 8	Number of VTs [if >10/day for two weeks]
Unrest-parameter 9	Detectable ground deformation (Y/N)
Magmatic unrest-parameter 1	Increase in C/S, or decrease after increase (Y/N)
Magmatic unrest-parameter 2	Detectable SO_2, HCl, HF (Y/N)
Magmatic unrest-parameter 3	Extreme increase in T [>300 °C]
Magmatic unrest-parameter 4	Any VLPs (Y/N)
Magmatic unrest-parameter 5	No. of LPs after significant VT swarms (#/day) (>5–10)
Magmatic unrest-parameter 6	Consistent increase in No. of VTs for 1 month (Y/N)
Magmatic unrest-parameter 7	Deep VTs [>8 km] (#/week) (4–5)
Magmatic unrest-parameter 8	Detectable radial deformation (localized-coherent signal) (Y/N)
Magmatic unrest-parameter 9	Surface deformation (island wide, >6 cm in over 6 months) (Y/N)
Magmatic eruption-parameter 1	Decreasing C/S after increase (Y/N)
Magmatic eruption-parameter 2	Increase in Cl, Br, F content in hot springs/pools (Y/N)
Magmatic eruption-parameter 3	Decrease in H_2O/CO_2 and/or H_2S/SO_2 and/or SO_2/HCl (Y/N)
Magmatic eruption-parameter 4	Phreatic activity (Y/N)
Magmatic eruption-parameter 5	Large thermal anomaly [incandescense] (Y/N)
Magmatic eruption-parameter 6	Landslides in hydrothermal areas (Y/N)
Magmatic eruption-parameter 7	Acceleration of VTs, LPs, hybrids [weekly] (Y/N)
Magmatic eruption-parameter 8	Presence of harmonic tremor (Y/N)
Magmatic eruption-parameter 9	Shallowing of VTs hypocenters in the ediffice or shallow depths [<3 km] (Y/N)
Magmatic eruption-parameter 10	Sudden reversal of activity (Y/N)
Hydrothermal unrest-parameter 1	Anomalous behavior of Boiling Lake [overflow, lower or higher T than usual, no return of lake, etc.] (Y/N)
Hydrothermal unrest-parameter 2	Changes in hydrothermal features (Y/N)
Hydrothermal unrest-parameter 3	Increase in B and/or NH_4 concentration in waters (Y/N)
Hydrothermal unrest-parameter 4	Increase in CH_4/CO_2 (fumaroles) (Y/N)
Hydrothermal unrest-parameter 5	Increase in T of fumaroles (Y/N)
Hydrothermal eruption-parameter 1	Increase in T of fumaroles (fuzzy 120–200 °C)
Hydrothermal eruption-parameter 2	riSe of water level in pools/overflow of BL (Y/N)
Hydrothermal eruption-parameter 3	Increase in extension of fumarolic field (Y/N)
Hydrothermal eruption-parameter 4	Muddy pools (Y/N)

(continued)

Table 7 (continued)

Node–parameter#	Parameter and threshold(s) (Y/N indicates a boolean observation)
Hydrothermal eruption-parameter 5	Boiling/bubbling of pools that previously didn't (Y/N)
Hydrothermal eruption-parameter 6	Inflation of fumarolic field (Y/N)
Hydrothermal eruption-parameter 7	Landslides in hydrothermal areas (Y/N)
Hydrothermal eruption-parameter 8	New/extension of alteration areas (Y/N)

Table 8 Resulting probabilities from real-time application of BET_UNREST at Dominica during VUELCO simulation exercise

		P(U)	P(MU)	P(MEr)	P(HU)	P(HEr)	P(TU)
Phase 1	mean	1	0.26	0.06	0.62	0.42	0.12
	10th prctile	1	0	0	0.05	0.01	0
	50th prctile	1	0.06	0	0.73	0.32	0
	90th prctile	1	0.85	0.22	1	0.95	0.5
Phase 2	mean	1	0.82	0.53	0.13	0.03	0.05
	10th prctile	1	0.29	0.01	0	0	0
	50th prctile	1	1	0.56	0.001	0	0
	90th prctile	1	1	1	0.54	0.06	0.06
Phase 3	mean	1	0.70 (0.24)	0.17 (0.07)	0.08	0.02	0.22
	10th prctile	1	0.09	0	0	0	0
	50th prctile	1	0.87	0.02	0	0	0.08
	90th prctile	1	1	0.68	0.27	0.03	0.80

In bracket estimates of mean values without including HCl anomaly in Phase 3

maps of the spatial probability of vent opening: one for the case of magmatic eruption, and one for hydrothermal eruption (Fig. 4). We believe this could be particularly useful, for example in a volcanic system like Dominica, where there are numerous areas showing hydrothermal activity, thus increasing the uncertainty on the position of a possible phreatic event.

The parameter "detectable SO_2, HCl, HF" created confusion and opened up a scientific discussion. For the sake of transparency, we provide the mean values of *P*(*MU*) and *P*(*MEr*) including, or not, the HCl anomaly (Table 8). Beyond the scientific implications of this issue, this concern reflected the sensitivity of BET_UNREST to the interpretation of some parameters. When relatively few monitoring parameters are provided, the weight of a single anomaly can be high: this is somehow a measure of the epistemic uncertainty.

4 Discussion and Implications for Unrest Tracking

This chapter presents the need for an updated BET model and tool that is able to account for the non-magmatic nature of some volcanic unrest episodes, which can often go under-estimated, if not totally neglected. The new model (BET_UNREST) and tool (PyBetUnrest) allow the tracking of unrest phases at volcanic systems and enables short-term volcanic forecasts. It has been fully developed within the VUELCO project, during which time it has been applied to some of the project's target volcanoes. In general, when we are able to distinguish magma-on-the-move (Rouwet et al. 2014) from the monitoring observations the new model basically "collapses" to BET_EF (or, better, the assessment of the probabilities related to

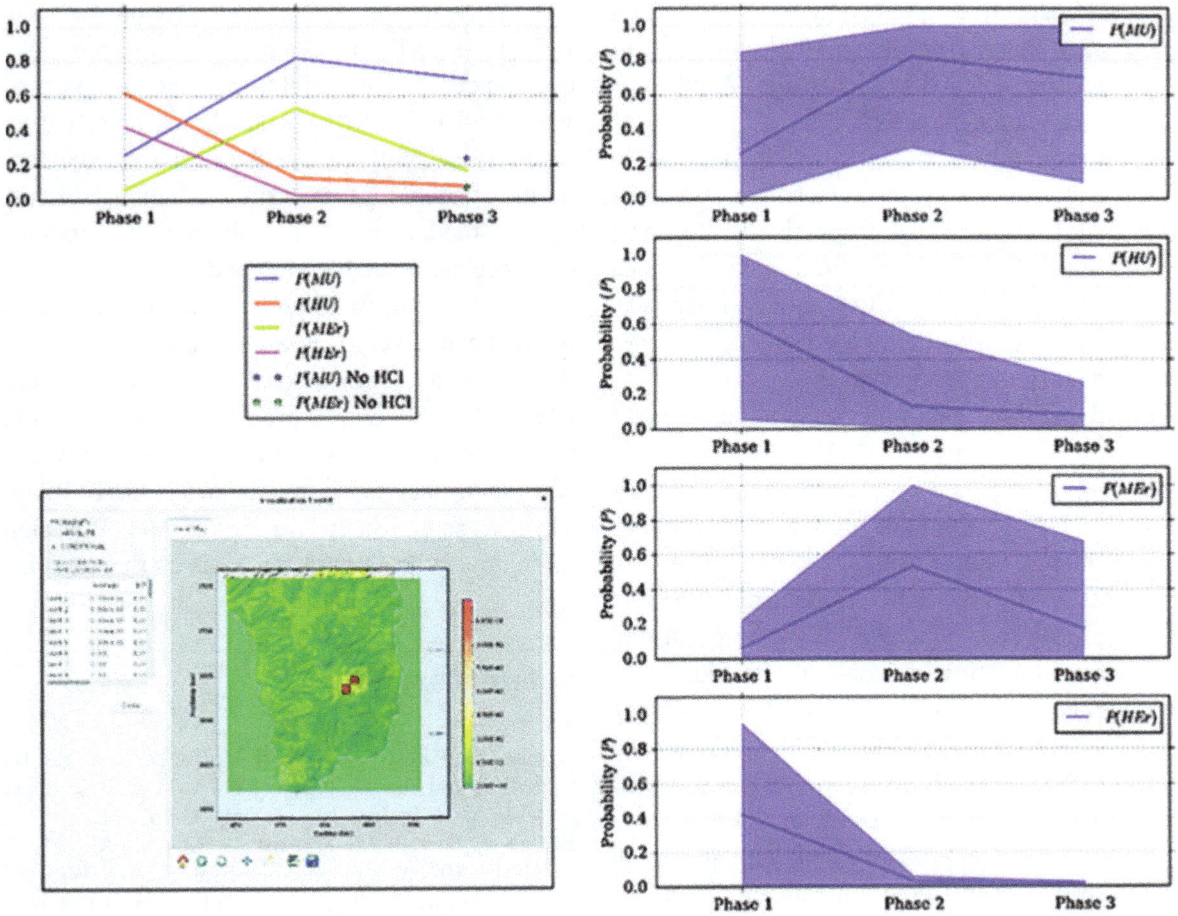

Fig. 4 Average values (*top left*) obtained by BET_UNREST during the three phases of Dominica exercise for *P*(*MU*), *P*(*HU*), *P*(*MEr*) and *P*(*HEr*). Asterisk points are the alternative average values for *P*(*MU*) *and P*(*MEr*) without considering HCl as detectable. On the *right column* the same probabilities are shown together with their confidence interval between 10th and 90th percentiles. On *bottom left*, a snapshot of PyBetUnrest tool shows the spatial probability of vent opening during Phase 1, localising the most probable position of the phreatic eruption

magmatic outcomes provided by the two models coincide). On the other hand, if we are not able to identify a magmatic "active role" in the unrest (from the available monitoring observations), BET_UNREST is still able to provide the probabilities of hazardous events that accompany non-magmatic volcanic unrest, rather than neglecting them. As discussed in Rouwet et al. (2014), a very difficult case is presented by phreatomagmatic eruptions that, sometimes, can occur without any precursors indicating magma movement. This is surely an important limit to overcome which requires further efforts to detect subtle changes in the very short-term (hours to minutes) by improving monitoring techniques.

The chapter illustrates the development and implementation of BET_UNREST model and PyBetUnrest tool through three different applications:

(i) the pure retrospective analysis at Popocatépetl volcano, where there is no compelling need for a hydrothermal branch due to the current magmatic nature of the unrest episodes. Popocatépetl has remained in unrest from December 1994 to present and, for this application, BET_UNREST and PyBetUnrest were run using the UNAM Data Base for the learning period 1997–2012, with a retrospective application aiming to forecast major eruptions (column heights greater than 8 km) for the April–June 2013 volcanic activity.

(ii) the application based on a simulation exercise at Cotopaxi. Here we tested the BET_UNREST retrospectively, but, this time, using the invented data provided during the VUELCO simulation exercise, in addition to data based on the real past history of the volcano.

(iii) the almost real-time simulation exercise organised by the VUELCO project in Dominica (May 2015). The volcanic system of Dominica presents a "prototype" setting for BET_UNREST due to its hydrothermal character. Phreatic/phreatomagmatic activity occurred during the simulation, coinciding with high associated probabilities from BET_UNREST (the average values $P(HU) = 0.73$ and $P(HEr) = 0.32$). We also positively tested the feasibility of providing different maps of the spatial probability of vent opening in case of magmatic or phreatic eruption.

As mentioned in previous sections, we implemented the BET_UNREST model into PyBetUnrest software tool using a graphical user interface aiming to provide a fast, open and user-friendly tool, which extends the usage of BET_UNREST to volcanologists with different expertise. The PyBetUnrest tool reached a mature and usable version during the Dominica simulation and its first stable release has been uploaded to Vhub cyber-infrastructure.

With these exercises we strongly believe we have brought BET a step closer to a full and proper implementation during a crisis situation. The PyBetUnrest tool eventually worked as expected, but it is important to take advantage of the lessons learned during these applications and pursue more tests that will improve its design and prove its usefulness in real-case scenarios.

As a final comment, we would like to remark that, as with any other event tree model (e.g. BET models by Marzocchi et al. 2004, 2008, 2010; HASSET model by Sobradelo et al. 2013), one can always apply and "populate" the BET_UNREST model in any "volcanic" circumstance. The uncertainty on the results provided by BET_UNREST, and consequently their practical use, will however be strongly dependent on the available information and data used to set up the models rules. If only a few pieces of evidence are available, the models results will be characterised by a large uncertainty, and thus might be not very helpful for decision-makers. As more and more knowledge is gathered, BET_UNREST output probabilities will become more attractive from a practical point of view, since their uncertainty will be increasingly small. This is an intrinsic feature of the Bayesian inferential procedure at the basis of the model.

References

Aguilera E, Pareschi MT, Rosi M, Zanchetta G (2004) Risk from Lahars in the Northern Valleys of Cotopaxi Volcano (Ecuador). Nat Hazards 33:161–189

Barberi F, Coltelli M, Frullani A, Rosi M, Almeida E (1995) Chronology and dispersal characteristics of recently (last 5000 years) erupted tephra of Cotopaxi (Ecuador): implications for long-term eruptive forecasting. J Volcanol Geotherm Res 69:217–239

Biass S, Bonadonna C (2011) A quantitative uncertainty assessment of eruptive parameters derived from tephra deposits: the example of two large eruptions of Cotopaxi volcano, Ecuador. Bull Volcanol 73:73–90. doi:10.1007/s00445-010-0404-5

Constantinescu R, Rouwet D, Gottsmann J, Sandri L, Tonini R (2015) Tracking volcanic unrest at Cotopaxi, Ecuador: the use of BET_EF tool during an unrest simulation exercise. Geophys Res Abs, 17-EGU 2015–2251

Constantinescu R, Robertson R, Lindsay JM, Tonini R, Sandri L, Rouwet D, Patrick Smith P, Stewart R (2016) Application of the probabilistic model BET_UNREST during a volcanic unrest simulation exercise in Dominica, Lesser Antilles, Geochem Geophys Geosyst 17:4438–4456, doi:10.1002/2016GC006485

De la Cruz-Reyna S, Tilling RI (2008) Scientific and public responses to the ongoing volcanic crisis at Popocatépetl Volcano, Mexico: importance of an effective hazards-warning system. J Volcanol Geotherm Res 170:121–134

Fournier N, Witham F, Moureau-Fournier M, Bardou L (2009) Boiling Lake of Dominica, West Indies: high-temperature volcanic crater lake dynamics. J Geophys Res 114(B02203). doi:10.1029/2008JB005773

Garcia-Aristazabal A (2010) Analysis of eruptive and seismic sequences to improve the short- and long-term eruption forecasting. PhD Università deli Studi di Bologna, pp 167

Hall M, Mothes P (2008) The rhyolitic–andesitic eruptive history of Cotopaxi volcano. Ecuador Bull Volcanol 70(6):675–702

Joseph EP, Fournier N, Lindsay JM, Fischer TP (2011) Gas and water geochemistry of geothermal systems in Dominica, Lesser Antilles island arc. J Volcanol Geotherm Res 206:1–14. doi:10.1016/j.jvolgeores.2011.06.007

Marzocchi W, Bebbington M (2012) Probabilistic eruption forecasting at short and long time scales. Bull Volcanol 74:1777–1805. doi:10.1007/s00445-012-0633-x

Marzocchi W, Woo G (2007) Probabilistic eruption forecasting and the call for an evacuation. Geophys Res Lett 34:L22310. doi:10.1029/2007GL031922

Marzocchi W, Woo G (2009) Principles of volcanic risk metrics: theory and the case study of Mount Vesuvius and Campi Flegrei. Italy J Geophys Res 114:B03213. doi:10.1029/2008JB005908

Marzocchi W, Sandri L, Gasparini P, Newhall CG, Boschi E (2004) Quantifying probabilities of volcanic events: the example of volcanic hazard at Mount Vesuvius. J Geophys Res 109:B11201 doi:10.1029/2004JB003155

Marzocchi W, Sandri L, Selva J (2008) BET_EF: a probabilistic tool for long- and short-term eruption forecasting. Bull Volcanol 70:623–632

Marzocchi W, Sandri L, Selva J (2010) BET_VH: a probabilistic tool for long-term volcanic hazard assessment. Bull Volcanol 72:705–716

Mendoza-Rosas AT, De la Cruz-Reyna S (2008) A statistical method linking geological and historical eruption time series for volcanic hazard estimations: applications to active Polygenetic volcanoes. J Volcanol Geotherm Res. doi:10.1016/j.jvolgeores.2008.04.005

Molina I, Kumagai H, García-Aristizábal A, Nakano M, Mothes P (2008) Source process of very-long-period events accompanying long-period signals at Cotopaxi Volcano, Ecuador. J Volcanol Geotherm Res 176:119–133

Newhall CG, Hoblitt RP (2002) Constructing event trees for volcanic crises. Bull Volcanol 64:3–20. doi:10.1007/s004450100173

Phillipson G, Sobradelo R, Gottsmann J (2013) Global volcanic unrest in the 21st century: an analysis of the first decade. J Volcanol Geotherm Res 264:183–196

Pistolesi M, Cioni R, Rosi M, Cashman KV, Rossotti A, Aguilera E (2013) Evidence for lahar-triggering mechanisms in complex stratigraphic sequences: the post-twelfth century eruptive activity of Cotopaxi Volcano. Ecuador Bull Volcanol 75:698. doi:10.1007/s00445-013-0698-1

Rouwet D, Sandri L, Marzocchi W, Gottsmann J, Selva J, Tonini R, Papale P (2014) Recognizing and tracking hazards related to non-magmatic unrest: a review. J Appl Volcanol 3:17. doi:10.1186/s13617-014-0017-3

Rouwet D, Hidalgo S, Joseph EP, González-Ilama G (2017) Fluid geochemistry and volcanic unrest: dissolving the haze in time and space. In: Gottsmann J, Neuberg, J, Scheu B (eds) Volcanic Unrest: from Science to Society—IAVCEI Advances in Volcanology, Springer, Berlin

Selva J, Costa A, Sandri L, Macedonio G, Marzocchi W (2014) Probabilistic short-term volcanic hazard in phases of unrest: a case study for tephra fallout. J Geophys Res 119:8805–8826

Simkin T, Siebert L (1994) Volcanoes of the world, 2nd edn. Geoscience Press for the Smithsonian Institution, Tucson, p 349

Sobradelo R, Bartolini S, Marti J (2013) HASSET: a probability event tree tool to evaluate future volcanic scenarios using Bayesian inference. Bull Volcanol 76:770. doi:10.1007/s00445-013-0770-x

Tonini R, Sandri L, Thompson MA (2015) PyBetVH: a Python tool for probabilistic volcanic hazard assessment and for generation of Bayesian hazard curves and maps. Comput Geosci 79:38–46

Tonini R, Sandri L, Rouwet D, Caudron C, Marzocchi W, Suparjan (2016) A new Bayesian Event Tree tool to track and quantify unrest and its application to Kawah Ijen volcano. Geochem Ge-ophys Geosyst 17:2539–2555, doi:10.1002/2016GC006327

Woo G (2008) Probabilistic criteria for volcano evacuation decisions. Nat Hazards 87–97. doi:10.1007/s11069-007-9171-9

Groundwater flow and volcanic unrest

Alia Jasim, Brioch Hemmings, Klaus Mayer and Bettina Scheu

Abstract

Hydrology around active volcanoes is strongly controlled by the interaction between groundwater, and the fluids, dissolved elements and heat associated with magmatic intrusion. The chemical and mechanical processes associated with magmatic unrest can result in observable changes in the hydrothermal system. Consequently, observations of chemical and physical hydrothermal variations may provide insights into the state of volcanic activity. Additionally, the interaction between hydrological and volcanic systems leads to the presence of high-temperature, pressurised, and often acidic fluids, which add to, and intensify, the volcanic hazard. In the following chapter we present the major components of, and controls on, magmatic hydrothermal systems focusing on the mutual perturbation between the groundwater flow system and the volcanic system. We explore how these conditions can be modified by volcanic unrest and we identify feedbacks between dynamic hydrothermal behaviour and on-going unrest. The interaction between these systems, and therefore the associated monitoring signals, are the result of complex groundwater-volcano coupling within multi-phase flow system in evolving lithologies. Nonetheless, detailed monitoring of hydrothermal and hydrological behaviour can provide insights into unrest and the evolution of hazards at restless volcanoes.

Keywords

Hydrothermal system · Groundwater
Fluid flow · Permeability · Unrest-monitoring

1 Resumen

Brevemente resumimos nuestra comprensión de los sistemas magmáticos hidrotermales y discutimos las mayores incógnitas y sus implicaciones en el vigilancia volcánica. También proveemos directrices adicionales para la recolección de datos a usarse en la calibración de la variabilidad del sistema de aguas subterráneas, alrededor de volcanes activos, como un paso crucial para desacoplar las señales magmáticas de las puramente hidrotermales.

A. Jasim (✉)
University of Bristol, Earth Science School,
Wills Memorial Building, Queen's Road,
Clifton BS8 1RJ, UK

K. Mayer · B. Scheu
Department of Earth and Environmental Sciences,
Ludwig-Maximilians-Universität München (LMU),
Theresienstrasse 41/III, 80333 Munich, Germany

B. Hemmings
GNS Science, 1 Fairway Drive, Lower Hutt,
New Zealand

Advs in Volcanology (2019) 83–99
DOI 10.1007/11157_2018_33

Published Online: 23 May 2018

La interacción entre los sistemas hidrológico y volcánico es un elemento importante durante reactivación volcánica. Los cambios en el comportamiento hidrológico de un volcán activo, como la elevación del nivel del agua subterránea, la descarga de manantiales, los cambios de temperatura y de la química, pueden ser indicadores preliminares de evolución de la actividad volcánica. Las interacciones hidrológicas pueden también alterar y aumentar el peligro volcánico existente. Las interacciones físicas y químicas entre la roca encajante y los diferentes tipos de fluido pueden modificar los caminos de desgasificación, generando distribuciones de presión dinámicas dentro del edificio volcánico. Aún los procesos lentos, como el desarrollo creciente de zonas de alteración permanentes, pueden manifestarse como un peligro dinámico asociado con una reactivación continua o futura, ya que las rocas altamente cristalinas son hidrotérmicamente alteradas produciendo arcillas débiles secundarias. Discutimos los principales parámetros que controlan las reacciones y sus efectos en la distribución de la alteración en ambientes volcánicos.

Debido a la introducción del calor de la fuente en el sistema del agua saturada, se presentan peligros adicionales. Esto frecuentemente conlleva a explosiones freáticas y freato-magmáticas. La presencia de paquetes de gases bajo la superficie además incrementa este peligro. El balance entre el ingreso de agua freca fría, la desgasificación y la disipación de calor, está críticamente relacionado con la abilidad del sistema para transmitir fluidos, el mismo que evoluciona en función tanto de los procesos químicos (ej., las reacciones de dilución/precipitación mineral) como físicos (ej., fracturamiento y compactación de la roca), produciendo así propiedades hidrológicas de la roca fuertemente dependientes de la escala (ej., porosidad, permeabilidad y conductividad térmica).

En resumen, las señales físicas y químicas, o la perturbación hidrológica asociada con la reactivación magmática, son complejas y dependientes del sitio. Las contribuciones de los diferentes componentes del fluido, y sus interacciones con caminos de flujo existentes, pueden determinar cómo un sistema evoluciona en períodos de calma. Esta evolución controla la posible respuesta a la perturbación termodinámica y química asociada con la iniciación de la reactivación volcánica. Dadas las intrincadas retroalimentaciones entre el magma, la hidrología, y los cambios repentinos de los sistemas involucrados, únicamente mediciones de alta frecuencia (de horas a semanas) de la temperatura, pH, conductividad eléctrica del agua, profundidad del nivel del agua subterránea, del contenido de REE (Elementos de Tierras Raras), RFEs (Elementos de Formación de Rocas) y gas disuelto, conjuntamente con mediciones geofísicas, pueden aclarar la evolución del sistema magmático, la apertura/cierre de fracturas y la dinámica estacional del agua subterránea.

2 Introduction

Much of the research relating to the interaction between hydrological and volcanic systems has focused on the role of hydrothermal systems in the development of economic mineral deposits. hydrothermal systems have formed vast ore-deposits around the world, most of them clearly result from the interaction between magmatic and meteoric fluids (Hedenquist and Lowenstern 1994).

The interaction between hydrological and volcanic systems is an important element in volcanic unrest. Changes in hydrological behaviour, such as water table elevation, spring discharge, temperature and chemistry, at an active volcano can provide early indications of changes in volcanic activity. Hydrological interactions can also alter and augment the existing volcanic hazard. Chemical and physical interactions between host rocks and different fluid types can modify fluid degassing pathways, generating dynamic pressure distributions within a volcanic edifice. Additional hazards are also presented by the introduction of a heat source into a water saturated system, this frequently results in dangerous phreatic and phreatomagmatic explosions. Understanding the controls on hydrological and hydrothermal behaviour in volcanic settings is essential for understanding the array of hazards presented by volcanic unrest. Continued development of this understanding is also providing new volcano monitoring opportunities. Despite the clear relevance and importance of hydrological and volcanic interactions in relation to volcanic unrest, the dynamics of this interaction remain poorly constrained.

3 Hydrothermal System

Although insulated or distal, cool groundwater aquifers can respond to volcanic perturbation, the clearest manifestation of volcanic and hydrological interactions is a hydrothermal system. Fumaroles, often visible on active volcanoes, represent the surface expression of this hydrothermal system.

A magmatic hydrothermal system is composed of three main elements: a **host rock** (or reservoir), which contains a circulating **fluid**, set in motion by an igneous **heat source** (Fig. 1). While the difference in relief between stratovolcanoes and calderas can lead to contrasting hydrological systems, the lower limit of any hydrological system is commonly defined as the brittle-ductile transition zone. Within this zone, fluid pressures transition from hydrostatic to lithostatic as rock permeability becomes severely reduced (Fournier 1999). When this region is subjected to high strain rates, fracturing may occur, leading to episodic influxes of mass and heat to the hydrothermal system (Bodnar et al. 2007). While agreement exists on the definition of lower limit of a volcanic hydrological system, the same is not true for the upper limit.

We consider the water table as the upper limit of the hydrological system. However, the earth's surface could equally be considered part of the system. This adds the further complexity of flow within the unsaturated (vadose) zone (Hemmings et al. 2015a). Furthermore, the upper limit of the hydrological system closely depends on the precipitation regime. Precipitation is a function of geography, including both latitude and elevation. Together with surface processes it determines the recharge dynamics of the aquifer and the depth and fluctuation of the water table. A number of studies suggest a correlation between the fluctuation of the water table and elevated seismicity. Both the reduction in effective stress due to a seasonal increase in hydrostatic pore pressure, and the snow unloading, lead to a seasonal peak of seismicity (Saar and Manga 2003; Christiansen et al. 2005). Furthermore, Mason et al. (2004) identify seasonal peaks in the eruption rate of volcanoes, which may be due to the load/unload seasonal stress cycle imposed by the hydrological cycle.

The definition of water table implies a water saturated medium below it. However, crater lakes (Fournier et al. 2009) and caldera settings (Bruno et al. 2007; Jasim et al. 2015) often have portion of the water table sustained by a two phases system (liquid and gas). Similarly, the condensation of magmatic gases (primarily vapour) often feeds the groundwater reservoir (Chiodini et al. 2001). While the location of the water table is important, the conventional definition does not really apply in volcanic settings, especially in high-relief stratovolcanoes, in which it is often unclear to what extent the edifice is water saturated. Specifically, the local water table may differ from the regional water table, with highly dynamic elevation changes controlled by both meteoric and volcanic processes.

Many studies suggest the presence of high elevation springs, however it is not clear whether they are fed by the regional water table or by perched saturated layers high up on the cone (Cabrera and Custodio 2004; Custodio 2007; Cruz and Oliveira Silva 2001; Hemmings et al. 2015a; Ingebritsen and Scholl 1993; Join et al. 2005; Peterson 1972). Access to wells on the flank of volcanoes and geophysical imaging methods can help resolve the hydrogeology behind such springs (Finn et al. 1987, 2001; Aizawa et al. 2008). However, in either scenario, the development of saturated flow units at high elevation has a particular relevance to forecasting volcanic hazards such as lahar, landslides and flank collapses, which can be sudden, unpredicted and deadly. Such mass wasting events all involve the displacement of material from the

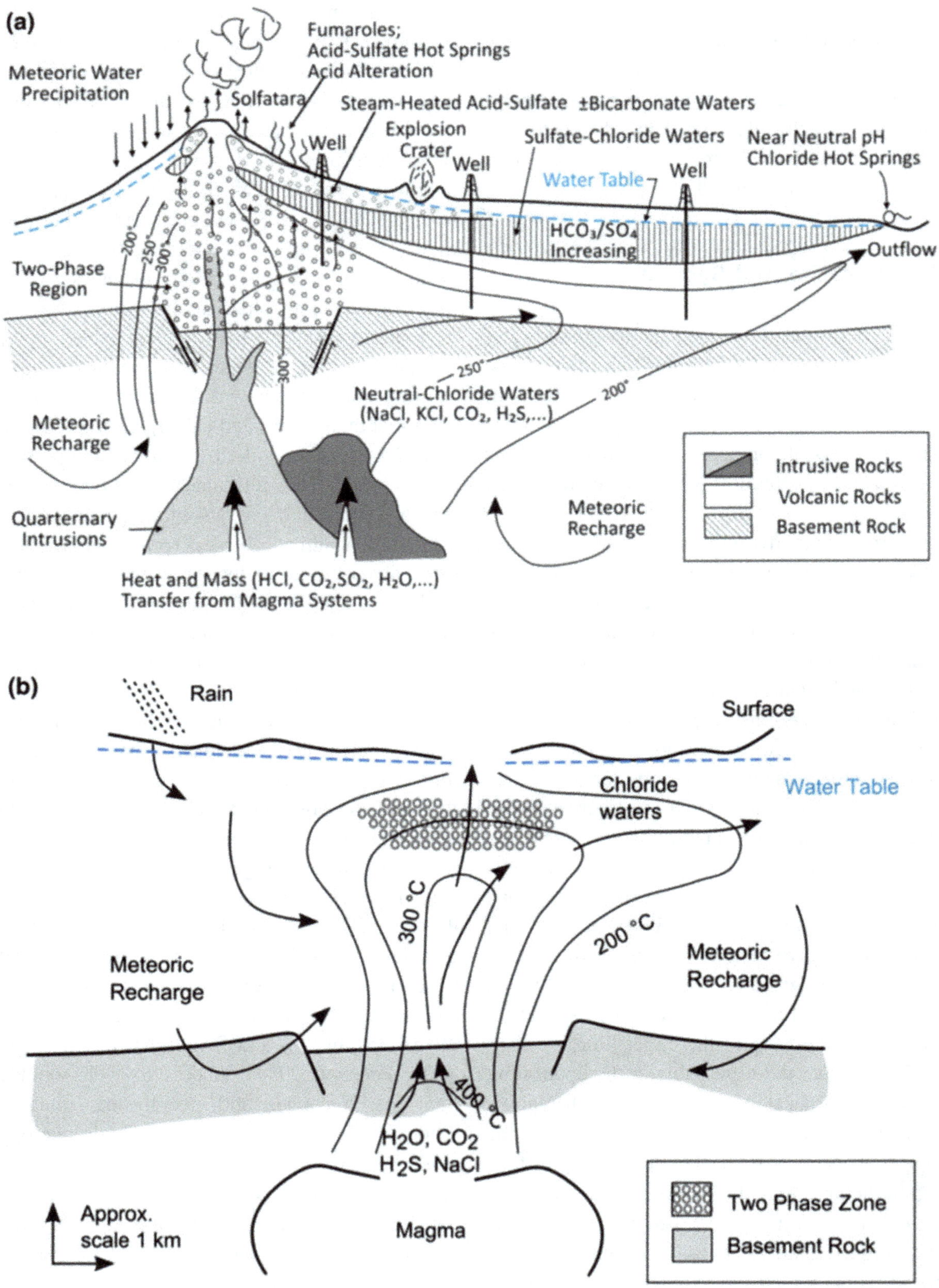

Fig. 1 Conceptual model of a hydrothermal system in, **a** high-relief volcano (modified after Goff and Janik 2000) and **b** caldera (modified after Kuhn 2004)

upper part of the volcano to the surrounding valleys. They can be triggered by gravitational instability (often in response to heavy rainfall), increase in pore pressure, reduction of rock strength, volcano-tectonic earthquake or intrusion of magma. The depressurisation induced by mass movement on the volcanic edifice may also result in the sudden reactivation of the magmatic system.

The waters circulating within volcanic systems are often high-temperature, sometimes supercritical fluids. They interact with the host rock through chemical reactions that result in hydrothermal alteration. The evolution and dynamics of hydrological and volcanic interactions are strongly controlled by fluid flow. This, in-turn, is a function of the pressure, temperature, fluid composition, and, critically, the system's ability to transmit fluids.

3.1 Fluid flow

Laminar flow through saturated porous media is described by Darcy's Law (Eq. 1),

$$q = -\frac{k}{\mu}\frac{dp}{dl} \tag{1}$$

where q is specific discharge (m/s), k is permeability of the porous media (m^2), μ is fluid viscosity (Pa s), and dP/dl is the hydraulic head, or pressure, gradient (Pa/m) along length l (m). The fluid viscosity is usually approximated as that of water (liquid or vapour). However, the combined effect of topography and the presence of a deep source of heat and fluids likely produce vast unsaturated (or two-phase) portions within the volcano. Dissolved air and gas near the surface, the phase transition from water (liquid) to vapour due to the temperature gradient, and the decompression of upwelling fluids are some of the processes that produce two-phase (liquid and gas) fluid flow regions within a volcanic system. To extend the Darcy's equation to two phase flow we define the liquid saturation (S_w) as the fraction of a representative bulk volume of the porous medium filled by water and, similarly, the gas saturation (S_g) as the fraction of a representative bulk volume of the porous medium filled by the gas phases, such that (Eq. 2)

$$S_g + S_w = 1, \tag{2}$$

The capillary pressure P_c (in Pa) is due to the pressure difference between the two phases (Eq. 3), hence

$$P_c = P_g - P_w \tag{3}$$

and is a unique function of water saturation (S_w). Finally, due to the competing flow of the two phases, the relative permeability (kr) is less than or equal to the single phase (usually water) permeability, k (m^2), of the medium. We define the relative permeability of the gas phase (kr_g) and the relative permeability of the liquid phase (kr_w) as (Eqs. 4 and 5)

$$kr_g = \frac{k_g}{k} \tag{4}$$

$$kr_w = \frac{k_w}{k} \tag{5}$$

where k_g and k_w are the effective permeabilities for each of the two fluids. Again the relative permeabilities are assumed to be a unique function of water saturation (S_w). Hence, we define the flow velocity vector (m/s) of the gas phase along a direction D (m) as (Eq. 6)

$$v_g = -k\frac{kr_g}{\mu_g}(\nabla P_g - \rho_g g \nabla D) \tag{6}$$

and similarly the velocity (m/s) of the water (Eq. 7)

$$v_w = -k\frac{kr_w}{\mu_w}(\nabla P_w - \rho_w g \nabla D) \tag{7}$$

where μ and ρ are respectively the viscosity (Pa s) and the density (kg/m^3) of the gas (g) and liquid (w) and g is the acceleration (m/s^2) due to gravity. Similarly, for non-laminar flow further terms (e.g., Forchheimer term) can be added to Darcy's equation of fluid flow to consider the inertial effects due to turbulence.

3.1.1 Permeability and Porosity

Permeability and porosity are the primary parameters controlling flow and storage of fluids in the subsurface (Manning and Ingebritsen 1999). The permeability also controls the heat regime of the hydrothermal system: high permeabilities ($\geq 10^{-14}$ m^2) favour advective transfer of heat away from the igneous source, resulting in low-temperature vapour-dominated systems. Contrastingly, low-permeabilities ($<10^{-16}$ m^2) favour slow heat conduction, also producing low-temperature systems. The hottest hydrothermal plumes reside in intermediate permeabilities, around 10^{-15} m^2 (Hayba and Ingebritsen 1997). Whilst permeability (k) is an important parameter, it is often one of the least well constrained. It can vary over 17 orders of magnitude, from $\sim 10^{-20}$ m^2 in intact crystalline rocks to $\sim 10^{-9}$ m^2 in porous and fractured basalt (Table 1).

Porosity (ϕ), the ratio of voids over total volume (voids and solid) in a given material, defines the storage capacity of the rocks; it also affects permeability and effective thermal conductivity. In many porous and fractured media, there is a positive correlation between ϕ and k, as permeability is simply the interconnected pore network. Clays and volcanic tuff are unusual in that they can have high porosity values but low permeabilities, at least in part due to their very small particle sizes, propensity to bridge pores and form aggregates (Neuzil 1994). Permeability in such lithology is greatly enhanced by the presence of discontinuities such as fissures, joints, shears and faults that can connect otherwise isolated pores.

Porosity and permeability of volcanic units are primarily controlled by the type of volcanic product (e.g., lava, pyroclastic density current, ash fall) and depositional environment. Inherent heterogeneities between deposits can be enhanced by subsequent compaction, fracturing and chemical alteration. The resulting hydrogeology is complex; hydrological rock properties (porosity, permeability and thermal conductivity) are strongly scale dependent, particularly when fluid flow is focussed along high permeability channels or fractures, as is common in volcanic settings. Such flow pathways themselves are an active component of a dynamic system. Their ability to transmit fluids and therefore the role they play in hydrothermal circulation can be modified by physical changes - the opening or closing of fractures in response to stress-field changes - and chemical alteration, which can both enhance and obstruct fluid flow, as discussed in the next section.

3.2 Chemical Reactions

Chemical alteration is an important process within a hydrothermal system. Dissolution can reduce cohesion and weaken a volcanic edifice. This can lead to catastrophic flank collapses and debris avalanches (Reid 2004). Such events can depressurise the magmatic system and trigger an eruption. Conversely, chemical precipitation and deposition processes can cause plugging in areas of intense mineralisation, this can promote pressurisation, which can also lead to flank collapse, as pore pressures increase within the edifice. Such pressurisation can also generate violent steam-driven explosions (Ingebritsen et al. 2010). Chemical dissolution and precipitation processes also alter the host-rock permeability. These processes are sensitive to the thermodynamic conditions and the proportions of different fluid components (Pirajno 2010). Feedbacks between physical and chemical flow behaviour within a hydrothermal system can modify the physical and chemical characteristics of its surface expression - hydrothermal fluid and fumarolic discharge. Therefore, monitoring the hydrothermal discharge can provide clues about the state of volcanic unrest and can even be used to help predict volcanic eruptions (Cronan et al. 1997).

To maximise the value of chemical analysis of hydrothermal discharge as a volcanic monitoring tool, and to fully understand the hazard presented by hydrological and magmatic interactions it is necessary to quantify the rates, spatial distribution and physical effects of chemical alteration within a hydrothermal system.

Table 1 Measured permeability and porosity ranges for various rock types

Rock type	Permeability (m^2)		Porosity (%)
	Min	Max	Range
Unconsolidated rocks			
Gravel	10^{-10}	10^{-7}	25–40
Clean sand	10^{-13}	10^{-9}	5–50
Silty sand	10^{-14}	10^{-10}	
Silt, loess	10^{-16}	10^{-12}	35–50
Unweathered clay[a]	10^{-20}	10^{-15}	40–80
Consolidated rocks			
Shale	10^{-20}	10^{-16}	0–10
Unfractured metamorphic and igneous	10^{-20}	10^{-17}	0–5
Sandstone	10^{-17}	10^{-13}	5–35
Limestone and dolomite	10^{-16}	10^{-13}	0–20
Fractured igneous and metamorphic	10^{-15}	10^{-13}	0–10
Permeable basalt	10^{-14}	10^{-9}	0–25
Karst limestone	10^{-13}	10^{-9}	5–50
Fractured basalt			5–50
Basalt near surface[b]	10^{-14}	10^{-12}	
Basalt at 1 km depth[b]	10^{-18}	10^{-10}	
Andesite[c]	10^{-20}	10^{-18}	0.2–0.3
Thermometamorphic[d]	10^{-18}	10^{-14}	2–17
Campi Flegrei trachy-phonolite			
Tuff, surface[e]	10^{-16}	10^{-15}	48–52
Tuff, depth[e]	10^{-17}	10^{-15}	19–52
Chaotic tuff/tuffites < 1 km depth[d]	10^{-18}	10^{-14}	6–40
Chaotic tuff/tuffites > 1 km depth[d]	10^{-18}	10^{-14}	5–36
Tuffites (HT altered)[f]	10^{-16}	10^{-16}	0.05–0.07
Lava[d]	10^{-18}	10^{-14}	7–25
Montserrat andesite			
Pyroclastic flow deposit[g]	10^{-18}	10^{-13}	
Lava[g]	10^{-17}	10^{-13}	
Lahar deposit[g]	10^{-14}	10^{-13}	

Sedimentary rocks are given for comparison. Hydrothermal systems in limestone usually develop as skarn deposit. Data from Freeze and Cherry (1979). [a]Neuzil (1994), [b]Ingebritsen et al. (2006), [c]Petrov et al. (2005), [d]Piochi et al. (2014), [e]Peluso and Arienzo (2007), [f]Giberti et al. (2006), [g]Hemmings et al. (2015a)

3.2.1 Reaction Controlling Parameters

Many factors may influence hydrothermal alteration, including temperature, pressure, rock type, fluid flux, fluid composition, and time. The relative importance of each of these has been much discussed in the literature (Gifkins et al. 2005; Pirajno 2010) and appears to vary between different case studies.

3.2.2 Fluid Composition

The dominant form of chemical alteration (dissolution and/or precipitation) is principally a function of the composition of the circulating hydrothermal fluids. Meteoric water and seawater are the main sources of fluid in hydrothermal systems with an additional and dynamic contribution of magmatic fluids. The composition of hydrothermal fluid critically affects the mineral-fluid equilibria and therefore the concentration of rock forming elements (RFEs) such as Silica (SiO_2), Sodium (Na), Potassium (K), Calcium (Ca) and Magnesium (Mg). The mineral-fluid equilibria and solubility of RFEs, as well as sulphate (SO_4^{2-}), chloride (Cl^-) and bicarbonate (HCO_3-), is affected by temperature, pressure and water/rock (W/R) ratio as well as the composition of the host rock. These factors also affect kinetic rate of chemical alteration. In addition, the relative mobility of elements depends on the characteristics of fluid flow, the number of phases (liquid and gas) and chemical condition along the flow path including pH, redox condition, sulfidation state, availability of ligands.

The chloride-sulphate-bicarbonate ternary diagram by Giggenbach and Soto (1992) provides a tool to identify water end-members (Fig. 2). It represents graphically the classification of thermal water suggested by Ellis and Mahon (1977) based on major ions, which identifies (i) neutral alkali-chloride waters which result from extensive interaction with the reservoir rocks and may cause silica or carbonate supersaturation at surface condition; (ii) acid-sulphate waters which result from the condensation of volcanic gases into the shallower groundwater system and are often depleted in alkali and Cl but enriched in metals; (iii) bicarbonate waters which usually show thermodynamic equilibrium with the reservoir rock and are common at the edge of magmatic-hydrothermal systems (Ellis and Mahon 1977; Giggenbach and Soto 1992; Goff and Janik 2000).

3.2.3 Acidity of Hydrothermal Fluids

Upper regions of hydrothermal systems are often characterized by steam-heated fumarolic alteration due to the presence of acidic, sulphate-rich fluids (Rye 2005). These fluids may cause leaching of the host rocks, resulting in an increase in both rock porosity and permeability. Eventually extreme acidic fluids (pH < 2) generate the development of vuggy silica and thereby facilitate faster gas escape in the shallow zone (Mayer et al. 2016 and references therein). In the presence of abundant sulphate ions and Al-rich host rocks, within a lesser acidic environment (pH > 2), the formation of alunite dominates (alunitic alteration, Pirajno 2010).

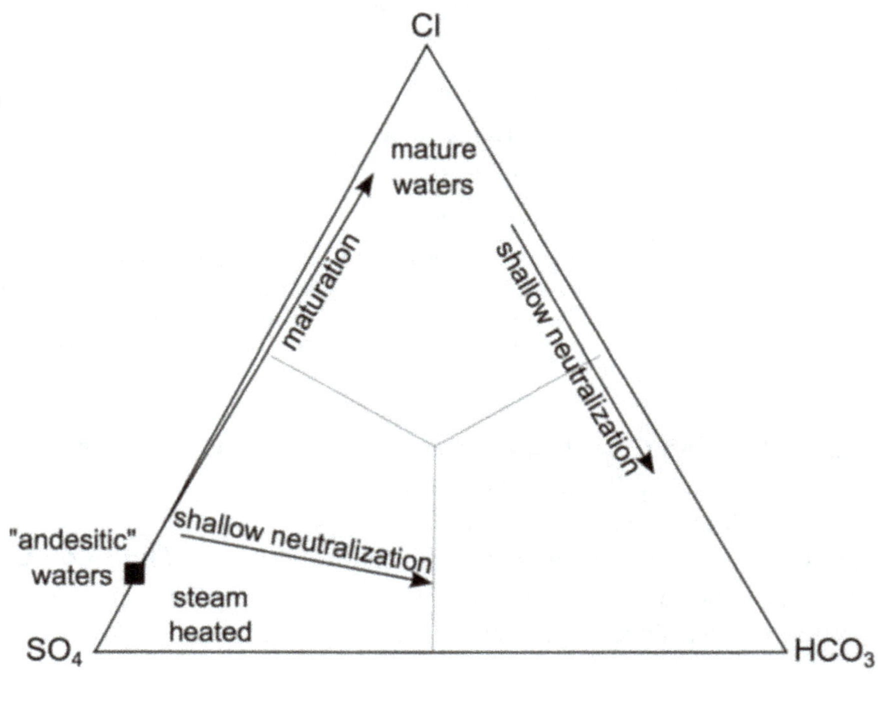

Fig. 2 Classification ternary diagram of thermal waters from Giggenbach and Soto (1992) based on the content of major solutes: Cl^-, SO_4^{2-} and HCO_3-. The vertexes represent the alkali-chloride, sulphate and bicarbonate water type end members, whilst the arrows show major differentiation processes

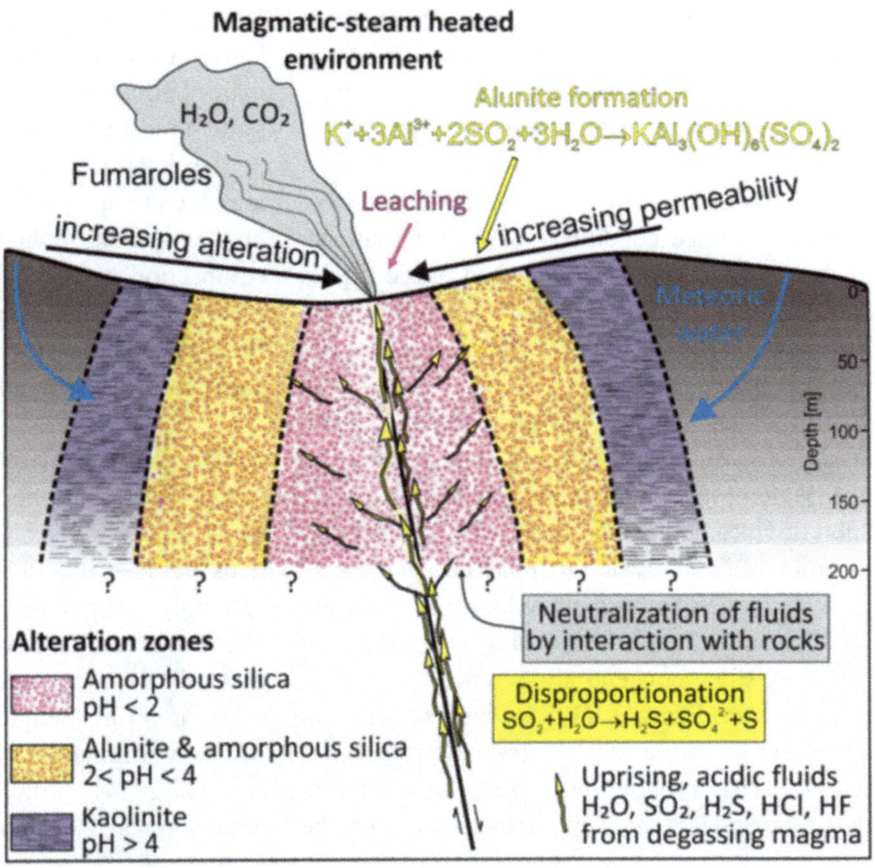

Fig. 3 Conceptual model for the formation of near surface high-sulfidation alteration. pH and composition control the development of alteration zones with increasing distance to the main fumarolic conduit. A highly permeable, acidic core characterized by amorphous silica is laterally replaced by a zone of alunite and amorphous silica. Successive neutralization of the fluids promotes the formation of kaolinite. Rock permeability as well as the degree of alteration increase toward the center of the hydrothermal activity (Mayer et al. 2016)

Often the fluids undergo progressive neutralization as they flow away from the degassing vents. This results in a sequence of alteration facies (Fig. 3) from silicic to advanced argillic to intermediate argillic (Fulignati et al. 1998).

In similar environments the distribution of kaolinite and alunite may also be affected by the presence of groundwater. Alunite preferentially forms at or above the groundwater table where atmospheric oxygen could oxidize H_2S to H_2SO_4, which is required for the formation of alunite (Mutlu et al. 2005).

3.2.4 Water/Rock Ratio

The amount of water and the rock surface area available for reactions are two of the primary controls on alteration. Therefore, the Water/Rock (W/R) ratio and rock porosity and permeability will determine the type and extent of alteration. W/R ratios range between 0.1 and 0.4 (Henley and Ellis 1983) while porosity can vary from 0 to ~80% (Freeze and Cherry 1979; Neuzil 1994). Static systems, with low porosity and low W/R ratios, are termed "closed systems", or "rock dominated systems". In this case, the secondary minerals depend nearly entirely on temperature and are of similar chemical composition to the original rocks, although possibly in a hydrated form (typical of propylitic alteration, see below). The alteration minerals often resemble those due to metamorphism (Giggenbach 1984).

Most hydrothermal systems are, however, "open systems", or "liquid dominated systems", which are characterised by high W/R ratios. In these cases, the fluid composition has a greater importance. The minerals that precipitate in open systems are the result of alteration by mobile fluids of constantly changing composition. In these systems permeability is highly influential and systematic spatial patterns of alteration zoning are common.

3.2.5 Rock Type

Many studies support the assumptions that the chemical and mineralogical composition of the original rock will change the composition of the equilibrium solution and therefore the rate of individual dissolution/precipitation reactions (e.g. Pirajno 2010). Consequently, various investigations have attempted to assess the most easily altered minerals in the host rocks (e.g. Browne 1984). Glass, followed by olivine are the least stable phases at surface conditions, as such basalts are likely to alter more rapidly than felsic rocks. Numerous studies of basalt dissolution have shown that the presence of glass can increase dissolution rates (Wolff-Boenisch et al. 2006; Berger et al. 1994; Stefansson and Gislason 2001; Zakharova et al. 2007; Hu et al. 2010; Gudbrandsson et al. 2011). However, experimental results on the alteration of volcanic materials are biased by the dominant use of basalt as starting material.

Even though basaltic glass dissolves relatively rapidly, basalts are still low-silica magmas, and therefore silica concentrations remain higher in felsic rocks. Browne (1978) showed that, at temperatures above 280 °C, the host rock composition has a negligible effect on alteration minerals. Indeed, he gathered evidence of the same stable alteration assemblages in basalts, sandstones, rhyolites and andesites. Most accessible hydrothermal systems, however, are at temperatures below 280 °C and in such systems Browne (1978) reports high-silica zeolites in rhyolitic volcanoes, and low-silica zeolites in basaltic and andesitic systems.

3.2.6 Pressure

Pressure generally has a secondary role in hydrothermal alteration (Robb 2005). An important exception is the role of pressure in controlling boiling in hydrothermal environments. Boiling at depth leads to low-salinity vapour and high-salinity brine. This phase separation is responsible for transport and deposition of elements that are key to ore mineralization (Henley and Berger 2013). In the upper ~400 m, pressure is due to the weight of the hot, possibly vapour rich, water column leading to pressure gradient below hydrostatic. At the margin of the hydrothermal system mixing between cold and hot water is enhanced by the pressure difference (Henley 1985). At greater depth, pressures exceed hydrostatic, feeding the upper reservoir (Henley 1985). Sharp pressure gradients can occur between high and low permeability portions of a hydrothermal system, for example within the flow system feeding fumaroles. Major processes occur at the interface of liquid-gas phases, such as massive precipitation of minerals (Lu and Kieffer 2009). In addition, high pressures can cause rock compaction, thus reduce permeability and drive pressure solution. Conversely, rapid increase in fluid pressure can promote fracturing leading to increases in permeability.

3.2.7 Temperature

Temperature, on the other hand, controls the general alteration patterns of hydrothermal systems because it is the main control on mineral solubility (Giggenbach 1988; Oelkers et al. 2009). For example, metal chlorides and alkaline minerals are more soluble at high temperatures, while gypsum, anhydrite, calcite and dolomite show retrograde solubility below ~100 °C (Frazer 2014). Silica solubility increases as temperatures rise, until ~300 °C. With further temperature increases, silica solubility decreases (Fournier 1985). These types of thermodynamic relationships in single-phase hydrothermal systems are relatively well constrained, and are reviewed in detail by Oelkers et al. (2009).

4 Hydrothermal Systems and Unrest

Many of the conditions that control fluid flow and chemical alteration are modified by the re-activation of the magmatic system, and evolve during volcanic unrest. For example, the introduction of fresh magma into the deep portions of an active hydrothermal system can critically change the pressure and temperature conditions within the system, thus leading to the development of gas pockets in the subsurface (Jasim et al. 2015). This can rapidly lead to phreatic eruptions. Thus it is crucial to expand unrest tracking to include monitoring of non-magmatic hazards (Sandri et al. 2017—this volume).

Seismicity, gases and ground deformation are usually monitored around active and restless volcanoes (Sparks 2003). Gravity anomaly studies (Gottsmann et al. 2008; Coco et al. 2016) and acoustic waves (Ferrazzini and Aki 1987) also provide insight into subsurface processes. Measurements of water chemistry composition from hydrothermal manifestation such as boiling pools, crater lakes and thermal springs are also routinely conducted (Varekamp et al. 2001, 2009; Federico et al. 2002; Tassi et al. 2003). The increase of Rare Earth Elements (REE)/Cl, RFEs (e.g., Ca, Mg, K)/Cl and the increase in SiO_2 concentration are indicative of either intrusion of fresh magma within the hydrothermal reservoir or exposure to water/rock interaction of fresh rock due to hydrofracturing (Varekamp et al. 2008). A pH drop and temperature increase in spring water can also be indicative of an increase in magmatic activity. However the majority of springs in volcanic environment are fed by the regional groundwater reservoirs, thus representing the cooler water inflow of the hydrothermal system (Jasim 2016). Thermal waters are focused on fluid upwelling pathways such as faults and fractures (Curewitz and Karson 1997; Hemmings et al 2015b; Jasim et al. 2015).

The influx of magmatic fluids can manifest as changes in chemical compositions of hydrothermal discharges. In particular the strong field ligands Cl^- and F^-, from magmatic degassing, can mobilize metals and F^- greatly enhances the dissolution rates of aluminium silicates (Oelkers and Gislason 2001; Wolff-Boenisch et al. 2004) with detrimental effect on rock mechanical properties. However, along the upwelling flow path, mixing with surface waters and chemical reactions with the rock often occur, overprinting the magmatic signature. Magmatic gases and vapour either mix with deep circulating water (Giggenbach 1988) or they condense to form in situ thermal waters (Rye 1993). In both cases, they rapidly dissociate and form a strong acidic solution which causes cation leaching of the host rock leading to advanced argillitic alteration (Giggenbach 1988; Symonds et al. 2001). Prior to mineral precipitation, the resulting waters are enriched in Si, Na^+, K^+, Mg^{2+} and Ca^{2+}, and other metals, proportionally to their concentrations in the host rock (rock congruent dissolution) until the solution is more or less neutralised. This is therefore called the "primary neutralisation zone". Continued acid leaching in K-feldspar systems leads invariably to the formation of alunite, an important component of high-sulfidation epithermal systems, where it can replace entire masses of rocks. For example, alunite can be formed indirectly from K-feldspar through the formation of K-mica and kaolinite (Eqs. 8a–8c).

$$\underset{\text{K-feldspar}}{3KAlSi_3O_8} + 2H^+ \rightarrow \underset{\text{K-mica}}{KAl_3Si_3O_{10}(OH)_2} + 2K^+ + 6SiO_2 \tag{8a}$$

$$\underset{\text{K-mica}}{KAl_3Si_3O_{10}(OH)_2} + 2H^+ + 3H_2O \rightarrow \underset{\text{Kaolinite}}{3Al_2Si_2O_5(OH)_4} + 2K^+ \tag{8b}$$

$$\underset{\text{Kaolinite}}{3Al_2Si_2O_5(OH)_4} + 2K^+ + 6H^+ + 4SO_4{}^{2-} \rightarrow \underset{\text{Alunite}}{2KAl_3(SO_4)_2(OH)_6} + 6SiO_2 + 3H_2O \tag{8c}$$

Most field and experimental observations imply that permeability in hydrothermal systems tends to decrease with time. However, most of these systems remain active for long periods of time (typically 10^3–10^6 years, Ingebritsen et al. 2010). Mechanisms must exist by which permeable pathways are maintained and or developed to allow continuing circulation of hydrothermal fluids and ongoing alteration. Such mechanisms include: (i) the periodic re-organisation of flow patterns related to spatial variations in dissolution and precipitation behaviour (Ritchie and Pritchard 2011); (ii) the dissolution of minerals because of pulses of acidic fluids (Plumlee 1999); (iii) feedback between permeability reduction, fluid pressure and rock mechanics resulting in hydrofracturing, shear dislocation, mineral dissolution and the opening of flow pathways (Barnes 2015; Weis 2015); (iv) pulsating volcanic activity causing fracturing, periodic variations in temperature, pressure and the composition of the circulating fluids (Bodnar et al. 2007); (v) Cooling of the magmatic sills and dikes may lead to thermal cracking (Cathles et al. 1997) and the thermal expansion of the rock; both of which cause increases in fracture density (Chen et al. 1999) enhancing rock permeability; and (vi) stressed induced fracturing (Tapponnier and Brace 1976). Furthermore, fluid pathways often ease the movement of magma towards the surface as shown by the 1975–1984 volcano-tectonic crisis at Krafla caldera (Iceland), which lead to the emplacement of fault-controlled pseudodikes at shallow depths (<100 m) and eruptive events (Opheim and Gudmundsson 1989).

The evolution of a hydrothermal system involves the interplay between a number of mechanisms, physical and chemical, that operate at very different timescales ranging from seconds to hundreds of years. Where volcanic unrest results in rapid modification of the hydrothermal system, hazards associated with unrest can manifest rapidly, with limited warning or precursory activity. Dynamic changes in permeability, related to rapid mineral precipitation or opening of fractures can immediately modify flow pathways. This can result in dramatic changes in heat and fluid flow, near surface pressurisation, hydrothermal outflow and phreatic explosions. Even slow processes such as the incremental development of pervasive alteration zones can manifest as a dynamic hazard in response to continued or future unrest, as strong crystalline rocks are hydrothermally altered into weak secondary clays.

5 Monitoring and Signals

The dynamic hydrological and hydrothermal response to volcanic unrest means that, boreholes, springs, fumaroles, crater lakes and geophysical imaging of the hydrological system can provide a rare window into the state of a volcano and the evolution of volcanic hazard. Hydrological monitoring itself is multi-parametric; insights can be gained from exploring physical and chemical patterns. For instance, the effect of groundwater on volcanic gases changes according to their solubility. As such, a larger proportion of SO_2, HCl and HF emitted from magma remain in solution in water compared to CO_2 and H_2S. Hence, SO_2, HCl and HF can be detected at the surface, only during intense magmatic activity or after drying of degassing pathways (Symonds et al. 2001). In the absence of active degassing, the isotopic ratio of the gases dissolved in groundwater2 such as $^3He/^4He$ (positive) and $\delta^{13}C$ (negative) can be indicative of a magmatic source (Sorey et al. 1998; Allard et al. 1997; Federico et al. 2002).

Fluctuations of the water table/spring discharge have also been frequently recorded before the onset of magmatic activity and are often interpreted as the effect of opening and closing of fractures during the intrusion of fresh magma (Tanguy 1994; Shibata and Akita 2001; Newhall et al. 2001). Alternatively, the effect of the water phase transition from liquid to gas at relatively shallow (<2 km) depth may also cause uplift of the water table (Jasim et al. 2015). Water levels in boreholes can be relatively easily monitored and have been observed to respond to tectonic and volcanic perturbations in a range of volcanic settings (e.g. Usu Volcano, Japan; Kilauea Volcano, Hawai'i; Koryajskii Volcano, Kamchatka).

Level changes have been attributed to thermal pressurisation, compression of water saturated rocks and opening of fractures in response to the intrusion of magma. However, the magnitude and even the sign of this hydrological response is a complex function of the nature of the thermal and mechanical perturbation, the orientation and connectivity of permeable pathways and even the design of the well itself. Thus, interpreting such signals in relation to magmatic unrest requires some prior understanding of the hydrological features involved.

Spring discharge fluctuations are harder to measure than well water level changes, especially on the flanks of volcanoes experiencing unrest, and are therefore less well documented. Decline in non-thermal spring discharge on Centre Hills, Montserrat, were observed prior to the onset of volcanic activity at the adjacent Soufrière Hills Volcano in 1995. This was followed by an increase after the cessation of the second eruptive phase in 2004 (Hemmings et al. 2015a). The mechanism behind such fluctuation is unclear, it may relate to fracture dynamics associated with magmatic pressurisation (and depressurisation). Regular temperature measurement and chemical analysis of spring systems and hydrological lakes in volcanic settings can provide insights into the differences and changes in flow pathways related to magmatic perturbation.

Chemical analysis of thermal springs and fumaroles are more common hydrological/hydrothermal monitoring strategies employed at active volcanoes. Changes in chemical composition and isotopic concentrations are often related to changes in the relative contribution of magmatic fluids to other groundwater species. Although there are general indicators for increased magmatic fluid contribution to discharging hydrothermal fluids, effective use of spring temperature, chemistry and discharge data as volcanic unrest monitoring tools requires a good understanding of the underlying composition of the hydrological and hydrothermal features and the likely sensitivity to different perturbation scenarios, in specific volcanic areas. For example, Taran et al. (2008) proposed that lower flowing, acidic springs at El Chichón volcano, Mexico would make a better monitoring target than near-neutral, high discharge springs. Potential chemical indicators of unrest in these springs include increase in relative concentration of Mg, and increase in Cl/B and Cl/Br ratios.

In summary, the physical and chemical signals or hydrological perturbation associated with magmatic unrest are complex and site dependant. Relative contributions of different fluid components and their interaction with existing flow pathways, can determine how a system evolves during quiescent periods. This evolution dictates the likely response to thermodynamic and chemical perturbation associated with the initiation of volcanic unrest. Given the intricate feedbacks between magma, hydrology and hazards and the sudden changes of the systems involved, only high-frequency (hour-week) monitoring of temperature, pH, electrical conductivity of water, depth of the water table, REE, rock forming elements and dissolved gas coupled with geophysical monitoring can untangle the evolution of the magmatic system, the opening/closure of fractures and the seasonal groundwater dynamic.

6 Open Questions—Important Unknowns

We have established that circulating hydrothermal fluids are highly reactive and may result in precipitation of alteration products or dissolution of the host rock, both of which may cause porosity, and permeability changes. However, the precise nature of this alteration varies with fluid chemistry, rock mineralogy and thermodynamic conditions. This uncertainty in alteration makes predicting the impact of water/rock interaction (WRI) on porosity and permeability, and therefore on fluid flow, particularly challenging. The background fluid flow regime is a critical part of the local expression of heat-flow as well as pressure distribution that surrounds a magmatic system. As such it may exercise an important control over the dynamics of the magmatic system that is currently poorly understood. Data constraining the time scales over which hydrothermal alteration occurs, related to data gathered from long term monitoring of coupled magmatic-hydrothermal systems, are thus crucial to inform ongoing interpretations and further predictions of areas experiencing magmatic-hydrothermal unrest.

Acknowledgements The authors thank Shaul Hurwitz for his constructive comments during the review of this manuscript and Pablo Palacios for carefully editing the extended abstract. This project has received funding from the European Union's Seventh Program for research, technological development, and demonstration under grant agreement no. 282759 (VUELCO). K. Mayer and B. Scheu also acknowledge the support of a PROCOPE grant (Hot Hydrothermal Volcanic Systems; project-ID 57130387), funded and implemented by the Deutscher Akademischer Austauschdienst (DAAD) in Germany, and the Ministry of Foreign and European Affairs (MAE) and the Ministry of Higher Education and Research (MESR) in France. K. Mayer and B. Scheu acknowledge the support of the ERC Advanced Investigator Grant (EVOKES—no. 247076).

Glossary

Hydrothermal system A groundwater system that has an area of recharge, an area of discharge, and a heat source. When a magma supplies the heat source and volatiles, the hydrothermal system is termed a magmatic hydrothermal system

Hydrothermal alteration hydrothermal alteration is a complex process involving chemical, mineralogical, and textural changes, due to the interaction of hot aqueous fluids and the host rocks through which they circulate

Permeability Connected pore space of a rock or lithology, controlling fluid flow within a reservoir

Porosity Ratio of voids over the total volume of the rock with respect to a reference rock-volume

Water/rock interaction (WRI) The set of chemical reactions between aqueous fluids and rocks. These reactions modify both the chemistry of the circulating fluid and the mineralogy of the host rock

Fluid generic term for either liquid or gas or both

Liquid liquid state of matter (e.g., water)

Gas gas state of matter (e.g., vapour)

Water table level below which water saturation occurs

References

Aizawa K (2008) Classification of self-potential anomalies on volcanoes and possible interpretations for their subsurface structure. J Volcanol Geoth Res 175:253–268

Allard P, Jean-Baptiste P, D'Alessandro W, Parello F, Parisi B, Flehoc C (1997) Mantlederived helium and carbon in groundwaters and gases of Mount Etna, Italy. Earth Planet Sci Lett 148:501–516

Barnes H (2015) Hydrothermal processes: the development of geochemical concepts in the latter half of the twentieth century. Geochem Perspect 4:1–93

Berger G, Claparols C, Guy C, Daux V (1994) Dissolution rate of a basalt glass in silica-rich solutions: implications for long-term alteration. Geochim Cosmochim Acta 58:4875–4886

Bodnar RJ, Cannatelli C, De Vivo B, Lima A, Belkin HE, Milia A (2007) Quantitative models for magma degassing and ground deformation (bradyseism) at Campi Flegrei, Italy: implications for future eruptions. Geology 35:791–794

Browne PRL (1978) Hydrothermal alteration in active geothermal fields. Annu Rev Earth Planet Sci 6: 229–250

Browne PRL (1984) Subsurface stratigraphy and hydrothermal alteration of Eastern section of the Olkaria geothermal field, Kenya. In: Proceedings of the 6th New Zealand geothermal workshop, vol 1, pp 33–41

Bruno PPG, Ricciardi GP, Petrillo Z, Di Fiore V, Troiano A, Chiodini G (2007) Geophysical and hydrogeological experiments from a shallow hydrothermal system at Solfatara Volcano, Campi Flegrei, Italy: response to caldera unrest. J Geophys Res: Solid Earth 112:1–17

Cabrera MC, Custodio E (2004) Groundwater flow in a volcanic-sedimentary coastal aquifer: Telde area, Gran Canaria, Canary islands, Spain. Hydrogeol J 12: 305–320

Cathles LM, Erendi AHJ, Barrie T (1997) How long can a hydrothermal system be sustained by a single intrusive event? Econ Geol 92:766–771

Chen Y, Xiaodong W, Fuqing Z (1999) Experiments on thermal fracture in rocks. C Sci Bull 17:1610–1612

Chiodini G, Frondini F, Cardellini C, Granieri D, Marini L, Ventura G (2001) CO_2 degassing and energy release at Solfatara volcano, Campi Flegrei, Italy. J Geophys Res 106:16213–16221

Christiansen LB, Hurwitz S, Saar MO, Ingebritsen SE, Hsieh P (2005) Seasonal seismicity at western United States volcanic centers. Earth Planet Sci Lett 240: 307–321

Coco A, Gottsmann J, Whitaker F, Rust A, Currenti G, Jasim A, Bunney S (2016) Numerical models for ground deformation and gravity changes during volcanic unrest: simulating the hydrothermal system dynamics of an active caldera. Solid Earth Discussions 7:557–577

Cronan DS, Johnson AG, Hodkinson RA (1997) Hydrothermal fluids may offer clues about impending volcanic eruptions. Eos 78:341–345

Cruz JV, Oliveira Silva M (2001) Hydrogeologic framework of Pico Island, Azores, Portugal. Hydrogeol J 9:177–189

Curewitz D, Karson JA (1997) Structural settings of hydrothermal outflow: fracture permeability maintained by fault propagation and interaction. J Volcanol Geoth Res 79:149–168

Custodio E (2007) Groundwater in volcanic hard rocks. In: Krasny and Sharp (eds) Groundwater in fractured rocks. IAH Selected Paper Series, vol 9, pp 95–108

Ellis AJ, Mahon WAJ (1977) Chemistry and geothermal systems. Academic Press (392)

Federico C, Aiuppa A, Allard P, Bellomo S, Jean-Baptiste P, Parello F, Valenza M (2002) Magma-derived gas influx and water-rock interactions in the volcanic aquifer of Mt. Vesuvius, Italy. Geochimica et Cosmochimica Acta 66:963–981

Ferrazzini V, Aki K (1987) Slow waves trapped in a fluid-filled infinite crack: implications for volcanic tremor. J Geophys Res 92:9215–9223

Finn C, Williams DL (1987) An aeromagnetic study of Mount St. Helens. J Geophys Res: Solid Earth 92:10194–10206

Finn CA, Sisson TW, Deszcz-Pan M (2001) Aerogeophysical measurements of collapse-prone hydrothermally altered zones at Mount Rainier volcano. Nature 409:600–603

Fournier RO (1985) The behavior of silica in hydrothermal solutions. In: Berger and Bethke (eds) Geology and geochemistry of epithermal systems. Reviews in Economic Geology, vol 2, pp 45–61

Fournier RO (1999) Hydrothermal processes related to movement of fluid from plastic into brittle rock in the magmatic-epithermal environment. Econ Geol 94:1193–1211

Fournier N, Witham F, Moreau-Fournier M, Bardou L (2009) Boiling Lake of Dominica, West Indies: high-temperature volcanic crater lake dynamics. J Geophys Res 114:1–17

Frazer AM (2014) Advances in understanding the evolution of diagenesis in carboniferous carbonate platforms: insights from simulations of palaeohydrology, geochemistry, and stratigraphic development. PhD Thesis, University of Bristol, UK (261)

Freeze R, Cherry J (1979) Groundwater. Prentice Hall (604)

Fulignati P, Gioncada A, Sbrana A (1998) Geologic model of the magmatic hydrothermal system of vulcano (Aeolian Islands, Italy). Mineral Petrol 62:195–222

Giberti G, Yven B, Zamora, M, Vanorio T (2006) Database on laboratory measured data on physical properties of rocks of Campi Flegrei volcanic area (Italy). In: Zollo, Capuano, Corciulo (eds) Geophysical exploration of the Campi Flegrei (Southern Italy) Caldera' interiors: data, methods and results, vol 1, pp 179–192

Gifkins CC, Herrmann W, Large RR (2005) Altered volcanic rocks: a guide to description and interpretation. Ph.D. thesis, University of Tasmania, Australia (275)

Giggenbach WF (1984) Mass transfer in hydrothermal alteration systems—A conceptual approach. Geochim Cosmochim Acta 48:2693–2711

Giggenbach WF (1988) Geothermal solute equilibria. Derivation of Na-K-Mg-Ca geoindicators. Geochim Cosmochim Acta 52:2749–2765

Giggenbach WF, Soto RC (1992) Isotopic and chemical composition of water and steam discharges from volcanic-magmatic-hydrothermal systems of the Guanacaste Geothermal Province, Costa Rica. Appl Geochem 7:309–332

Goff F, CJ Janik (2000) Geothermal systems. In: Sigurdsson, Houghton, McNutt, Rymer, Stix (eds) Encyclopedia of volcanoes, vol 1, pp 817–834

Gottsmann J, Camacho AG, Martí J, Wooller L, Fernández J, Garcia A, Rymer H (2008) Shallow structure beneath the Central Volcanic Complex of Tenerife from new gravity data: implications for its evolution and recent reactivation. Phys Earth Planet Inter 168:212–230

Gudbrandsson S, Wolff-Boenisch D, Gislason SR, Oelkers EH (2011) An experimental study of crystalline basalt dissolution from $2 < pH < 11$ and temperatures from 5 to 75 C. Geochim Cosmochim Acta 75:5496–5509

Hayba D, Ingebritsen S (1997) Multiphase groundwater flow near cooling plutons. J Geophys Res 102:12235–12252

Hedenquist J, Lowenstern J (1994) The role of magmas in the formation of hydrothermal ore deposits. Nature 370:519–527

Hemmings B, Whitaker F, Gottsmann J, Hughes A (2015a) Hydrogeology of montserrat, review and new insights. J Hydrol: Reg Stud 3:1–30

Hemmings B, Gooddy D, Whitaker F, Darling GW, Jasim A, Gottsmann J (2015b) Groundwater recharge and flow on Montserrat, West Indies: insights from groundwater dating. J Hydrol: Reg Stud 4:611–622

Henley RW (1985) The geothermal framework of epithermal deposits. Rev Econ Geol 2:1–24

Henley RW, Berger BR (2013) Nature's refineries metals and metalloids in arc volcanoes. Earth Sci Rev 125:146–170

Henley R, Ellis A (1983) Geothermal Systems Ancient and Modern: a Geochemical Review. Earth Sci Rev 19:1–50

Hu SM, Zhang RH, Zhang XT, Huang WB (2010) Experimental study of water-basalt interactions in Luzong volcanic basin and its applications. Acta Petrologica Sinica 26:2681–2693

Ingebritsen SE, Scholl MA (1993) The hydrogeology of Kilauea volcano. Geothermics 22:255–270

Ingebritsen S, Ward S, Neuzil C (2006) Groundwater in geologic processes. Cambridge University Press (564)

Ingebritsen SE, Geiger S, Hurwitz S, Driesner T (2010) Numerical simulation of magmatic hydrothermal systems. Rev Geophys 48:1–33

Jasim A (2016) Exploring the complexity of groundwater flow in volcanic terrains: a combined numerical, experimental and field data approach. Ph.D. thesis, University of Bristol, UK (199)

Jasim A, Whitaker FF, Rust AC (2015) Impact of channelized flow on temperature distribution and fluid flow in restless calderas: insight from Campi Flegrei caldera, Italy. J Volcanol Geoth Res 303:157–174

Join JL, Folio JL, Robineau B (2005) Aquifers and groundwater within active shield volcanoes. Evolution of conceptual models in the Piton de la Fournaise volcano. J Volcanol Geoth Res 147:187–201

Kuhn M (2004) Reactive flow modeling of hydrothermal systems. Springer, Berlin (264)

Lu X, Kieffer S (2009) Thermodynamics and mass transport in multicomponent, multiphase H_2O systems of planetary interest. Annu Rev Earth Planet Sci 37:449–477

Manning C, Ingebritsen S (1999) Permeability of the continental crust: Implications of geothermal data and metamorphic systems. Rev Geophy 37:127–150

Mason BG, Pyle DM, Dade WB, Jupp T (2004) Seasonality of volcanic eruptions. J Geophys Res: Solid Earth 109:1–12

Mayer K, Scheu B, Montanaro C, Yilmaz TI, Isaia R, Aßbichler D, Dingwell DB (2016) Hydrothermal alteration of surficial rocks at Solfatara (Campi Flegrei): Petrophysical properties and implications for phreatic eruption processes. J Volcanol Geoth Res 320:128–143

Mutlu H, Sariiz K, Kadir S (2005) Geochemistry and origin of the Şaphane alunite deposit, Western Anatolia, Turkey. Ore Geol Rev 26:39–50

Neuzil CE (1994) How permeable are clays and shales? Water Resour 30:145–150

Newhall CG, Albano SE, Matsumoto N, Sandoval T (2001) Roles of groundwater in volcanic unrest. J Geol Soc Philippines 56:69–84

Oelkers EH, Gislason SR (2001) The mechanism, rates and consequences of basaltic glass dissolution: I. An experimental study of the dissolution rates of basaltic glass as a function of aqueous Al, Si and oxalic acid concentration at 25 C and pH = 3 and 11. Geochim Cosmochim Acta 65:3671–3681

Oelkers EH, Benezeth P, Pokrovski GS (2009) Thermodynamic databases for water-rock interaction. In: Oelkers and Schott (eds) Thermodynamics and kinetics of water-rock interaction. Reviews in Mineralogy and Geochemistry, 70, 1–37

Opheim JA, Gudmundsson A (1989) Formation and geometry of fractures, and related volcanism, of the Krafla fissure swarm, northeast Iceland. Geol Soc Am Bull 101:1608–1622

Peluso F, Arienzo I (2007) Experimental determination of permeability of Neapolitan Yellow Tuff. J Volcanol Geoth Res 160:125–136

Peterson FL (1972) Water development on tropic volcanic islands-type example: Hawaii. Ground Water 10:18–23

Petrov VA, Poluektov VV, Zharikov AV, Velichkin VI, Nasimov RM, Diaur NI, Terentiev VA, Shmonov VM, Vitovtova VM (2005) Deformation of metavolcanics in the Karachay Lake area, Southern Urals: petrophysical and mineral-chemical aspects. Geol Soc London, Spec Publ 240:307–322

Piochi M, Kilburn CRJ, Di Vito MA, Mormone A, Tramelli A, Troise C, De Natale G (2014) The volcanic and geothermally active Campi Flegrei caldera: an integrated multidisciplinary image of its buried structure. Int J Earth Sci 103:401–421

Pirajno F (2010) Hydrothermal processes and mineral systems. Springer, Berlin (1250)

Plumlee GS (1999) The environmental geology of mineral deposits. In: Plumlee and Logsdon (eds) The Environmental Geochemistry of Mineral Deposits, Part A. Processes, Techniques, and Health Issues: Society of Economic Geologists. Reviews in Economic Geology, 6, 71–116

Reid ME (2004) Massive collapse of volcano edifices triggered by hydrothermal pressurization. Geology 32:373–376

Ritchie LT, Pritchard D (2011) Natural convection and the evolution of a reactive porous medium. J Fluid Mech 673:286–317

Robb L (2005) Introduction to ore-forming processes. Blackwell Science Ltd (384)

Rye RO (1993) The evolution of magmatic fluids in the epithermal environment: the stable isotope perspective. Economica Geologica 88:733–753

Rye RO (2005) A review of the stable-isotope geochemistry of sulphate minerals in selected igneous environments and related hydrothermal systems. Chem Geol 215:5–36

Saar MO, Manga M (2003) Seismicity induced by seasonal groundwater recharge at Mt. Hood, Oregon. Earth and Planetary Science Letters 214:605–618

Sandri L, Tonini R, Rouwet D, Constantinescu R, Mendoza-Rosas AT, Andrade D, Bernard B (2017) The need to quantify hazard related to non-magmatic unrest: from BET_EF to BET_UNREST. This volume

Shibata T, Akita F (2001) Precursory changes in well water level prior to the March, 2000 eruption of Usu volcano, Japan. Geophys Res Lett 28:1799–1802

Sorey ML, Evans WC, Kennedy BM, Farrar CD, Hainsworth LJ, Hausback B (1998) Carbon dioxide and helium emissions from a reservoir of magmatic gas beneath Mammoth Mountain, California. J Geophy Res: Solid Earth 103:15303–15323

Sparks RSJ (2003) Forecasting volcanic eruptions. Earth Planet Sci Lett 210:1–15

Stefansson A, Gislason SR (2001) Chemical weathering of basalts, Southwest Iceland: effect of rock crystallinity and secondary minerals on chemical fluxes to the ocean. Am J Sci 301:513–556

Symonds RB, Gerlach TM, Reed MH (2001) Magmatic gas scrubbing: implications for volcano monitoring. J Volcanol Geoth Res 108:303–341

Tanguy JC (1994) The 1902-1905 eruptions of Montagne Pelee, Martinique: anatomy and retrospection. J Volcanol Geoth Res 60:87–107

Tapponnier P, Brace WF (1976) Development of stress-induced microcracks in Westerly granite. Int J Rock Mech Min Sci 13:103–112

Taran Y, Rouwet D, Inguaggiato S, Aiuppa A (2008) Major and trace element geochemistry of neutral and acidic thermal springs at El Chichon volcano, Mexico. J Volcanol Geoth Res 178:224–236

Tassi F, Vaselli O, Capaccioni B, Macias JL, Nencetti A, Montegrossi G, Magro G (2003) Chemical composition of fumarolic gases and spring discharges from El Chichon volcano, Mexico: Causes and implications of the changes detected over the period 1998-2000. J Volcanol Geoth Res 123:105–121

Varekamp JC (2008) The volcanic acidification of glacial Lake Caviahue, Province of Neuquen, Argentina. J Volcanol Geoth Res 178:184–196

Varekamp JC, Ouimette AP, Herman SW, Bermudez A, Delpino D (2001) Hydrothermal element fluxes from Copahue, Argentina: A "beehive" volcano in turmoil. Geology 29:1059–1062

Varekamp JC, Ouimette AP, Herman SW, Flynn KS, Bermudez A, Delpino D (2009) Naturally acid waters from Copahue volcano, Argentina. Appl Geochem 24:208–220

Weis P (2015) The dynamic interplay between saline fluid flow and rock permeability in magmatic-hydrothermal systems. Geofluids 15:350–371

Wolff-Boenisch D, Gislason SR, Oelkers EH, Putnis CV (2004) The dissolution rates of natural glasses as a function of their composition at pH 4 and 10.6, and temperatures from 25 to 74 C. Geochim Cosmochim Acta 68:4843–4858

Wolff-Boenisch D, Gislason SR, Oelkers EH (2006) The effect of crystallinity on dissolution rates and CO_2 consumption capacity of silicates. Geochim Cosmochim Acta 70:858–870

Zakharova EA, Pokrovsky OS, Dupre B, Gaillardet J, Efimova LE (2007) Chemical weathering of silicate rocks in Karelia region and Kola peninsula, NW Russia: assessing the effect of rock composition, wetlands and vegetation. Chem Geol 242:255–277

Experimental Simulations of Magma Storage and Ascent

C. Martel, R.A. Brooker, J. Andújar, M. Pichavant, B. Scaillet and J.D. Blundy

Abstract

One of the key issues in utilizing precursor signals of volcanic eruption is to reliably interpret geophysical and geochemical data in terms of magma movement towards the surface. An important first step is to identify where the magma is stored prior to ascent. This can be studied through phase-equilibrium experiments designed to replicate the phase assemblage and compositions of natural pyroclasts or by measuring volatiles in melt inclusions from previous eruptions. The second crucial step is to characterize the magmatic conditions and processes that will guide the eruption style. This may be addressed through controlled dynamic decompression or deformation experiments to examine the different rates that govern the kinetics of syn-eruptive degassing, crystallization, and strain. Comparing the compositional and textural characteristics of these experimental products with the natural samples can be used to retrieve magma ascent conditions. These experimental simulations allow interpretation of direct observations and *in situ* measurements of syn-eruptive processes leading to more accurate forecasting of future eruptive scenarios.

C. Martel (✉) · J. Andújar · M. Pichavant · B. Scaillet
Institut Des Sciences de La Terre D'Orléans, Université D'Orléans-CNRS-BRGM, Orléans, France
e-mail: caroline.martel@cnrs-orleans.fr

R.A. Brooker · J.D. Blundy
School of Earth Sciences, University of Bristol, Wills Memorial Building, Queens Road, Bristol BS8 1RJ, UK

Advs in Volcanology (2019) 101–110
DOI 10.1007/11157_2017_20

Published Online: 02 August 2017

1 Linking Geophysical and Geochemical Warning Signals to Magmatic Processes

A key objective in volcanology is to forecast eruptions, i.e. to establish when, where, and how an eruption will occur and what magnitude it will be. The prerequisite of such forecasting is to (i) detect reliable precursory signals of magma ascent to subsurface and (ii) anticipate the eruption style in order to inform the crisis management strategy. To reach these objectives, the intensification of the geophysical and geochemical signals associated with an unrest episode has to be interpreted in terms of magma movements. This is a far from trivial task because (i) seismicity has to distinguish signals related to magma movements from those of rock fracturing and/or gas percolation (*see Chapter "Volcano Seismology: detecting unrest in wiggly lines"*), (ii) ground deformation has to precisely track magma motion towards the surface (*see Chapter "Volcano geodesy and multiparameter investigations"*), and (iii) the flux and the speciation of emitted gas at the surface has to be interpreted in terms of magma ascent and degassing (*see Chapter "Volcanic gases and low temperature volcanic fluids"*). For any of these monitoring signals, their interpretation in terms of impending eruption requires knowing at what depth beneath the volcano magma is stored (magma storage conditions) and how it progresses toward the surface (magma ascent conditions). A pertinent approach to investigate the conditions of magma storage and ascent consists of comparing petrological and textural studies of previously erupted products to the results of experimental simulations carried out under realistic magma conditions. During the last decades the development of powerful analytical and experimental tools has led to great advances in this type of investigation. Of course, this is an a posteriori approach (using previously erupted products) that relies on considering past eruptive behaviour of a given volcanic system as a guide to future activity. For this reason, it is necessary to understand the fundamental magmatic processes at any particular volcano and in the long term, build up a record that links the pre-eruptive signals with eruptive products. In this way we can successfully use the warning signals to forecast or even start to predict the timing and style of an imminent eruption.

2 Magma Storage

How and where magma is stored before an eruption are enduring and complex questions, particularly given the range of hypotheses covering single versus multiple storage regions or dyke feeder systems. Key parameters in interpreting the precursory geophysical and geochemical signals are the depth of storage and the volatile content dissolved in the magma, as the exsolution of these provides an important driving force for explosive eruptions. Magma consists principally of phenocrysts (crystals larger than 50–100 μm) coexisting with a silicate melt containing dissolved volatile species (e.g. H_2O, CO_2, S species, F, Cl, etc.…). To assess the magma storage conditions, one has to determine the parameters (i.e. the pressure, temperature, redox state, volatile content) that govern equilibrium between melt and the phenocrysts. This is accomplished by comparing the phase assemblage and compositions of the natural products to those obtained by phase-equilibrium experiments where all these parameters are controlled.

2.1 Decoding Natural Pyroclasts

Phenocrysts. Mineral compositions are sensitive to various intensive (pressure, temperature, fO_2) and extensive (host liquid composition and volatile content) variables. In some cases, the compositions can be used as geothermobarometers (or hygro/oxy/chemo-meters) to retrieve first-order parameters of crystallization. For instance, we can calculate crystallization temperature from the composition of orthopyroxene-clinopyroxene pairs (e.g. Lindsley and Andersen 1983) or amphibole-plagioclase pairs (e.g. Holland and Blundy 1994), temperature and oxygen fugacity from Fe–Ti oxide pairs (e.g. Ghiorso and Evans 2008) or pressure from amphibole composition (e.g. Ridolfi and Renzulli 2012). Such calculations require two main criteria: (i) the thermobarometers must have been calibrated experimentally for conditions and compositions relevant to those of the target samples and (ii) the analysed phenocrysts must be identified as part of the phenocryst assemblage in chemical equilibrium with the melt in the reservoir under pre-eruptive conditions. This can achieved by comparison with experiments that simulate a range of variables. Establishing different 'equilibrium' events also becomes crucial as magma mixing (and associated reheating prior to eruption or further crystallization outside the storage zone) very often blur the pre-intrusion/eruption equilibrium conditions. Moreover, many eruptions involve the entrainment of crystals that did not grow from the erupted magma and are consequently out of equilibrium with the melt in which they occur. These are often termed xenocrysts or antecrysts and increasingly recognised as important aspects of magma's crystal cargo (e.g. Streck 2008; Kilgour et al 2013). The processes of mixing, reheating, and potentially decompression recorded in these crystals, may be used to provide timescales of reservoir dynamics, such as residence times between the last magma recharge and eruption (e.g. Saunders et al. 2012). Indeed, the compositional zoning of some phenocrysts witness cycles of recharge events prior to the final eruption (e.g. Druitt et al. 2012). Diffusion chronometry provides a means to infer the crystal residence time in the reservoir prior to eruption (e.g. Costa and Morgan 2010), but again requires experimental calibration of the rates involved under various conditions.

Glass inclusions. Glass (or "melt") inclusions are aliquots of melt that are trapped in crystals, usually during stages of rapid crystal growth (Fig. 1a). If trapped in phenocrysts crystallizing in the magma chamber, these inclusions will be the only witness of the composition of the melt in equilibrium with the phenocrysts prior to eruption, provided that they remained sealed after entrapment (no volatile leak, no crystallization or post-entrapment interaction with the host mineral). Indeed, after leaving the storage region, the melt initially surrounding the phenocrysts is likely to degas and crystallize microlites upon ascent, therefore deviating significantly from its

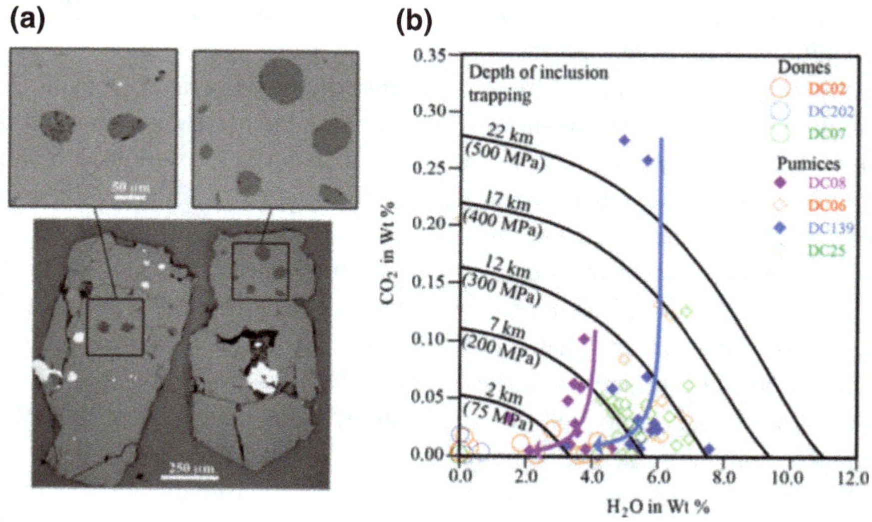

Fig. 1 Melt inclusion (*MI*) volatile contents from the Morne aux Diables complex (*MAD*), northern Dominica, as determined by ion-microprobe measurements. **a** Close up of inclusions shows they were either slightly vesiculated, during capture (fluid-saturated) or became so during ascent. The *right* hand crystal contains a very large melt inclusion around the bright oxide inclusions that has vesiculated more extensively, possibly due to cracks developing and exposing the melt to the full decompression effect during eruption. In **b** the maximum amount of H_2O and CO_2 that can be dissolved in a melt at any given depth follows pressure dependent isopleths that can be determined experimentally (in this case from Tamic et al. 2001). This represents the minimum pressure (depth of entrapment of a MI). These trends are controlled by the lower solubility of CO_2 compared to H_2O, such that there is a rapid drop in CO_2 before H_2O starts to be lost (*coloured arrows*)

pre-eruptive composition and character. The two main objectives of studying glass inclusions in phenocrysts are (i) mineral-melt thermobarometry (clinopyroxene-melt or plagioclase-melt; e.g., Putirka 2008; Water and Lange 2015) and (ii) pre-eruption volatile contents and speciation that can be converted into gas saturation pressure using experimentally-derived volatile solubility laws and which represent an end-member in the calculations of the volcanic degassing budget (dissolved vs. released gases). This latter approach requires that the magma be demonstrably volatile-saturated at the time of melt inclusion entrapment, for example by looking at relationships between dissolved volatile contents and trace elements in inclusions (Blundy and Cashman 2008). If the magma was not saturated, the calculated pressure generally represents a minimum depth prior to volatile volatile-saturated, the calculated pressure generally represents a minimum pressure estimate. As an example, the glass inclusions in phenocrysts from pyroclasts of the last eruption of Morne aux Diables, Dominica, have been analysed by ion microprobe. The data show about 6 to 8 wt% dissolved H_2O, up to 3000 ppm CO_2, together with some chlorine and fluorine. Available H_2O–CO_2 solubility models based on experiments for comparable compositions (e.g. Tamic et al. 2001) indicate melt entrapment during phenocryst crystallization at pressures as high as 400–500 MPa (depth of <22 km) for sample DC139 and shallower depth for sample DC08 although it is also possible this magma originally contained even more CO_2 than is recorded by any of the analysed melt inclusions. (Figure 1b). Small vesicles in the inclusions of Fig. 1a suggest there was exsolution of volatiles during ascent. The possible disequilibrium between such vesicles and melt produced during very rapid ascent is discussed in *Chapter "Magma degassing: the diffusive fractionation model and beyond"*.

Complexity of open-systems. An eruption is often triggered by the injection of new magma into the reservoir, which reheats, mingles, and mixes with the resident magma. Upon ascent to the surface, the two batches may interact to varying degrees and can further crystallize and cool. Therefore, imprudent use of geothermobarometers may yield large pressure and temperature ranges that cannot be easily reconciled with a single/specific episode of equilibrium crystallization. In order to retrieve the storage conditions of the resident magma, i.e. prior to deep magma mixing or before possible modification within the volcanic conduit, a detailed petrological study is necessary to identify precisely the different stages of perturbation and their characteristics in terms of phase assemblage and chemical composition. Experimental petrology is one of the tools that helps to unravel the various magmatic processes at work and their relative impact on magma chemistry and magmatic evolution.

2.2 Phase-Equilibrium Experiments

Phase-equilibrium experiments use natural (or analogue) products as starting material that are subjected to high-pressure (HP) and high-temperature (HT) in various devices under controlled conditions of pressure, temperature, oxygen fugacity, and volatile content. Such experiments are powerful tools to simulate realistic magmatic conditions for the crustal reservoirs that feed volcanic systems. Experimental equipment ranges from cold-seal pressure vessels, internally-heated pressure vessels, and piston-cylinder apparatus, depending on the investigated conditions. The principle is to reproduce the natural assemblage, proportion, and chemical compositions of the phenocrysts and equilibrium coexisting melt in the magma storage region. This is then compared with the natural samples in order to retrieve the pre-eruptive crystallization conditions (Fig. 2).

The first prerequisite for such an approach is a detailed petrological and mineralogical study of the erupted samples in order to identify the magmatic processes potentially perturbing equilibrium crystallization in the reservoir. Indeed, the relevance of the experimental study relies on accurate petrological knowledge that dictates the choice of the starting material and run procedure (Pichavant et al. 2007). The second prerequisite

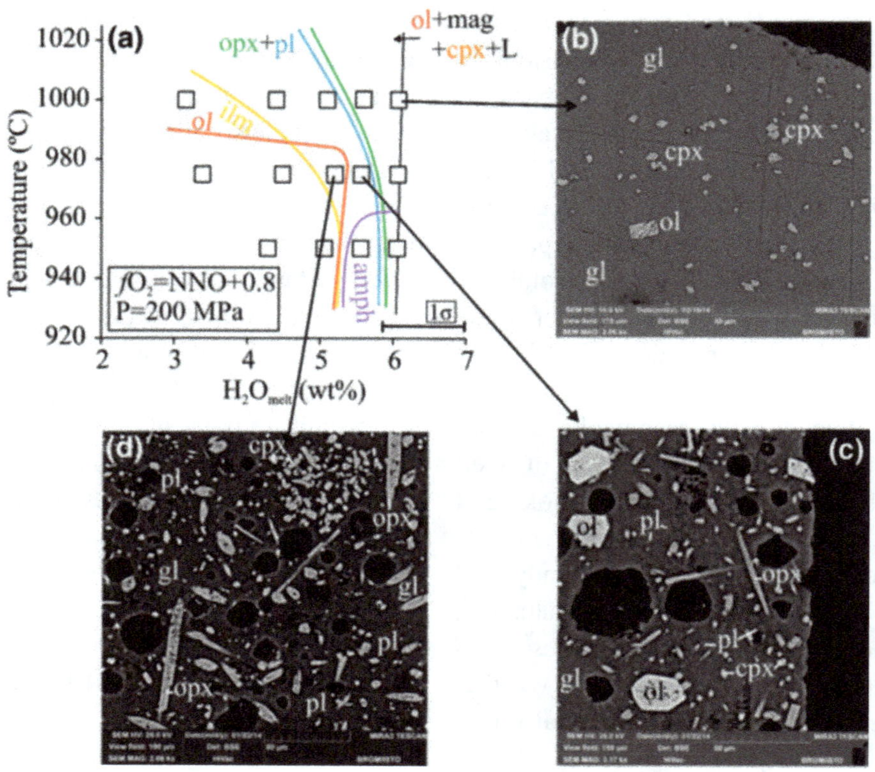

Fig. 2 Phase equilibrium experiments for the Tungurahua 2006 andesite, showing **a** mineral stability fields as a function of temperature and H_2O content at 200 MPa and oxidizing conditions (fO_2 = NNO + 0.8 log unit). SEM images of the experimental charges are shown in **b** for 1000 °C and H_2O saturation (~6.1 wt% H_2O dissolved in melt), **c** 975 °C and ~5.7 wt% H_2O, and **d** 975 °C and 5.2 wt% H_2O. *Note* the drastic increase of crystal content with cooling or dehydration; *gl* for glass (*L* for silicate liquid), *ol* for olivine, *cpx* for clinopyroxene, *opx* for orthopyroxene, *pl* for plagioclase, *amph* for amphibole, *mag* for magnetite, and *ilm* for ilmenite

is the appropriate choice of the volatile species (H_2O, CO_2, sulphur, etc.) and contents to be added to the starting material. These dissolved volatiles can impact crystallization (sequence, mineral stability fields, and phase compositions; Scaillet and Pichavant 2003; Riker et al. 2015). Where volatile measurements on melt inclusions are available, these can be used, although it is possible that the melt inclusions may no longer be representative of the initial volatile content (e.g. due to possible leakage or recrystallization). Consequently, volatile species and contents become experimental parameters that have to be varied within a range that is first inferred from the study of the glass inclusions (when available), and/or based on previous work carried out on similar bulk rock compositions. It should be remembered that entrapment of melt inclusions requires crystallisation to occur and magmas may undergo substantial volatile loss (degassing) prior to any crystallisation. For instance, Blundy et al. (2010) speculate that many magmas had original CO_2 contents significantly higher than those recorded by any melt inclusions.

It is clear that an approach involving the combination of petrological study of the natural products and phase-equilibrium experiments can help to retrieve the storage conditions of magmas of a wide range of compositions. This is a prerequisite step for the interpretation of unrest signals and construction of eruptive scenarios.

3 Magma Ascent

During ascent in the volcanic conduit the silicate melt around the phenocrysts degasses by exsolving its dissolved volatiles as gas bubbles. This may lower the liquidus triggering the crystallization of microlites (i.e. crystals smaller than about 50–100 μm). The residual melt is transformed both chemically, by degassing and differentiation as microlites crystallize and physically, by an increase in melt viscosity and a change from a single liquid phase to a three-phase suspension (i.e. liquid, gas bubble, and microcrystals). Both types of transformations have drastic effects on the bulk magma flow conditions (rheology) that control ascent rate and the ductile versus brittle behaviour of the magma.

The rate of magma decompression/ascent is the key parameter that controls the kinetics of degassing and crystallization, and ultimately, the eruptive style. In silicic to intermediate systems, slow ascent rates typical of effusive eruptions such as lava flow or dome growth, i.e. cm/s to mm/s (Gardner and Rutherford 2000) yield timescales long enough for extensive degassing and crystallization. In contrast, the high ascent rates prevalent during paroxysmal Strombolian or Plinian eruptions (i.e. of the order of m/s; Gardnerand and Rutherford 2000) are able to generate physico-chemical disequilibria of both degassing and crystallization processes, driving gas overpressures that may be released explosively.

3.1 Textures of Natural Pyroclasts

Decompression-induced degassing of the magma creates gas bubbles, the number density of which has been demonstrated to correlate with the decompression rate simulated by experiments (Mourtada-Bonnefoi and Laporte 2004; Mangan et al. 2004). The growing bubbles can rapidly coalesce and form gas escape channels. In this case, the bubble number densities in the erupted/quenched pyroclasts may no longer be representative of the initial ascent rates under which nucleation was triggered. The timescale of outgassing by magma foam collapse varies from a couple of hours to about 1000 h for magmas having bulk viscosities of $\sim 10^4$ to $10^{5.5}$ Pa.s, respectively (Martel and Iacono-Marziano 2015). This restricts the use of the degassing process to simulations of rapid magma ascent rates such as those during Plinian events. To investigate longer transit times in the conduit, one requires information from magmatic processes with timescales longer than degassing. Microlite crystallization is one of those processes, because diffusion in the melt, that controls crystal growth, occurs on timescales ranging from hours in mafic melts, to days in silicic melts. Microlite number density, volume proportion, size and shape have all been used as markers of the undercooling (liquidus temperature minus magma temperature) that drives crystallization (e.g. Hammer et al. 1999); the higher the undercooling, the more numerous, smaller, and irregularly-shaped the crystals. With decompression (and dehydration of the melt), liquidus temperature increases (as does undercooling), so that it becomes possible to infer the depth of crystallization in the conduit by relating the textural characteristics of the microlites to undercooling and pressure (Fig. 3). This approach has been used by Melnik et al. (2011) to constrain both magma flow and reservoir shape for the 1980–86 dome-forming eruptions of Mount St. Helens (USA). Such modelling requires an accurate determination of the dependence of undercooling on pressure, which can be achieved through decompression experiments (e.g. Riker et al. 2015).

3.2 Dynamic Experiments

Dynamic experiments, such as decompression, deformation or shock-wave experiments, are valuable tools to investigate degassing, crystallization, strain, mixing, or fragmentation of a magma. They provide information on the physics, chemistry, and kinetics of syn-eruptive magmatic processes, which can be used in turn to decode natural pyroclast formation in order to better identify geophysical and geochemical precursory signals of an eruption. Shock-wave experiments dedicated to the fragmentation process are covered in *Chapter "From unrest to eruption: Conditions for phreatic versus magmatic activity"*; the discussion below concentrates on decompression and deformation experiments.

Decompression experiments performed at elevated pressure (HP) and temperature (HT) have proved useful in accurately simulating magma ascent in volcanic conduits (e.g. Hammer and Rutherford 2002). In particular, experimental decompression rates can cover the most of the perceived range of magma ascent rates during

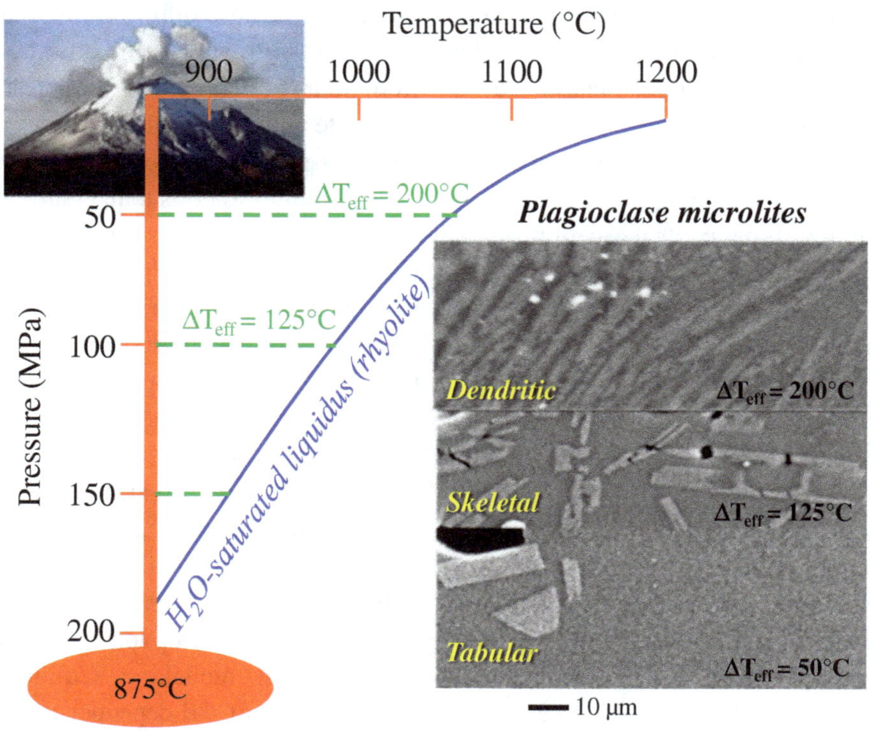

Fig. 3 Relationships between effective undercooling (ΔT_{eff}; *in green*) and microlite textural characteristics in a H_2O-saturated rhyolitic melt (modified after Mollard et al. 2012). Plagioclase microlites crystallized at 150 MPa after an isothermal quasi-instantaneous decompression from 200 MPa, i.e. ΔT_{eff} = 50 °C, are represented by scarce large tabular crystals. With decreasing pressure and increasing ΔT_{eff}, microlites become more numerous, but smaller in size, and they show more complex shapes ranging from skeletal (*hollow*) to dendritic

volcanic eruptions. For instance, Plinian ascent rates of the order of m/s can be simulated experimentally by decompression durations from seconds to hours whereas the slow ascent rates recorded for dome eruptions can be reproduced by decompression durations of several days or weeks. More generally, decompression experiments can simulate different natural eruptive scenarios depending on the applied decompression rate, final pressure, and dwell time at final pressure. In basaltic H_2O- and CO_2-bearing magmas, experimental decompression in the duration range of <1–10 h has provided information on degassing processes leading to either regular or paroxysmal Strombolian eruptions (*Chapter 15 "Magma degassing: the diffusive fractionation model and beyond"*). In silicic melts, decompression pathways and durations from ten seconds to forty days have been investigated experimentally to evaluate the lifetime of rhyolitic foams as a function of bulk viscosity (Martel and Iacono-Marziano 2015). Deformation experiments performed in vessels equipped with torsion or coaxial deformation modules have shown that the lifetime of such magmatic foams is drastically reduced when a differential stress field prevails, because it enhances bubble coalescence (e.g. Okumura et al. 2009). The recent implementation of HP-HT devices that allow magma deformation at pressure coupled with in situ measurements of permeability, represents a considerable step forward for investigating the explosive-effusive transition of volcanic eruptions in the laboratory under realistic conditions (Kushnir et al. 2017).

Figure 4 illustrates how timescales of degassing and crystallization during decompression can be used to decipher eruption style. Magmas from both, Plinian and dome-forming eruptions (dome, block-and-ash flows, surges), degas during ascent. However, gases in dome-forming magmas escape from the melt (leading to dense pyroclasts)

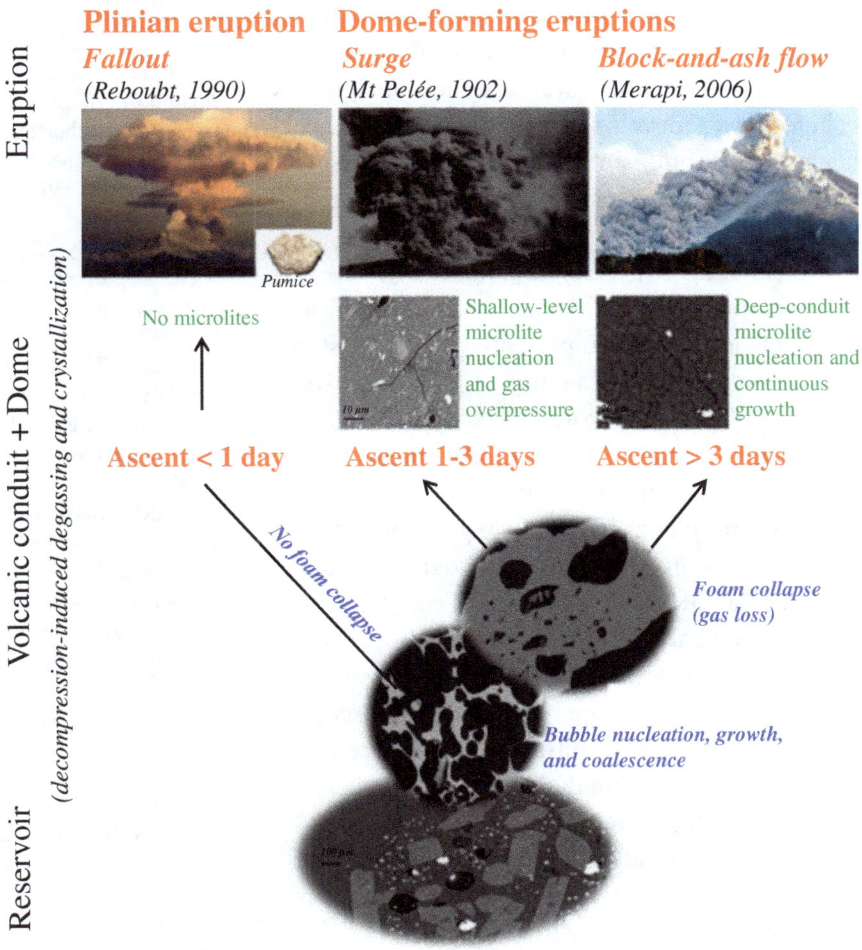

Fig. 4 Deciphering eruption explosivity from decompression-induced degassing and crystallization experiments (see text) [*Photograph of Redoubt is from R. Clucas (USGS), Mt. Pelée is from* Lacroix (1904), *and Merapi comes from the website:* gunungmerapi. weebly.com]

whereas Plinian foams have no time to collapse through gas escape (leading to pumiceous pyroclasts) which suggests Plinian ascent durations are limited to a couple of hours (Martel and Iacono-Marziano 2015). Furthermore, in contrast to Plinian magmas, dome-forming magmas have time to crystallize during ascent. At Mt. Pelée, the moderately-explosive block-and-ash flows in 1929–1932 may have degassed and crystallized continuously during an ascent lasting more than 3–6 days, so that little gas overpressure remains at dome level. In contrast, the devastating surges in 1902 may have resulted from rapid ascent (i.e. <3 days) that did not allow crystallization in the conduit, followed by extensive microlite crystallization at dome level (due to large effective undercooling). The exsolving gas and high overpressurization resulting from this extensive crystallization may have triggered the violent surges (Martel 2012).

This application of decompression experiments highlights the possibility of forecasting the style of an eruption provided the magma ascent rate towards the surface can be determined by some remote means (seismology, gravity, geodesy, gas discharge).

4 Future Directions

The combination of equilibrium and dynamic experiments can simulate many of the conditions relevant to magma eruption, because realistic pressures, temperatures and rates of decompression or shear are now accessible in the laboratory. The one parameter that remains impossible to simulate is an extended timescale. Experiments last typically a maximum of weeks, or occasionally months. In practical terms, this leads to small crystal sizes compared with nature, occasional difficulties in establishing equilibrium, or very short diffusion profiles for controlled 'disequilibrium' experiments. However, as new analytical techniques are developed we can start to make nanoscale measurements that allow us to measure profiles developed on laboratory timescale, providing access to ever faster natural processes (Saunders et al. 2014; Lloyd et al. 2014).

Development of *in situ* observation or measurement techniques represents a major step forward in the understanding of the magmatic processes linked to volcanic eruptions. It is becoming possible to make *in situ* observations using cameras coupled to HP-HT vessels equipped with transparent windows (e.g. Gondé et al. 2011) or 4D *in situ* X-ray tomography (e.g. Pistone et al. 2015). HP-HT vessels coupled with *in situ* analytical techniques are already capable of measuring the volatile species either dissolved in the melt under pressure by *in situ* spectroscopy techniques (Raman or Infrared) or exsolved as vapour. Laboratory simulation of magma degassing and crystallization with these new *in situ* approaches will allow the identification of potential geophysical and geochemical signals that may be used as unrest precursors.

References

Blundy JD, Cashman KV (2008) Petrologic reconstruction of magmatic system variables and processes. Miner Inclusions Volcanic Proc 69:179–239

Blundy JD, Cashman KV, Rust AC, Witham F (2010) A case for CO_2-rich arc magmas. Earth Planet Sci Lett 290(3–4):289–301

Costa F, Morgan D (2010) Time constraints from chemical equilibration in magmatic crystals. In Dosseto A, Turner SP, Van Orman JA (eds) Timescales of magmatic processes: from core to atmosphere. Wiley-Blackwell, pp 125–159

Druitt TH, Costa F, Deloule E, Dungan M, Scaillet B (2012) Decadal to monthly timescales of magma transfer and reservoir growth at a caldera volcano. Nature 482:77–80

Ghiorso MS, Evans BW (2008) Thermodynamics of rhombohedral oxide solid solutions and a revision of the Fe-Ti Two-oxide geothermometer and oxygen-barometer. Am J Sci 308:957–1039

Gondé C, Martel C, Pichavant M, Bureau H (2011) In situ bubble vesiculation in silicic magmas. American Mineralogy 96:111–124

Hammer JE, Cashman KV, Hoblitt RP, Newman S (1999) Degassing and microlite crystallization during pre-climactic events of the 1991 eruption of Mt. Pinatubo Philippines. Bull Volcanol 60:355–380

Hammer JE, Rutherford MJ (2002) An experimental study of the kinetics of decompression-induced crystallization in silicic melts. J Geophys Res 107 (ECV8):1–23

Holland T, Blundy JD (1994) Non-ideal interactions in calcic amphiboles and their bearing on amphibole-plagioclase thermometry. Contrib Miner Petrol 116(4): 433–447

Kilgour GN, Blundy JD, Cashman KV, Mader HM (2013) Small volume andesite magmas and melt-mush interactions at Ruapehu, New Zealand: evidence from melt inclusions. Contrib Minerology Petrol 166 (2):371–392

Kushnir ARL, Martel C, Champallier R, Arbaret L (2017) In situ confirmation of permeability development in shearing bubble-bearing melts and implications for volcanic outgassing. Earth Planet Sci lett 458:315–326

Lacroix A (1904) La Montagne Pelée et ses éruptions. Masson Paris, p 662

Lindsley DH, Andersen DJ (1983) A two-pyroxene thermometer. J Geophys Res 88(S02):001. doi:10.1029/JB088iS02p0A887

Lloyd AS, Ruprecht P, Hauri EH, Rose W, Gonnerman HM, Plank T (2014) NanoSIMS results from olivine-hosted melt embayments: magma ascent rate during explosive basaltic eruptions. J Volcanol Geoth Res 283:1–18

Mangan MT, Sisson TW, Hankins WB (2004) Decompression experiments identify kinetic controls on explosive silicic eruptions. Geophys Res Lett 31: L08605. doi:10.1029/2004GL019509

Martel C (2012) Eruption dynamics inferred from microlite crystallization experiments: application to Plinian and dome-forming eruptions of Mt. Pelée (Martinique, Lesser Antilles). J Petrol 53:699–725

Martel C, Iacono-Marziano G (2015) Bubble coalescence, outgassing, and foam collapsing in decompressed rhyolitic melts. Earth Planet Sci Lett 412:173–185

Melnik OE, Blundy JD, Rust AC, Muir DD (2011) Subvolcanic plumbing systems imaged through crystal size distributions. Geology 39(4):403–406

Mollard E, Martel C, Bourdier J-L (2012) Decompression-induced crystallization in hydrated silica-rich melts: Empirical models of experimental plagioclase nucleation and growth kinetics. J Petrol 53:1743–1766

Mourtada-Bonnefoi CC, Laporte D (2004) Kinetics of bubble nucleation in a rhyolitic melt: an experimental study of the effect of ascent rate. Earth Planet Sci Lett 218:521–537

Okumura S, Nakamura M, Takeuchi S, Tsuchiyama A, Nakano T, Uesugi K (2009) Magma deformation may induce non-explosive volcanism via degassing through bubble networks. Earth Planet Sci Lett 281:267–274

Pichavant M, Costa F, Burgisser A, Scaillet B, Martel C, Poussineau S (2007) Equilibration scales in silicic to

intermediate magmas—implications for experimental studies. J Petrol 48:1955–1972
Pistone M, Arzilli F, Dobson KJ, Cordonnier B, Reusser E, Ulmer P, Marone F, Whittington AG, Mancini L, Fife JL, Blundy JD (2015) Gas-driven filter pressing in magmas: insights into in-situ melt segregation from crystal mushes. Geology 43:699–702
Putirka KD (2008) Thermometers and barometers for volcanic systems. Rev Mineral Geochim 69:61–120
Ridolfi F, Renzulli A (2012) Calcic amphiboles in calc-alkaline and alkaline magmas: thermobarometric and chemometric empirical equations valid up to 1130 °C and 2.2 GPa. Contrib Mineral Petrol 163:877–895
Riker JM, Blundy JD, Rust AC, Botcharnikov RE, Humphreys MCS (2015) Experimental phase equilibria of a Mount St. Helens rhyodacite: a framework for interpreting crystallization paths in degassing silicic magmas. ContribMineral Petrol 170(6):535–560. doi:10.1007/s00410-015-1160-5
Rutherford MJ, Gardner JE (2000) Rates of magma ascent. In: Sigurdsson H (ed) Encyclopedia of volcanoes. Academic San Diego, California, pp 207–217
Saunders KE, Blundy JD, Dohmen RG, Cashman KV (2012) Linking petrology and seismology at an active volcano. Science 336(6084):1023–1027
Saunders K, Buse B, Kilburn MR, Kearns S, Blundy JD (2014) Nanoscale characterisation of crystal zoning. Chem Geol 364:20–32
Scaillet B, Pichavant M (2003) Experimental constraints on volatile abundances in arc magmas and their implications for degassing processes. In: Oppenheimer C, Pyle D, Barclay J (eds) Volcanic degassing. Geological Society, London, Special Publications 213, 23–52
Streck MJ (2008) Mineral textures and zoning as evidence for open system processes. Rev Mineral Geochem 69:595–622
Tamic N, Behrens H, Holtz F (2001) The solubility of H_2O and CO_2 in rhyolitic melts in equilibrium with a mixed CO_2-H_2O fluid phase. Chem Geol 174: 333–347
Waters LE, Lange RA (2015) An updated calibration of the plagioclase-liquid hygrometer-thermometer applicable to basalts through rhyolites. Am Miner 100:2172–2184

Magma Chamber Rejuvenation: Insights from Numerical Models

C.P. Montagna, P. Papale, A. Longo and M. Bagagli

Abstract

Most volcanic systems on Earth are characterized by chemically different magmas that can be found in the erupted products throughout their history. The reasons are multiple, including variations in the mantle source and/or crustal assimilation, as well as shallower processes such as fractional crystallization or mixing and mingling. Magma chamber rejuvenation indicates the processes that happen whenever a magma intrudes from the mantle to shallower depths and encounters an already established storage zone (i.e. a magma chamber or reservoir). Magmas rising from depth are typically characterized by higher temperatures, larger volatile contents and more primitive, mantle-like compositions than those residing in the shallow crust. The interaction with magmas that have already resided at shallower depths for a while (years to thousands of years) varies the physical and chemical properties of both the involved magmatic end-members. Typically, volatile-rich magmas coming from depth are lighter than degassed shallow magma; therefore, a gravitational instability sets in as the two come into contact, which generates convection and thus intense mingling and mixing among the two. These dynamic interactions cause variations in the physical and chemical properties of the magmas themselves, as well as in the stress conditons both inside the reservoir and in the host rock. The volcanic system as a whole enters an unrest scenario, that can evolve to eruption or not depending on the specific conditions. Numerical simulations of the dynamics within magmatic systems can shed light on the features of magma chamber rejuvenation, providing the time

C.P. Montagna (✉) · P. Papale · A. Longo · M. Bagagli
Istituto Nazionale di Geofisica e Vulcanologia, via U. della Faggiola 32, Pisa, Italy
e-mail: chiara.montagna@ingv.it

Present Address:
M. Bagagli
D-ERDW, ETH-Zuerich, Zürich, Switzerland

Advs in Volcanology (2019) 111–122
DOI 10.1007/11157_2017_21

Published Online: 17 August 2017

scales of mixing processes and possibly of the evolution towards eruption. Coupling with models for the visco-elastic response of the host rock allows the identification of the onset of recharge processes from the analysis of geophysical signals observed at the surface.

Keywords
Magma chamber · Magma dynamics · Magma mixing

1 Extended English/Spanish abstract

Most volcanic systems on Earth are characterized by chemically different magmas that can be found in the erupted products throughout their history, either in synchronous eruptive episodes, or in different epochs of volcanic activity. This chemical heterogeneity can have multiple reasons, and it originates from deep in the mantle, due to variations in the source and/or crustal assimilation during ascent, to shallower crustal regions, where processes such as fractional crystallization or mixing and mingling take place. Magma chamber rejuvenation comprises some of the aforementioned processes at shallow level. Whenever a magma intrudes from the mantle to shallower depths and encounters an already established storage zone (a magma chamber or reservoir), the magma already emplaced gets rejuvenated by the incoming more primitive magma. Magmas rising from the deep regions of a volcano feeding system are typically characterized by higher temperatures, larger volatile contents and more primitive, mantle-like compositions. On the other hand, magmas that have resided at shallower depths for a while (years to thousands of years) have evolved by fractional crystallization, thus they have changed their composition towards a more felsic one, and have lost most of their gaseous phase, that can escape towards the surface. The interaction between primitive and evolved magmas varies the physical and chemical properties of both the end-members involved. Typically, volatile-rich magmas coming from depth are lighter than degassed shallow magma, albeit having a higher liquid density: they are characterized by a much larger gas content. The light magma tends to rise inside the denser reservoir; a gravitational instability sets in as the two magmatic mixtures come into contact, and generates convection inside the reservoir. As a consequence, intense mingling and mixing are generated among the two end-members. These dynamic interactions cause variations in the physical and chemical properties of the magmas themselves, that loose their identity as initial end-members and become a more homogeneous mixture. The volcanic system as a whole enters an unrest scenario, that can evolve to eruption or not depending on the specific conditions. Numerical simulations of a magmatic system representing magma injection into a shallow reservoir show that mixing is very intense at the time of contact, and can be efficient on time scales of hours to day in homogeneizing the system. Depending on the geometry of the volcano feeding system, and even more on the volatile content of the incoming and resident magmas, the process can be suppressed or enhanced. Sills favour mixing, while more vertically elongated, dike-like reservoirs slow the dynamical interactions. As the presence of a gaseous phase is the engine of the gravitational instability that triggers the dynamics, a higher volatile content, which translates into a higher gas content, in the deep regions of the feeding system strongly accelerates the rejuvenation process. As mixing patterns are found almost ubiquitously in products from volcanoes around the world, comparison of the observed features to the model predictions can provide insights on the features of magma chamber rejuvenation, including the time scales over which mixing processes are efficient and possibly the timings for the evolution towards an eruption or not. Coupling to models that describe the visco-elastic response of the host rock to stress variations within the magmatic system provides hints as to how to identify recharge processes at depth from the analysis of geophysical signals observed at the surface. Characteristic features of ground deformation associated to convection and mixing is the appearance of oscillation of exremely long period, on the order of hours (Ultra-Long-Period, ULP), that can be detected by instruments such as continuous tiltmeters and dilatometers. Their records can identify the onset of the interaction among different magmas, thus provide time scales for unrest duration and evolution.

2 Introduction

Magmas evolve in many ways during their residence time within the crust, determining whether they are going to be erupted or not. Magma chamber rejuvenation takes place whenever a magma intruding from the mantle to shallower depths encounters an already established storage zone (i.e. a magma chamber or reservoir). It can involve many different processes such as reheating and melting of the residing magmas, fractional crystallization due to changes in the pressure and temperature conditions, mingling and mixing among the different components; typically, it takes place at shallow crustal depths. Magmas rising from depth are often characterized by higher temperatures, larger volatile contents and more primitive, mantle-like compositions with respect to those that have been residing at shallower levels for a while (months, years to thousands of years). This general scenario can have a variety of declinations, depending on the specific setting and physico-chemical characteristics of the magmatic mixtures involved. The shallow magma can be highly crystalline, a mush, that can be rejuvenated by the heat from the incoming component (Bachmann and Bergantz 2003, 2008; Girard and Stix 2009; Bain et al. 2013; Till et al. 2015); or, at the other end, it can still be hot and more fluidal, especially if injection episodes are frequent (Voight et al. 2010), giving rise to mixing and mingling phenomena (Montagna et al. 2015). The interaction among the deep and shallow components changes the physical and chemical properties of both the involved magmatic end-members, triggering an unrest phase that can evolve to eruption or not depending on the specific conditions. Evidence of chamber rejuvenation both in igneous and in intrusive rocks, manifested mostly by mingling and mixing patterns, is almost ubiquitous at volcanic systems worldwide, and it is often invoked as eruption trigger.

Magma movement at depth implies mass re-distribution, pressure changes, and pressure transients which translate into variations in the gravity field, shape and slope of the volcano flanks, and seismic signals registered at the surface. Understanding the complex relationships between quantities measured by volcano monitoring networks and shallow magma processes is a crucial step for the comprehension of volcanic processes and in evaluating more realistic hazard forecast. The ability to detect the onset of magma recharge at depth is fundamental as it can provide hints to unrest duration and evolution, and possibly eruption timings.

In this work we describe a forward-modeling approach to describe magma chamber dynamics, specifically for what concerns rejuvenation episodes, and link it to the geophysical observables that are expected as a consequence. This provides a framework for the consistent interpretation of geological and geophysical records of unrest periods at active volcanoes. This methodology allows for identification of rejuvenation episodes in ground deformation records, and possibly discrimination between those episodes that lead to eruption or not.

3 Numerical Simulations of Magma Chamber Rejuvenation

3.1 Magmatic System

We refer as an archetypal case to the Phlegraean Fields magmatic system, where seismic imaging and attenuation tomographies have identified a huge (probably around 10 km wide) magma reservoir at a depth of around 8 km (Zollo et al. 2008; De Siena et al. 2010), while a variety of geophysical and geochemical evidence suggests that smaller (probably less than 1 km^3), shallower batches of magma have been forming throughout the caldera history at virtually any depth smaller than 9 km (Arienzo et al. 2010; Di Renzo et al. 2011). These shallow magma bodies have been identified as actively involved in past eruptions, which at least in some cases shortly followed the arrival of volatile-rich, less differentiated magmas from the deep feeding system (Arienzo et al. 2009; Fourmentraux et al. 2012). Chemical compositions of erupted magmas range from shoshonitic to trachytic to phonolitic; geochemical analyses on melt inclusions suggest a variety of processes contributing to this variability, such as recharge from depth, intra-chamber mixing, syn-eruptive mingling (Arienzo et al. 2010; Fourmentraux et al. 2012). The same analyses show that deep magmas are typically rich in gas, especially CO_2 (Mangiacapra et al. 2008), while shallow magmas are unusually crystal-poor, down to less than 3 wt% (Arienzo et al. 2009). To study the magmatic dynamics occurring as a consequence of a recharge event, we simplify the magmatic system retaining its most peculiar features. We model the injection of CO_2-rich shoshonitic magma coming from a deep reservoir into a shallower, much smaller chamber, containing more evolved and partially degassed phonolitic magma (see Table 1 for compositions). The two chambers are connected by a dyke. This idealized layout captures several first-order characteristics of prototype magmatic systems, including a composite structure, vertical extension, and heterogeneous composition, and it approximates systems composed by long-lived, interconnected multiple reservoirs believed to exist at many active volcanoes (Elders et al. 2011).

Table 1 Composition of the phonolite and shoshonite magma types employed in the simulations

Composition	SiO_2 (wt%)	TiO_2 (wt%)	Al_2O_3 (wt%)	Fe_2O_3 (wt%)	FeO (wt%)	MnO (wt%)	MgO (wt%)	CaO (wt%)	Na_2O (wt%)	K_2O (wt%)
Phonolite	53.5	0.6	19.8	1.6	3.2	0.1	1.8	6.8	4.7	7.9
Shoshonite	52.5	0.9	17.6	1.9	5.7	0.1	3.6	7.9	3.4	4.3

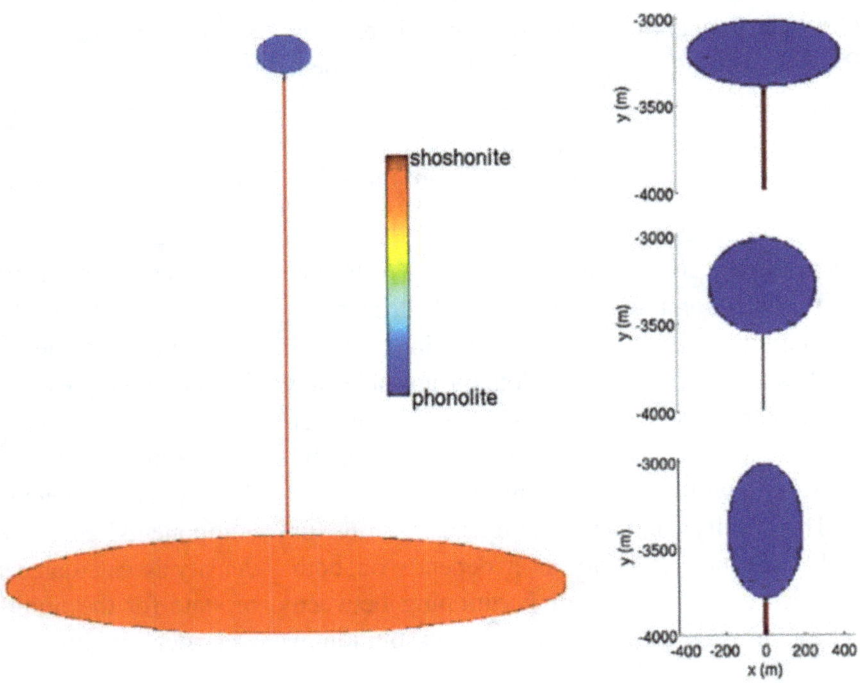

Fig. 1 Initial conditions for the numerical simulations of the magmatic system. On the *left*, the whole domain is shown, indicating the two magmatic end-members. On the *right*, the *upper* portion of the domain shows the three different geometries explored

Figure 1 shows the system domain for the numerical simulations. We assume one of the horizontal dimensions of the magmatic system to be much larger than the other, so that our domain is two-dimensional. The deep chamber is elliptical, 1 km thick and 8 km wide; its top is at 8 km depth. The geometry of the shallow chamber has been varied as shown in Fig. 1, keeping its surface area fixed. In the elliptical cases, the semi-axes measure 400 and 800 m, respectively, while the circular chamber has a radius of 283 m.

The initial conditions of the system are also shown in Fig. 1. The shallow chamber hosts a differentiated, volatile-poor phonolitic magma. Its volatile content has been varied from 0.3 wt% CO_2 and 2.5 wt% H_2O to 0.1 wt% CO_2 and 1 wt % H_2O. In the feeding dyke and deep reservoir is a less evolved, basaltic shoshonite, containing 1 wt% CO_2 and 2 wt% H_2O. Such a low water content in the phonolitic end-member derives from melt inclusion data from Phlegrean Fields (Arienzo et al. 2010). Typically, more evolved magmas are expected to have a relatively larger water content (Signorelli et al. 2001; Cannatelli et al. 2007; Pappalardo et al. 2007; Mollo et al. 2015), resulting possibly in smaller density contrasts at the interface among the two magmas thus less efficient mixing dynamics.

Volatiles partitioning between gaseous and liquid phases is computed following Papale et al. (2006) as a function of composition and pressure. Pressure at time 0 consists of a depth-dependent magmastatic contribution superimposed to the host rock confining pressure. The interface between the two magmas, at the inlet of the shallow chamber, is gravitationally unstable, the lower magma being less dense due to its higher gas content. The dynamics is solely driven by buoyancy, without any external forcing.

The density contrast at the interface varies for each simulated scenario, as it depends on both volatiles content and their partitioning between liquid and gaseous phases; volatiles exsolution in turn depends on pressure, thus on the depth at which the interface is placed, which is different for each geometry of the shallow chamber (Fig. 1). Temperature differences between interacting magmas are often negligible (Sparks et al. 1977), particularly at Phlegraean Fields (Mangiacapra et al. 2008; Arienzo et al. 2010), thus the system is assumed isothermal. As a result, there is no need to speculate on the thermal status of the surrounding rock, thus reducing model uncertainties. Moreover, heat transfer effects are expected to play a minor role on the short simulated time scales (hours; Di Renzo et al. 2011).

3.2 Magma Dynamics

Interaction among the two magmas develops as a consequence of the initial gravitational instability at the interface. We solve numerically the two-dimensional space-time evolution of the system, consisting of a mixture of two different magmatic components, each of them including a liquid (silicate melt and dissolved volatiles) and a gaseous (exsolved volatiles) fractions. The equations of motion for the mixture express conservation of mass for each component k = 1, 2, and momentum for the whole mixture (Longo et al. 2012a):

$$\frac{\partial(\rho y_k)}{\partial t} + \nabla\cdot(\rho \mathbf{u} y_k) = -\nabla\cdot(\rho D_k \nabla y_k), \quad \sum_k y_k = 1 \tag{1}$$

$$\frac{\partial(\rho \mathbf{u})}{\partial t} + \mathbf{u}\cdot\nabla(\rho\mathbf{u}) = -\nabla p + \nabla\cdot\left\{\mu\left[\nabla\mathbf{u} + (\nabla\mathbf{u})^{\mathrm{T}} - \frac{2}{3}\nabla\cdot\mathbf{u}\right]\right\} + \rho\mathbf{g}. \tag{2}$$

In the above, t is time; ρ is mixture density, y_k is mass fraction of component k, $\mathbf{u}$ is fluid velocity, D_k is the k-th coefficient of mass diffusion, p is pressure, μ is viscosity and $\mathbf{g}$ is gravity acceleration.

The magmatic mixture is considered ideal. Its density is evaluated as weighted sum of the components' densities; for each component, density is calculated using a non-ideal equation of state for the liquid phase, real gas properties and ideal mixture laws for multiphase fluids.

Mixture viscosity is computed through standard rules of mixing for one phase mixtures and with a semi-empirical relation in order to account for the effect of non-deformable gas bubbles. Liquid viscosity is modeled as in Giordano et al. (2008), and it depends on liquid composition and dissolved water content. The assumption of Newtonian rheology is justified by the very low strain rates and the crystal-free nature of the magmas. The generalized Fick's law is used to describe mass diffusion. Volatile partitioning between gaseous and liquid phases is evaluated at every point in the space-time domain as function of mixture composition and pressure as in Papale et al. (2006). All the physical properties of the two magmas are evaluated at every point in the space-time domain depending on the local conditions of pressure, velocity and mass fractions, which are the unknowns in Eqs. (1) and (2). The equations are solved numerically using GALES, a finite element C++ code specifically designed for volcanic fluid dynamics (Longo et al. 2012a).

The evolution in space and time of the system is complex and presents a number of interesting features. Figure 2 summarizes the results regarding magma dynamics, showing the evolution of composition in time in the shallow chamber for the five different simulation scenarios.

The initial inverse density contrast at the contact interface between the two magmas gives rise to convective mass transfer from the deeper parts of the system to shallower depths and vice versa. The unstable density contrast is solely due to the different volatile content of the two mixtures: the shoshonitic melt has an higher density than the phonolitic. The role played by volatiles is crucial, and it is exsolved gases that ultimately determine the buoyant dynamics. A Rayleigh-Taylor instability develops, which acts to bring the system to gravitational equilibrium by overturning it. The instability develops starting from the perturbed interface, with a first plume of light material that rises into the chamber. Depending on the initial density contrast as well as on the geometry of the shallow chamber, the initial plume starts developing at different times. The dynamics is strongly enhanced by higher density contrasts; geometry also plays an important role when density contrasts are similar, with horizontally elongated, sill-like chambers favouring convection with respect to more dyke-like setups (see also Fig. 2).

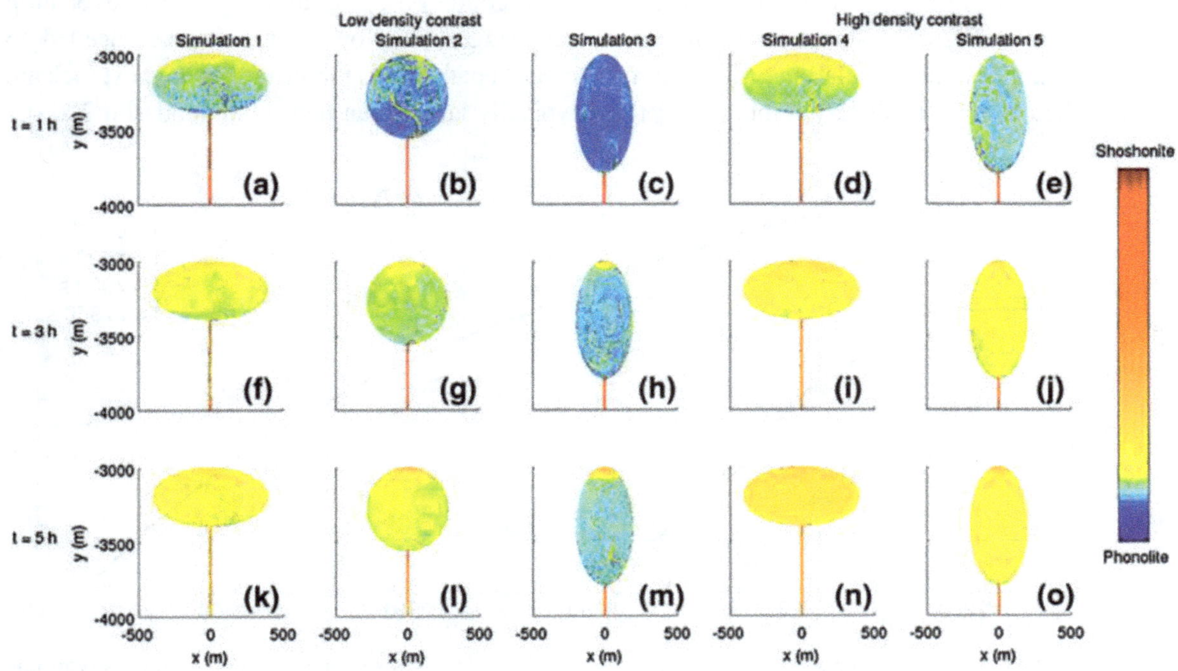

Fig. 2 Snapshots of variation of composition with time in the shallower parts of the system for the different simulations. *Columns* correspond to different simulations; *rows* correspond to different times

Plumes of light magma coming from depth keep entering the shallow reservoir as discrete filaments, following irregular trajectories and showing typical convective patterns. The lighter material tends to rise into the chamber, thereby decreasing more and more its density as volatiles exsolve in lower-pressure environments; on the other hand, the denser magmatic mixture initially residing in the chamber sinks into the feeder dyke, increasing its density by the reverse process of volatile dissolution at higher pressures. The plumes thus progressively increase their buoyancy, enhancing their expansion and acceleration. During the rise, vortexes form at the head of the plumes and subsequent plumes interact among themselves, further favouring mixing. The dynamics creates complicated patterns that maximize the interaction among the two different magmatic mixtures (Petrelli et al. 2011). Mingling is evident for all simulated conditions both within the chamber itself and even more in the feeding dyke (Fig. 2), and it is strongly intensified by the chaotic patterns that form as a consequence of deep magma injection.

Independently from system geometry or density contrast at the interface, mingling is very efficient in the feeding dyke, more than inside the upper chamber. Figure 2 shows that since the very beginning of the simulations, the magma entering the chamber is already a mixture of the two initial end-members, and not the pure shoshonitic composition.

As the dynamics proceeds, faster for higher density contrasts and sill-like setups, the gas-rich mixture tends to accumulate at the top of the chamber, thereby originating a stable density stratification that has indeed been testified at various magmatic systems (Arienzo et al. 2009). The stratification is more prominent in vertically elongated, dyke-like reservoirs (Fig. 2). The density profile along the vertical direction, evaluated averaging along horizontal planes (Fig. 3), illustrates that a quasi-stable profile is reached after some hours of simulated time.

As time proceeds, convection slows down due to smaller buoyancy of the incoming already mixed component, and the instability proceeds in time asymptotically: the more the two end-members have mingled, the less intense is convection.

The evolution of pressure in the system is highly heterogeneous in space and time. Alternating phases dominated by buoyancy and sinking at chamber inlet result in pressure fluctuations with periods of hundreds of seconds and amplitudes decreasing with time (Fig. 4). Typically pressure variations are smaller than 1 MPa; under these conditions, it is unlikely that rejuvenation can trigger eruption, as the stresses needed to create a pathway to the surface in the host rock are typically larger than that (Gudmundsson 2006).

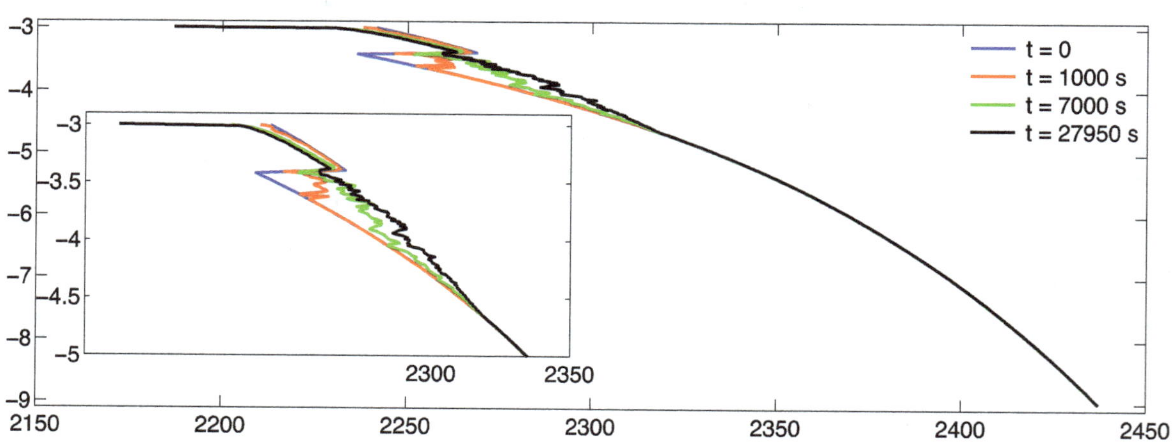

Fig. 3 Total mixture density averaged over horizontal planes as function of depth for simulation 1, at different times. The inset shows the upper 5 km of the domain; the *black line* represents the quasi-equilibrium density profile at the end of the simulation

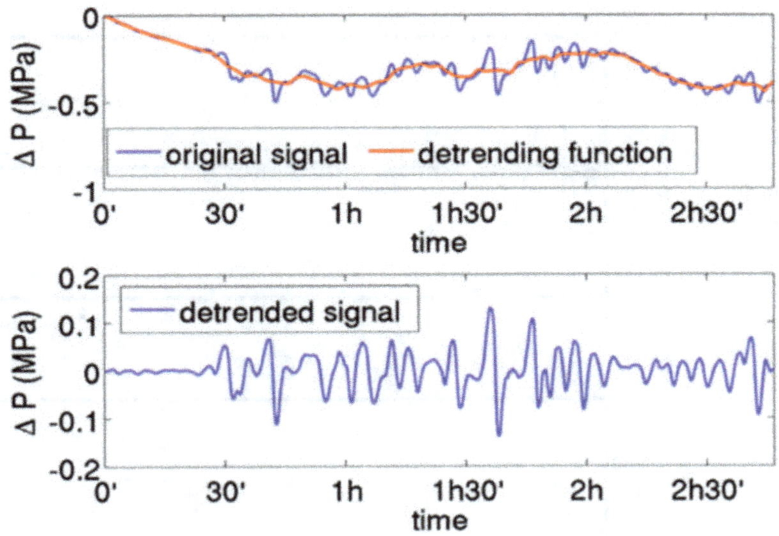

Fig. 4 Pressure variations as a function of time at a point on the boundary of the upper chamber, for simulation #1. The *upper* diagram shows the difference between the local pressure at current time and at time zero, while the *bottom* diagram shows the same quantity after subtraction of a detrending function (*red curve* in the *upper* diagram)

3.3 Ground Deformation

Determining the time–space-dependent ground displacement requires modeling the magma–rocks boundary conditions and the mechanical response of rocks, the latter depending on heterogeneous rock properties, presence and distribution of faults, interfaces, fluids, and volcano topography (e.g., O'Brien and Bean 2004). A first-order analysis performed here assumes magma–rock one-way coupling and adopts the Green's functions formulation for a homogeneous, infinite medium (Aki and Richards 2002).

We consider as point sources the fluid dynamics computational grid nodes located at the reservoir walls. As source time functions, we use the respective temporal evolutions of magmatic forces computed from pressures and stresses provided at those nodes by the numerical simulations of magma convection and mixing dynamics. Ground displacement at a series of virtual receivers is finally obtained by integrating, over all sources, the Green's functions associated with individual sources.

Continuity of pressure and stress is taken as the boundary condition along the non moving magma–rock interface. Physical properties of rocks are homogeneous averages that describe the volcanic edifices within the range of considered depths (<10 km, $v_P = 3000$ m/s; $v_P/v_S = 1/\sqrt{3}$, $\rho = 2500$ kg/m^3).

Propagation of pressure disturbances in the host rock medium reveals that the computed pressure oscillations, originated by the ingression of buoyant magma in the magma chamber, translate into Ultra Long Period ground displacement dynamics with amplitudes of millimeter to micrometer order (Fig. 5; Longo et al. 2012b). ULP ground movements like those predicted by the present modeling could not be detected by classical broadband seismometers (although more recent seismometers extend their working range up to 100–200 s periods), while they are visible in the records from other instruments, especially

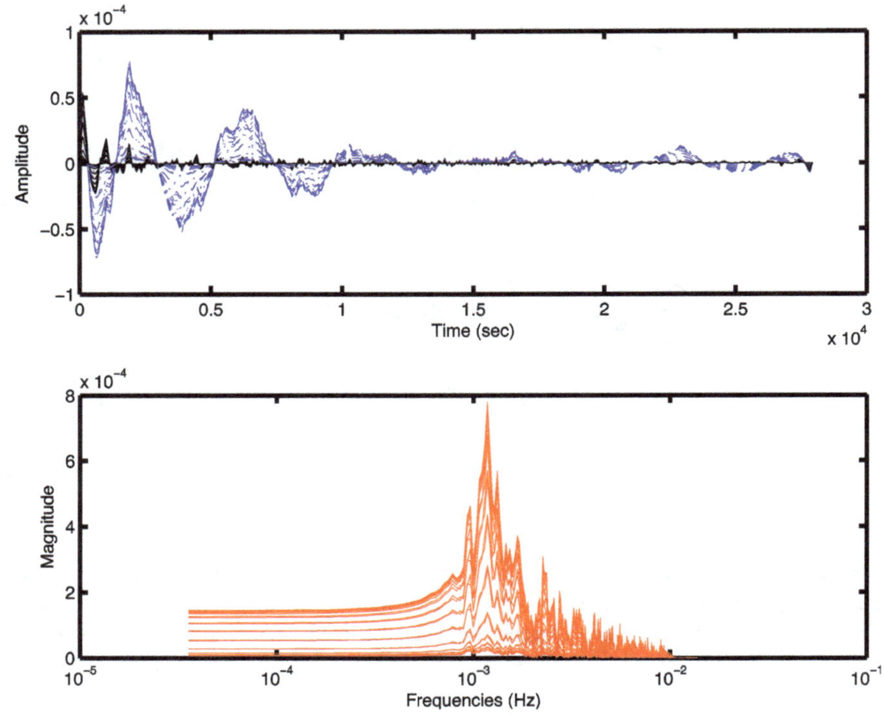

Fig. 5 Synthetic seismic signal (*blue* detrended as to represent an instrumental record, *black* filtered [0.001,0.01] Hz) and corresponding frequency spectrum for the vertical component at all synthetic stations. Example from simulation #1

borehole dilatometers characterized by high signal-to-noise ratio (Sacks et al. 1971).

4 Discussion and Conclusions

The arrival of fresh magma into an already emplaced reservoir and the consequent internal dynamics have often been invoked as possible eruption triggers, especially at Phlegraean Fields (Arienzo et al. 2010); on the other hand, footprints of magma chamber rejuvenation are often found in intrusive granites as well, testifying that it does not necessarily lead to eruption. Magma chamber rejuvenation can thus be regarded as a prototypical volcanic unrest process, that can lead to eruption or not depending on the specific system conditions (e.g. host rock compliance, volume and volatile content of injected magma).

The time scales for rejuvenation processes to be effective in magmatic reservoirs are relatively short, on the order of hours. This is consistent with what has been observed from the analyses of erupted products (Fourmentraux et al. 2012) as well as from experiments on diffusive fractionation (Perugini et al. 2010), and opens a completely new perspective in terms of unrest duration: indications of mixing in erupted products suggest that recharge events can happen within a very short time frame from eruption, otherwise the evidence would be wiped out by the efficient mixing process (Montagna et al. 2015).

In terms of geophysical observables, convective mingling dynamics in magmatic reservoirs is associated with ultra-long-period seismic signals, characterized by frequencies in the range $10^{-2}-10^{-4}$ Hz (Longo et al. 2012b).

The results obtained by our modeling specifically refer to the Phlegraean system. They can be extended to many other volcanoes where evidence of rejuvenation has been observed in similar magmatic settings, characterized by relatively primitive magmas that are not very different in composition and temperature, such as e.g. Mount Etna (Viccaro et al. 2006) or Stromboli (La Felice et al. 2011). For more evolved magmas, reservoirs can be dominated by crystal-rich regions (Marsh 1981; Koyaguchi and Kaneko 1999; Bachmann and Bergantz 2004; Hildreth 2004; Huber et al. 2009; Cooper and Kent 2014), and they are often at a lower temperature. The approach described above must be applied with caution in such cases, as the dynamics of the incoming primitive magma is more likely to be described as flow through a porous medium (mush) than as fluid mingling and mixing. Nonetheless, there is some evidence for crystal-poor silicic magma reservoirs to be reactivated as well (Bachmann et al. 2002; Deering et al. 2011; Huber et al. 2012; Sliwinski et al. 2015; Wolff et al. 2015).

Given the short time scales over which the dynamical processes described here can be effective and lead to eruption, it would be beneficial to be able to routinely detect the signals described above for eruption forecasting and mitigation actions. This is especially true for long-dormant volcanoes such as Phlegraean Fields, one of the highest-risk volcanic areas in the world given the large population living within the caldera borders (Arienzo et al. 2010), for which there is still no widely accepted means of discriminating the precursors of an impending eruption (Druitt et al. 2012).

Acknowledgements This work has received funds from the European Union's Seventh Programme for research, technological development and demonstration under grant agreements No. 282769 VUELCO and No. 308665 MED-SUV. The manuscript has largely benefited from reviews by Fabio Arzilli and Olivier Bachmann.

Glossary

Magma chamber A storage volume for magma in the crust. Typically, during their rise from the mantle magmas accumlate in regions where there are geological discontinuities.

Rejuvenation The process by which magmas coming from depth modify chemical and physical properties of more evolved magmas already emplaced at shallower levels.

Primitive magma A magmatic melt that has composition similar to that of the mantle (smaller silica content).

Evolved magma A magmatic melt that has undergone processes within the crust that modified its chemical properties, such as fractional crystallization and crustal assimilation. Its composition is characterized by higher silica content.

Convection Exchange of mass and energy by means of cell patterns.

Green's functions Source to receiver transfer function; in this context through the volcanic rock medium.

Index

Volcanic unrest
Magma chamber dynamics
Magma mingling: magma mixing
Eruption precursors
Volcanic unrest duration
Magma evolution
Ground deformation
Ultra-Long-Period seismicity
Volcano seismicity

References

Aki K, Richards PG (2002) Quantitative seismology. University Science, Sausalito

Arienzo I, Civetta L, Heumann A, Woerner G, Orsi G (2009) Isotopic evidence for open system processes within the Campanian Ignimbrite (Campi Flegrei—Italy) magma chamber. Bull Volcanol 71(3):285–300

Arienzo I, Moretti R, Civetta L, Orsi G, Papale P (2010) The feeding system of Agnano—Monte Spina eruption (Campi Flegrei, Italy): Dragging the past into present activity and future scenarios. Chem Geol 270 (1–4):135–147

Bachmann O, Bergantz GW (2003) Rejuvenation of the Fish Canyon magma body: a window into the evolution of large-volume silicic magma systems. Geology 31(9):789–792

Bachmann O, Bergantz GW (2004) On the origin of crystal-poor rhyolites: Extracted from batholithic crystal mushes. J Petrol 45:1565–1582

Bachmann O, Bergantz GW (2008) The Magma Reservoirs That Feed Supereruptions. Elements 4(1):17–21

Bachmann O, Dungan MA, Lipman PW (2002) The Fish Canyon magma body, San Juan volcanic field, Colorado: Rejuvenation and eruption of an upper-crustal batholith. J Petrol 43:1469–1503

Bain AA, Jellinek AM, Wiebe RA (2013) Quantitative field constraints on the dynamics of silicic magma chamber rejuvenation and overturn. Contrib Min Petr 165(6):1275–1294

Cannatelli C, Lima A, Bodnar RJ, De Vivo B, Webster JD, Fedele L (2007) Geochemistry of melt inclusions from the Fondo Riccio and Minopoli 1 eruptions at Campi Flegrei (Italy). Chem Geol 237:418–432

Cooper KM, Kent AJR (2014) Rapid remobilization of magmatic crystals kept in cold storage. Nature 506:480–483

Deering CD, Bachmann O, Vogel TA (2011) The Ammonia Tanks Tuff: erupting a melt-rich rhyolite cap and its remobilized crystal cumulate. Earth Planet Sc Lett 310:518–525

De Siena L, Del Pezzo E, Bianco F (2010) Seismic attenuation imaging of Campi Flegrei: evidence of gas reservoirs, hydrothermal basins, and feeding systems. J Geophysi Res 115(B9):1–18

Di Renzo V, Arienzo I, Civetta L, D'Antonio M, Tonarini S, Di Vito MA, Orsi G (2011) The magmatic feeding system of the Campi Flegrei caldera: architecture and temporal evolution. Chem Geol 281(3–4):227–241

Druitt TH, Costa F, Deloule E, Dungan M, Scaillet B (2012) Decadal to monthly timescales of magma transfer and reservoir growth at a caldera volcano. Nature 482(7383):77–80

Elders WA, Friðleifsson GÓ, Zierenberg RA, Pope EC, Mortensen AK, Guðmundsson Á, Lowenstern JB, Marks NE, Owens L, Bird DK, Reed M (2011) Origin of a rhyolite that intruded a geothermal well while drilling at the Krafla volcano Iceland. Geology 39 (3):231–234

Fourmentraux C, Metrich N, Bertagnini A, Rosi M (2012) Crystal fractionation, magma step ascent, and syn-eruptive mingling: the Averno 2 eruption (Phlegraean Fields, Italy). Contrib Min Petr 163(6):1121–1137

Giordano D, Russell JK, Dingwell DB (2008) Viscosity of magmatic liquids: a model. Earth Planet Sci Lett 271:123–143

Girard G, Stix J (2009) Magma recharge and crystal mush rejuvenation associated with early post-collapse Upper Basin Member rhyolites, Yellowstone caldera. Wyoming. J Petrol 50(11):2095–2125

Gudmundsson A (2006) How local stresses control magma-chamber ruptures, dyke injections, and eruptions in composite volcanoes. Earth Sci Rev 79:1–31

Hildreth WS (2004) Volcanological perspectives on Long Valley, Mammoth Mountain, and Mono Craters: several contiguous but discrete systems. J Vol Geoth Res 136:169–198

Huber C, Bachmann O, Manga M (2009) Homogenization processes in silicic magma chambers by stirring and mushification (latent heat buffering). Earth Planet Sci Lett 283:38–47

Huber C, Bachmann O, Dufek J (2012) Crystal-poor versus crystal-rich ignimbrites: a competition between stirring and reactivation. Geology 40:115–118

Koyaguchi T, Kaneko K (1999) A two-stage thermal evolution model of magmas in continental crust. J Petrol 40:241–254

La Felice S, Landi P (2011) The 2009 paroxysmal explosions at Stromboli (Italy): magma mixing and eruption dynamics. Bull Volc 73(9):1147–1154

Longo A, Barsanti M, Cassioli A, Papale P (2012a) A finite element Galerkin/least-squares method for computation of multicomponent compressible in compressible flows. Comp Fluids 67:57–71

Longo A, Papale P, Vassalli M, Saccorotti G, Montagna CP, Cassioli A, Giudice S, Boschi E (2012b) Magma convection and mixing dynamics as a source of Ultra-Long-Period oscillations. Bull Volcanol 74:873–880

Mangiacapra A, Moretti R, Rutherford MJ, Civetta L, Orsi G, Papale P (2008) The deep magmatic system of the Campi Flegrei caldera (Italy). Geophys Res Lett 35

Marsh BD (1981) On the crystallinity, probability of occurrence, and rheology of lava and magma. Contrib Min Petr 78:85–98

Mollo S, Masotta M, Forni F, Bachmann O, De Astis G, Moore G, Scarlato P (2015) A K-feldspar-liquid hygrometer specific to alkaline differentiated magmas. Chem Geol 392:1–8

Montagna CP, Papale P, Longo A (2015) Timescales of mingling in shallow magmatic reservoir. In: Caricchi L, Blundy J D (eds) Chemical, physical and temporal evolution of volcanic systems. Gelogical Society, London, Special Publications 422

O'Brien GS, Bean CJ (2004) A 3D discrete elastic lattice method for seismic wave propagation in heterogeneous media with topography. Geophys Res Lett 31, L14608

Papale P, Moretti R, Barbato D (2006) The compositional dependence of the multicomponent volatile saturation surface in silicate melts. Chem Geol 229:78–95

Pappalardo L, Ottolini L, Mastrolorenzo G (2007) The Campanian Ignimbrite (southern Italy) geochemical zoning: insight on the generation of a super-eruption from catastrophic differentiation and fast withdrawal. Contrib Min Petr 156:1–26

Perugini D, Poli G, Petrelli M, Campos CP, Dingwell DB (2010) Time-scales of recent Phlegrean Fields eruptions inferred from the application of a diffusive fractionation model of trace elements. Bull Volcanol 72(4):431–447

Petrelli M, Perugini D, Poli G (2011) Transition to chaos and implications for time-scales of magma hybridization during mixing processes in magma chambers. Lithos 125(1–2):211–220

Sacks IS, Selwyn S, Evertson DW (1971) Sacks-Evertson strainmeter, its installation in Japan and some preliminary results concerning strain steps. P Jpn Acad 47 (9):707–712

Signorelli S, Vaggelli G, Carroll C, Romano MR (2001) Volatile element zonation in Campanian Ignimbrite magmas (Phlegrean Fields, Italy): evidence from the study of glass inclusions and matrix glasses. Contrib Min Petr 140:543–553

Sliwinski JT, Bachmann O, Ellis BS, Dávila-Harris P, Nelson BK, Dufek J (2015) Eruption of shallow crystal cumulates during caldera-forming events on Tenerife. Canary Islands, J Petrol 56(11):2173–2194

Sparks R, Sigurdsson H, Wilson L (1977) Magma mixing: a mechanism for triggering acid explosive eruptions. Nature 267:315–318

Till CB, Vazquez JA, Boyce JW (2015) Months between rejuvenation and volcanic eruption at Yellowstone caldera. Wyoming. Geology 43(8):695–698

Viccaro M, Ferlito C, Cortesogno L, Cristofolini R, Gaggero L (2006) Magma mixing during the 2001 event at Mount Etna (Italy): Effects on the eruptive dynamics. J Vol Geo Res 149(1–2):139–159

Voight B, Widiwijayanti C, Mattioli G S, Elsworth D, Hidayat D, Strutt M (2010) Magma-sponge hypothesis and stratovolcanoes: Case for a compressible reservoir and quasi-steady deep influx at Soufriere Hills Volcano, Montserrat. Geophys Res Lett 37(19): L00E05

Wolff JA, Ellis BS, Ramos FC, Starkel WA, Boroughs S, Olin PH, Bachmann O (2015) Remelting of cumulates as a process for producing chemical zoning in silicic tuffs: A comparison of cool, wet and hot, dry rhyolitic magma systems. Lithos 236–237:275–286

Zollo A, Maercklin N, Vassallo M, Dello Iacono D, Virieux J, Gasparini P (2008) Seismic reflections reveal a massive melt layer feeding Campi Flegrei caldera. Geophys Res Lett 35(12):L12306

Magma Mixing: History and Dynamics of an Eruption Trigger

Daniele Morgavi, Ilenia Arienzo, Chiara Montagna, Diego Perugini and Donald B. Dingwell

Abstract

The most violent and catastrophic volcanic eruptions on Earth have been triggered by the refilling of a felsic volcanic magma chamber by a hotter more mafic magma. Examples include Vesuvius 79 AD, Krakatau 1883, Pinatubo 1991, and Eyjafjallajökull 2010. Since the first hypothesis, plenty of evidence of magma mixing processes, in all tectonic environments, has accumulated in the literature allowing this natural process to be defined as fundamental petrological processes playing a role in triggering volcanic eruptions, and in the generation of the compositional variability of igneous rocks. Combined with petrographic, mineral chemistry and geochemical investigations, isotopic analyses on volcanic rocks have revealed compositional variations at different length scales pointing to a complex interplay of fractional crystallization, mixing/mingling and crustal contamination during the evolution of several magmatic feeding systems. But to fully understand the dynamics of mixing and mingling

D. Morgavi · D. Perugini
Department of Physics and Geology, University of Perugia, Piazza Università, 06100 Perugia, Italy

D. Morgavi (✉) · D.B. Dingwell
Department Earth and Environmental Sciences, Ludwig-Maximilian-University (LMU), Theresienstrasse 41/III, 80333 Munich, Germany
e-mail: daniele.morgavi@unipg.it

I. Arienzo
Istituto Nazionale Di Geofisica E Vulcanologia, Osservatorio Vesuviano, Via Diocleziano, 328, 8124 Naples, Italy

C. Montagna
Istituto Nazionale Di Geofisica E Vulcanologia, Via Della Faggiola, 32, Pisa 56126, Italy

Advs in Volcanology (2019) 123–137
DOI 10.1007/11157_2017_30

Published Online: 05 December 2017

processes, that are impossible to observe directly, at a realistically large scale, it is necessary to resort to numerical simulations of the complex interaction dynamics between chemically different magmas.

Keywords
Magma mixing · Mingling · Isotope · Modelling

1 Magma Mixing: A Brief Historical Overview

One of the first investigations on magma mixing recorded in the literature is the work of the chemist Bunsen (1851), a scholar at the University of Heidelberg who published research on the chemical variation of igneous rocks from the western region of Iceland. Through chemical analyses Bunsen highlighted that the linear correlation between pairs of chemical elements in binary plots in those Icelandic rocks was the consequence of "simple" binary mixing between two magmas with different chemical composition. Bunsen published this data and, for the first time, magma mixing was taken into account to explain the chemical variation of a suite of igneous rocks. This idea triggered a strong critical reaction from the geological community; the strongest opposition coming from Wolfgang Sartorius Freiherr von Waltershausen an expert on the Iceland and Etna volcanic areas at that time. He mostly argued against the method used by Bunsen of averaging rock analyses to calculate the starting end-members that eventually took part in the mixing process. Sartorius criticized not only the arbitrary choice of the end-members but also disliked the idea of Bunsen of an extensive layering of felsic/mafic rocks and magmas beneath Iceland.

Since the beginning of the 20th century, the experimental and thermodynamic work of Norman L. Bowen (e.g., Bowen 1928) has had a profound influence upon the way petrological processes and igneous differentiations are conceived. The conceptual model of fractional crystallization was firmly established as the most fundamental petrological process for generating the diversity of igneous rocks and remained so for many decades. Although Bowen did not explicitly deny the possibility of magma mixing, he reinterpreted field evidence of magma mixing rather as immiscibility of liquids (e.g., Bowen 1928). In 1944, Wilcox published a work on the Gardner River complex (Yellowstone, USA; Wilcox 1944) which is now considered a milestone for evidence of magma mixing, even if at that time it received strong comments from Fenner and remained one of the few papers on the topic. Only in the 1970's geoscientists started to deeply investigate magma mixing, recorded as a plethora of unequivocal evidence in both plutonic and volcanic rocks, as a major petrogenetic process (e.g., Eichelberger 1978, 1980; Blundy and Sparks 1992; Wiebe 1994; Wilcox 1999). Since the first hypotheses about the origin of mixed igneous rocks (e.g., Bunsen 1851), plenty of evidence of magma mixing processes, in all tectonic environments, throughout geological time, has accumulated in the literature allowing this natural process to be defined as a fundamental petrological process playing a key role in the generation of the compositional variability of igneous rocks and as a major process for planetary differentiation (e.g., Eichelberger 1978, 1980; Blundy and Sparks 1992; Wiebe 1994; De Campos et al. 2004; Perugini and Poli 2012; Morgavi et al. 2016).

2 Magma Mixing: Field Evidence

It is common practice in the petrological community to split magma mixing into two separate physico/chemical processes: (i) mechanical mixing (also referred to as "magma mingling"), by which two or more batches of magma mingle without chemical exchanges between them, and (ii) a chemical mixing (also referred to as "magma mixing") triggered by chemical exchanges between the interacting magmas in which elements move from one magma to the other according to compositional gradients continuously generated by the mechanical dispersion of the two magmas (e.g., Flinders and Clemens 1996). Physically, "magma mingling" is mainly controlled by the viscosity contrast between the two magmas; decreasing of the viscosity contrast results in progressively more efficient mingling dynamics (e.g., Sparks and Marshall 1986; Grasset and Albarede 1994; Bateman 1995; Poli et al. 1996; Perugini and Poli 2005). Chemically, "magma mixing" is driven by the mobility of chemical elements in the melt fractions of the two magmas (e.g., Lesher 1990; Baker 1990). Linear variations in inter-elemental plots for a set of rock samples have long been considered as the sole evidence for the occurrence of magma mixing (e.g., Fourcade and Allegre 1981).

The adoption of the above conceptual models led to the common practice of applying the term magma mingling to indicate the process acting to physically disperse (no chemical exchanges are involved) two or more magmas, whereas the term magma mixing indicates that the mingling process is also accompanied by chemical exchanges. Although such a conceptual approach may allow us to simplify the complexity of the magma mixing process and make it more tractable from the petrological point of view, unfortunately such terminology is not consistently used in the literature and this causes some misunderstanding. Although it is not always easy to clearly discriminate between the two processes, mingling is quite a rare process in nature as physical dispersion and chemical exchanges must occur in tandem during magma mixing processes (e.g., Wilcox 1999; Perugini and Poli 2012).

The most common evidence for magma mixing in igneous rocks is the occurrence of textural heterogeneity; the processes responsible for this have been discussed extensively in many works in the last decades (e.g., Eichelberger 1975; Anderson 1976; Bacon 1986; Didier and Barbarin 1991; Wada 1995; De Rosa et al. 1996; Ventura 1998; Smith 2000; Snyder 2000; De Rosa et al. 2002; Perugini et al. 2002, 2007; Perugini and Poli 2005, 2012; Pritchard et al. 2013; Morgavi et al. 2016).

In order to provide possible classification of magma mixing structures, the evidence of mechanical mixing in igneous rocks can be roughly divided into three different groups: (i) flow structures, (ii) magmatic enclaves and (iii) physico-chemical disequilibria in melts and crystals (e.g., Walker and Skelhorn 1966; Didier and Barbarin 1991; Hibbard 1995; Flinders and Clemens 1996; Wilcox 1944, 1999; Perugini et al. 2002, 2003; Streck 2008; Perugini and Poli 2012; Morgavi et al. 2016). Flow structures can be readily recognized in field outcrops as they show alternating light and dark coloured bands constituted by magmas with different compositions. Figure 1a, b shows some examples of fluid structures occurring in volcanic rocks from Grizzly Lake outcrop in Yellowstone National Park (USA) (Pritchard et al. 2013; Morgavi et al. 2016) and from Soufrière Hills volcano (Island of Montserrat, UK) (Plail et al. 2014). In particular, Fig. 1a shows flow bands of rhyolitic magma (white) intruding in a basaltic/hybrid magma (red to dark grey) whereas Fig. 1b shows an alternation of flow bands of rhyolitic (white) and hybrid magma (dark grey) across which basaltic enclaves (light grey) occur. The latter are surrounded by flow bands of hybrid filaments (blue).

Magmatic enclaves are probably the structural evidence that, according to common thinking, mostly characterize magma mixing processes. The term magmatic enclave is used to identify a discrete portion of a magma occurring within a host magma with a different composition (e.g., Wilcox 1944; Walker and Skelhorn 1966; Bacon 1986, Didier and Barbarin 1991). Generally, enclaves display quite sharp contacts with the

Fig. 1 Detailed images of the mixing features present at Grizzly Lake (Yellowstone) (**a**, **b**) and Soufrière Hills volcano (Montserrat) (**c**, **d**). Figure 1a shows the rhyolitic magma (white) has apparently intruded into the basaltic magma and the hybrid portions are present at the contact between the two end-members. Two large basaltic enclaves are visible at the bottom left and at the centre right. Figure 1b shows the stretching and folding of hybrid magma (dark grey) into a rhyolitic portion (white) with the presence of several basaltic enclaves. Figure 1c shows a basaltic enclave surrounded by andesitic magma (Soufrière Hills, from the 2010 eruption). Figure 1d exhibits at the centre a basaltic enclave in and andesitic host from Soufrière Hills volcano

host rock, although it is not rare to observe that some enclaves display engulfment and disruption of their boundaries due to infiltration of the host magma. Some examples of enclaves found in the volcanic rocks from Soufrière Hills are shown in Fig. 1c, d.

Disequilibrium textures in minerals (Fig. 2a–c) can be viewed as recorders of the thermal and compositional disequilibria operating in the magmatic system during the development of magma mixing processes. As the zoning pattern can be well preserved in minerals from both the plutonic and volcanic rocks, crystal populations from both environments can be used to reconstruct the time evolution of thermal and compositional exchanges between the two magmas during mixing. Recent studies highlighted the importance of detailed investigations of crystal compositional variability not only to reconstruct the fluid-dynamic regime governing the evolution of the igneous body, but also to understand the length-scale of the compositional variability induced by the mixing process, the latter being considered as a proxy to estimate the residence time of magmas in sub-volcanic reservoirs prior to eruption (e.g., Costa and Chakraborty 2004; Martin et al. 2008; Chamberlain et al. 2014; Perugini et al. 2003).

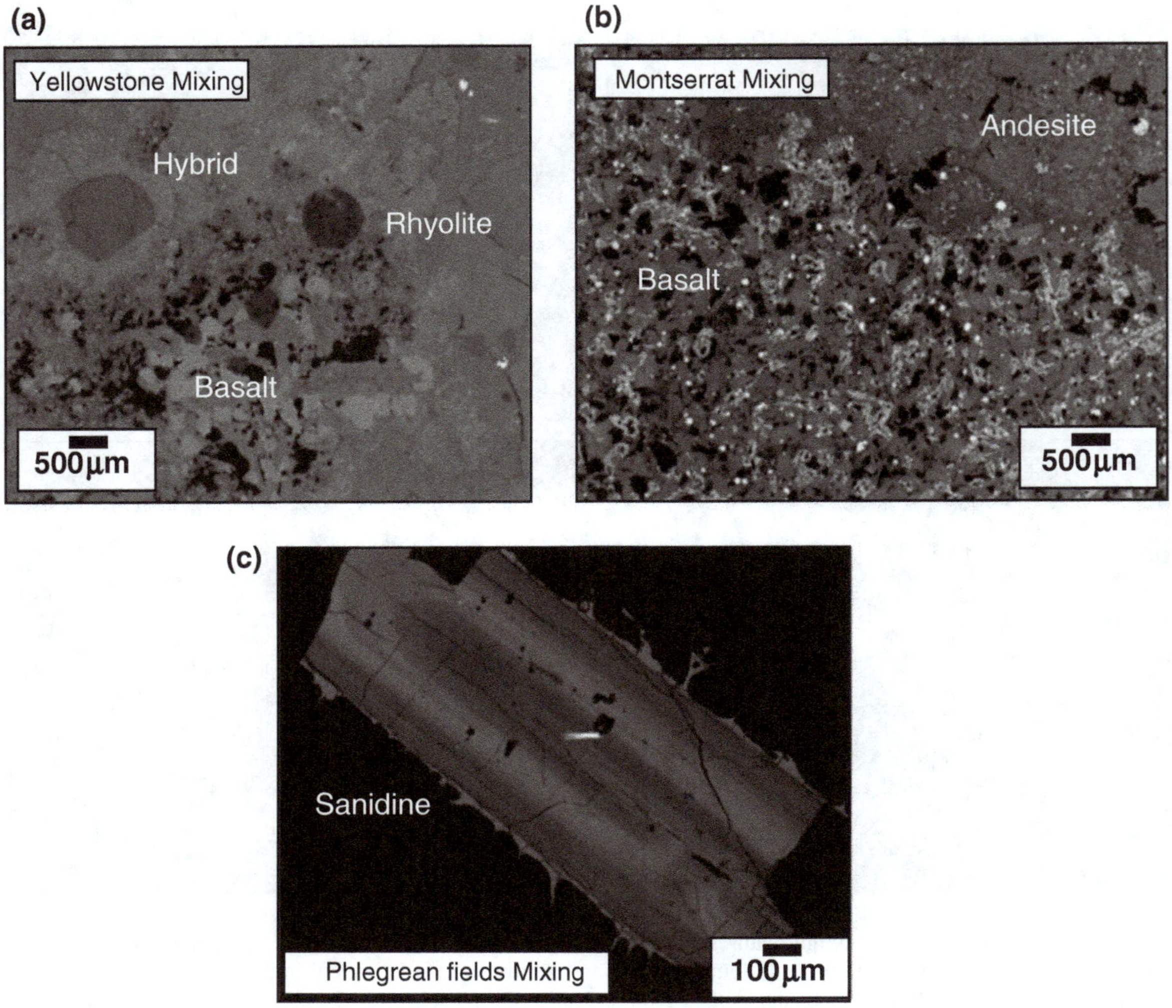

Fig. 2 Backscattered electron (BSE) image of a section of disequilibrium textures in rock and minerals from Yellowstone, Montserrat and Phlegrean Fields. **a** Mixed rock from Grizzly Lake Complex (Yellowstone) showing the interaction between the basaltic portion, the hybrid portion and the rhyolitic portion. **b** Mixed rock section from the 2010 Soufrière Hill eruption (Montserrat) showing disequilibrium texture in the basaltic and the andesitic rock. **c** Crystal from the 4.67 ± 0.09 cal ka Agnano Monte-Spina eruption (Phlegrean fields) occurred from a vent in the Agnano-San Vito area, has a darker (i.e., Ba-poorer) resorbed core and an inner rim with "swirly" zonation textures that indicate crystallization and dissolution. The outer rim is characterized by small-scale wavy oscillatory zoning that results from high-frequency growth and resorption events

3 Numerical and Experimental Studies: New Ideas for Deciphering the Complexity of Magma Mixing

Studies focused on numerical and experimental investigation of magma mixing dynamics (e.g., Perugini et al. 2003, 2008, 2015; De Campos et al. 2004, 2008, 2011; Petrelli et al. 2011; Montagna et al. 2015; Morgavi et al. 2015; Laumonier et al. 2014) can provide additional tools for a better understanding of the complexity of the mixing process, the evolution of which is governed by a continuous exchanges. One of the most striking results arising from these studies is that, during mixing, chemical elements experience a diffusive fractionation process due to the development in time of chaotic mixing dynamics (Perugini et al. 2006, 2008). This process is considered the source of strong deviations in many chemical elements from the linear variations in inter-elemental plots that would otherwise be expected, based on a conceptual model classically adopted in the geochemical modelling of magma mixing processes (e.g., Fourcade and Allegre 1981; Perugini and Poli 2012 and references therein). Recent studies on the mineralogical and geochemical features of mixed rocks (e.g., Hibbard 1981, 1995; Wallace and Bergantz 2002; Costa and Chakraborty 2004; Perugini and Poli 2005; Slaby et al. 2010), as well as those focused on quantitative analyses of morphologies related to textural heterogeneity (e.g., Wada 1995; De Rosa et al. 2002; Perugini and Poli 2005; Perugini et al. 2002, 2003) have highlighted the dominant role played by chaotic mixing dynamics in producing the substantial complexity of geochemical variations and textural patterns found in the resultant rocks (e.g., Flinders and Clemens 1996; De Campos et al. 2011; Morgavi et al. 2013a, b, c, 2016). Despite significant attention in the past, however, few works have focused on the understanding of the relationship between the morphology of the mixing patterns and the geochemical variability of the system using experimental devices (e.g., De Rosa et al. 2002).

Based upon the combination of field observations and the outcome from numerical simulations, a new experimental apparatus has been developed to perform mixing experiments using high viscosity silicate melts at high temperature (De Campos et al. 2011; Morgavi et al. 2013a, c, 2015). This device has been used to study the mixing process between natural melts, enabling the investigation of the influence of chaotic dynamics on the geochemical evolution of the system of mixing magmas (Morgavi et al. 2013a, b, c, 2015).

Preliminary results indicate that the time evolution of compositional exchanges between magmas from the experiments can be effectively modelled, leading to the prospect that the record of magma mixing processes may serve as chronometers to estimate the time interval between mixing and eruption (Perugini et al. 2010; Perugini and Poli 2012; Morgavi et al. 2013a, b, c, 2015; Perugini et al. 2015).

4 Geochemical Evidence of Magma Mixing/Mingling: An Example from the Campi Flegrei Volcanic Area

In some volcanic areas chemically and isotopically distinct magmas have been erupted, and their composition identified by analyzing the chemical composition of the erupted products (e.g., Pantelleria (Italy), Gedemsa and Fanta 'Ale (Main Ethiopian Rift), Gorely Eruptive Center (Kamchatka); Civetta et al. 1997; Giordano et al. 2014; Seligman et al. 2014). However, in other volcanic complexes the majority of the erupted products are chemically rather homogeneous, displaying a dominant composition (e.g., trachybasalt at Mt. Etna; trachyte at Campi Flegrei, Italy; basalt at Réunion Island, Indian Ocean; andesite at Popocatepetl, Mexico). Despite the roughly homogenous composition of products from these volcanic areas, their isotopic features suggest that complex open system processes occurred and superposed the main fractional crystallization trend. In fact, isotopic analyses (e.g., Sr, Nd, Pd, B) have been proven to be an important tool for discriminating between closed-system fractional crystallization and open-system magma mixing/mingling or crustal contamination (e.g., James 1982; Knesel et al. 1999; Turner and Foden 2001). Combined with petrographic, mineral chemistry and chemical investigations, isotopic analyses on volcanic rocks have revealed compositional variations at different length-scales (bulk rock, minerals, single crystals) pointing to a complex interplay of fractional crystallization, mixing/mingling and crustal contamination during the evolution of several magmatic feeding systems (e.g., Di Renzo et al. 2011 and references therein; Melluso et al. 2012; Corsaro et al. 2013 and references therein; Di Muro et al. 2014 and references therein; Brown et al. 2014). Furthermore, together with conventional isotopic analyses, current technologies permit high precision, in situ determination of Sr isotopic ratios of portions of phenocryst and glasses. In fact, microsampling by MicroMill™, coupled with isotopic measurement by Thermal

Ionization Mass Spectrometry (TIMS), allows for the performance of high precision determination of Sr isotopic composition of single crystals or portions of them. This information, unobtainable from bulk samples, has been used successfully to gather information on the time- and length-scales of the pre-eruptive magmatic processes, for identifying mantle sources and/or magmatic end-members and for tracking the time evolution of magma differentiation (e.g., Davidson et al. 1990; Davidson and Tepley 1997; Davidson et al. 1998; Knesel et al. 1999; Font et al. 2008; Kinman et al. 2009; Francalanci et al. 2012; Braschi et al. 2012; Jolis et al. 2013; Arienzo et al. 2015).

Among the active volcanic areas worldwide the volcanic hazard posed by the Campi Flegrei caldera is extremely high, due to its explosive character. Both the high volcanic hazard and the intense urbanization result in an extreme volcanic risk in this area, leading to a considerable interest in understanding which processes might contribute to triggering of eruptions and controlling? eruptive dynamics. The Campi Flegrei caldera is a nested and resurgent structure in the Campania Region, South Italy (Orsi et al. 1996), possibly formed after two large caldera forming eruptions: the Campanian Ignimbrite eruption (39 ka, Fedele et al. 2008) and the Neapolitan Yellow Tuff (15 ka, Deino et al. 2004). Its magmatic system is still active as testified by the occurrence of the last eruption in 1538 AD, as well as the present widespread fumaroles and hot springs activity, and the persistent state of unrest (Del Gaudio et al. 2010; Chiodini et al. 2003, 2012, 2015; Moretti et al. 2013). For compositionally homogenous magmas such as those extruded at the Campi Flegrei caldera (trachytes and phonolites being by far the most abundant rocks), major oxide and trace element variations cannot be used to unequivocally establish which magma evolution processes operated. Thus, together with petrographic, mineral chemistry and chemical data, isotopic investigations on volcanic rocks spanning the history of the volcano have been performed in recent decades in order to define the role of variable magmatic processes in the evolution of its feeding system up to eruption (e.g., Civetta et al. 1997; D'Antonio et al. 1999, 2013; de Vita et al. 1999; Pappalardo et al. 2002; Fedele et al. 2008, 2009; Tonarini et al. 2004, 2009; Arienzo et al. 2010, 2011; Perugini et al. 2010; Di Vito et al. 2011; Melluso et al. 2012; Arienzo et al. 2015).

In particular, detailed investigations of the geochemical and isotopic (Sr, Nd, Pb, and B) features of the younger than 15 ka Campi Flegrei volcanic products gave understanding to how many variable magmatic components, rising from large depth and/or stagnating in middle crustal reservoir(s), recharged the shallowest reservoir(s) and interacted with magma batches left from previous eruptions (Di Renzo et al. 2011). One identified magmatic component, geochemically similar to magma from the Neapolitan ca. 0.70750, 143Nd/144Nd ratio of ca. 0.51247, 206Pb/204Pb of ca. 19.04 and d11B of ca. −7.8‰), has been the most prevalent component over the past 15 ka. A second magmatic component, having geochemical features similar to the Minopoli 2 magma (D'Antonio et al. 1999; Di Renzo et al. 2011), first erupted 10 ka ago, is shoshonitic in composition. It is the most enriched in radiogenic Sr (87Sr/86Sr of ca. 0.70850) and unradiogenic Nd and Pb (143Nd/144Nd ratio of ca. 0.51238, 206Pb/204Pb of ca. 18.90), and it is characterised by the lowest d11B value of ca. −7.4‰. The third component is trachytic in composition and is characterized by lower 206Pb/204Pb (ca. 19.08), 87Sr/86Sr (ca. 0.70720) and d11B (−9.8‰) and higher 143Nd/144Nd (ca. 0.51250), with respect to the Neapolitan Yellow Tuff component (Tonarini et al. 2009; Di Renzo et al. 2011; Arienzo et al. 2015). This third composition is known as the Astroni 6 component due to the fact that it best recognized in the Astroni 6 erupted products (Di Renzo et al. 2011). During the past 5 ka of activity, this new component has been suggested to have mixed in variable proportions with the Neapolitan Yellow Tuff and Minopoli 2 magmatic components, which dominated the Campi Flegrei volcanic activity mostly in the time span from 15 to 5 ka (Fig. 3).

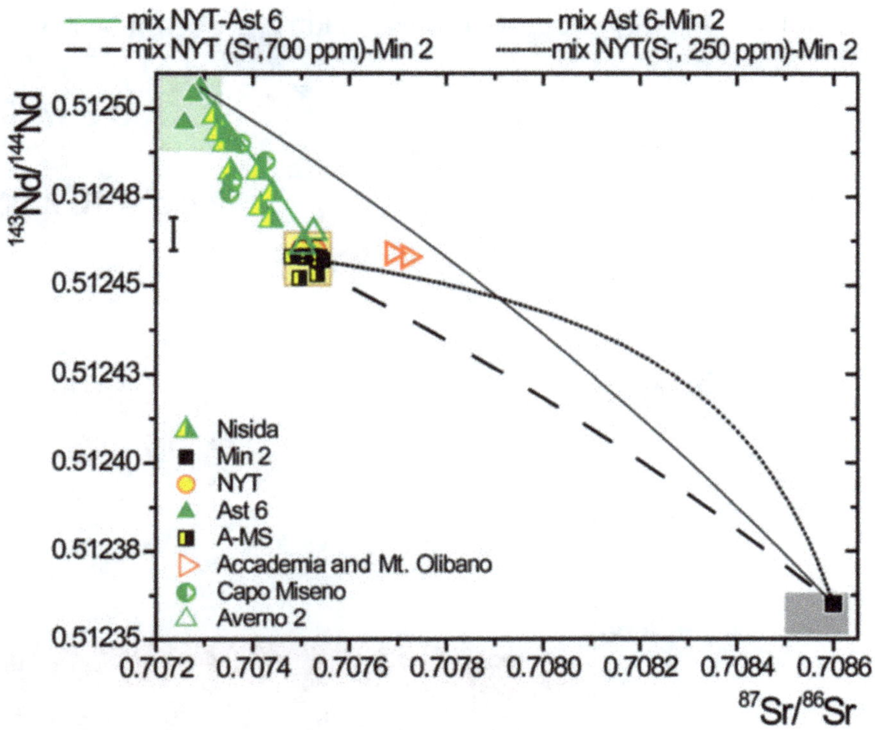

Fig. 3 Modelling of the Sr-Nd isotopic features of some of the Campi Flegrei volcanics of the past 5 ka, by assuming mixing among the Astroni 6 (Ast-6)-, Neapolitan Yellow Tuff (NYT)- and Minopoli 2 (Min 2)-like magmatic components. The green, yellow and black boxes represent the range of Sr and Nd isotopes of the products erupted during the Astroni 6, Neapolitan Yellow Tuff and the Minopoli 2 eruptions, respectively. Symbols inside the plot represent volcanic products belonging to the listed eruptions. The vertical error bar is the uncertainty in 143Nd/144Nd determination at the 2σ level of confidence; that for 87Sr/86Sr is included in the symbols. Modified after Arienzo et al. (2016)

Based on isotope investigations and melt inclusions studies, Arienzo et al. (2016) suggested that the Astroni 6 component, although undergoing differentiation during uprising, had a deep origin (larger than 8 km depth). Indeed, this magma rose not only inside and along the margins of the caldera, but also at the intersection between SE-NW and NE-SW regional fault systems mixing with the NYT-like magma component at shallower depth, and possibly entrapping crystals accumulated during older eruptions.

This detailed study of the Campi Flegrei volcanic system highlights that Sr isotopic microanalysis and, in general, more conventional isotopic analyses, coupled with petrographic, mineral chemistry and geochemical data can provide a better knowledge of the mixing/mingling processes and of the mixing end-members. In turn, they provide (i) information for evaluating the volcanic hazards and mitigating the related risks and (ii) the basic geochemical and petrologic knowledge for the numerical simulations.

5 Numerical Simulation of Magma Mingling and Mixing

To understand mixing and mingling processes at a realistically large scale, it is necessary to resort to numerical simulations of the complex interaction dynamics between chemically different magmas. Referring to the archetypal case of the Campi Flegrei magmatic system as described above, the interaction of a shoshonite and a more evolved phonolite has been investigated in detail to provide constraints on the time and length scales of the mixing dynamics. The simulated system consists of a very large, deep (8 km) reservoir connected by a dike to a shallower,

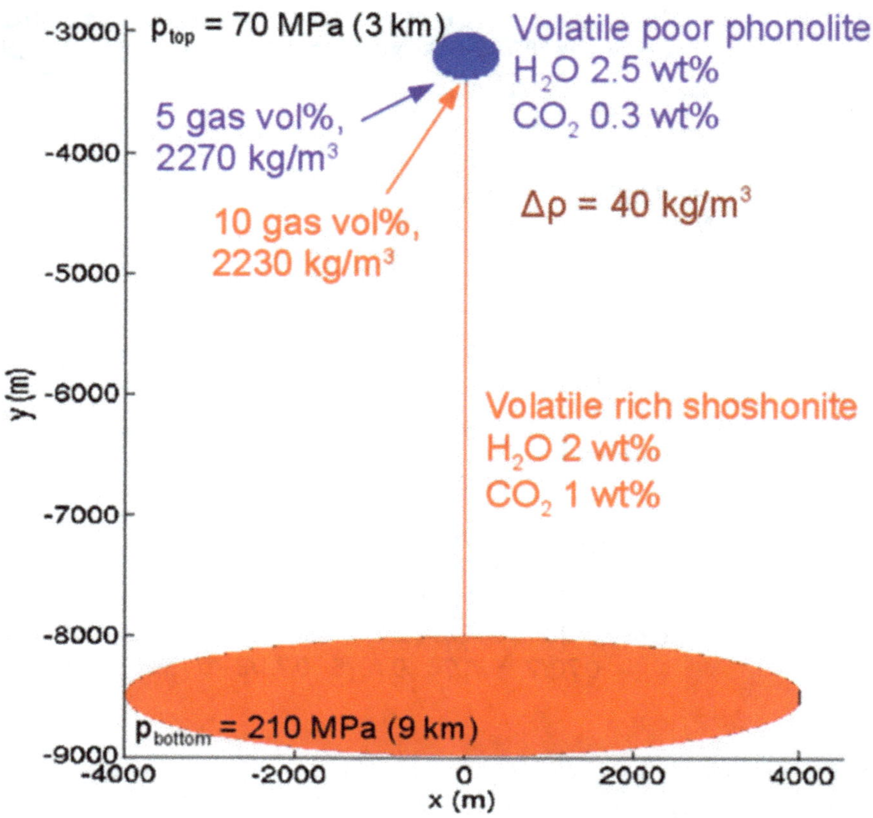

Fig. 4 Initial setup for the numerical simulations of magma chamber replenishment at Campi Flegrei caldera. The two interacting magmas are characterized by different compositions and volatile contents, thus densities and gas volume fractions

smaller chamber (Fig. 4). The chemical interactions between the two magmas cannot be resolved on the simulated large scale, as the computational costs required would be too high. The shoshonitic magma, being richer in volatiles than the resident phonolitic melt, rises into the phonolitic chamber by buoyancy, generating convection and mixing within the reservoir and in the feeding dike. Both magmatic components include a liquid (silicate melt) and a gaseous phase, that cannot decouple from the host melt. Space-time varying volatile exsolution is computed as a function of local composition, pressure and temperature following Papale et al. (2006). More details on magma chamber dynamics can be found in Chap. 8 'Magma chamber rejuvenation: insights from numerical models' of this book.

The main results show that the chaotic patterns observed in the products and in recently developed experimental setups (Morgavi et al. 2013a, b, c, 2015) are reproducible (Fig. 5), and that the two magmas mingle very efficiently from the beginning of their interaction. As time progresses, convection slows down due to smaller buoyancy of the incoming mixed component, and the instability proceeds in time asymptotically: the more the two end-members have mingled, the less intense the convection. A time-dependent mixing efficiency η_C can be defined as:

$$\eta_C = \frac{|m_R(t) - m_R(0)|}{m_R(0)}. \quad (1)$$

In the equation above, $m_R(t)$ is the mass of the resident phonolitic magma at time t, thus $m_R(0)$ is the initial phonolite mass. The mixing efficiency η_C represents the relative variation of the mass of the initially resident magma in a certain region of the domain. Figure 6 shows the time evolution of mixing efficiency in the shallow chamber, for different simulated setups in terms of chamber geometry and volatile content. It clearly shows

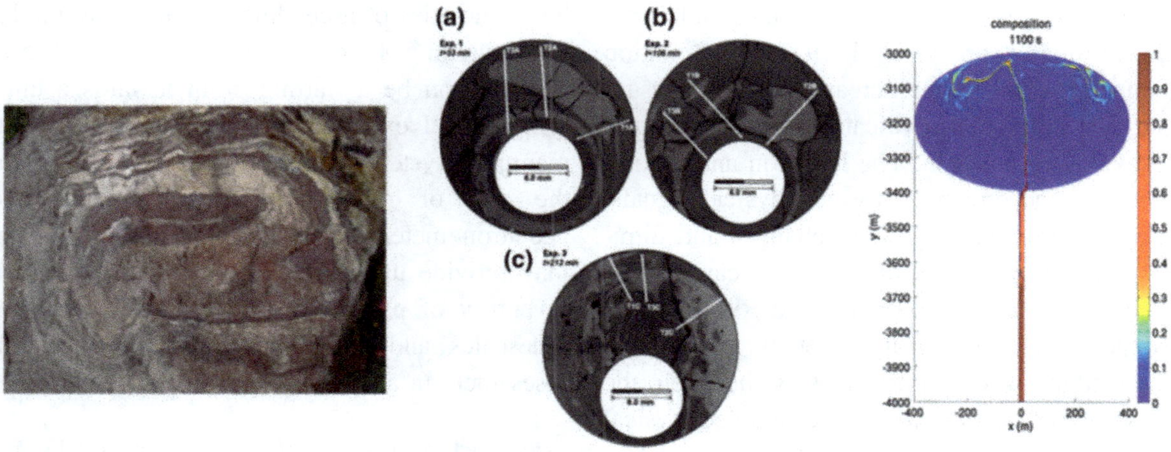

Fig. 5 Magma mixing in natural samples, experimental setups 684 and numerical modelling. Left: Lesvos (Greece) lava flow, from Perugini and Poli (2012); centre: experimental setup, from Morgavi et al. (2013b); right: simulations of magma chamber replenishment

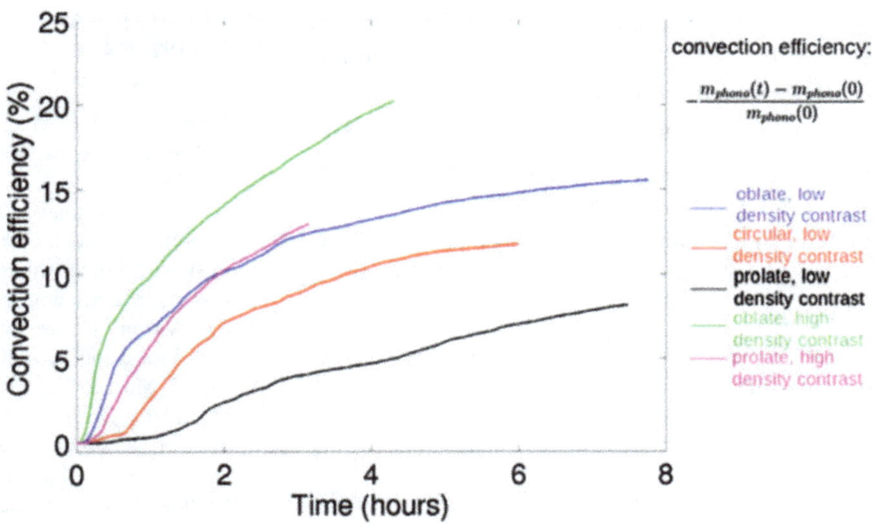

Fig. 6 Variation of convection efficiency with time in the shallow chamber for different simulated scenarios, characterized by varying geometry and total volatile content in the shallow chamber

that mingling is very effective on relatively short time scales, on the order of hours, in agreement with mixing timescales derived from the geochemical modelling of magma mixing experiments (Perugini et al. 2010). When buoyancy drives fast dynamics, more than 40% of the original end-member magmas have mingled in the feeder dyke within 4 h from the arrival of the gas-rich magma from depth. The asymptotic behaviour seems to be reached earlier for less efficient setups: after 4–5 h from the onset of the instability, the system seems to have reached a quasi-steady state. In these cases, a much smaller part of the two magmas has mingled.

6 Magma Mixing Time Scale and Eruption Trigger

The most violent and catastrophic volcanic eruptions on Earth have been triggered by the refilling of a felsic volcanic magma chamber by a hottest and more mafic magma (Kent et al. 2010; Murphy 1998). Examples include Vesuvius 79 AD (Cioni et al. 1995), Krakatau 1883 (Self 1992), Pinatubo 1991 (Kress 1997), the Campanian Ignimbrite (Arienzo et al. 2009) and Eyjafjallajökull 2010 (Sigmundsson et al. 2010). Injection of the more mafic magma into the felsic magma triggers convection dynamics and widespread mixing (Sparks et al. 1977). Vesiculation induced by convection increases magma pressure and may fracture the volcanic edifice triggering an explosive eruption. The injection and mixing process is accompanied by geophysical signals, such as earthquakes, gravity changes and ultra-long-period ground oscillations, that can now be accurately detected (Williams and Rymer 2002; Longo et al. 2012; Bagagli et al. 2017). The knowledge of the time elapsing between the beginning of mixing (and associated geophysical signals) and eruption is thus of greatest importance in forecasting the onset of a volcanic eruption.

Recent studies highlighted that in order to preserve magma mixing structures (i.e., filaments, swirls, bandings) in the rocks, the time elapsed between the beginning of mixing and the subsequent eruption must be very short; on the order of hours or days (Perugini et al. 2010). Preserved structures would indicate mixing was the last process recorded by the magmatic system and its study can unravel unprecedented information on pre-eruptive behaviour of volcanoes.

The compositional heterogeneity produced by magma mixing, and subsequently frozen in time in the volcanic products, can hence be viewed as a broken clock at a crime scene; it can potentially be used to determine the time of the incident. Following this idea and combining numerical simulations with magma mixing experiments using natural compositions and statistical analyses it was shown that for three volcanic eruptions from the Campi Flegrei volcanic system (Astroni, Averno and Agnano Monte Spina) the mixing-to-eruption timescale are of the order of a few minutes (Perugini et al. 2015). These timescales indicate that very little time elapsed from the moment mixing started until eruption. These results are in agreement with recent numerical simulations of magma mixing (Montagna et al. 2015) that highlight mixing timescales of a few hours to attain complete hybridization of magmas for the Campi Flegrei magmatic systems.

These results have implications for civil protection planning of future volcanic crisis as the high velocities of ascending magmas may imply little warning time in volcanic crises. These findings can be a starting point towards a unifying model explaining chemical exchanges in magmatic systems and supplying information on the use of chemical element mobility as geochronometers for volcanic eruptions. This may provide unparalleled clues for building an inventory of past and recent volcanic eruption timescales and could be decisive for hazard assessment in active volcanic areas.

Acknowledgements This research was funded by the European Union's Seventh Programme for research technological development and demonstration under grant agreement No 282759—VUELCO and by the ERC Consolidator Grant 612776—CHRONOS.

References

Anderson AT (1976) Magma mixing: petrological process and volcanological tool. J Volcanol Geotherm Res 1:3–33

Arienzo I, D'Antonio M, Di Renzo V, Tonarini S, Minolfi G, Orsi G, Carandente A, Belviso P, Civetta L (2015) Isotopic microanalysis sheds light on the magmatic endmembers feeding volcanic eruptions: the Astroni 6 case study (Campi Flegrei, Italy) J Volcanol Geoth Res 304:24–37

Arienzo I, Mazzeo FC, Moretti R, Cavallo A, D'Antonio M (2016) Open-system magma evolution and fluid transfer at Campi Flegrei caldera (Southern Italy) during the past 5 ka as revealed by geochemical and isotopic data: the archetype of Nisida eruption. Chem Geol 427:109–124

Bacon CR (1986) Magmatic inclusions in silicic and intermediate volcanic rocks. J Geophys Res 91:6091–6112

Bagagli M, Montagna CP, Papale P, Longo A (2017) Signature of magmatic processes in strainmeter records at Campi Flegrei (Italy). Geophys Res Lett 44(2):718–725

Baker DR (1990) Chemical Interdiffusion of Dacite and Rhyolite—Anhydrous Measurements at 1 Atm and 10 Kbar, Application of Transition-State Theory, and Diffusion in Zoned Magma Chambers. Contrib Mineral Petr 104:407–423

Bateman R (1995) The interplay between crystallization, replenishment and hybridisation in large felsic magma chambers. Earth Sci Rev 39:91–106

Blundy J, Sparks SRJ (1992) Petrogenesis of mafic enclaves in granitoids of the Adamello massif, Italy. J Petrol 33:1039–1104

Bowen N (1928) The evolution of the igneous rocks. Princeton University Press, Princeton, N.J

Braschi E, Francalanci L, Vougioukalakis GE (2012) Inverse differentiation pathway by multiple mafic magma refilling in the last magmatic activity of Nisyros Volcano, Greece. Bull Volcanol 74:1083–1100

Brown RJ, Civetta L, Arienzo I, D'Antonio M, Moretti R, Orsi G, Tomlinson EL, Albert PG, Menzies MA (2014) Geochemical and isotopic insights into the assembly, evolution and disruption of a magmatic plumbing system before and after a cataclysmic caldera Dcollapse eruption at Ischia volcano (Italy). Contrib Mineral Petrol 168, 1035. doi:https://doi.org/10.1007/s00410-014-1035-1

Bunsen W (1851) Uber die Processe der vulkanischen Gesteinsbildungen Islands. Ann Phys Chem (Dritte Reihe) 83:197–272

Chamberlain KJ, Morgan DJ, Wilson CJN (2014) Timescales of mixing and mobilisation in the Bishop Tuff magma body: perspectives from diffusion chronometry. Contrib Mineral Petr 167:1034. https://doi.org/10.1007/s00410-014-1034-2

Chiodini G, Caliro S, De Martino P, Avino R, Gherardi F (2012) Early signals of new volcanic unrest at Campi Flegrei caldera? Insights from geochemical data and physical simulations. Geology 40:943–946

Chiodini G, Todesco M, Caliro S, Del Gaudio C, Macedonio G, Russo M (2003) Magma degassing as a trigger of bradyseismic events; the case of Phlegrean Fields (Italy). Geophys Res Lett 30(8):1434. https://doi.org/10.1029/2002GL01679

Chiodini G, Vandemeulebrouck J, Caliro S, D'Auria L, De Martino P, Mangiacapra A, Petrillo Z (2015) Evidence of thermal-driven processes triggering the 2005–2014 unrest at Campi Flegrei caldera. Earth Planet Sci Lett 414:58–67

Cioni R, Civetta L, Marianelli P, Metrich N, Santacroce R, Sbrana A (1995) Compositional layering and syn-eruptive mixing of a periodically refilled shallow magma chamber: the AD 79 Plinian eruption of Vesuvius. J Petrol 36:739–776

Civetta L, Orsi G, Pappalardo L, Fisher RV, Heiken G, Ort M (1997) Geochemical zoning, mingling, eruptive dynamics and depositional processes—the Campanian Ignimbrite, Campi Flegrei Caldera, Italy. J Volcanol Geotherm Res 75:183–219

Corsaro RA, Di Renzo V, Di Stefano S, Miraglia L, Civetta L (2013) Relationship between magmatic processes in the plumbing system of Mt. Etna and the dynamics of the eastern flank: inferences from the petrologic study of the products erupted from 1995 to 2005. J Volcanol Geotherm Res 251:75–89

Costa F, Chakraborty S (2004) Decadal time gaps between mafic intrusion and silicic eruption obtained by chemical zoning patterns in olivine. Earth Planet Sci Lett 227:517–530

D'Antonio M, Civetta L, Orsi G, Pappalardo L, Piochi M, Carandente A, de Vita S, Di Vito MA, Isaia R (1999) The present state of the magmatic system of the Campi Flegrei caldera based on a reconstruction of its behavior in the past 12 ka. J Volcanol Geotherm Res 91:247–268

D'Antonio M, Tonarini S, Arienzo I, Civetta L, Dallai L, Moretti R, Orsi G, Andria M, Trecalli A (2013) Mantle and crustal processes in the magmatism of the Campania region: inferences from mineralogy, geochemistry, and Sr–Nd–O isotopes of young hybrid volcanics of the Ischia island (South Italy). Contrib Mineral Petr 165:1173–1194. https://doi.org/10.1007/s00410-013-0853-x

Davidson JP, McMillan NJ, Moorbath S, Worner G (1990) The Nevados de Payachata volcanic region (18S, 69 W, N. Chile) II. Evidence for widespread crustal involvement in Andean magmatism. Contrib Mineral Petr 105:412–432

Davidson JP, Tepley FJ III (1997) Recharge in volcanic systems; evidence from isotopic profiles of phenocrysts. Science 275:826–829

Davidson JP, Tepley FJ III, Knesel KM (1998) Crystal isotopic stratigraphy; a method for constraining magma differentiation pathways. Eos 79:185–189

De Campos CP, Dingwell DB, Fehr KT (2004) Decoupled convection cells from mixing experiments with alkaline melts from Phlegrean Fields. Chem Geol 213:227–251

De Campos CP, Dingwell DB, Perugini D, Civetta L, Fehr TK (2008) Heterogeneities in magma chambers: insight from the behaviour of major and minor elements during mixing experiments with natural alkaline melts. Chem Geol 256:131–145

De Campos CP, Perugini D, Ertel-Ingrisch W, Dingwell DB, Poli G (2011) Enhancement of Magma mixing efficiency by Chaotic Dynamics: an experimental study. Contrib Mineral Petr 161:863–881

Deino AL, Orsi G, de Vita S, Piochi M (2004) The age of the Neapolitan Yellow Tuff caldera-forming eruption (Campi Flegrei caldera—Italy) assessed by $^{40}Ar/^{39}Ar$ dating method. J Volcanol Geotherm Res 133:157–170

De Rosa R, Donato P, Ventura G (2002) Fractal analysis of min- gled/mixed magmas: an example from the Upper Pollara eruption (Salina Island, Southern Tyrrhenian Sea, Italy). Lithos 65:299–311

De Rosa R, Mazzuoli R, Ventura G (1996) Relationships between deformation and mixing processes in lava flows: a case study from Salina (Aeolian Islands, Tyrrhenian Sea). Bull Volcanol 58:286–297

de Vita S, Orsi G, Civetta L, Carandente A, D'Antonio M, Deino A, di Cesare T, Di Vito MA, Fisher RV, Isaia R, Marotta E, Necco A, Ort M, Pappalardo L, Piochi M, Southon J (1999) The Agnano-Monte Spina eruption (4100 years BP) in the restless Campi Flegrei caldera (Italy). J Volcanol Geotherm Res 91:269–301

Del Gaudio C, Aquino I, Ricciardi GP, Ricco C, Scandone R (2010) Unrest episodes at Campi Flegrei: a reconstruction of vertical ground movements during 1905–2009. J Volcanol Geotherm Res 195:48–56

Di Muro A, Metrich N, Vergani D, Rosi M, Armienti P, Fourgeroux T, Deloule E, Arienzo I, Civetta L (2014) The shallow plumbing system of Piton de la Fournaise volcano (La Reunion island, Indian Ocean) revealed by the major 2007 caldera forming eruption. J Petrol

55(7):1287–1315. https://doi.org/10.1093/petrology/egu025

Di Renzo V, Arienzo I, Civetta L, D'Antonio M, Tonarini S, Di Vito MA, Orsi G (2011) The magmatic feeding system of the Campi Flegrei caldera: architecture and temporal evolution. Chem Geol 281:227–241

Di Vito MA, Arienzo I, Braia G, Civetta L, D'Antonio M, Orsi G (2011) The Averno 2 fissure eruption: a recent small size explosive event at the Campi Flegrei caldera. Bull Volcanol 73:295–320. https://doi.org/10.1007/s00445-010-0417-0

Didier J, Barbarin B (1991) Enclaves and Granite Petrology, vol 13. Elsevier, Amsterdam

Eichelberger JC (1975) Origin of andesite and dacite: evidence of mixing at Glass Mountain in California and other circum-Pacific volcanoes. Geol Soc Amer Bull 86:1381–1391

Eichelberger JC (1978) Andesitic volcanism and crustal evolution. Nature 275:21–27

Eichelberger JC (1980) Vesiculation of mafic magma during replenishment of silicic magma reservoirs. Nature 288:446–450

Fedele L, Scarpati C, Lanphere M, Melluso L, Morra V, Perrotta A, Ricci G (2008) The Breccia Museo formation, Campi Flegrei, southern Italy: geochronology, chemostratigraphy and relationship with the Campanian Ignimbrite eruption. Bull Volcanol 70:1189–1219

Fedele L, Zanetti A, Morra V, Lustrino M, Melluso L, Vannucci R (2009) Clinopyroxene/liquid trace element partitioning in natural trachyte-trachyphonolite systems: insights from Campi Flegrei (southern Italy). Contrib Mineral Petr 158:337–356. https://doi.org/10.1007/s00410-009-0386-5

Flinders J, Clemens JD (1996) Non-linear dynamics, chaos, complexity and enclaves in granitoid magmas. Trans R Soc Edinburgh: Earth Sci 87:225–232

Font L, Davidson JP, Pearson DG, Nowell GM, Jerram DA, Ottley CJ (2008) Sr and Pb isotopic micro-analysis of plagioclase crystals from Skye lavas: an insight into open-system processes in a flood basalt province. J Petrol 49(8):1449–1471

Fourcade S, Allegre CJ (1981) Trace element behaviour in granite genesis: a case study the calc-alkaline plutonic association from the Querigut Complex (Pyrenees France). Contrib Mineral Petr 76:177–195

Francalanci L, Avanzinelli R, Nardini I, Tiepolo M, Davidson JP, Vannucci R (2012) Crystal recycling in the steady-state system of the active Stromboli volcano: a 2.5-ka story inferred from in situ Sr-isotopic and trace element data. Contrib Mineral Petr 163:109–131

Grasset O, Albarede F (1994) Hybridisation of mingling magmas with different densities. Earth Planet Sci Lett 121:327–332

Hibbard MJ (1981) The magma mixing origin of mantled feldspar. Contrib Mineral Petrol 76:158–170

Hibbard MJ (1995) Petrography to petrogenesis. MacMillan Publishing Company, London

Jolis EM, Freda C, Troll VR, Deegan FM, Blythe LS, McLeod CL, Davidson JP (2013) Experimental simulation of magma-carbonate interaction beneath Mt. Vesuvius, Italy. Contrib Mineral Petr 166:1335–1353

Kent AJR, Darr C, Koleszar AM, Salisbury MJ, Cooper KM (2010) Preferential eruption of andesitic magmas through recharge filtering. Nat Geosci 3:631–636

Kinman WS, Neal CR, Davidson JP, Font L (2009) The dynamics of Kerguelen Plateau magma evolution: new insights from major element, trace element and Sr isotopic microanalysis of plagioclase hosted in Elan Bank basalts. Chem Geol 264:247–265

Knesel KM, Davidson JP, Duffield WA (1999) Evolution of silicic magma through assimilation and subsequent recharge: evidence from Sr isotopics in sanidine phenocrysts Taylor Creek Rhyolites. J Petrol 40:773–786

Kress V (1997) Magma mixing as a source for Pinatubo sulphur. Nature 389:591–593

Lesher CE (1990) Decoupling of chemical and isotopic exchange during magma mixing. Nature 344:235–237

Longo A, Papale P, Vassalli M, Saccorotti G, Montagna CP, Cassioli A, Giudice S, Boschi E (2012) Magma convection and mixing dynamics as a source of Ultra-Long-Period oscillations. Bull Volcanol 74(4):873–880. doi:https://doi.org/10.1007/s00445-011-0570-0, https://doi.org/10.1007/s00445-011

Martin VM, Morgan DJ, Jerram DA, Caddick MJ, Prior DJ, Davidson JP (2008) Bang! Month-scale eruption triggering at Santorini Volcano. Science 321:1178

Melluso L, De' Gennaro R, Fedele L, Franciosi L, Morra V (2012) Evidence of crystallization in residual, Cl-F-rich, agpaitic, trachyphonolitic magmas and primitive Mg-rich basalt-trachyphonolite interaction in the lava domes of the Phlegrean Fields (Italy). Geol Mag 149(3):532–550

Montagna CP, Papale P, Longo A (2015) Timescales of mingling in shallow magmatic reservoir. In: Caricchi L, Blundy JD (eds) Chemical, Physical and Temporal Evolution of Volcanic Systems. Gelogical Society, London, Special Publications 422, pp 131–140

Morgavi D, Perugini D, De Campos CP, Ertl-Ingrisch W, Dingwell DB (2013a) Time evolution of chemical exchanges during mixing of rhyolitic and basaltic melts. Contrib Mineral Petr 166(2):615–638

Morgavi D, Perugini D, De Campos CP, Ertl-Ingrisch W, Dingwell DB (2013b) Morphochemistry of patterns produced by mixing of rhyolitic and basaltic melts. J Volcanol Geotherm Res 253:87–96

Morgavi D, Perugini D, De Campos CP, Ertl-Ingrisch W, Lavallee Y, Morgan L, Dingwell DB (2013c) Interactions Between Rhyolitic and Basaltic Melts Unraveled by Chaotic Magma Mixing Experiments. Chem Geol 346:119–212

Morgavi D, Petrelli M, Vetere FP, Gonzalez-Gartia D, Perugini D (2015) High temperature apparatus for chaotic mixing of natural silicate melts. Rev Sci Instrum 86(10):105108

Morgavi D, Arzilli F, Pritchard C, Perugini D, Mancini L, Larson P, Dingwell BD (2016) The Grizzly Lake complex (Yellowstone Volcano, USA): mixing between basalt and rhyolite unraveled by microanalysis and X-ray microtomography. Lithos 260:457–474

Murphy MD (1998) The role of Magma Mixing in triggering the current eruption at the Soufriere Hills Volcano, Montserrat, West Indies. Geophys Res Lett 25:3433–3436

Laumonier M, Scaillet B, Pichavant M, Champallier R, Andujar J, Arbaret L (2014) On the conditions of magma mixing and its bearing on andesite production in the crust. Nature communications, volume 5

Papale P, Moretti R, Barbato D (2006) The compositional dependence of the saturation surface of H_2O + CO_2 fluids in silicate melts. Chem Geol 229(1–3):78–95

Pappalardo L, Piochi M, D'Antonio M, Civetta L, Petrini R (2002) Evidence for multistage magmatic evolution during the past 60 kyr at Campi Flegrei (Italy) deduced from Sr, Nd and Pb isotopic data. J Petrol 43(8):1415–1434

Perugini D, De Campos CP, Dingwell DB, Petrelli M, Poli G (2008) Trace element mobility during magma mixing: preliminary experimental results. Chem Geol 256:146–157

Perugini D, de Campos CP, Petrelli M, Dingwell DB (2015) Concentration variance decay during magma mixing: a volcanic chronometer. Sci Rep 5:14225

Perugini D, Petrelli M, Poli G (2006) Diffusive fractionation of trace elements by chaotic mixing of magmas. Earth Planet Sci Lett 243:669–680

Perugini D, Petrelli M, Poli G, De Campos C, Dingwell DB (2010) Time-scales of recent Phlegrean fields eruptions inferred from the application of a 'Diffusive Fractionation' model of trace elements. Bull Volcanol 72:431–447

Perugini D, Poli G (2005) Viscous fingering during replenishment of felsic magma chambers by continuous inputs of mafic magmas: field evidence and fluid-mechanics experiments. Geology 33:5–8

Perugini D, Poli G (2012) The mixing of magmas in plutonic and volcanic environments: analogies and differences. Lithos. https://doi.org/10.1016/j.lithos.2012.02.002

Perugini D, Poli G, Gatta G (2002) Analysis and simulation of Magma mixing processes in 3D. Lithos 65:313–330

Perugini D, Poli G, Mazzuoli R (2003) Chaotic advection, fractals and diffusion during mixing of Magmas: evidence from Lava Flows. J Volcanol Geotherm Res 124:255–279

Perugini D, Valentini L, Poli G (2007) Insights into Magma chamber processes from the analysis of size distribution of Enclaves in Lava Flows: a case study from Vulcano Island (Southern Italy). J Volcanol Geotherm Res 166:193–203

Petrelli M, Perugini D, Poli G (2011) Transition to Chaos and implications for Time-scales of Magma Hybridization during mixing processes in Magma chambers. Lithos 125:211–220

Plail M, Barclay J, Humphreys MCS, Edmonds M, Herd RA, Christopher TE (2014) Characterization of mafic enclaves in the erupterd products of Soufriere Hills Volcano, Montserrat, 2009–2010. In Wadge G, Robertson REA, Voight B, (eds) The eruption of Soufriere Hills Volcano, Montserrat from 2000 to 2010. Geological Society, London, Memoirs, 39, 343–360

Poli G, Tommasini S, Halliday AN (1996) Trace elements and isotopic exchange during acid-basic magma interaction processes. Trans Royal Soc Edinburgh: Earth Sci 87:225–232

Pritchard CJ, Larson PB, Spell TL, Tarbert KD (2013a) Eruption-triggered mixing of extra-caldera basalt and rhyolite complexes along the East Gallatin-Washburn fault zone, Yellowstone National Park, WY, USA. Lithos 175–176:163–177. doi:https://doi.org/10.1016/j.lithos.2013.04.022

Self S (1992) Krakatau revisited: the course of events and interpretation of the 1883 eruption. Geoj Lib 28:109–121

Sigmundsson F, Hreinsdottir S, Hooper A, Arnadottir T, Pedersen R, Roberts JM, Oskarsson N, Auriac A, Decriem J, Einarsson P, Geirsson H, Hensch M, Ofeigsson BG, Sturkell E, Sveinbjornsson H, Feigl LK (2010) Intrusion triggering of the 2010 Eyjafjallajokull explosive eruption. Nature 468:426–430

Slaby E, Gotze J, Worner G, Simon K, Wrzalik R, Smigielski M (2010) K-feldspar phenocrysts in microgranular magmatic enclaves: a cathodoluminescence and geochemical study of crystal growth as a marker of magma mingling dynamics. Lithos 105:85–97

Smith JV (2000) Structures on interfaces of mingled magmas, Stewart Island, New Zealand. J Struct Geol 22:123–133

Sparks SRJ, Marshall LA (1986) Thermal and mechanical constraints on mixing between mafic and silicic magmas. J Volcanol Geotherm Res 29:99–124

SparksSRJ Sigurdsson H, Wilson L (1977) Magma mixing: a mechanism for triggering acid explosive eruptions. Nature 267:315–318

Tonarini S, D'Antonio M, Di Vito MA, Orsi G, Carandente A (2009) Geochemical and isotopical (B, Sr, Nd) evidence for mixing and mingling processes in the magmatic system feeding the Astroni volcano (4.1–3.8 ka) within the Campi Flegrei caldera (South Italy). Lithos 107:135–151

Tonarini S, Leeman WP, Civetta L, D'Antonio M, Ferrara G, Necco A (2004) B/Nb and 5^{11}B systematics in the Phlegrean Volcanic District (PVD). J Volcanol Geotherm Res 113:123–139

Ventura G (1998) Kinematic significance of mingling-rolling structures in lava flows: a case study from Porri volcano (Salina Southern Tyrrhenian Sea). Bull Volcanol 59:394–403

Wada K (1995) Fractal structure of heterogeneous ejecta from the Meakan volcano, eastern Hokkaido, Japan: implications for mixing mechanism in a volcanic conduit. J Volcanol Geotherm Res 66:69–79

Walker GPL, Skelhorn RR (1966) Some associations of acid and basic igneous rocks. Earth Sci Rev 2:9–109
Wallace G, Bergantz G (2002) Wavelet-based correlation (WBC) of crystal populations and magma mixing. Earth Planet Sci Lett 202:133–145
Wiebe RA (1994) Silicic magma chambers as traps for basaltic magmas: the Cadillac mountain intrusive complex, Mount Desert island, Maine. J Geol 102:423–427
Wilcox RE (1944) Rhyolite-basalt complex of Gardiner River, Yellowstone Park, Wyoming. Geol Soc Am Bull 55:047–1080
Wilcox RE (1999) The idea of Magma Mixing: history of a struggle for Acceptance. J Geol 107(4):421–432
Williams-Jones G, Rymer H (2002) Detecting volcanic eruption precursos: a new method using gravity and deformation measurements. J Volcanol Geotherm Res 113:379–389

Gases as Precursory Signals: Experimental Simulations, New Concepts and Models of Magma Degassing

M. Pichavant, N. Le Gall and B. Scaillet

Abstract

Volatile release during magma ascent in volcanic conduits (magma degassing) forms the basis for using volcanic gases as precursory signals. Recent high temperature high pressure experimental simulations have yielded results that challenge key assumptions related to magma degassing and are important for the interpretation of glass inclusion and gas data and for using volcanic gas as precursory signals. The experimental data show that, for ascent rates expected in natural systems, pure H_2O basaltic melts will evolve mostly close to equilibrium when decompressed from 200 to 25 MPa. In the same way, degassing of H_2O–S species evolves at near equilibrium, although this conclusion is limited by the number of S solubility data available for basaltic melts. However, degassing of CO_2 is anomalous in all studies, whether performed on basaltic or rhyolitic melts. CO_2 stays concentrated in the melt at levels far exceeding solubilities. The anomalous behaviour of CO_2, when associated with near equilibrium H_2O losses, yields post-decompression glasses with CO_2 concentrations systematically higher than equilibrium degassing curves. Therefore, there is strong experimental support for disequilibrium degassing during ascent of CO_2-bearing magmas. The existence of volatile concentration gradients around nucleated gas bubbles suggests that degassing is controlled by the respective mobilities (diffusivities) of volatiles within the melt. The recently formulated diffusive fractionation model reproduces the main characteristics, especially the volatile concentrations, of experimental glasses. The model also shows that the gas phase is more H_2O-rich than expected at equilibrium because CO_2 transfer toward the gas phase is hampered by its retention within the melt. However, only integrated gas compositions are calculated. Similarly, only bulk experimental fluid compositions are determined in recent experiments. Thus, constraints on the local gas phase are becoming necessary for the application to volcanoes. This stresses the need for the direct analysis of gas bubbles nucleated in decompression experiments. Pre-eruptive changes in volcanic CO_2/SO_2 and H_2O/CO_2 gas ratios are interpreted to reflect different pressures of gas-melt segregation in the conduit, an approach that assume gas-melt equilibrium. However, if disequilibrium magma degassing is accepted, the use of volatile saturation codes is no longer possible and caution must be exercised with the application of local equilibrium to volcanic gases. Future developments in the interpretation of gas data require progress

M. Pichavant (✉) · N. Le Gall · B. Scaillet
CNRS, Orléans, France
e-mail: michel.pichavant@cnrs-orleans.fr

Advs in Volcanology (2019) 139–154
DOI 10.1007/11157_2018_35

Published Online: 09 June 2018

from both sides, experimental and volcanological. One priority is to reduce the gap in scales between experiments and gas measurements.

Keywords
Magma ascent · Decompression experiments
Disequilibrium degassing · Volatile diffusion
Gas compositions

Extended Abstract

Volcanic gases are one of the main tools to monitor changes in the activity of volcanoes and forecast their eruption. Magma ascent toward the surface is associated with the exsolution of volatiles initially dissolved in the melt, a process designated as "magma degassing". Classically, the interpretation of volcanic gases relies on the assumption that degassing takes place at equilibrium. However, several observations (CO_2 contents of basaltic seafloor glasses, H_2O and CO_2 concentrations in glass inclusions, explosive basaltic volcanism) do not fit easily in such a model. Recently, decompression, ascent and degassing of magmas in volcanic conduits have been simulated by high temperature high pressure experiments. Results from these simulations stress the need to critically reconsider the whole mechanism of degassing in basaltic but also rhyolitic magmas. The new experimental data show that, at the decompression rates tested, pure H_2O basaltic melts will evolve mostly close to equilibrium when decompressed from 200 to 25 MPa. In the same way, degassing of H_2O–S species evolves at near equilibrium, although this conclusion is limited by the number of S solubility data available for basaltic melts. Degassing of CO_2 is anomalous in all studies, whether performed on basaltic or rhyolitic melts. CO_2 stays concentrated in the melt at levels far exceeding solubilities. The anomalous behaviour of CO_2, when associated with near equilibrium H_2O losses, yields post-decompression glasses with CO_2 concentrations systematically higher than equilibrium degassing curves. Therefore, there is strong experimental support for disequilibrium degassing during ascent of CO_2-bearing magmas. The existence of volatile concentration gradients around nucleated gas bubbles suggests that degassing is controlled by the respective mobilities (diffusivities) of volatiles within the melt. The contrasted diffusivities of dissolved volatile species (in particular H_2O and CO_2) selectively limit their transfer toward the gas phase for timescales typical for magma ascent. The diffusive fractionation model recently formulated reproduces the main characteristics, especially the volatile concentrations, of experimental glasses. It provides a framework to interpret the new experimental observations and the systematic deviations from equilibrium observed in CO_2-bearing systems, although coupling between volatile diffusion and vesiculation requires a more elaborate treatment. The model also shows that the gas phase is more H_2O-rich than expected at equilibrium because CO_2 transfer toward the gas phase is hampered by its retention within the melt. However, only integrated gas compositions are calculated. In the same way, only bulk experimental fluid compositions are determined in recent experiments. Since the gas phase is essential for the application to volcanoes, constraints on the local gas phase are becoming necessary. Compositions of gas bubbles in decompression experiments must be linked not only with pressure but also with volatile concentrations of local melts, specific degassing textures and mechanisms. As a way in this direction, local gas-melt equilibrium assumes that chemical equilibrium persists at the local scale, despite evidence for disequilibrium at larger scales. However, there are alternative ways to constrain the composition of nucleated gas bubbles, thus stressing the need for their direct analysis in decompression experiments. Pre-eruptive changes in CO_2/SO_2 and, in some cases, H_2O/CO_2 gas ratios observed at several basaltic volcanoes are generally interpreted to reflect different pressures of gas-melt segregation in the conduit, an approach that assume gas-melt equilibrium. However, if disequilibrium magma degassing is accepted, volatile saturation codes can no longer be directly used. Caution also must be exercised with the application of local gas-melt equilibrium to volcanic gases which are probably closer to integrated rather than to local compositions. Future developments in the interpretation of gas data require progress from both sides, experimental and volcanological. One priority is to reduce the gap in scales between experiments and gas measurements to refine interpretations of gas compositions as unrest signals.

1 Magma Degassing and Volcanic Gases as Precursory Signals

Volcanic gases are one of the main tools used to monitor changes in the activity of volcanoes and forecast their eruption. This approach is rooted in the strong pressure dependence of the solubility of volatiles (mainly H_2O, CO_2, SO_2, H_2S, Cl) in silicate melts. Accordingly, magma ascent toward the surface is associated with the exsolution of volatiles initially dissolved in the melt, a process designated as "magma degassing". The different volatiles have contrasted solubilities in silicate melts and, therefore, are expected to react differently to decompression. This forms the basis for using volcanic gas ratios to infer magma ascent and depth of gas segregation in volcanic conduits. For example, the sudden increase of gas CO_2/SO_2 ratio has been used as an indication for deep magma recharge at Stromboli (Aiuppa et al. 2010). At Soufriere Hills volcano (Montserrat), a correlation has been noted between gas HCl/SO_2 and the level of shallow activity as marked by the rate of lava extrusion and dome growth (Christopher et al. 2010; Edmonds et al. 2010).

Classically, the interpretation of volcanic gases relies on the assumption that degassing takes place at equilibrium. In the case of basaltic magmas, this assumption is supported by the high temperatures, low viscosities and high volatile diffusivities (Sparks et al. 1994). Vesiculation (i.e., the combined processes of bubble nucleation, growth and coalescence) is thought to be relatively easy in basaltic melts and degassing of basaltic magma is classically viewed as an equilibrium process. However, several observations do not fit easily in such a model. They include (1) the existence of basaltic seafloor glasses often supersaturated in CO_2 (e.g., Aubaud et al. 2004), (2) the occurrence of glass inclusions with H_2O and CO_2 concentrations inconsistent with closed system equilibrium degassing (e.g., Metrich et al. 2010) and (3) the occurrence of explosive basaltic volcanism (e.g., Head and Wilson 2003) which implies sudden rather than gradual release of volatiles.

Recently, decompression and ascent of basaltic magmas in volcanic conduits has been simulated by high temperature high pressure petrological experiments. These simulations stress the need to critically reconsider the whole mechanism of degassing in basaltic but also more silicic magmas. In particular, the assumption of equilibrium degassing is now becoming increasingly challenged. This has major implications for the interpretation of glass inclusion and gas data and, more generally, for the use of volcanic gas as precursory signals. In this Chapter, first, the recent experimental simulations are reviewed. We show that they all demonstrate an anomalous behaviour for CO_2 which tends to stay dissolved within the melt at concentrations too high for equilibrium. Second, the diffusive fractionation model which has been proposed to account for the new experimental observations is described and critically discussed. Finally, the implications of disequilibrium degassing for experimental fluid compositions and the interpretation of volcanic gas data as precursory signals are explored.

2 Experimental Simulations

2.1 Basaltic Systems

Following early work on systems with only pure CO_2 (Lensky et al. 2006), decompression experiments on hydrous basaltic melts have been carried out recently by Pichavant et al. (2013) at 1150–1180 °C, for initial pressures of 200–250 MPa, final pressures of 100, 50 and 25 MPa and for decompression rates between $\sim$1.5 down to 0.25 m/s. Melts from Stromboli, pre-synthesized to incorporate dissolved H_2O (2.7–3.8 wt%) and CO_2 (600–1300 ppm), were used as starting materials. The experiments were of continuous decompression type, and both constant (one ramp) and variable (two ramps) decompression rates were imposed. Final melt H_2O concentrations were homogeneous and always close to equilibrium solubility values. In contrast, the rate of vesiculation was found to control the final melt CO_2 concentration. High vesicularity charges had glass CO_2 concentrations that follow theoretical equilibrium degassing paths whereas glasses from low vesicularity charges showed marked deviations from equilibrium, with CO_2 concentrations up to one order of magnitude higher than equilibrium solubilities (Fig. 1a). The experimental results were interpreted in light of the slower diffusivity of CO_2 relative to H_2O in basaltic melts.

Yoshimura (2015) decompressed a natural evolved basaltic melt containing dissolved H_2O and CO_2 at 1200 °C and between 1000 and 500 MPa. The short decompression duration of 10 min over this pressure interval simulates a very fast ascent rate ($\sim$32 m/s for a rock density of 2650 kg/m^3). A vesiculated glass was produced and Fourier Transform Infrared Spectroscopy (FTIR) profiles revealed large CO_2 concentration gradients in the melt adjacent to gas bubbles. In contrast, the melt H_2O content was almost constant throughout the sample. The glass volatile concentration data cover a near vertical trend in the H_2O–CO_2 diagram (Fig. 1a).

Le Gall and Pichavant (2016a) extended the decompression experiments performed by Pichavant et al. (2013), using essentially the same procedures and materials. Three starting volatile compositions were investigated: series #1 (4.91 wt% H_2O, no CO_2), series #2 (2.41 $\pm$ 0.04 wt% H_2O, 973 $\pm$ 63 ppm CO_2) and series #3 (0.98 $\pm$ 0.16 wt% H_2O, 872 $\pm$ 45 ppm CO_2). The volatile-bearing glasses were synthesized at 1200 °C and 200 MPa, then continuously decompressed at a fast decompression rate of 3 m/s in the pressure range 150–25 MPa and then rapidly quenched. Post-decompression glasses were characterized texturally by X-ray microtomography. Volatile equilibrium was reached or approached during decompression in all series #1 melts with just water. In contrast, disequilibrium degassing occurred systematically in series #2 and #3 melts which retained elevated CO_2 concentrations (Fig. 1a). In similar experiments performed on the same three glass series but at a slower decompression rate of 1.5 m/s, Le Gall and Pichavant (2016b) found that series #1 (CO_2-free) melts followed equilibrium degassing until 100 MPa final pressure (P_{fin}). But at both 60 and 50 MPa P_{fin}, a slight H_2O-supersaturation was recognized, associated with a second bubble nucleation event that occurred at 25 MPa. In comparison, in series #2 and #3 (CO_2-bearing) melts, disequilibrium degassing was systematic, glasses retaining high non-equilibrium CO_2 concentrations (Fig. 1a).

The behavior of H_2O–, CO_2– and S-bearing basaltic melts during decompression was investigated by Le Gall et al. (2015a). Stromboli melts with 2.72 $\pm$ 0.02 wt% H_2O, 1291 $\pm$ 85 ppm CO_2 and 1535 $\pm$ 369 ppm S were synthesized at 1200 °C and 200 MPa and then decompressed to final pressures (P_{fin}) ranging from 150 to 25 MPa, followed by rapid quenching. The continuous decompressions were conducted at rates of 1.5 and 3 m/s. During decompression, S (and H_2O) were lost slightly more from the melt than expected from equilibrium degassing models, whilst significant CO_2 was retained at elevated concentrations in the melt (Fig. 1a). It was found that the degassing trend recorded by Stromboli glass inclusions could be closely reproduced by the experiments (Fig. 1b; Le Gall et al. 2015a). For andesitic melts, Fiege et al. (2014) observed that the fluid/melt partition coefficient for sulfur increases with the decompression rate. However, the influence of decompression rate on S degassing was marked only for oxidizing conditions, corresponding to sulfate as the only S species, thus making necessary to consider the different behaviour of S^{2-} and S^{6+} during degassing.

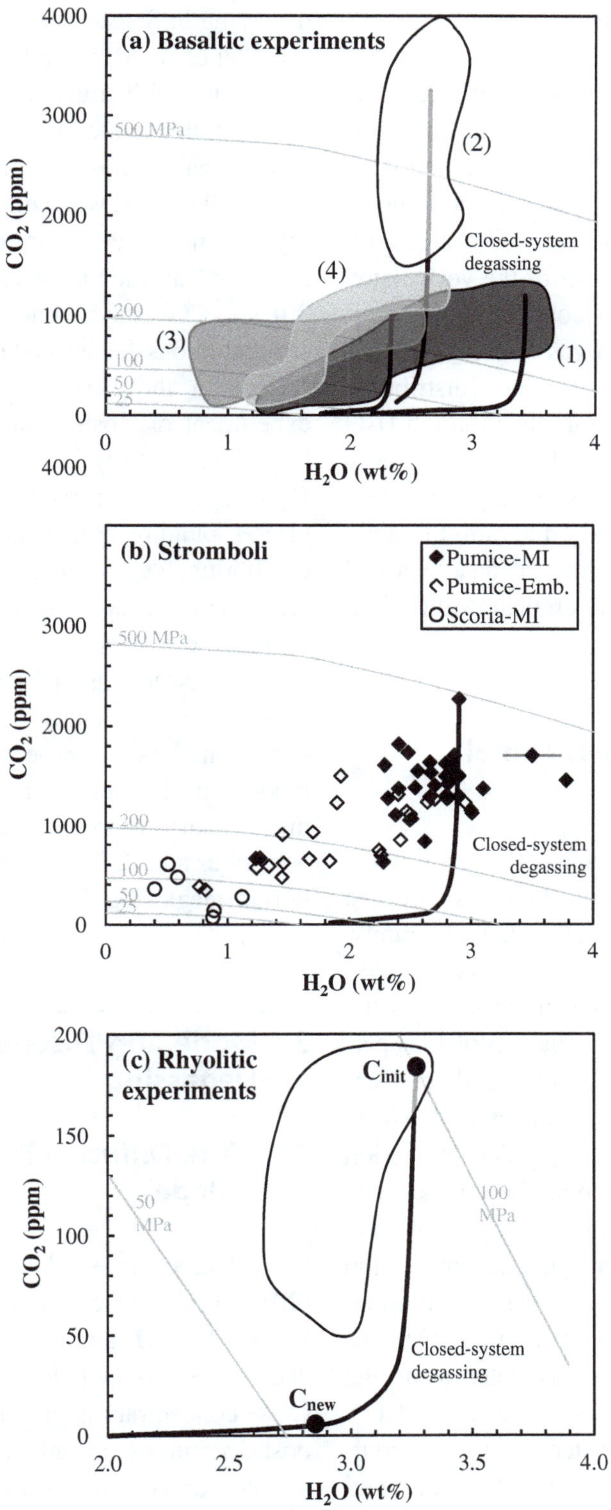

Fig. 1 H_2O–CO_2 glass concentration diagrams for **a** basaltic decompression experiments, **b** Stromboli glass inclusions and **c** rhyolitic decompression experiments. In **a**, **b** and **c**, light grey curves are isobars labelled with pressure in MPa. In **a**, fields for post-decompression glasses are distinguished with (1) referring to Pichavant et al. (2013), (2) to Yoshimura (2015), (3) to Le Gall and Pichavant (2016a, b) and (4) to Le Gall et al. (2015a). Black curves are closed-system equilibrium degassing trajectories redrawn from the original figures (Pichavant et al. 2013; Yoshimura 2015; Le Gall and Pichavant 2016a, b; Le Gall et al. 2015a). In **b**, the glass inclusion data are from Metrich et al. (2010). MI: glass inclusions, Emb: embayments. In **c**, the glass data field and the bold theoretical equilibrium closed-system degassing curve are redrawn from Yoshimura (2015). C_{init} and C_{new} are the composition of the pre-decompression melt and of the estimated post-decompression melt at the gas-melt interface, respectively (Yoshimura 2015; see also Figs. 5 and 6)

2.2 Rhyolitic Systems

Yoshimura (2015) decompressed a natural rhyolitic melt containing dissolved H_2O and CO_2 from 100 to 50 MPa at 800 °C. The duration of the decompression was 5000 s corresponding to a decompression rate of 0.38 m/s (for a density of 2650 kg/m^3). FTIR analysis of the vesiculated glass sample showed CO_2 concentration gradients in the melt away from gas bubbles. In contrast, H_2O was found to be distributed homogeneously within the sample although H_2O concentrations decreased significantly relative to the pre-decompression melt showing it had reequilibrated. On the H_2O–CO_2 diagram, the glass volatile concentrations define a near vertical array located left of the theoretical equilibrium degassing curve (Fig. 1c).

2.3 Summary of Experimental Evidence

In all experiments above, melt vesiculation is the result of decompression, in most cases single-step (constant decompression rate) and, more rarely, multi-step (variable decompression rates, Pichavant et al. 2013). Vesiculation leads to the generation of a gas (or fluid) phase. Volatiles partition between melt and gas, and volatile concentrations in post-decompression glasses evolve from those initially dissolved in pre-decompression glasses. The evaluation of equilibrium vs. disequilibrium degassing is performed by comparing volatile concentrations of post-decompression glasses with theoretical closed-system equilibrium degassing trajectories (Fig. 1). For pure H_2O melts, this equilibrium trajectory is calculated using the experimental solubility data of Lesne et al. (2011). For CO_2– and S-bearing systems, gas-melt equilibrium thermodynamic models (Newman and Lowenstern 2002; Papale et al. 2006; Burgisser et al. 2015) are used. Results show that pure H_2O basaltic systems evolve close to equilibrium when decompressed from 200 to 25 MPa with ascent rates of 1.5 and 3 m/s, although small levels of H_2O supersaturation are observed below 100 MPa (Le Gall and Pichavant 2016a, b). In the same way, degassing of S species evolves at near equilibrium (Le Gall et al. 2015a) although the reference equilibrium model (Burgisser et al. 2015) is somewhat uncertain due to the limited number of S solubility data for calibration. Degassing of CO_2 is anomalous in all studies, whether performed on basaltic or rhyolitic melts (Pichavant et al. 2013; Yoshimura 2015; Le Gall and Pichavant 2016a, b; Le Gall et al. 2015a). CO_2 stays concentrated in the melt at concentrations far exceeding solubilities (Fig. 1). Except in the very fast basalt decompression experiment of Yoshimura (2015), the anomalous behaviour of CO_2 is associated with significant H_2O losses which results in post-decompression glasses plotting systematically left of theoretical equilibrium degassing trajectories in Fig. 1. We conclude that recent experimental studies strongly support the possibility of disequilibrium degassing, i.e., that ascending melts can keep volatile concentrations (particularly CO_2) significantly different from those expected from equilibrium modelling. The questions thus arise of (1) the mechanisms responsible for this disequilibrium behaviour and (2) of the consequences of disequilibrium degassing for the composition of the gas phase.

3 Modelling Disequilibrium Degassing

3.1 The Diffusive Fractionation Model

Both Pichavant et al. (2013) and Yoshimura (2015) observed decoupling between the behaviour of H_2O and CO_2 during experimental decompression and degassing. In both studies, CO_2 concentration gradients were found in post-decompression glasses, either around gas bubbles or near the gas-melt interface. In contrast, no such diffusion profiles were identified for H_2O, despite concentrations being lower (in most cases) than in pre-decompression glasses (Pichavant et al. 2013; Yoshimura 2015; Le Gall et al. 2015a; Le Gall and Pichavant 2016a, b).

Pichavant et al. (2013) suggested that two characteristic distances, the gas interface distance (either the distance between two bubbles in the melt or the distance to the gas-melt interface) and the volatile diffusion distance (a function of respective diffusivities of volatiles in the melt) control the degassing process. Yoshimura (2015) quantitatively formulated a diffusive fractionation model to describe the ascent and degassing of volatile-bearing magmas. The reader is referred to this work for details about the calculations. The model is based on a diffusivity of CO_2 being one log unit lower than for H_2O (e.g., Zhang and Ni 2010). Decompression trajectories computed from the model are shown on Fig. 2 for different ascent rates, from 0.1, 1, 10, 100 to ∞ m/s. Although very high ascent rates (e.g., Peslier et al. 2015) are necessary for degassing trajectories to shift significantly left to the equilibrium reference curve, the modelling results qualitatively reproduce the main characteristics of experimental post-decompression glasses, i.e., the elevated CO_2 glass concentrations, the significant H_2O losses and the melt concentration trends in H_2O–CO_2 diagrams (Fig. 1).

Yoshimura (2015) emphasized the relative simplicity of his model. For example, bubble growth was not considered as in other more elaborated theoretical treatments (e.g., Gonnermann and Manga 2005). Rather than continuously varying boundary (gas-melt) interface volatile concentrations and bubble-bubble distances as in a natural ascending magma, the calculations were performed step-by-step (i.e., at different pressures) along the decompression ramp, with fixed boundary concentrations and bubble-bubble distance (Yoshimura 2015). It is also important to note that the volatile concentrations on Fig. 2 correspond to averages computed by integrating the concentrations in the melt along the diffusion profiles (*distance integrated compositions*).

Gas phase compositions were calculated by mass balance using the initial volatile concentrations and the average volatile concentrations left in the melt after decompression and degassing (Yoshimura 2015). Results are shown on Fig. 3 and they correspond to compositions

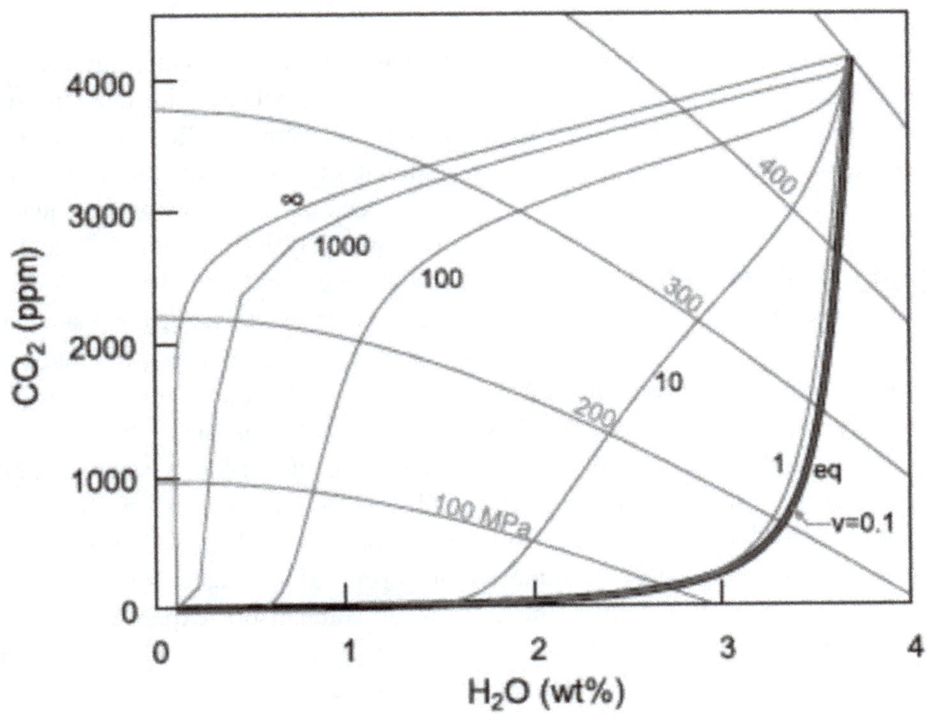

Fig. 2 H_2O and CO_2 melt volatile concentrations computed with the diffusive fractionation model for different decompression/ascent rates (v from 0.1 to ∞, in m/s). Isobars (light curves) are labelled with pressure in MPa. The heavy curve labelled "eq" is the equilibrium closed-system degassing trajectory as calculated by Yoshimura (2015). Figure redrawn from Yoshimura (2015). See text for details

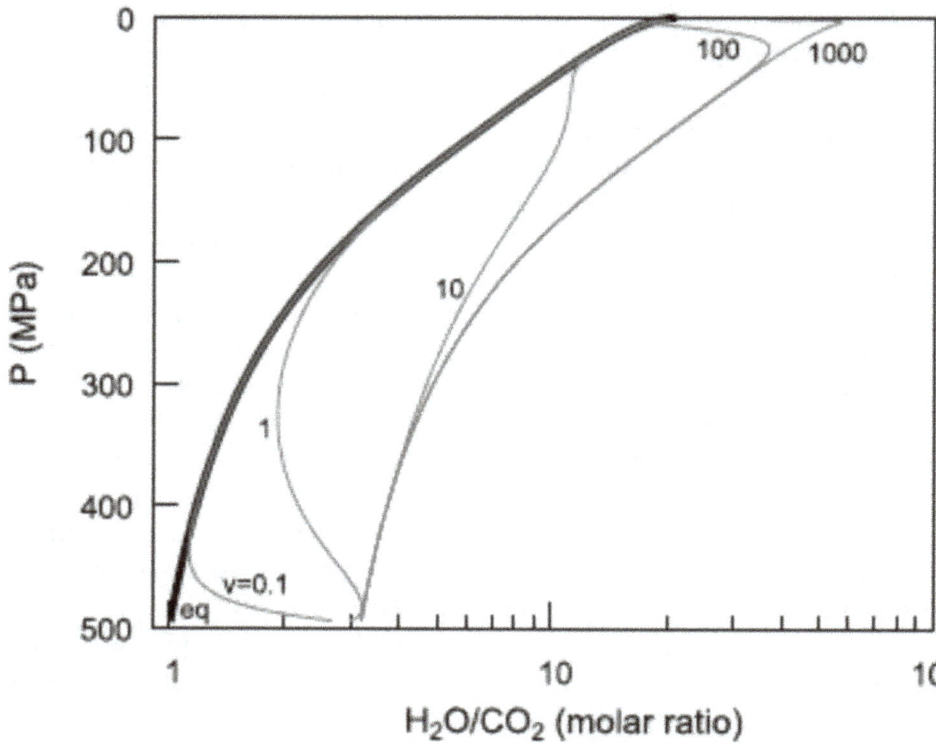

Fig. 3 Variations of the gas H_2O/CO_2 molar ratio (integrated compositions) with pressure computed from the diffusive fractionation model using different decompression/ascent rates (v from 0.1 to 1000, in m/s). The heavy curve labelled "eq" is the equilibrium closed-system degassing trajectory. Figure redrawn from Yoshimura 2015. See text for details

integrated along decompression (*pressure integrated compositions*). These compositions are more H_2O-rich (higher H_2O/CO_2 ratios) than gases generated under equilibrium degassing.

3.2 Coupling Between Diffusion and Vesiculation

Coupling between volatile diffusion and vesiculation is a necessity in diffusive degassing models because vesiculation defines the density of bubbles, their sizes and the distances between them (e.g., Pichavant et al. 2013; Le Gall et al. 2016a, b). This issue was addressed by Yoshimura (2015), although in a relatively simplified manner. The distance between bubbles was defined as being a function of only two variables, the distance between bubbles at the bottom of the decompression column (arbitrary value) and the vesicularity. Vesicularity must change along with decompression and degassing. So, the vesicularity term should embody the textural variations associated with magma ascent. In the model of Yoshimura (2015), the vesicularity was computed from the amount of volatiles exsolved upon decompression, using an equation of state for H_2O–CO_2 gas mixtures to calculate the density of the gas and assuming a constant density for the melt. In so doing, it is apparent that only a vesicularity corresponding to equilibrium degassing is considered. Thus, for a given initial bubble-bubble distance, the distance between bubbles in the decompression column depends only on the equilibrium vesicularity. Degassing trajectories (Fig. 2) and integrated gas compositions (Fig. 3) were calculated on this basis.

For comparison, experimental vesicularities, bubble diameters and bubble number densities are shown on Fig. 4 for three series of basaltic melts decompressed from 200 to 25 MPa final pressure (P_{fin}) at 3 m/s (Le Gall and Pichavant 2016a). Systematic variations within the three glass series are observed depending on P_{fin}. In most cases, the vesicularity data plot intermediate between the two equilibrium vesicularity curves, which were computed in a similar way than Yoshimura (2015) but only for two end-member cases corresponding to pure H_2O and pure CO_2 gas. The vesicularity data for the series #1 melts (with pure H_2O) are in general much lower than the theoretical vesicularities calculated for pure H_2O gas. The data also show large changes in bubble sizes and bubble number densities that do not directly correlate with vesicularity. Le Gall and Pichavant (2016a) emphasized that

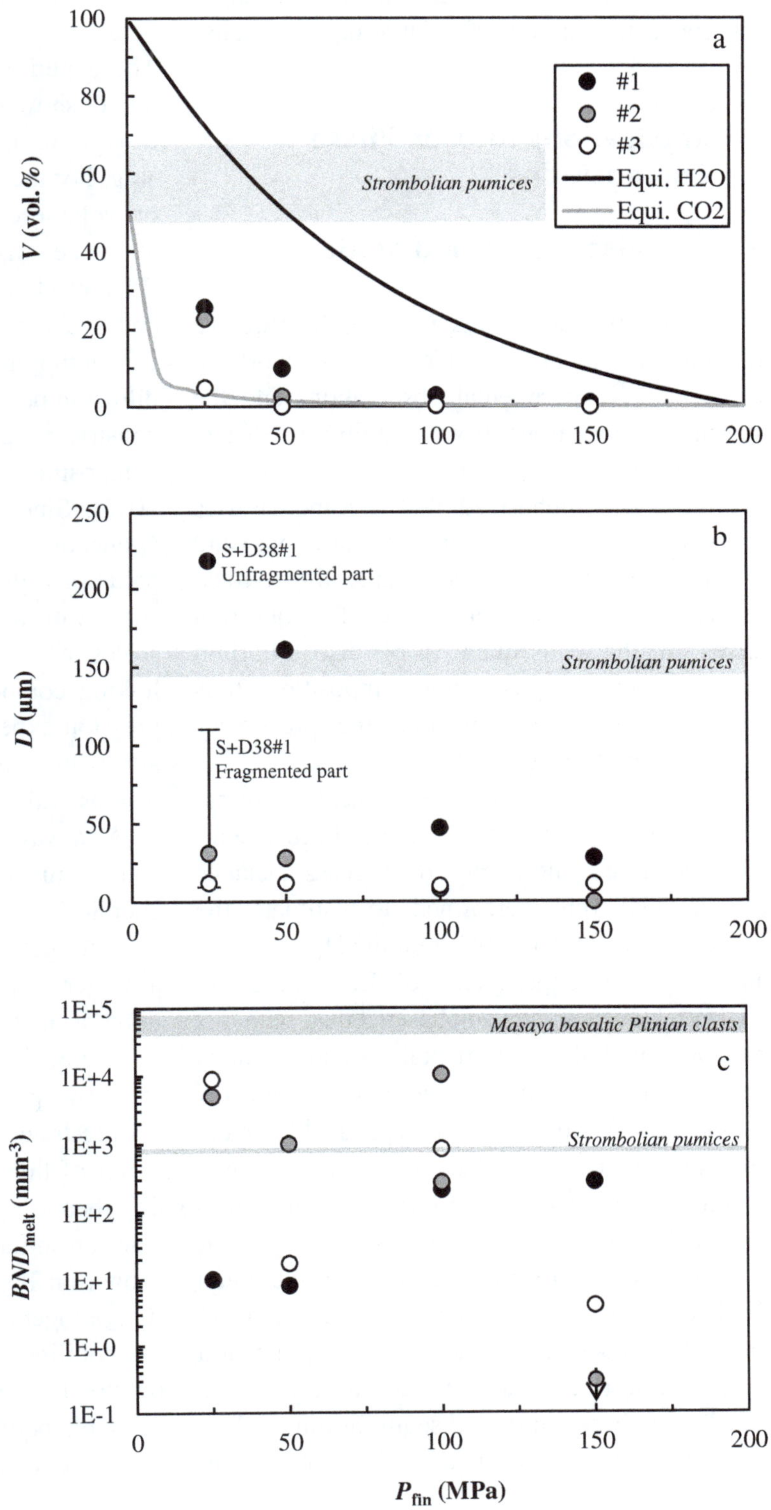

Fig. 4 Textural data for post-decompression experimental glasses plotted as a function of final pressure (P_{fin}) and comparison with data for natural basaltic pumices (Stromboli, Masaya). Vesicularities (**a**), bubble diameters (**b**) and bubble number densities (**c**) for three series of basaltic melts decompressed from 200 to 150, 100, 50 and to 25 MPa P_{fin} at 3 m/s (Le Gall and Pichavant 2016a). Pre-decompression melt concentrations, series #1: 4.91 wt% H_2O (no CO_2), series #2: 2.41 ± 0.04 wt% H_2O, 973 ± 63 ppm CO_2 and series #3: 0.98 ± 0.16 wt% H_2O, 872 ± 45 ppm CO_2. Note that, for charge S + S38#1 which was partially fragmented, the vesicularity (**a**) and BND (**c**) data concern the unfragmented part. The bubble diameter data (**b**) are for both the unfragmented (black symbol) and the fragmented (minimum and maximum values) parts. Figure redrawn from Le Gall and Pichavant (2016a)

degassing textures result from several processes including bubble nucleation, growth, coalescence, plus buoyancy-driven bubble migration. We conclude that, although the diffusive fractionation model of Yoshimura (2015) provides a basis for coupling volatile diffusion calculations and vesiculation processes, more work is needed to incorporate the complex textural changes associated with ascent of volatile-bearing melts.

4 Implications for Gas Phase Compositions

4.1 Available Data and Models

Despite the limitations noted above, the diffusive fractionation model provides a framework to interpret the experimental observations and the systematic deviations from equilibrium degassing observed in CO_2-bearing systems. However, it should be emphasized that the model uses analytical data (glass volatile concentrations) and physicochemical properties (volatile diffusivities) related only to the melt phase. The question arises of the consequences of disequilibrium degassing for the gas phase composition. It is worth remembering here that the precursory signals come from gas data.

In the decompression experiments summarized above, the gas phase has not been chemically analysed although some mass balance calculations were performed to estimate the composition of the gas phase in the H_2O–, CO_2– and S-bearing experiments of Le Gall et al. (2015a). However, it is emphasized that, with this method, only bulk experimental gas compositions are provided (*charge and pressure integrated compositions*). No information is available on the composition of individual bubbles generated during decompression. The gas calculations performed by Yoshimura (2015) also use a similar mass balance approach, i.e., pressure integrated fluid compositions are given. However, the local gas at the gas-melt interface has an equilibrium composition (local gas-melt equilibrium). The differences between the disequilibrium (calculated with the model) and the equilibrium (calculated assuming equilibrium degassing) gases (Fig. 3) is the consequence of CO_2 degassing being hampered by its retention within the melt. Therefore, disequilibrium is evidenced in the compositions of the pressure integrated fluids.

4.2 Composition of Gas Bubbles

The experiments and the diffusive fractionation model show that melt and gas both evolve under disequilibrium during magma ascent and degassing. For the melt, this conclusion is based either on volatile concentration measurements in glass at some distance of the gas/melt interface (Pichavant et al. 2013; Yoshimura 2015; Le Gall and Pichavant 2015a, 2016a, b) or on average concentrations calculated by integration along diffusion profiles (Yoshimura 2015). For the gas, constraints are available only on integrated compositions (Le Gall et al. 2015a; Yoshimura 2015). Since the gas phase is essential for the application to volcanoes, and given the interpretations proposed for the melt phase, constraints on the gas phase composition at smaller scales are becoming necessary. This requires linking compositions of gas bubbles in decompression experiments not only with pressure but also with volatile concentrations of local melts as well as with degassing textures and mechanisms.

As a way toward this direction, local gas-melt equilibrium can be assumed. This implies that chemical equilibrium persists locally between gas and melt, despite evidence for disequilibrium at larger scales. Therefore, the volatile compositions of melt and gas at the interface are defined by equilibrium partitioning of volatiles between these two phases (e.g., Dixon and Stolper 1995). To illustrate this concept, a schematic representation of the gas-melt interface for a H_2O– and CO_2-bearing melt decompressed isothermally from an initial (P_{init}) to a final (P_{fin}) pressure is shown on Fig. 5a. Initial volatile concentrations (C_{init}), together with the P_{init} and P_{fin} isobars and the equilibrium degassing trajectory are shown on the H_2O–CO_2 diagram of Fig. 5b. If local gas-melt equilibrium is assumed, the interface melt H_2O and CO_2 concentrations at P_{fin} (C_{new})

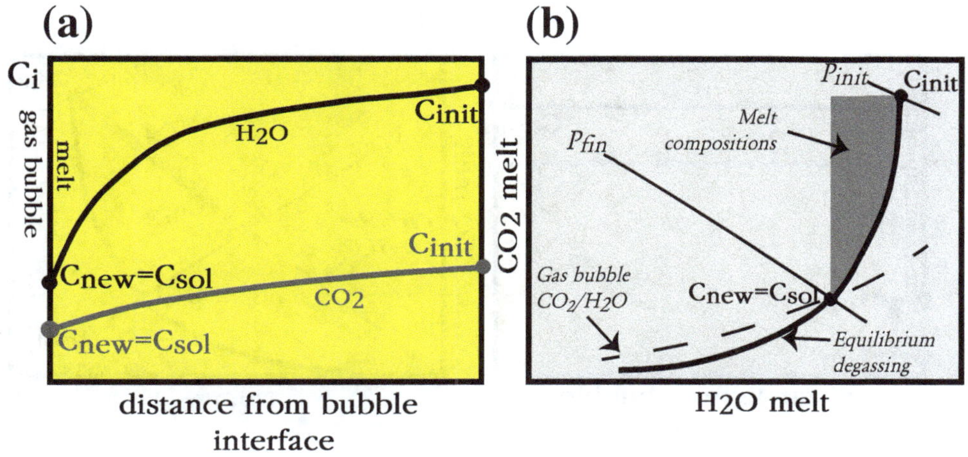

Fig. 5 Schematic illustration of local gas-melt equilibrium. **a** detail of the gas-melt interface region in a concentration (C_i) versus distance diagram where C_i refers to the volatile concentration in the melt. The gas bubble is on the left. The two curves are melt volatile concentration profiles for H_2O (black) and CO_2 (grey) respectively, generated as a result of diffusion in the melt during decompression from P_{init} to P_{fin}. C_{init} give volatile concentrations of the pre-decompression melt, C_{new} gas/melt interface volatile concentrations at P_{fin} and C_{sol} volatile solubilities at P_{fin}. Black lettering is used for H_2O and grey for CO_2. **b** H_2O–CO_2 diagram illustrating the evolution during decompression and degassing. The black bold curve is the equilibrium degassing trajectory. The two black lines are isobars labelled with initial (P_{init}) and final (P_{fin}) pressures along the decompression path. The dashed curve is the CO_2/H_2O isopleth passing through C_{sol} and it defines the composition of the gas bubble in local equilibrium with the interface melt. The shaded domain gives the range of possible melt compositions generated upon decompression from P_{init} down to P_{fin}

are equal to their solubilities (C_{sol}) at $P = P_{fin}$ (intersection of the equilibrium degassing curve with the P_{fin} isobar, Fig. 5b). Note that diffusive fractionation generates H_2O and CO_2 concentration gradients within the melt (Fig. 5a), the range of possible melt compositions during decompression being represented by the dark grey domain in Fig. 5b. The interface melt is the only melt at equilibrium with the local gas at P_{fin} which has a CO_2/H_2O corresponding to the fluid isopleth on Fig. 5b (e.g., Dixon and Stolper 1995). For a low pressure (e.g., 25 MPa), the local gas (e.g., a gas bubble nucleated at P_{fin}) is relatively H_2O-rich. In comparison, the pressure integrated gas assuming bulk *equilibrium* degassing from P_{init} to P_{fin} would be necessarily less H_2O-rich since most of the CO_2 must have been outgassed from the melt. This gas is less H_2O-rich than the pressure integrated gas produced by *disequilibrium* degassing from P_{init} to P_{fin} (Fig. 3). Thus, individual bubbles nucleated at P_{fin} can have CO_2/H_2O different from the composition of integrated gases generated continuously during decompression.

An alternative way to constrain the composition of gas bubbles is illustrated on Fig. 6. It starts from the observation that bubble nucleation is, from a kinetic point of view, an instantaneous process (e.g., Mourtada-Bonnefoi and Laporte 2002, 2004). Nucleation of a gas bubble draws volatiles from the local melt and the possibility that the initial CO_2/H_2O of the gas bubble is the same as the local melt should be considered. According to this hypothesis, represented schematically on Fig. 6, the local melt next to the nucleated bubble (C_{new}) is volatile-depleted but its CO_2/H_2O (r, Fig. 6a) is the same than the initial melt (C_{init}). Melt and gas bubble compositions are thus both located on a mixing line between C_{init} and C_{new} which passes through the origin of the H_2O–CO_2 diagram (Fig. 6b). The net result is the nucleation of individual gas bubbles more CO_2-rich than expected from local gas-melt equilibrium (Fig. 5b).

The previous discussion emphasizes the compositional variability of nucleated gas bubbles and the need for their direct analysis in decompression experiments. Comparison between the

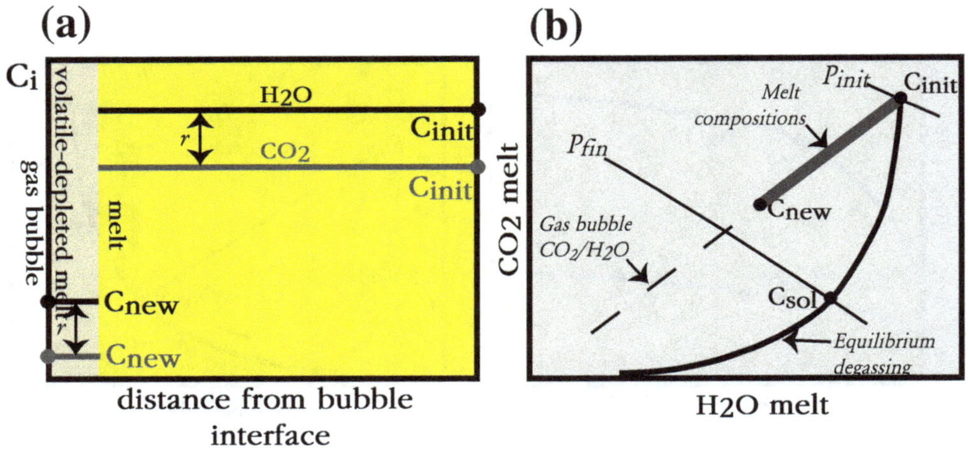

Fig. 6 Hypothetical model for the composition of a gas bubble nucleated during decompression of H_2O–, CO_2-bearing basaltic melts. **a** detail of the gas-melt interface region in a concentration (C_i) versus distance diagram where C_i refers to the volatile concentration in the melt. The gas bubble is on the left. The two horizontal lines are melt volatile concentrations for H_2O (black) and CO_2 (grey) drawn as straight lines because bubble nucleation is an instantaneous event. The narrow zone near the interface is the melt region depleted in volatiles drawn to form the bubble. C_{new} are volatile concentrations in the depleted melt region. Notice that the H_2O/CO_2 ratio (r) is identical in both the depleted and non-depleted melt regions because during nucleation volatiles are drawn from the local melt and the initial CO_2/H_2O of the gas bubble is the same as the local melt. C_{init} give volatile concentrations of the pre-decompression melt, C_{new} gas/melt interface volatile concentrations left after bubble nucleation at P_{fin}. Black lettering is used for H_2O and grey for CO_2. **b** H_2O–CO_2 diagram illustrating the evolution during decompression and degassing. The black bold curve gives the schematic location of the theoretical equilibrium degassing trajectory. The two black lines are isobars labelled with initial (P_{init}) and final (P_{fin}) pressures along the decompression path. Melts produced as a result of decompression and bubble nucleation plot on the straight line joining C_{init} and C_{new}. This line passes through the origin of the diagram because both C_{init} and C_{new} have the same CO_2/H_2O ratio. Note that the location of C_{new} along this line is arbitrary. The CO_2/H_2O ratio of the gas bubble (dashed line) is also the same as C_{init} and C_{new}. It is higher than the gas bubble controlled by local gas-melt equilibrium (Fig. 5)

CO_2/H_2O of nucleated bubbles and results of gas-melt volatile partitioning models (e.g., Dixon and Stolper 1995; Papale et al. 2006) would provide a crucial test of the local gas-melt equilibrium model (Fig. 5). If the nucleated bubbles prove to be CO_2-rich, then alternative models of control of gas composition would be supported (Fig. 6). One important aspect is that, in CO_2-bearing systems, bubble nucleation during decompression is continuous, occurring over a large pressure range (Le Gall and Pichavant 2016a, b). This is because CO_2-bearing melts are volatile-supersaturated (Fig. 1) and, so, the driving force for nucleation of new bubbles is always present. Thus, decompression of CO_2-bearing melts continuously leads to the nucleation of new bubbles, which increases the relevance of the hypothetical mechanism illustrated in Fig. 6.

5 Discussion and Perspectives for Gas Monitoring

5.1 Degassing Processes

Experimental simulations show that, for ascent rates expected in natural systems, equilibrium degassing occurs in pure H_2O melts. In contrast, results for CO_2-bearing melts conclusively demonstrate that degassing generates melt volatile concentrations out of equilibrium. The experimental database supporting this conclusion has recently expanded. It now includes basaltic and rhyolitic melts, and S-bearing as well as S-free systems. Several of those experimental decompression studies have been scaled to natural systems so that results are realistic and applicable. The decompression experiments on Stromboli basalt cover ascent rates ranging from 0.25 to 3 m/s (Pichavant et al. 2013; Le Gall and Pichavant 2016a, b; Le Gall et al. 2015a), well in the range of current estimates for basaltic magmas (e.g., Rutherford 2008; Peslier et al. 2015). An ascent rate approximately 10 times faster was used by Yoshimura (2015). Therefore, in the case of basaltic magmas, equilibrium degassing should be viewed more as a reference situation rather than as a general mechanism. This is quite a change in paradigm which has major implications for how gas signals are interpreted.

One remaining issue concerns the role of crystals on bubble nucleation, heterogeneous rather than homogeneous. All decompression studies considered in this paper were performed on very crystal-poor, if not totally crystal-free, melts and bubble nucleation appears to be mostly homogeneous. Crystals present in experimental basaltic products include Fe–Ti oxides (Le Gall and Pichavant 2015b) and rare Fe sulphides (Le Gall et al. 2015a). Le Gall and Pichavant (2015b) have documented heterogeneous nucleation of bubbles on Fe–Ti oxide crystals (and also on Fe sulphides, Le Gall et al. 2015a). Recently, Shea (2017) has stressed the importance of magnetite as a key mineral phase promoting heterogeneous bubble nucleation in natural magmas. However, Fe oxide phenocrysts and sulphides are uncommonly present in amounts exceeding a few vol.% in natural magmas. This led Le Gall and Pichavant (2015b) to conclude that heterogeneous bubble nucleation is not an important mechanism in basaltic melts if driven by Fe oxides. Yet, heterogeneous nucleation on silicate phases is still an open question. For example, olivine, clinopyroxene and plagioclase are typical phenocrysts and microlites in Stromboli basalts (e.g., Pichavant et al. 2011). On the basis of limited textural evidence, Pichavant et al. (2013) ruled out the possibility of heterogeneous nucleation of gas bubbles on clinopyroxene and olivine crystals. However, additional investigations seem warranted to guarantee full applicability of the decompression experiments above.

Disequilibrium degassing, as documented in the experiments, is the consequence of the anomalous behaviour of CO_2. CO_2-supersaturated melts are systematically generated during decompression. The interpretation suggested by Pichavant et al. (2013) and quantitatively formulated by Yoshimura (2015) is that, because of its restricted diffusive mobility within the melt, CO_2 has limited access to the gas phase for timescales typical of magma ascent. However, our knowledge of volatile diffusivities in silicate melts is still very fragmentary. There are very few diffusivity data for H_2O and CO_2 on the same melt. S is another volatile which reputedly has a slow diffusivity in silicate melts. Yet, the behaviour of S during degassing differs from that of CO_2, although we are still short of S solubility data for basaltic melts (e.g., Lesne et al. 2015). Acquisition of fundamental data (especially volatile diffusivity and solubility data for basaltic melts) is needed for the elaboration of more detailed interpretations of the decompression experiments. Future magma ascent models should also incorporate the textural complexities associated with the vesiculation process.

5.2 Gases as Unrest Signals

Pre-eruptive changes in gas ratios have been observed at several basaltic volcanoes such as Stromboli (Burton et al. 2007; Aiuppa et al. 2010), Etna (Aiuppa et al. 2007) and Villarrica (Aiuppa et al. 2017) among others. Transition from passive degassing to more explosive paroxysmal eruption regimes is marked by temporal increases of the CO_2/SO_2 gas ratio in the volcanic plume. In some cases, the CO_2/SO_2 variations are correlated with a decrease of the H_2O/CO_2 gas ratio (e.g., Aiuppa et al. 2017). These variations in volcanic gas ratios have been generally interpreted to reflect different pressures of gas-melt segregation in the conduit, high CO_2/SO_2 (and low H_2O/CO_2) indicating deep conditions and low CO_2/SO_2 (and high H_2O/CO_2) shallow conditions (e.g., Edmonds 2008; Burton et al. 2007; Allard 2010; Aiuppa et al. 2017). In this approach, the pressure-dependent evolution of the gas phase exsolved upon magma ascent and decompression is calculated by using volatile saturation codes (Newman and Lowenstern 2002; Moretti and Papale 2004; Papale et al. 2006; Burgisser et al. 2015). This implicitly assumes chemical equilibrium between gas and melt, an assumption which, as shown above, is now largely questioned. If disequilibrium magma degassing is accepted, then the consequences for the interpretation of gas signals need to be examined.

Firstly, one might argue that gas-melt equilibrium can persist at local scale, despite disequilibrium at larger scales. Thus, volatile saturation codes could still be used and applied to local gas and melt compositions, for example to model the composition of unconnected bubbles nucleated within the melt. In contrast, volcanic gases necessarily require, to be sampled, that the magma is permeable and, so, that the gas phase is connected. It is quite possible that the gases sampled are mixtures of different components, either integrated from several discrete degassing events along ascent or issued from different parts of the plumbing system. Therefore, volcanic gases are probably more representative of integrated compositions as discussed above than to compositions of local gases. We have shown previously that individual bubbles with compositions defined by local gas-melt equilibrium at a given pressure (Fig. 5) can have CO_2/H_2O different from integrated gases generated continuously during decompression (Fig. 3). We conclude to the limited applicability of local gas melt equilibrium to interpret volcanic gas ratios.

Secondly, disequilibrium gas-melt degassing due to CO_2 retention within the melt implies that CO_2/SO_2 and H_2O/CO_2 gas ratios can no longer be directly related to pressures of gas-melt segregation. Calculations using the diffusive fractionation model (Fig. 3) show that the pressure integrated gases have a higher H_2O/CO_2 (and also presumably a lower CO_2/SO_2 because CO_2 is retained within the melt) than the same gases calculated assuming equilibrium with the melt (Fig. 3). This demonstrates the possibility of changing the gas ratios depending on the degassing mechanism (equilibrium vs. disequilibrium). It is worth emphasizing that disequilibrium degassing associated with CO_2 retention produces integrated fluids that are less, not more, CO_2-rich (Fig. 3).

The CO_2-rich gases observed on basaltic volcanoes have been generally attributed to deep-seated processes such as fluxing of CO_2 or arrival of CO_2-rich magmas (e.g., Aiuppa et al. 2010, 2017; Allard 2010). In contrast, the degassing mechanism of Fig. 6 (although it needs validation from direct analysis of gas bubbles in decompression experiments) allows CO_2-rich gas bubbles to be generated at low pressures. It also provides an example of how gas ratios can be changed at constant pressure depending on the degassing mechanism. The initially CO_2-rich bubbles (Fig. 6) will probably shift rapidly with time toward lower CO_2/H_2O because of preferential diffusion of H_2O from the melt. However, nucleation is a continuous process in CO_2-bearing basaltic melts (Le Gall and Pichavant 2016a, b) and reequilibration of previously nucleated bubbles by diffusion will be accompanied by the nucleation of new CO_2-rich bubbles.

We conclude that future developments in the interpretation of gas data require progress from both sides, experimental and volcanological. Some crucial experimental information at small scale is still missing such as the composition of individual gas bubbles nucleated in the decompression experiments and the influence of crystals on bubble nucleation. In parallel, at larger scales, the representativity and the significance of the gas phase sampled on active basaltic volcanoes needs to be better demonstrated, for example by combining gas measurements with detailed textural studies of eruption products. It is expected that future work will narrow the gap in scales between experiments and gas measurements to refine interpretations of gas compositions as unrest signals.

Acknowledgements This paper has benefited from discussions with P. Allard, C. Martel, N. Metrich, A. Bertagnini, R. Moretti, P. Papale and M. Pompilio, reviews by R. Brooker and F. Wadsworth and from editorial comments by B. Scheu. Discussion with S. Yoshimura was helpful. The VUELCO consortium provided a scientifically demanding and interdisciplinary forum for the elaboration of ideas developed in this study. The Ph.D. thesis of NLG was supported by the VUELCO project.

References

Aiuppa A, Moretti R, Federico C, Giudice G, Gurrieri S, Liuzzo M, Papale P, Shinohara H, Valenza M (2007) Forecasting Etna eruptions by real-time observation of volcanic gas composition. Geology 35:1115–1118

Aiuppa A, Bertagnini A, Metrich N, Moretti R, Di Muro A (2010) A model of degassing for Stromboli volcano. Earth Planet Sci Lett 295:195–204

Aiuppa A, Bitetto M, Francofonte V, Velasquez G, Bucarey Parra C, Giudice G, Liuzzo M, Moretti R, Moussallam Y, Peters N, Tamburello G, Valderrama OA, Curtis A (2017) A CO_2-gas precursor to the March 2015 Villarrica volcano eruption. Geochem Geophys Geosyst 18:2120–2132

Allard P (2010) A CO_2-rich gas trigger of explosive paroxysms at Stromboli basaltic volcano, Italy. J Volcanol Geotherm Res 189:363–374

Aubaud C, Pineau F, Jambon A, Javoy M (2004) Kinetic disequilibrium of C, He, Ar and carbon isotopes during degassing of mid-ocean ridge basalts. Earth Planet Sci Lett 222:391–406

Burgisser A, Alletti M, Scaillet B (2015) Simulating the behavior of volatiles belonging to the C-O–H–S system in silicate melts under magmatic conditions with the software D-Compress. Comput Geosci 79:1–14

Burton M, Allard P, La Spina A, Murè F (2007) Magmatic gas composition reveals the source depth of slug-driven strombolian explosive activity. Science 317:227–230

Christopher T, Edmonds M, Humphreys MCS, Herd RA (2010) Volcanic gas emissions from Soufrière Hills Volcano, Montserrat 1995–2009, with implications for mafic magma supply and degassing. Geophys Res Lett 37: L00E04. https://doi.org/10.1029/2009gl041325

Dixon JE, Stolper EM (1995) An experimental study of water and carbon dioxide solubilities in mid-ocean ridge basaltic liquids. Part II. Applications to degassing. J Petrol 36:1633–1646

Edmonds M (2008) New geochemical insights into volcanic degassing. Philos Trans R Soc A 366:4559–4579

Edmonds M, Aiuppa A, Humphreys M, Moretti R, Giudice G, Martin RS, Herd RA, Christopher T (2010) Excess volatiles supplied by mingling of mafic magma at an andesite arc volcano. Geochem Geophys Geosyst 11:Q04005. https://doi.org/10.1029/2009GC002781

Fiege A, Behrens H, Holtz F, Adams F (2014) Kinetic vs. thermodynamic control of degassing of H2O–S ± Cl-bearing andesitic melts. Geochim Cosmochim Acta 125:241–264

Gonnermann HM, Manga M (2005) Non-equilibrium magma degassing: results from modelling of the ca. 1340 AD eruption of Mono craters, California. Earth Planet Sci Lett 238:1–16

Head JW III, Wilson L (2003) Deep submarine pyroclastic eruptions: theory and predicted landforms and deposits. J Volc Geotherm Res 121:155–193

Le Gall N, Pichavant M (2015b) Heterogeneous bubble nucleation on Fe–Ti oxides in H_2O- and H_2O-CO_2-bearing basaltic melts. (In preparation)

Le Gall N, Pichavant M (2016a) Homogeneous bubble nucleation in H2O- and H_2O-CO_2-bearing basaltic melts: results of high temperature decompression experiments. J Volcanol Geotherm Res 327:604–621

Le Gall N, Pichavant M (2016b) Effect of ascent rate on homogeneous bubble nucleation in the system basalt-H_2O-CO_2: Implications for Stromboli volcano. Am Mineral 101:1967–1985

Le Gall N, Pichavant M, Di Carlo I, Scaillet B (2015a) Sulfur partitioning between melt and fluid during degassing of ascending C-O-H-S-bearing basaltic magma: an experimental study. In preparation

Lensky NG, Niebo RW, Holloway JR, Lyakhovsky V, Navon O (2006) Bubble nucleation as a trigger for xenolith entrapment in mantle melts. Earth Planet Sci Lett 245:278–288

Lesne P, Scaillet B, Pichavant M, Iacono-Marziano G, Bény J-M (2011) The H_2O solubility of alkali basaltic melts: an experimental study. Contrib Mineral Petrol 162:133–151

Lesne P, Scaillet B, Pichavant M (2015) The solubility of sulphur in hydrous basaltic melts. Chem Geol 418:104–116

Metrich N, Bertagnini A, Di Muro A (2010) Conditions of magma storage, degassing and ascent at Stromboli: new insights into the volcanic plumbing system with inferences on the eruptive dynamics. J Petrol 51:603–626

Moretti R, Papale P (2004) On the oxidation state and volatile behaviour in multicomponent gas–melt equilibria. Chem Geol 213:265–280

Mourtada-Bonnefoi CC, Laporte D (2002) Homogeneous bubble nucleation in rhyolitic magmas: an experimental study of the effect of H_2O and CO_2. J Geophys Res 107(B4):2066. https://doi.org/10.1029/2001jb000290

Mourtada-Bonnefoi CC, Laporte D (2004) Kinetics of bubble nucleation in a rhyolitic melt: an experimental study of the effect of ascent rate. Earth Planet Sci Lett 218:521–537

Newman S, Lowenstern JB (2002) VolatileCalc: a silicate melt–H_2O–CO_2 solution model written in Visual Basic for Excel. Comput Geosci 28:597–604

Papale P, Moretti R, Barbato D (2006) The compositional dependence of the saturation surface of $H_2O + CO_2$ fluids in silicate melts. Chem Geol 229:78–95

Peslier AH, Bizimis M, Matney M (2015) Water disequilibrium in olivines from Hawaiian peridotites: recent metasomatism, H diffusion and magma ascent rates. Geochim Cosmochim Acta 154:98–117

Pichavant M, Pompilio M, D'Oriano C, Di Carlo I (2011) The deep feeding system of Stromboli, Italy: insights from a primitive golden pumice. Eur J Miner 23:499–517

Pichavant M, Di Carlo I, Rotolo SG, Scaillet M, Burgisser A, Le Gall N, Martel C (2013) Generation of CO_2-rich melts during basalt magma ascent and degassing. Contrib Mineral Petrol 166:545–561

Rutherford MJ (2008) Magma ascent rates. In: Putirka KD, Tepley F (eds) Minerals, inclusions and volcanic processes. Mineralogical Society of America Reviews in Mineralogy vol 69, pp 241–271

Shea T (2017) Bubble nucleation in magmas: A dominantly heterogeneous process? J Volcanol Geotherm Res 343:155–170

Sparks RSJ, Barclay J, Jaupart C, Mader HM, Phillips JC (1994) Physical aspects of magmatic degassing I. Experimental and theoretical constraints on vesiculation. In: Carroll MR, Holloway JR (eds) Volatiles in Magmas. Mineralogical Society of America Reviews in Mineralogy, vol 30, pp 413–445

Yoshimura S (2015) Diffusive fractionation of H_2O and CO_2 during magma degassing. Chem Geol 411:172–181

Zhang Y, Ni H (2010) Diffusion of H, C, and O components in silicate melts. In: Zhang Y, Cherniak DJ (eds) Diffusion in minerals and melts. Mineralogical Society of America Reviews in Mineralogy vol 72, pp 171–225

Crystals, Bubbles and Melt: Critical Conduit Processes Revealed by Numerical Models

M. E. Thomas, J. W. Neuberg and A. S. D. Collinson

Abstract

Understanding how magma moves within a conduit is an important question that is still poorly understood. In particular, estimation of the magma ascent rate is key for interpreting monitoring signals and therefore, predicting volcanic activity. This relies on understanding how strongly different magmatic processes occurring within the conduit control the ascent rate. These processes are controlled by changes in magmatic parameters such as the water content or temperature and understanding/linking changes of such parameters to monitoring data is an essential step in the use of these data as a predictive tool. The results presented here are from a suite of conduit flow models based on Soufrière Hills Volcano, Montserrat, that assesses the influence of individual model parameters. By systematically changing these parameters, the results indicate that changes in conduit diameter and excess pressure in the magma chamber are amongst the dominant controlling variables. However, the single most important parameter controlling variations in the magma ascent rate is the volatile content. Therefore, understanding the processes controlling the volatile content within the conduit system and the outgassing of these volatiles is crucial to understanding and predicting potential unrest or eruption scenarios.

Keywords

Numerical modelling · Conduit processes
Low frequency earthquakes · Magma flow
Magma ascent rate

1 Introduction

A volcanic conduit provides the pathway for transport of magma and magmatic fluids within a volcano. It is possible to detect both this movement and the occurrence of conduit processes through geophysical monitoring techniques as discussed in Chap. 2 of this book. However, the extent to which changes in magma flow properties affect the data recorded on volcanoes is not well understood. Is it possible that a small change in magma temperature or water content could alter the processes or flow within the conduit enough to be recorded by geophysical monitoring instruments or simple visual observation? What effect does the size of gas bubbles within

M. E. Thomas (✉) · J. W. Neuberg
A. S. D. Collinson
School of Earth and Environment, Institute of Geophysics and Tectonics, University of Leeds, Leeds LS12 9JT, UK
e-mail: m.e.thomas@leeds.ac.uk

Advs in Volcanology (2019) 155–169
DOI 10.1007/11157_2018_36

Published Online: 30 August 2018

the magma have on the overall flow dynamics, and how big do these changes need to be to alter the eruption style? These types of question are addressed within this chapter in an attempt to identify the crucial parameters that cause changes in observed volcanic behaviour.

We use conduit flow models to analyse the key input parameters that control magma flow properties, such as the magma water content, crystal content and conduit geometry, to assess their relative importance to the overall magma flow dynamics. A list of all input parameters is presented in Table 1 along with the range of values studied. We focus on evolved silicic magmatic systems because of the wealth of relevant monitoring information and previous numerical modelling attempts relating to Soufrière Hills Volcano, Montserrat—a long lived andesitic dome forming eruption (Sparks et al. 2000; Wadge et al. 2014) and excellent natural laboratory. While these initial models are based on extrusive eruptions, the results of changing the model parameters have the potential to alter the eruption style to either more violent or gentle forms and it is noted that the underlying principles discussed here are applicable to other volcanic systems, including those exhibiting signs of unrest, yet to develop into a full blown eruption. If we can develop the findings presented in this chapter into the creation of threshold levels for recorded geophysical data, it may become possible to begin to predict when a volcano will evolve from a state of unrest to eruption.

Table 1 Parameters used in the reference model and range of parameter variations

Symbol/abbreviation	Variable	"Reference" model value	Range of modelled values
–	The melt composition	Rhyolitic (>71% SiO_2) (Barclay et al. 1998)	See Table 2
b_{ni}	Bubble number density	10^{10} m^{-3} (Cluzel et al. 2008)	10^7–10^{11} m^{-3}
D_{TBL}	Thickness of thermal boundary layer over which T_{diff} is lost	0.3 m (Collier and Neuberg 2006)	0.3–0.5 m
Γ	Bubble surface tension	0.06 N m^{-1} (Lyakhovsky et al. 1996)	0.05–0.25 N m^{-1}
χ_c	Magma chamber crystal volume fraction	40% (Barclay et al. 1998)	40–50%
Ls	Slip length of brittle failure of melt	0.01	0.01–1.0 m
P_e	Excess chamber pressure above lithostatic	0 MPa	0–20 MPa
P_{top}	Pressure at conduit exit	0.09 MPa	0.09–4.5 MPa
ρ_c	Average density of crystal assemblage	2700 kg m^{-3} (Burgisser et al. 2010)	2550–3200 kg m^{-3}
ρ_m	Density of pure melt	2380 kg m^{-3} (Burgisser et al. 2010)	–
T	Magma temperature	1150 K(Devine et al. 2003)	1100–1150 K
T_{diff}	Amount of cooling at conduit wall	200 K (Collier and Neuberg, 2006)	100–200 K
τ_s	Melt shear strength	–	10^5–10^7 Pa
$W_{\%}$	Initial dissolved water content of magma	4.5 wt% (Barclay et al. 1998)	3–8 wt%
w, d, r	Variables that define the conduit shape and size	See Fig. 1	See Fig. 2

2 The Model

In order to assess the effect of altering the model parameters, a standard or "reference" model is defined. This reference model is based on data available in the literature that refers to Soufrière Hills Volcano, and is outlined in Fig. 1 and Table 1. The general dimensions of the modelled conduit, shown in Fig. 1 are inferred from geochemical and observational data from Soufrière Hills Volcano (Barclay et al. 1998; Sparks et al. 2000), placing minimum depth constraints of 5–6 km for the position of the magma chamber and width estimates of 30–50 m for the conduit.

2.1 Governing Equations

Conduit flow is computed with a finite element approach within the code COMSOL Multiphysics®, and modelled in an axial symmetric domain space through the compressible formulation of the Navier-Stokes equation:

$$\rho\frac{\partial \boldsymbol{u}}{\partial t}+\rho\boldsymbol{u}\cdot\nabla\boldsymbol{u}=-\nabla p+\nabla\cdot\{\eta\left[\nabla\boldsymbol{u}+(\nabla\boldsymbol{u})^{T}\right]-\frac{2}{3}\eta[\nabla\cdot\boldsymbol{u}]\boldsymbol{I}\}+\boldsymbol{F} \tag{1}$$

and the continuity equation:

$$\frac{\partial\rho}{\partial t}+\nabla\cdot(\rho\boldsymbol{u})=0 \tag{2}$$

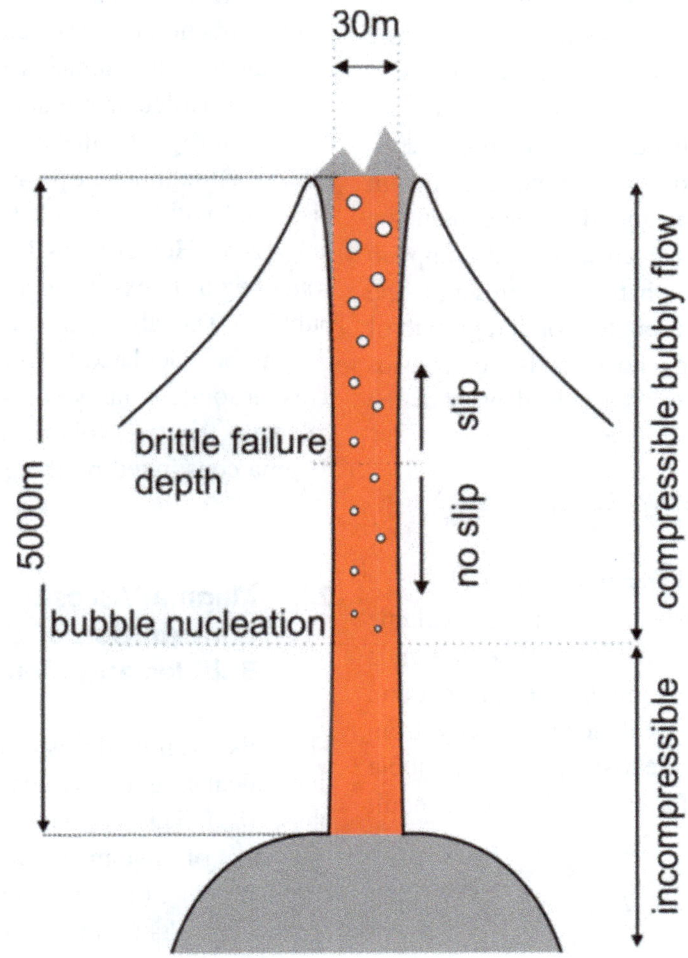

Fig. 1 Cartoon of the modelled volcanic system. Bubble nucleation and brittle failure depth vary with the model parameters considered

where ρ is density, $\boldsymbol{u}$ the velocity vector, p the pressure, η the dynamic viscosity and $\boldsymbol{F}$ the volume force vector (gravity). There is no time dependency in the model as they are solved to a steady state, so the terms $\rho\frac{\partial u}{\partial t}$ and $\frac{\partial p}{\partial t}$ in Eqs. (1) and (2) are neglected.

2.2 Magma Composition

The properties of the magma are modelled as the averaged properties of its constituents: melt, crystals and gas. For the reference model, the general composition of the melt is taken as rhyolitic, using the groundmass analysis of Montserrat dome rocks undertaken by Barclay et al. (1998). However, several melt compositions, were ultimately considered to assess the effect of melt composition on the modelled eruption dynamics (Table 2). Crystal content (χ_c) and density (ρ_c) are fixed as we assume a constant temperature and that the conduit ascent times are orders of magnitudes faster than the time required for crystal growth by decompression, meaning only the phenocrysts present in the magma chamber are accounted for and growth of microlites and microphenocrysts is not considered. The expression for the bulk density of the magma is given by:

$$\rho = \rho_m\chi_m(1-\chi_g) + \rho_g\chi_g + \chi_c\rho_c(1-\chi_g), \quad (3)$$

where χ_m is the initial fraction of melt ($1 - \chi_c$), ρ_m is the melt density and χ_g is the gas volume fraction (Table 1). For the gas phase, water is assumed as the only volatile species present and the gas density (ρ_g) is calculated directly from the ideal gas law with the assumption that bubble growth in is equilibrium with the conduit pressure:

$$pV = nRT, \quad (4)$$

where V is the volume of gas, R the ideal gas constant and T the temperature. The number of moles of water, n, is related to density by:

$$n = \frac{\mathrm{M}}{m}, \quad (5)$$

where M is the molar mass of water and m is the mass of water present. Thus, combining Eqs. (7) and (8) and considering a unit volume we get:

$$\rho_g = \frac{mp}{RT} \quad (6)$$

In the reference model, a single magma temperature is used with the exception of the temperature across a thermal boundary layer (TBL) defined adjacent to the conduit wall. A linear temperature drop is applied across the TBL, to simulate the cooling of the magma abutting the country rock in a well-established conduit (Collier and Neuberg 2006). The gas volume fraction (χ_g) is calculated by determining how much water remains dissolved within the melt at a particular pressure using the solubility of H_2O in rhyolitic melts presented by Liu et al. (2005). At significant pressures, all the water is dissolved within the melt fraction and χ_g is initially zero. However, as the pressure decreases, water begins to exsolve out of the melt and forms bubbles. The absolute volume of exolved gas (V) can be calculated from the ideal gas law (4). This absolute volume of gas is then used to calculate the gas volume fraction of the bulk magma constituted by the gas phase.

2.3 Magma Viscosity—The Contribution of Crystals, Bubbles and Melt

The bulk magma viscosity (η) is determined by first calculating the viscosity of the pure melt phase (η_m). This is done using a model for the viscosity of magmatic liquids presented by Giordano et al. (2008), that predicts the viscosity of silicate melts as a function of temperature and melt composition. It is important to note that the composition used in the viscosity model is that of the pure melt phase (rhyolitic) not the overall

Table 2 Compositions of melt used in the numerical simulations

Composition[a]	SiO_2	Al_2O_3	TiO_2	FeO	MgO	MnO	CaO	Na_2O	K_2O
a	71.41	13.58	0.28	2.78	1.64	0.13	4.86	3.73	1.6
b	76.97	11.21	0.29	1.89	0.26	0.12	1.29	4.07	2.37
c	77.10	9.83	0.18	1.17	0.22	0.10	1.52	4.14	1.72
d	78.66	11.20	0.39	1.93	0.30	0.10	1.48	3.57	2.38

[a]Compositions determined through, (a) rastered electron microprobe analysis of groundmass (Barclay et al. 1998); (b) Matrix glass composition (Rutherford and Devine 2003); (c) Quartz hosted melt inclusion (Devine et al. 1998); (d) Cameca SX50 microprobe analysis of interstitial glass (Burgisser et al. 2010). All melts are rhyolitic and composition (a) is used in the defined reference model

magma composition. The whole rock composition of recent Soufrière Hills Volcano magma is andesitic (Edmonds et al. 2010), but this includes the contribution of the crystals. The viscosity model is only used to calculate the actual viscosity of the liquid component (the melt), on which the crystals (the solid) have no bearing. When the effect of crystals within the melt *is* considered, the effective viscosity of the melt (liquid) and crystal (solid) mixture (η_{mc}) increases, and can be represented by the Einstein-Roscoe equation:

$$\eta_{mc} = \eta_m \left(1 - \frac{\chi_c}{\chi_c^{\max}}\right)^{-2.5}, \tag{7}$$

where $\chi_c^{\max}$ is the volume fraction of crystals at which the maximum packing is achieved and a commonly adopted value for this is 0.6 (*Marsh*, 1981), which is used within this study. Although this value was proposed for randomly packed spheres, and it has been shown by Marti et al. (2005) that $\chi_c^{\max}$ tends to decrease as the particle (crystal) shape becomes less isotropic. Ishibashi (2009) demonstrated that this value is a good approximation as the effect of particle shape on $\chi_c^{\max}$ is offset by effects of size heterogeneity and crystal alignment.

The presence of bubbles also affects the viscosity. If the bubbles within the magma remain un-deformed they act to increase viscosity, whilst if deformed (elongated in the direction of flow), they act to decrease *visosity* (Llewellin and Manga, 2005). Whether a bubble is in an un-deformed or deformed state is represented by the capillary number:

$$\mathrm{Ca} = \frac{\eta_m r E}{\Gamma} \tag{8}$$

where r is the un-deformed bubble radius, Γ, the bubble surface tension and E, a function of the strain rate within the magma flow defined below. If Ca > 1 then the bubbles can be considered deformed. The value of Ca will vary as a function of shear strain rate and elongation strain rate (Thomas and Neuberg 2012), meaning bubbles can be deformed within the model through either shear or extension. To account for strain acceleration or deceleration the dynamic capillary number (Cd) is required (Llewellin and Manga 2005). This compares the timescale over which the bubbles can respond to changes in their strain environment with the timescale over which the strain environment changes. If this value is large, the flow is termed unsteady and the bubbles are unable to deform independently in response to the flow. However, for the models considered here, conditions of unsteady flow are found only in a very small area near the exit of the conduit. Accounting for this within the models resulted in no noticeable change in the derived flow parameters, hence the computation of Cd is not considered.

Depending on the value of Ca, η is calculated using the suggested 'minimum variation' of Llewellin and Manga (2005):

$$\mathrm{Ca} = \begin{cases} <1 & \eta = \eta_{mc}(1-\chi_g)^{-1} \\ >1 & \eta = \eta_{mc}(1-\chi_g)^{5/3} \end{cases} \quad (9)$$

By assuming the homogeneous nucleation of a number of bubbles in a unit volume of melt as a single event, which is determined experimentally through the initial bubble number density (b_{ni}) (e.g. Hurwitz and Navon 1994), the bubble radius (Lensky et al. 2002) is given by:

$$r = \left[\frac{S_0{}^3 \rho_m (C_0 - C_m)}{\rho_g}\right]^{1/3}, \quad (10)$$

where C_0 and C_m are the initial and remaining amount of water dissolved in the melt respectively and S_0 is the initial size of the melt shell from which each bubble grows. S_0 is related to the instantaneous bubble number density (b_n) through the expression:

$$S_0{}^3 = \frac{3}{4\pi b_n}. \quad (11)$$

b_n is used rather than the initial value (b_{ni}) because homogeneous nucleation is assumed. Therefore, the bubble number density must remain constant with respect to the volume of the melt fraction in Eq. 3. This also accounts for bubble coalescence and b_n is given by:

$$b_n = \frac{b_{ni}}{\chi_m}\left[\chi_m - (1 - \chi_g)\right] \quad (12)$$

2.4 Brittle Failure of Melt

It is now well established that magma, or more specifically the melt component of a magma can fail in a brittle manner (e.g. Goto 1999). This is likely to generate low-frequency (LF) earthquakes (e.g. Neuberg et al. 2006) and effect the overall flow dynamics. In order to account for these effects, it is necessary to define conditions under which the melt may fracture. Shear failure of melt occurs when the shear stress ($\eta\dot{\varepsilon}$) exceeds the shear strength (τ_s), and has been represented as a brittle failure criterion (e.g. Tuffen et al. 2003):

$$\frac{\eta\dot{\varepsilon}}{\tau_s} > 1 \quad (13)$$

where $\dot{\varepsilon}$ is the shear strain rate. This criterion holds true under the assumption that during un-relaxed deformation the accumulation of shear stress in the melt obeys the Maxwell model:

$$\sigma_s = \frac{\eta}{\mu}\frac{\partial \sigma_s}{\partial t} = \eta\dot{\varepsilon} \quad (14)$$

where σ_s is the shear stress and μ the shear modulus.

The magma composition as discussed in Sects. 2.2 and 2.3 is considered without the effects of microlite growth, the reasons for doing so are outlined in Sect. 2.2. However, it is worth noting, that if considered, the influence of microlite growth would possibly increase the bulk viscosity significantly.

2.5 Boundary Conditions

Flow within the system is driven by a pressure gradient defined by boundary conditions at the top and bottom of the conduit. The top boundary is set to atmospheric pressure at the altitude of the conduit exit plus any overburden load from an emplaced lava dome. The bottom boundary is set to lithostatic pressure (assuming a homogeneous country rock density of 2600 kg m^{-3}) plus any imposed overpressure (P_e). Both the top and bottom pressure conditions are held constant throughout the model run. Initial boundary conditions along the length of the conduit are defined as no slip. When brittle failure of melt is considered within a model run, at the regions of the conduit wall where the brittle failure criterion was exceeded, the boundary conditions are changed to a tangential slip velocity ($\Delta \boldsymbol{u}$) defined by:

$$\Delta \boldsymbol{u} = \frac{1}{\beta}\sigma_s, \quad (15)$$

where σ_s is the tangential shear stress to the conduit wall and the coefficient β is a function of the slip length (L_s):

$$\beta = \frac{\eta}{L_s}, \tag{16}$$

The model is then re-run to account for the effect of changing boundary conditions at the conduit walls. Where this results in an increase in the predicted failure depth, an iterative approach is used and the model is re-run with the new depth until the depth at which brittle failure of the melt stabilises. For the purposes of this study, the failure depth is considered to have converged if the depth increase between iterative runs is less than 10% of the previous observed increase.

The size and shape of the conduit is also an important factor in influencing the accent rate, so for the purpose of assessing the potential magnitude of this influence, several possible conduit shapes were modelled. While the types of boundary conditions discussed above do not change, the relative locations of the boundaries do (Fig. 2). Case (a) is the simplest geometry change and represents just a change of the conduit radius (r). Case (b) represents a narrowing of the conduit. Case (c) represents a widening of the conduit. The extent to which the geometry of the conduit is changed within the models is discussed further in the next Section.

3 Critical Conduit Processes

3.1 Using Magma Ascent Rates to Assess Model Sensitivity

The ascent rate is a key parameter in understanding volcanic hazard because it has been directly linked to eruptive behaviour (e.g. Gonnerman and Manga 2007). By gaining a better understanding of which model parameters have the greatest effect on ascent rates, we can achieve an insight into which are the most important parameters controlling explosivity, and the likely severity of the volcanic hazard. For the purpose of comparing the various models we use two velocities, defined as $\overline{V}$ and V_{2500}, where $\overline{V}$ is the average accent velocity taken along a vertical profile through the centre of the conduit, and V_{2500} is the average accent velocity taken along a horizontal profile at a depth of 2500 m within the conduit.

The ascent rate has also been linked to monitoring data such as seismicity (e.g. Thomas and Neuberg 2012) or deformation (e.g. Zobin et al. 2011), therefore, it is possible to link the changes in model parameters to recorded monitoring data. In addition, there are physically observed variations in ascent rate estimated from a variety of methods, ranging from studying mineral reaction rims around phenocrysts within erupted magma

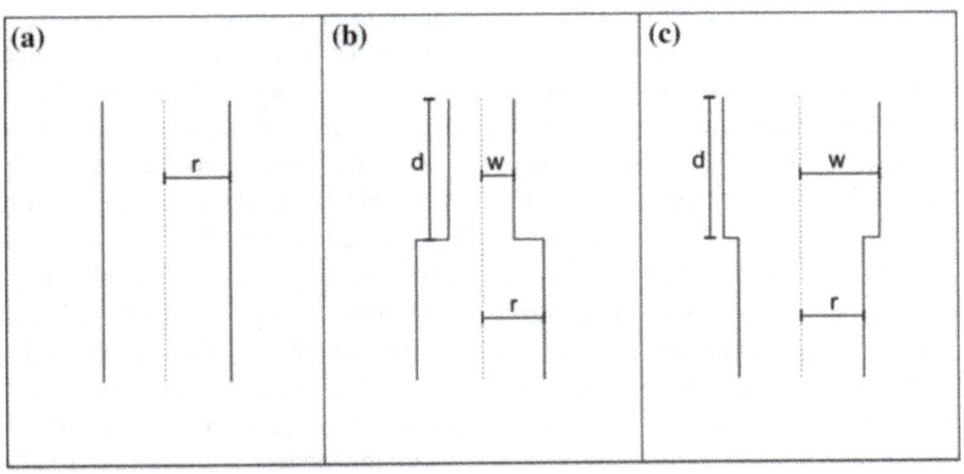

Fig. 2 Schematic diagram of the alternative conduit geometries modelled, showing **a** a constant conduit radius; **b** a narrowing conduit and **c** a widening conduit

(Rutherford and Devine 2003) to interpreting lava dome morphology (Sparks et al. 2000). This places constraints on the magnitude of changes to the modelled ascent rate engendered by altering the model input parameters we can consider realistic.

Matching the absolute values of physically observed and calculated ascent rates is currently beyond the scope of the model, however we can use the magnitude of the observed variations to provide upper and lower bounds to the extent to which the model input parameters are varied. Any changes that produce increases in ascent rate greater than two orders of magnitude over the reference model are not considered realistic in this work. This may seem at first an arbitrary discrimination, but there is a good reason that the observed or calculated ascent rates presented in the literature (e.g. Rutherford and Devine 2003; Castro and Gardner 2008) are "slow" ($<5 \times 10^{-2}$ m s^{-1}). Faster ascent rates, while likely to exist in nature, would almost certainly result in substantial fragmentation of the magma, making it very difficult to observe or calculate the actual magma ascent rate below the initial point of fragmentation. Fragmentation dynamics are not considered within the current model, hence no valid inferences or conclusions can be gained from studying the model runs that exhibit extremely fast ascent rates.

3.2 The Critical Model Parameters

Figure 3 summarises the sensitivity of ascent rate to the different model parameters presented in Table 1. The single parameter (within the modelled ranges) which has the strongest effect on the ascent velocities is the initial dissolved water content of the magma. This parameter affected both $\overline{\mathrm{V}}$ and V_{2500} to a large degree. In contrast there are several model parameters which have little effect on the modelled ascent velocities. These include the thermal boundary layer thickness and the temperature drop across it, as well as the bubble number density and bubble surface tension. Modifying the parameters involved in the brittle failure of the melt (magma shear strength and slip length) has a negligible effect on ascent rates and these results have not been plotted on Fig. 3. However, the contribution of the brittle failure of the melt to observed geophysical signals is considered very important, and will be discussed in Sect. 3.3.

It is unsurprising that the group of model parameters that appear to have the greatest effect on the magma ascent velocity, as seen in Fig. 3b (water content, temperature, crystal content, and chemical composition), also have the greatest effect on the magma viscosity (Sect. 2.3). Ultimately, modelling the ascent of magma is a fluid flow problem, and the properties that have the biggest effect on the fluid (magma) properties will have the biggest effect on the overall dynamics of the system. All other parameters have a much smaller direct effect on the fluid properties, and although they may be important to specific small scale magmatic processes when considered in isolation, with respect to the overall magma ascent they appear insignificant. For example, altering the properties of the bubbles within the magma, b_{ni} and Γ, the effect is to change the shape and the number of bubbles. Previous work has heavily focused on this area (e.g. Llewellin and Magna 2005) but the effect on the overall flow modelled here is minimal. The indication from this is that it is the total volatile content (water in this case) which is available that is more important to governing the overall flow dynamics, rather than how exactly it is stored in the magma. This particular observation is a key point as new estimates from Cassidy et al. (2015) suggest that basaltic South Soufrière Hills magmas (and by extension, possibly other basaltic arc magmas) have the potential to be extremely volatile-rich, containing up to >6 wt% H_2O prior to eruption. Firstly, this validates the use of high initial water contents used in the range of parameters modelled, and secondly, given the range of accent velocities generated within the models as a result of just changing the dissolved water content (the dark blue bars in

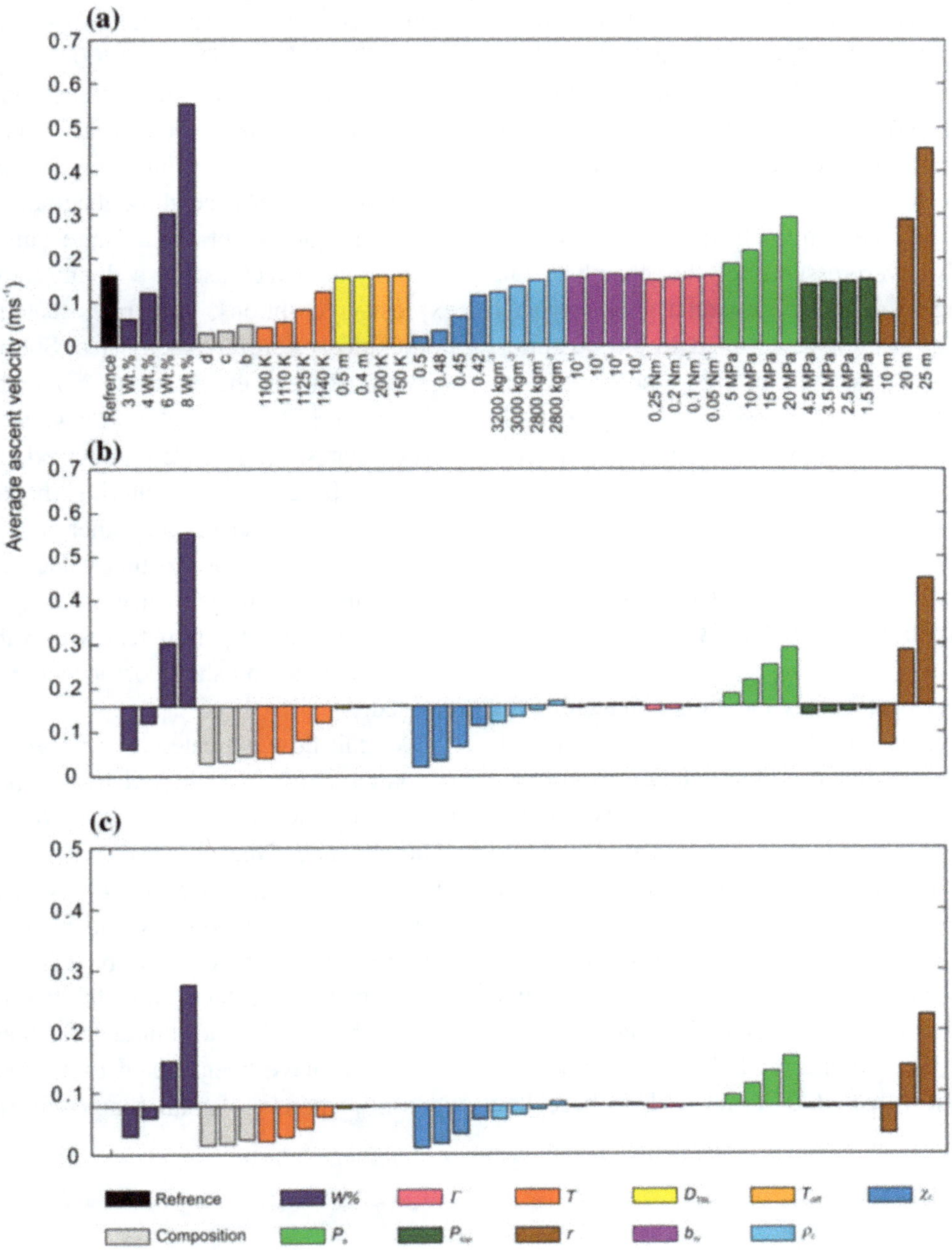

Fig. 3 **a** The ascent rate, $\overline{V}$ for each value of the parameters altered. The black bar represents the reference model with parameters as listed in Table 1. **b** The same data as (**a**) plotted relative to the reference model, which is represented by the horizontal black line. Parameter changes that caused increases in the ascent rates plot upward from the horizontal line, while parameter changes that caused decreases in ascent rates plot downwards. **c** V_{2500} for each value of the parameters altered, plotted relative to the reference model

Fig. 3), demonstrates that it is vitally important to obtain an accurate understanding of the magma components at the volcano of interest, rather than assuming "typical" values representative of a broad compositional category.

Outside of water content, temperature, crystal content, and chemical composition, the two parameters modelled which have the largest influence on the modelled ascent rate are the chamber overpressure and the conduit geometry. These are particularly important points when considering volcanoes entering periods of unrest following long periods of quiescence. It is problematic to achieve an accurate understanding of the magma components highlighted above at all volcanoes under these circumstances due to a likely lack of monitoring (a problem highlighted in Parts 1 and 2 of this book). Unless there has been long-term measurement of deformation occurring at the volcano now exhibiting signs unrest, it will be extremely difficult to estimate any likely overpressures in the chamber, and attempting to define the conduct geometry of a system that has not yet erupted would be almost impossible. It is therefore paramount that as much information as possible of all potentially active volcanic systems is routinely gathered before signs of unrest are detected.

3.3 Matching Observations—Explosivity and Seismicity

Although, as previously mentioned, matching the absolute values of physically observed and calculated ascent rates is currently beyond the scope of the model, key to giving the models real significance is determining whether the changes to important parameters highlighted in Sect. 3.1 can be theoretically linked to physical observations at real volcanic systems. Figure 4 shows values of V_{2500} for all of the modelled parameters in a manner similar to that presented in Fig. 3c, but in this case, the data are plotted relative to a baseline accent value of 0.02 ms^{-1}. This base line value was chosen because it has been highlighted by Rutherford and Devine (2003) as an ascent rate which may indicate a transition between effusive and explosive behaviour. This value has been obtained from quantifying the breakdown of hornblende in ascending magma, and while this technique is not an accurate barometer for defining an exact ascent velocity required for explosive eruptions, the rates calculated for non-explosive eruptive activity at Soufrière Hills volcano between the period of November 1995–September 2002 were below this value. Figure 4 shows that several model runs produced ascent rates of <0.02 ms^{-1} (by altering the melt composition or χ_c) and several other runs produced ascent rates very close to this value (by altering $W_{\%}$, T, and conduit geometry), indicating that by altering just single parameters within the system this theoretical threshold of accent rate can be crossed.

Conduit flow is treated as a closed system, so no outgassing is considered in the model, as a result the ascent velocities are overestimated (Thomas and Neuberg 2014). It is therefore predictable that if this process was included, far more of the model runs would result in ascent velocities that straddle the baseline in Fig. 4. This suggests that the ability for the ascent rate within the conduit to fluctuate, either side of values that have been linked to explosive eruptions in response to small changes in the system

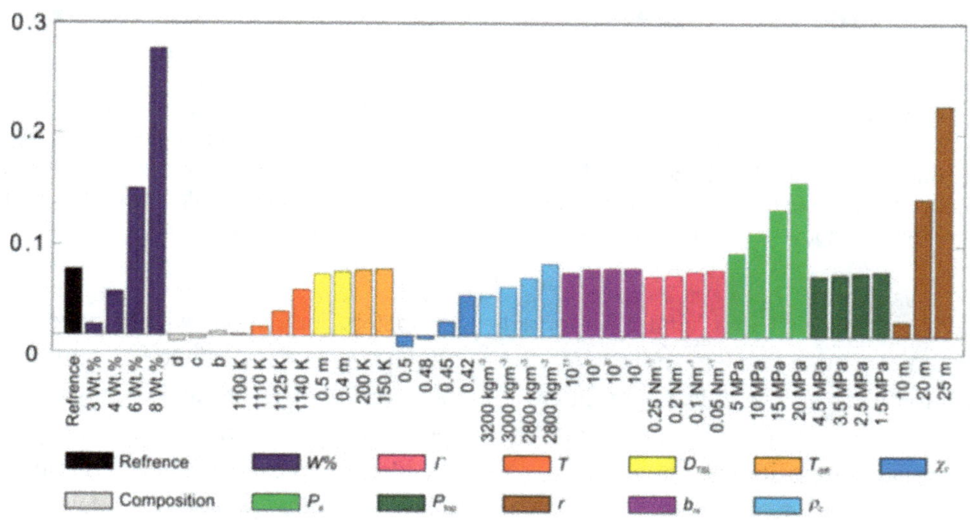

Fig. 4 The ascent rate, V_{2500} for each value of the parameters altered, plotted relative to an ascent velocity of 0.02 ms^{-1}

parameters is genuine. The requirement to accurately understand and model outgassing processes (see Sect. 4) is therefore an important capacity that is currently lacking.

One major discrepancy between physical observations and the model is the model results suggested that the brittle failure of the melt (Sect. 2.4) and the related LF seismicity would occur as a shallow process (in agreement with the work of Holland et al. (2011)).The physical observations place the location of this type of seismicity at Soufrière Hills at depths of ~1500 m below the conduit exit (Neuberg et al. 2006). This is because under normal conditions the shear stress required to break the melt (Eq. 14) can only be reached where the melt is extremely viscous, which occurs near the surface. In order to reach the higher shear stresses required to break the melt at greater depths, where the viscosity is lower, the shear strain rate ($\dot{\varepsilon}$) needs to increase. Since $\dot{\varepsilon}$ is equal to the lateral velocity gradient within a cylindrical conduit or dyke:

$$\dot{\varepsilon} = \frac{dv}{dx} \tag{17}$$

the simplest way to increase $\dot{\varepsilon}$ is to increase the velocity of the magma flowing within the conduit, or reduce the area through which it flows, which since mass must be conserved also has the effect of increasing the flow velocity.

To resolve this discrepancy between model and observations we introduce a constriction within the conduit as a plausible explanation for brittle failure at greater depths. We test its effect within the reference model by including a bottleneck region at a depth of 1500 m, reducing the conduit diameter from 15 to 10 m. This bottleneck is 100 m in length, which equates to only 1/50 of the total conduit length. Figure 5 shows ascent velocity and shear strain rate profiles from the bottleneck region compared to values from the same location of the conduit in the unmodified reference model. By altering this relatively small region of the conduit, the shear strain rate increases by a factor of four. Crucially, with the exception of small changes in the magma rheology caused by the induced pressure gradients within the bottleneck, the magma viscosity has not been altered. Due to the increased value of shear strain rate the brittle failure ratio (13) will increase by the same factor. By introducing such asperities into the conduit and increasing the strain rate it is possible to drive the brittle fracture of the melt to deeper levels in the conduit that match the location of recorded LF seismicity at Soufrière Hills volcano. This further

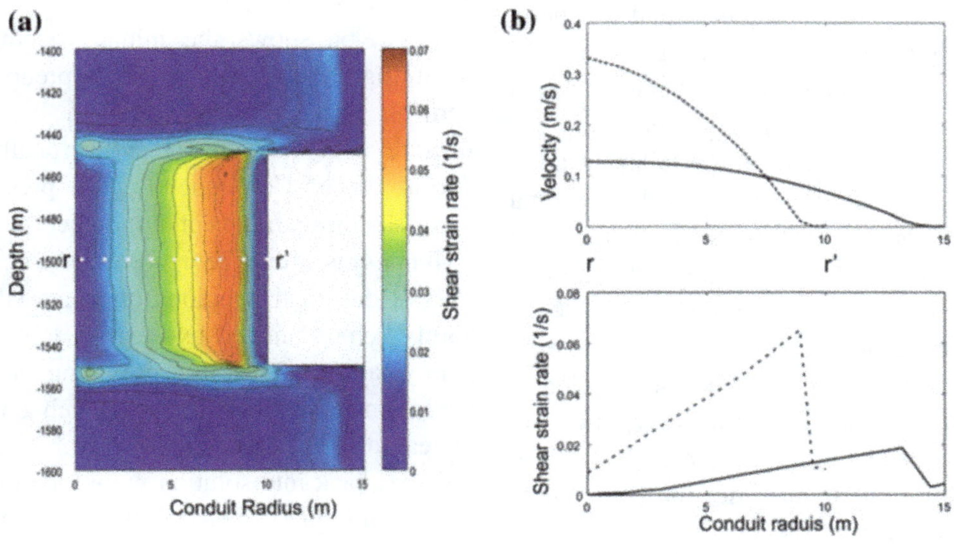

Fig. 5 **a** Plot of the shear strain rate within a simple bottleneck of 100 m length at intervals of 0.005. The values of shear strain rate are seen to be increases over the entire length of the bottleneck. **b** Cross conduit profiles taken at the same depth for velocity and shear strain rate for the case of the unmodified reference model (solid line) and a conduit containing a bottleneck (dashed line)

emphasises the importance of understanding the possible conduit geometry previously highlighted in Sect. 3.2. This is just one possible solution of generating seismicity deeper in the conduit. Any process that acts to increase the sear strain rate, such as strain localisation between crystals in a magma with a high crystal fraction (as discussed in Chap. 10 of this book) would be a possible explanation.

4 Pathways for Outgassing

The role of volatiles has been identified as the primary controlling parameter governing ascent rate within the conduit, but what was not considered was that the gas, once exsolved from the melt has the ability move independently to the melt. This is a commonly assumed process in basaltic systems, but not in silicic magmas of a relatively high viscosity as it is assumed that bubbles of gas cannot rise with significant speed within the magma. However fractures generated by brittle failure of the melt (Sect. 3.3) may provide ideal outgassing pathways for exsolved gas.

The behaviour of this exsolved gas is separated from the problem of magma ascent, and considered independent of any other parameters, through additional numerical modelling in Comsol Multiphysics (Collinson and Neuberg, 2012). In order to consider all possible outgassing pathways, including vertically through the conduit, or laterally through the walls, we model the gas response to brittle failure using a simplistic "block-style" model, with a central conduit and adjacent wall-rocks. Brittle failure is explicitly modelled as an increase in the permeability within narrow regions either side of the conduit. The problem is simplified to considering permeable flow through a static media. Consequently, the equations for Darcy's law (18), and the continuity equation (19) are amalgamated to derive a partial differential equation (20), which is solved for pressure (P):

$$\boldsymbol{u} = -\frac{k}{\eta_g}(\nabla P + \rho g \nabla z) \tag{18}$$

$$\frac{\partial}{\partial t}(\rho \varepsilon) + \nabla \cdot (\rho \boldsymbol{u}) = 0 \tag{19}$$

$$\frac{\partial}{\partial t}(\rho \varepsilon) + \nabla \cdot \rho \left[-\frac{k}{\eta_g}(\nabla P + \rho g \nabla z) \right] = 0 \tag{20}$$

The gas velocity (u) is then determined by using the pressure gradient within Darcy's law (18).

In contrast, with the models for ascent rate, time dependency is included in this model in order to understand the changes in the system through time, in response to a permeability (k) increase in response to brittle failure at the conduit-wall margin. Due to its abundance in volcanic systems, water is the only volatile considered here. The gas density (ρ) is calculated using the mean molar mass within the ideal gas law (6), and the gas viscosity (μ) assumed constant at 1.5×10^{-5} Pa (Collinson and Neuberg 2012).

Bulk permeabilities are set such that the conduit has a higher permeability (10^{-10} m^2) than the wallrocks (10^{-12} m^2), and initially, the conduit-wall margin is "sealed" with a very low permeability of 10^{-16} m^2. The fracturing is initiated at 1500 m, in accordance with measurements by Neuberg et al. (2006), and propagate vertically towards the surface, as an increase in permeability from 10^{-16} to 10^{-6} m^2.

Figure 6a shows the initial system, before fracturing, where the gas loss is predominantly vertical, through the conduit. In Fig. 6b, the fracture zone has propagated vertically up to 700 m depth. Consequently, the pressure has increased through the conduit and corresponding wall margins where the fractures have developed. This is due to the regions of increased permeability being confined by areas of lower permeability above. Thereby, providing a suitable environment for gas storage, which due to the increased pressurisation, may force exsolved volatiles back into solution within the melt. This change in pressurisation has resulted in a

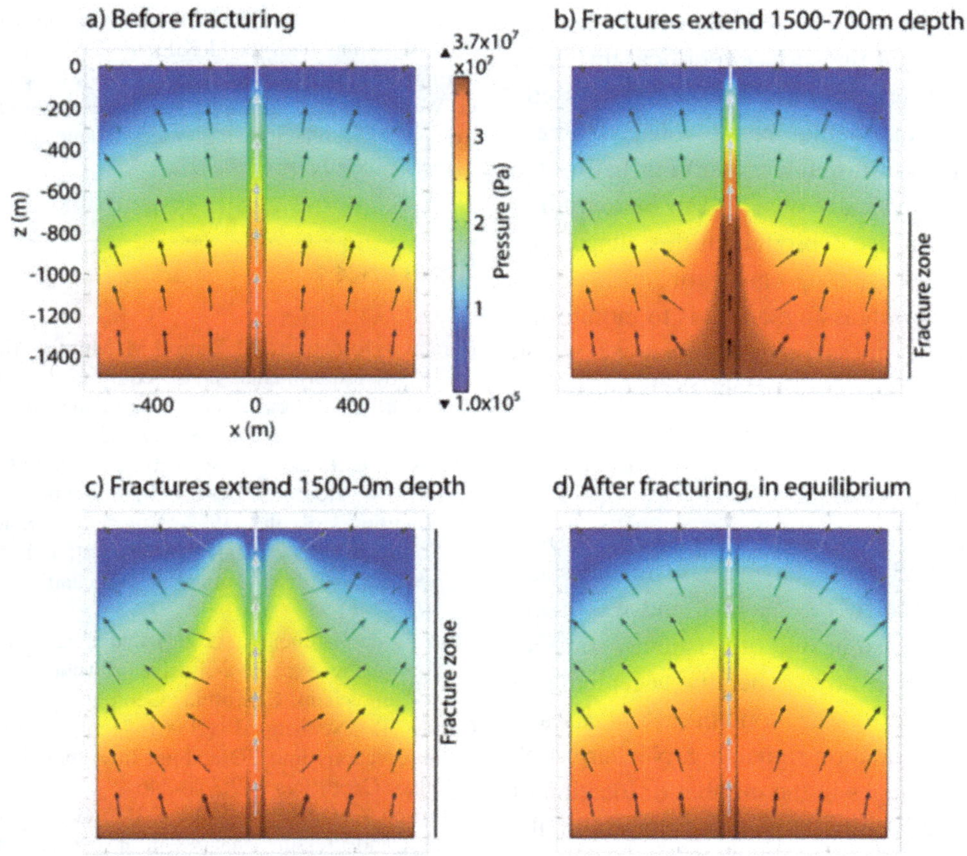

Fig. 6 **a** The initial system in equilibrium, before fracturing has commenced. **b** The systems of fractures has propagated upwards to a depth of 700 m. **c** The fracture zones have reached the surface. **d** The system has resumed equilibrium. Velocity range of arrows (ms^{-1}): **a** max: 0.30, min: 1.3×10^{-3}; **b** max: 0.38, min: 8.8×10^{-5}; **c** max: 0.33, min: 6.8×10^{-4}; **d** max: 0.30, min: 9.5×10^{-4}

corresponding change to the gas velocity pattern, which the lower conduit having a very low gas velocity due to the low pressure gradient. However, at shallow levels, the higher pressure gradient forces a much higher gas velocity towards the surface. On reaching the surface (Fig. 6c), the high permeability fractures result in a decrease in pressure throughout the conduit and a rapid expulsion of the stored gas. Equilibrium conditions resume approximately 4 h after the fracture zones reached the surface (Fig. 6d). This model shows a propagating fracture zone is an effective mechanism for degassing the conduit and wall margins. A key observation is the possibility for this mechanism to produce periods of cyclic activity which are observed at many silicic volcanoes (e.g. Holland et al. (2011)), which can be directly related to observed degassing patterns or through controlling the ascent rate and the associated geophysical signals (e.g. LF seismicity) within the conduit through moderating the amount of gas stored in the system.

5 Summary and Implications

In the introduction to this chapter, we asked the question of whether it is possible that small changes in the composition of properties of the magma could cause changes significant enough to be recorded in monitoring data or even visual observations. It is clear that these small changes, particularly when considering changes in the water content or conduit diameter, can have large effects of the ascent velocity of the magma. These effects are large enough that conceivable they could be simply observed as an increased extrusion rate at the surface. At this point however it may be too late. If increased extrusions rates are being observed at the surface a critical threshold of ascent rate may have already been surpassed.

More useful would be to observe these potential changes through monitoring before the magma was physically observed to be extruding faster at the surface. Throughout this chapter we have discussed the importance of shear stress, seismicity and outgassing, and it is to monitoring data relating to these processes that we would look for an indication of potential changes.

Increases in ascent velocity may lead to a change in eruption style, but they also potentially cause changes in monitoring data. As discussed in the chapter, any increase in ascent velocity of the magma will cause an increase in the shear stress experience by the magma, and will likely lead to an increase in rate of seismicity, or alter its location. It is now generally accepted that shear stress within the conduit also causes a significant deformation signal (e.g. Neuberg et al. 2018). Any changes in the number or location of low frequency earthquakes, or changes in the near-field deformation around the conduit could therefore be inferred as changes in the magma properties. In addition, the single most important parameter identified within this study was changes in the water content. The fractures generated by the seismicity, as a result of the increases in ascent velocity, have also been shown to be important outgassing pathways. An increase in water content would lead to an increase in volatiles and these would be more easily outgassed along the created fractures. Conceptually, a clear signal of an increase in ascent velocity (and a potential early warning of a change in eruption style) caused by increase in water content of the magma, could be an increase and deepening of low frequency earthquakes, accompanied by an increase in the deformation signal followed by significant outgassing event.

In this chapter we look at a single volcanic setting, but by changing the model parameter these effects could be assessed at any volcano, and regardless of the volcano studied the relative importance of the parameters considered should remain unaltered.

Acknowledgements The authors are grateful for inspiring discussions with Geoff Kilgour (GNS New Zealand) during J. Neuberg's study leave at GNS Research Centre in Wairakei, and contributions to research ideas form former Ph.D. students within the Volcanic Studies Group at the University of Leeds.

References

Barclay J, Rutherford M, Carroll M, Murphy M, Devine J (1998) Experimental phase equilibria constraints on pre-eruptive storage conditions of the Soufrière Hills magma. Geophys Res Lett 25:3437–3440

Burgisser A, Poussineau S, Arbaret L, Druitt TH, Giachetti T, Bourdier J-L (2010) Pre-explosive conduit conditions of the 1997 Vulcanian explosions at Soufrière Hills Volcano, Montserrat: I. Pressure and vesicularity distributions. J Volcanol Geoth Res 194:27–41

Castro JM, Gardner JE (2008) Did magma ascent rate control the explosive-effusive transition at the Inyo volcanic chain, California? Geology 36:279–282

Collier L, Neuberg J (2006) Incorporating seismic observations into 2D conduit flow modelling. J Volcanol Geoth Res 152:331–346

Collinson ASD, Neuberg JW (2012) Gas storage, transport and pressure changes in an evolving permeable volcanic edifice. J Volcanol Geoth Res 243–244:1–13

Cluzel N, Laporte D, Provost A, Kannewischer I (2008) Kinetics of heterogeneous bubble nucleation in rhyolitic melts: implications for the number density of bubbles in volcanic conduits and for pumice textures. Contrib Miner Petrol 156:745–763

Devine JD, Murphy MD, Rutherford MJ, Barclay J, Sparks RSJ, Carroll MR, Young SR, Gardner JE (1998) Petrologic evidence for pre-eruptive pressure-temperature conditions, and recent reheating, of andesitic magma erupting at the Soufrière Hills Volcano, Montserrat, W.I. Geophys Res Lett 25:3669–3672

Devine JD, Rutherford MJ, Norton GE, Young SR (2003) Magma storage region processes inferred from geochemistry of Fe–Ti oxides in andesitic magma, Soufrière Hills Volcano, Montserrat, WI. J Petrol 44:1375–1400

Edmonds M, Aiuppa A, Humphreys M, Moretti R, Giudice G, Martin RS, Herd RA, Christopher T (2010) Excess volatiles supplied by mingling of mafic magma at an andesite arc volcano. Geochem Geophys Geosyst 11:Q04005

Giordano D, Russell JK, Dingwell DB (2008) Viscosity of magmatic liquids: a model. Earth Planet Sci Lett 271:123–134

Gonnermann HM, Manga M (2007) The fluid mechanics inside a volcano. Annu Rev Fluid Mech 39:321–356

Gotto A (1999) A new model for volcanic earthquake at Unzen Volcano: melt rupture model. Geophys Res Lett 26:2541–2544

Holland ASP, Watson IM, Phillips JC, Caricchi L, Dalton MP (2011) Degassing processes during lava dome growth: Insights from Santiaguito lava dome, Guatemala. J Volcanol Geoth Res 202:153–166

Hurwitz S, Navon O (1994) Bubble nucleation in rhyolitic melts: experiments at high pressure, temperature, and water content. Earth Planet Sci Lett 122:267–280

Ishibashi H (2009) Non-Newtonian behavior of plagioclase-bearing basaltic magma: subliquidus, viscosity measurement of the 1707 basalt of Fuji volcano, Japan. J Volcanol Geoth Res 181:78–88

Lensky NG, Lyakhovsky V, Navon O (2002) Expansion dynamics of volatile-supersaturated liquids and bulk viscosity of bubbly magmas. J Fluid Mech 460:39–56

Liu Y, Zhang Y, Behrens H (2005) Solubility of H_2O in rhyolitic melts at low pressures and a new empirical model for mixed H_2O-CO_2 solubility in rhyolitic melts. J Volcanol Geoth Res 143:19–235

Llewellin E, Manga M (2005) Bubble suspension rheology and implications for conduit flow. J Volcanol Geoth Res 143:205–217

Lyakhovsky V, Hurwitz S, Navon O (1996) Bubble growth in rhyolitic melts: experimental and numerical investigation. Bull Volc 58:19–32

Marsh B (1981) On the crystallinity, probability of occurrence, and rheology of lava and magma: contributions to mineralogy and petrology. Bull Volcanol 78:85–98

Marti I, Höfler O, Fischer P, Windhab EJ (2005) Rheology of concentrated suspensions containing mixtures of spheres and fibers. Rheol Acta 44:502–512

Neuberg JW, Tuffen H, Collier L, Green D, Powell T, Dingwell D (2006) The trigger mechanism of low-frequency earthquakes on Montserrat. J Volcanol Geoth Res 153:37–50

Neuberg JW, Collinson ASD, Mothes PA, Ruiz CM, Aguaiza S (2018) Understanding cyclic seismicity and ground deformation patterns at volcanoes: Intriguing lessons from Tungurahua volcano. Ecuador, Earth Planet Sci Lett 482:193–200

Rutherford MJ, Devine JD (2003) Magmatic conditions and magma ascent as indicated by hornblende phase equilibria and reactions in the 1995–2002 Soufrière hills magma. J Petrol 44:1433–1454

Sparks RSJ, Murphy MD, Lejeune AM, Watts RB, Barclay J, Young SR (2000) Control on the emplacement of the andesite lava dome of the Soufrière Hills volcano, Montserrat by degassing-induced crystallization. Terra Nova 12:14–20

Thomas ME, Neuberg JW (2012) What makes a volcano tick—A first explanation of deep multiple seismic sources in ascending magma. Geology 40:351–354

Thomas ME, Neuberg JW (2014) Understanding which parameters control shallow ascent of silicic effusive magma. Geochem Geophys Geosyst 15:4481–4506

Tuffen H, Dingwell DB, Pinkerton H (2003) Repeated fracture and healing of silicic magma generate flow banding and earthquakes? Geology 31:1089–1092

Wadge G, Robertson REA, Voight B (eds) (2014) The eruption of Soufrière Hills Volcano, Montserrat from 2000 to 2010. Geol Soc, London, Memoirs, p 39

Zobin VM, Ramirez JJ, Santiago H, Alatorre E, Navarro C (2011) Relationship between tilt changes and effusive-explosive episodes at an andesitic volcano: the 2004–2005 eruption at Volcan de Colima, Mexico. Bull Volc 73:91–99

When Does Magma Break?

Fabian B. Wadsworth, Taylor Witcher, Jérémie Vasseur, Donald B. Dingwell and Bettina Scheu

Abstract

Geophysical signals arriving at the Earth's surface originate from a source mechanism at depth but are not necessarily directly observable. Therefore, well-posed experiments can provide insights into source mechanics and, importantly, the parameters required to model aspects of the sources of unrest signals. In this Chapter we detail one such example of how experimental laboratory work has improved our understanding of unrest signals. We focus on the failure of single- and multi-phase magmas, demonstrating that the liquid viscosity, and therefore the temperature and volatile content of a magma of a given composition, is the limiting parameter in determining whether a magma will ascend viscously or whether it can fracture during ascent. This critical threshold is characterized by a Deborah number, the ratio of the timescale of relaxation to the timescale of local flow. We show that for single-phase magmatic liquids and for vigorously vesiculating magmas, a local Deborah number of 10^{-2} is the limit above which mixed viscoelastic behaviour including fracture propagation can be expected, and a Deborah number of 1 is the limit above which magma is dominantly elastic and responds in a brittle manner to applied stresses. These thresholds can be understood in terms of the onset and peak of the Debye relaxation process for viscoelastic liquids. The apparent validity of a Maxwell model permits us to predict the maximum stress that can be supported by a volcanic liquid deforming in the high Deborah number range. We use these constraints to provide a map of timescales on which we contour dominant system responses from viscous to purely brittle; valid for all magmatic liquids. Finally, we explore the scaling necessary to extend these conceptual insights to crystal- and bubble-bearing magmas valid under specific conditions.

F.B. Wadsworth (✉) · T. Witcher · J. Vasseur
D.B. Dingwell · B. Scheu
Ludwig-Maximilians-Universität, Theresienstr.41, 80333 Munich, Germany
e-mail: fabian.wadsworth@min.uni-muenchen.de

Advs in Volcanology (2019) 171–184
DOI 10.1007/11157_2017_23

Published Online: 18 October 2017

The competing timescales of deformation and relaxation in magma are relevant to unrest source mechanisms that originate from magma deformation, such as long-period seismic signals that are used to predict eruption timing.

Keywords
Rheology · Strain rate · Experimental volcanology · Glass transition · Failure forecasting · Low frequency earthquakes · Viscous dissipation

Glossary

Rheology	The study of the response of a material to an applied stress or deformation. In the volcano-sciences this typically refers to magma-rheology which is dominated by the rheology of the liquid phase (a viscous or *viscoelastic* melt) and the additional effect of suspended phases (crystals and gases).
Viscoelasticity	A material response to an applied stress in which both elastic and viscous components of the deformation are observed. In magma, as the temperature decreases or the local strain rate increases, the elastic component becomes more dominant. When the viscous component is negligible, purely elastic behaviour and material-rupture can readily occur. See *Rheology*.

1 Introduction

There are a wide variety of observable unrest signals at active volcanoes. The key unrest signals are those that are diagnostic of a new regime of behaviour or of how the system may evolve in the future. One such family of events are the low-frequency earthquakes at volcanoes, which are thought to directly represent magma movement at moderate to shallow depths (Chouet et al. 1994; Chouet 1996; Neuberg et al. 2006; Salvage and Neuberg 2016). Such events are powerful tools particularly because they are closely associated with surface activity (Miller et al. 1998). Accelerating event rates of low frequency earthquakes can be used to retrospectively predict an eruption or dome collapse event with reasonable accuracy (Salvage and Neuberg 2016). The utility of the low-frequency event type has been solidified since physical hypotheses for the source mechanism have been put forward (Neuberg et al. 2006) and tested in scaled laboratory expeirments (Benson et al. 2008; Tuffen et al. 2008). The phenomenological observation of the use of certain signal types over others is useful and has led to accurate retrospective eruption forecasts (Voight 1988; Chouet 1996). However, knowledge of the physics of the source leads to a more diverse range of tools being deployed to understand the evolution of volcanic systems into the future. In the context of low-frequency earthquakes at volcanoes, once a source mechanism is identified (Tuffen et al. 2003; Neuberg et al. 2006), numerical models of magma ascent in the conduit beneath volcanoes can be used to reproduce the depths of the source from first principles (Thomas and Neuberg 2012) and could eventually make forecasts of other observables at the surface that would be consistent with impending eruption.

In this Chapter we explore the source mechanism of low-frequency earthquakes at volcanoes from a physical perspective using a compilation of data from scaled laboratory experiments. We use these datasets to demonstrate that the critical threshold for fracture propagation in a viscoelastic fluid such as magma is universal. We propose that this potentially simplifies the inputs required for effective modelling of source mechanics of low-frequency events. More than this, we hope that this provides a good example of how laboratory work can provide valuable insight into the physical feasibility of models proposed to explain unrest signals at volcanoes.

2 Scaling the Viscous-to-Brittle Transition in Magmas

2.1 Single-Phase Magmatic Liquids

Newtonian viscous liquids deform and flow under applied stresses such that the rate of shear strain $\dot{\gamma}$ resulting from an applied shear stress τ is proportional to the viscosity μ via $\tau = \mu\dot{\gamma}$. However, as large shear stresses are applied to viscous liquids, they can exhibit viscoelastic behaviour, which manifests as an apparent non-Newtonian relationship between τ and $\dot{\gamma}$ and can result in fracture propagation locally in the liquid. Of interest here is the transition from simple, apparently Newtonian viscous behaviour to complex viscoelastic behaviour.

Maxwell described the simplest viscoelastic model in which a liquid has a characteristic time required to relax an applied shear stress, termed the Maxwell relaxation timescale λ_r, which is given by $\lambda_r = \mu/G_\infty$ where G_∞ is the shear modulus in the purely elastic regime. The Maxwell model of viscoelasticity permits us to predict that if a shear strain is imposed on a viscoelastic liquid, the resultant shear stress will rise rapidly to a peak value τ_i and will relax over time t according to $\tau(t) = \tau_i \exp(-t/\lambda_r)$. This simple concept has proved powerful in describing the transition between Newtonian and viscoelastic behaviour in volcanic liquids (Dingwell 1995, 1996). Indeed, this model is invoked to explain the transition between volcanic liquids and volcanic glasses on cooling (e.g. Stevenson et al. 1995), the fragmentation of fluid and bubbly magma undergoing decompression (Alidibirov and Dingwell 1996; Kameda and Kuribara 2008) and the conditions under which extensive shear fracture networks can form at or near volcanic conduit margins (Gonnermann and Manga 2003; Tuffen et al. 2003; Kendrick et al. 2014; Hornby et al. 2015). In all cases, it is useful to define a Deborah number De, which is the dimensionless ratio between the Maxwell relaxation time and the timescale of deformation λ. For simple (viscometric) shearing flow, the latter timescale is $1/\dot{\gamma}$ and thus

$$\mathrm{De} = \frac{\lambda_r}{\lambda} = \frac{\mu\dot{\gamma}}{G_\infty} \tag{1}$$

Our hypothesis is that a Deborah number of unity separates the regime between viscous ($\mathrm{De} \ll 1$) and elastic ($\mathrm{De} \gg 1$) responses to shear stresses, or equivalently, between a system that can relax shear stresses efficiently from one in which shear stresses can accumulate elastically. In two different experimental types, both Webb and Dingwell (1990a) and Cordonnier et al. (2012a, b) showed that in fact the first evidence of behaviour that is not purely viscous occurs at $\mathrm{De} \geq 10^{-2}$ and not exactly at unity. This observation will be discussed later.

The power of Eq. 1 is that simple parameters that can be measured in the laboratory can be used to predict where this transitional point of $\mathrm{De} = 1$ will be. This critical Deborah number can be termed De'. In what follows, we give examples of how μ and G_∞ have been found for a range of volcanic liquid compositions at magmatic temperatures and volatile contents before exploring the validity and applicability of the scaling in Eq. 1. The dimensionless nature of Eq. 1 means that it can be assessed for any system so long as λ can be defined.

At the core of the experimental toolkit is the determination of the fundamental quantities that are required to model volcanic processes involved in unrest. The most variable physical quantity in volcanic systems is the viscosity of the liquid phase. Using a range of techniques, this viscosity can be determined with prodigious accuracy and significant effort has been expended in mapping the full range of composition, temperature (see Giordano et al. 2008) and volatile content (Hess and Dingwell 1996) relevant to shallow magmatic settings. Multicomponent models have been proposed such that the viscosity of any composition of silicate volcanic liquid on Earth can now be predicted as a function of temperature including the range of conditions between shallow magma storage and the surface (Hess and Dingwell 1996; Giordano et al. 2008). This provides parameterization of $\mu(T)$ for use in Eq. 1. The most commonly used form for $\mu(T)$ is the non-Arrhenian Vogel-Fulcher-Tammann equation of the form

$$\mu = A\exp\left(\frac{B}{T-C}\right), \quad (2)$$

where A, B, and C are coefficients that are experimentally determined and then parameterized as a function of families of oxides in the liquid structure (Giordano et al. 2008) or as a function of the dissolved water content (for calc-alkaline rhyolites: Hess and Dingwell 1996), to give two examples.

To give examples of the range of viscosities of interest in volcanic scenarios, we give end members in Fig. 1 for a calc-alkaline rhyolite using the model from Hess and Dingwell (1996), and for the basaltic liquid composition provided in Zhang et al. (1991) using the model from Giordano et al. (2008), both contoured for a range of water contents.

Unlike viscosity, the shear modulus in the elastic regime does not vary significantly in siliciate liquids and is not strongly dependent on composition or temperature. Indeed, a short survey of the values for G_∞ in silicate glasses and liquids at a range of temperature and a huge range of composition, provided by Dingwell and Webb (1989) shows that $G_\infty = 10^{10\pm0.5}$ Pa. Here we use this range in order to fully parameterize De (by predicting λ_r) as a function of temperature (via Eq. 2) for any silicate liquid in the shallow crust.

2.2 Extensions to Multiphase Magmas

Except in rare circumstances such as obsidian-forming eruptions, volcanic liquids are not often erupted without some proportion of suspended pore space (either as isolated bubbles or connected networks) and rigid crystals. In either case, the utility of Eq. 1 requires additional attention. We posit that for the liquid phase between the pores or the crystals, De given by Eq. 1 holds. However, we acknowledge that estimation or measurement of the rate of shear strain locally between pores or crystals would be difficult (Deubelbeiss et al. 2011). Therefore, a more robust criterion for the viscous-to-brittle transition in crystal- or bubble-bearing magma would require that we scale the bulk strain rate $\dot{\gamma}_b$ on the system for the effect of the suspended load.

For crystals, we show some first-order scaling attempts for their effect on the critical threshold for the onset of brittle behaviour in multiphase magma. The simplest view of the local flow of liquid in crystal-bearing magma under constant shear stress is that the rate of shear strain between the crystals should scale approximately with $\dot{\gamma} = \dot{\gamma}_b(1-\phi_x/\phi_m)^{-1}$ where ϕ_x is the suspended crystal volume fraction and ϕ_m is a jamming fraction above which no more crystals can be

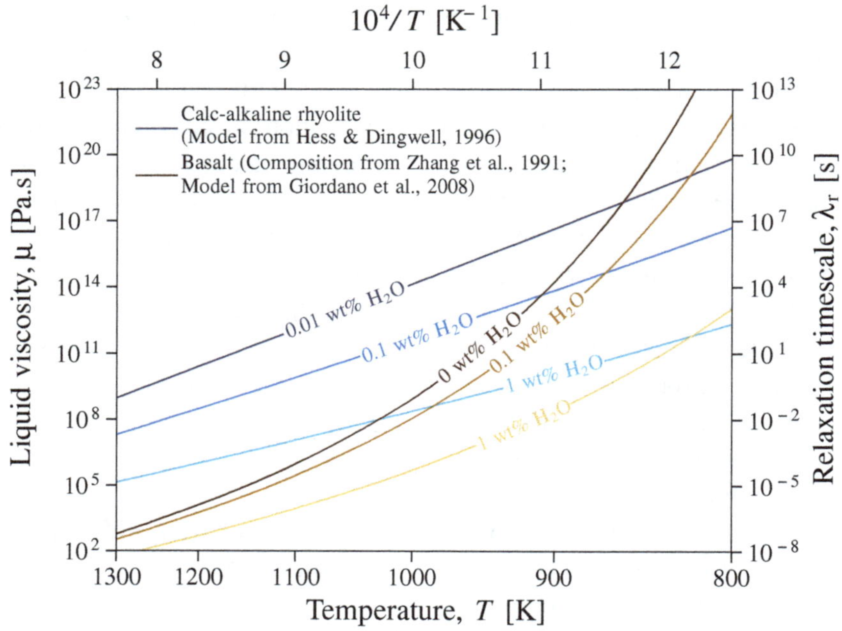

Fig. 1 The viscosity of end-member magmatic liquids with variable dissolved water concentrations. Plotted are the results for calc-alkaline rhyolite from the general viscosity model for hydrous silicic liquids (Hess and Dingwell 1996) for 0.01–1 wt% water, and a typical basaltic composition (composition from Zhang et al. 1991) calculated using the multicomponent viscosity model (Giordano et al. 2008) for 0–1 wt% water. The relaxation timescale is calculated assuming a composition independent value for G_∞ of 10^{10} Pa (Dingwell and Webb 1989)

added to the flowing system. This scaling would imply that the Deborah number for a crystal-bearing magma $\mathrm{De_x}$ would be

$$\mathrm{De_x} = \frac{\mu}{G_\infty}\dot{\gamma}_b\left(1 - \frac{\phi_x}{\phi_m}\right)^{-1}, \quad (3)$$

where ϕ_m is a function of crystal shape and roughness (Mueller et al. 2010; Mader et al. 2013). A critical value of $\mathrm{De_x}$, termed De'_x, is 10^{-2}, consistent with the limiting De' at $\phi = 0$. A new definition of the bulk failure criterion De' can now be made, which decreases as $\phi \to \phi_m$ so that $\mathrm{De}' = \mathrm{De}'_x(1 - \phi_x/\phi_m)$. In Fig. 2 we demonstrate how this concept predicts a linear relationship between $\mathrm{De_x}$ and ϕ_x/ϕ_m, which in turn shows that lower bulk strain rates are required to induce brittle behaviour in crystal-bearing magmas.

This hypothesis was found to hold for simple two-phase systems by Cordonnier et al. (2012a), but where those authors required that μ be replaced by the suspension viscosity of the whole system, which is at odds with the scaling of strain rate for local liquid effects only. Nonetheless, this approach described their data satisfactorily and the conceptual insight that crystals locally increase the liquid rate of shear strain relative to the bulk value is robust, with the implication is that the whole suspension will begin fracturing at lower bulk rates of strain relative to a single-phase liquid of the same composition as crystals are added. It may be that additional second-order effects are important at high ϕ_x, which are not accounted for in this simple analysis.

For the case of porous magmas, there are two considerations: (1) The growth of bubbles exerts a rate of shear strain in the liquid concentrated at the bubble walls, which is broadly independent of any bulk shearing deformation, and (2) like for the crystal case, the presence of bubbles changes the partitioning of the bulk rate of shear strain in the liquid between the bubbles. In the case of scenario (1), the Rayleigh-Plesset equation can be used to relate the bubble growth rate to the gas

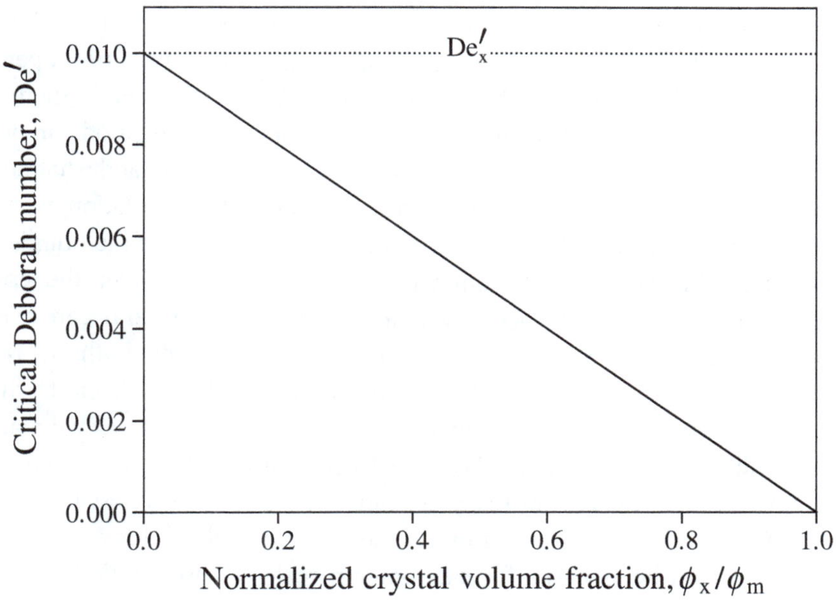

Fig. 2 The critical Deborah number required for fracturing De′ is reduced when crystals are suspended in magma. To a first-order, the De′ value is reduced proportional to $\mathrm{De}' = \mathrm{De}'_x(1 - \phi_x/\phi_m)$

pressure in the bubble, relative to the hydrostatic pressure (Sparks 1978). This can be augmented for the diffusion-controlled gradient of viscosity in the immediate liquid shell around a growing bubble (Prousevitch et al. 1993; Lensky et al. 2001). The component of the rate of shear strain tangential to the bubble wall $\dot{\gamma}_\theta$ can then be computed throughout bubble growth (Ichihara et al. 2002). If we were then to input $\dot{\gamma}_\theta$ into our computation of De, we would predict under which conditions the system would meet the criterion of $\mathrm{De}' = 10^{-2}$ locally at the bubble rim. These conditions would be best cast in terms of the critical rate of pressure or temperature change, or bulk initial volatile content that would allow the bubble to grow sufficiently fast that fractures could be propagated at the bubble wall.

In case (2), the effect of bubbles on the bulk Deborah number is less clear. We speculate that at $\mathrm{De} \gg 1$, the stress required to fracture a bubbly system will be analogous to that required to fracture a vesicular glass. In this case, models for the effect of spherical cavities on the stress required for fracturing are valid and they predict that the stress is reduced significantly as the bulk gas volume fraction increases (Sammis and Ashby 1986). This has been confirmed in the high Deborah number regime for porous liquids analogous to volcanic systems (Vasseur et al. 2013) but remains untested in the low Deborah number regime where bubble deformation may be important.

2.3 Apparent Non-newtonian Effects

In single-phase liquids at Deborah numbers below the critical threshold at which fracturing is observed, there is evidence that during steady state shearing flow, there is a non-Newtonian relationship between applied shear stress and resultant shear rate of strain (Simmons et al. 1982; Dingwell and Webb 1989; Webb and Dingwell 1990a; Cordonnier et al. 2012b). In the onset of this non-Newtonian, the onset of this non-Newtonian behaviour appears to be well described by the value above which viscous dissipation of heat is active on the system length scale of interest (Costa and Macedonio 2005; Hess et al. 2008). This can be scaled by the Brinkman number Br

$$\mathrm{Br} = \frac{\Phi_g}{\Phi_l} = \frac{\mu\dot{\gamma}^2}{kq}, \tag{4}$$

where Φ_g and Φ_l are the gain and loss power densities, respectively, q is the areal heat flux out of the system and k is the thermal conductivity of the material. $\Phi_g = \mu\dot{\gamma}^2$ represents the amount of energy produced in a system of a given volume by viscous dissipation of heat, while $\Phi_l = kq$ represents the energy lost due to diffusive thermal equilibration. When $\mathrm{Br} \gg 1$, heat is efficiently produced and inefficiently lost from the system, resulting in a bulk temperature increase in the liquid. This would manifest itself as an apparent shear thinning rheology if the temperature increase where not locally accounted for, and would be most likely to be operative at high viscosities and high shear strain rates. We note that unlike the Deborah number, the Brinkman number is scale dependent (as q depends on the area available for heat transfer out of the system) and so should be assessed for each system scale separately.

3 The Universal Breaking Timescales of Volcanic Liquids

Laboratory data related to the viscous-to-brittle transition in magmas has been collected in a variety of geometries. Using single-phase liquids, there are two dominant geometries: (1) thin fibers of silicate liquid of basaltic, andesitic, phonolitic and rhyolitic composition were stretched under constant load in tension until the fibers snapped in a singular fracture event (Webb and Dingwell 1990b), and (2) cylinders of synthetic borosilicate liquid were compressed under constant load and the bulk temporal evolution of the rate of axial shortening was determined as viscous (relaxed) if it continuously increased to a steady or near-steady value, or brittle (unrelaxed) if the axial shortening rate jumped due to fracturing events (Cordonnier et al. 2012b). In another type of experiment, analogue vesicular liquids were decompressed at different rates from pressure (Kameda and Kuribara 2008; Kameda et al. 2013). In this type of decompression experiment, if the dominant response was viscous (relaxed), then the sample was seen to grow due to decompression-driven bubble growth, and if the dominant response was brittle (unrelaxed), then smooth sample inflation was punctuated by visible fractures opening. In these decompression experiments there additionally was a violent rupture mode in which the fracturing was pervasive and shattered the sample in a vigorous fragmentation event. These data are selected here because the liquids used have a known relaxation time under the conditions used in the experiments and care was taken by the authors who originated the work to investigate the shear rates of strain local to the bubbles (discussed above).

In Fig. 3 we map these experimental results as a ratio of the deformation timescale to the relaxation timescale, which permits us to contour the plot for critical Deborah numbers. We colour-code the data according to the bulk mode of response of the sample to the deformation using *green* to represent purely viscous relaxed behaviour, *orange* to represent brittle unrelaxed behaviour and *red* to represent the complete violent rupture of the sample. These data suggest that the viscous relaxed behaviour transitions to unrelaxed brittle behaviour at a Deborah number of 10^{-2}, and are consistent across a huge range of experimental conditions and across both the single-phase compression and vesicular decompression experiment types. Furthermore, we see that a Deborah number of unity consistently separates the experiments for which the bulk response was unrelaxed and brittle from those for which the response was violent rupture, fragmentation or complete failure. It appears that these two thresholds, $\mathrm{De} = 10^{-2}$ and $\mathrm{De} = 1$ are universal even when comparing analogue room temperature liquids with high temperature silicate liquids deformed in a variety of ways. This lends power to the scaling provided by the dimensionless Deborah number and implies that we need only to define the liquid viscosity and a characteristic rate of deformation in order to predict whether a system will flow viscously or rupture violently. For example, the working viscosity of the analogue fluid used in Kameda and Kuribara (2008) is $10^0 \leq \mu \leq 10^{10}$ Pa.s, which extends to much lower used in Cordonnier et al. (2012b), and yet the scaling with the Deborah number

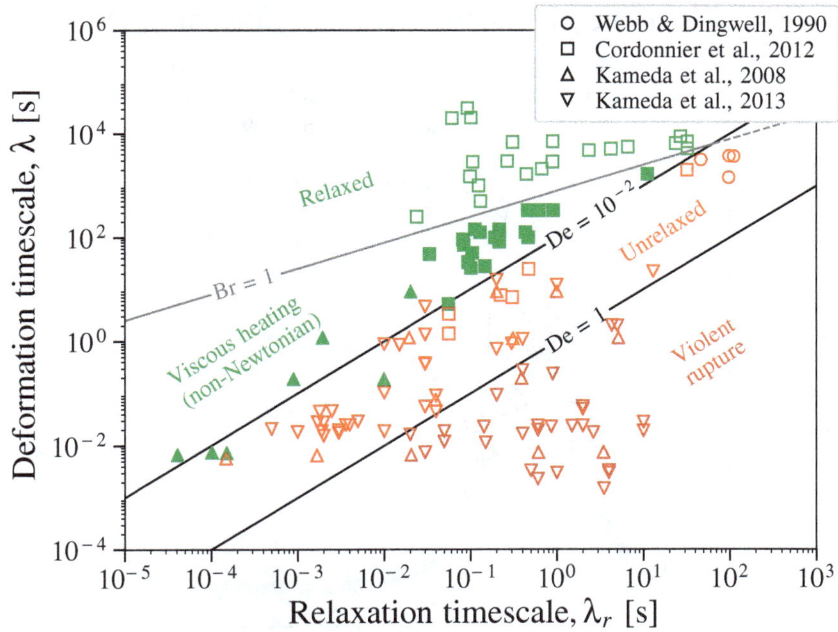

Fig. 3 A map of behaviour arising from deformation experiments on natural (Webb and Dingwell 1990b) and synthetic (Cordonnier et al. 2012b) silicate liquids and on analogues of magmatic liquids undergoing vesiculation (Kameda and Kuribara 2008; Kameda et al. 2013). The experimental results were obtained by applying a characteristic timescale of deformation λ of a liquid with a characteristic timescale of stress relaxation λ_r. Marked are ratios λ_r/λ, equivalent to Deborah numbers De, of 10^{-2} and 1, which separate experiments with a purely viscous response from those with an unrelaxed brittle response or a violent rupture response, respectively. Additionally marked is the Brinkman number Br of 1, which separates the boundary between isothermal viscous behaviour (*above the line*) and viscous behaviour in which the material heats up due to viscous dissipation as heat (*below the line*)

holds across a broad range simply because the deformation timescale was also smaller in the former example.

In the viscous field, Cordonnier et al. (2012b) additionally recorded whether samples hosted a measurable temperature increase due to viscous dissipation of heat during the experiment, or not. In Fig. 3, we plot those with a measureable temperature increase as filled symbols and those without this feature as unfilled symbols. To explain this, we plot the threshold dimensionless Brinkman number of unity Br = 1, using an estimation of the loss power density for the furnace and sample size used $\Phi_l = 10^{4.5}\text{W.m}^{-2}$ (Cordonnier et al. 2012b). We note that the Brinkman number curve consistently divides the regimes of purely isothermal experiments and those with measurable heat gain. On this map, the position of Br = 1 is non-unique and depends on sample size, such that on the scale of a volcanic conduit, for example, Br = 1 would occur at much higher deformation timescales for the same relaxation timescale (the curve would shift up in Fig. 3). This implies that in the natural case, viscous dissipation of heat may be far more important than shown here for the sample length scale (Costa and Macedonio 2005; Mastin 2005; Costa et al. 2007). Nevertheless, the Deborah number limits discussed appear to be universal and, importantly, are scale-independent.

To gain helpful physical insight into why $10^{-2} \leq \text{De} \leq 10^{0}$ is the transitional window between a purely viscous and a purely elastic response of a liquid to a deformation, we provide low-strain, high frequency oscillatory rheological measurement data for similar liquids (Fig. 4). Here, rods of single-phase liquid are subjected to a low-amplitude oscillatory strain with a forcing frequency ω at a range of temperatures similar to those used in experiments presented in Fig. 2. Here, ω is normalized with λ_r, yielding a dimensionless frequency or, equivalently, a

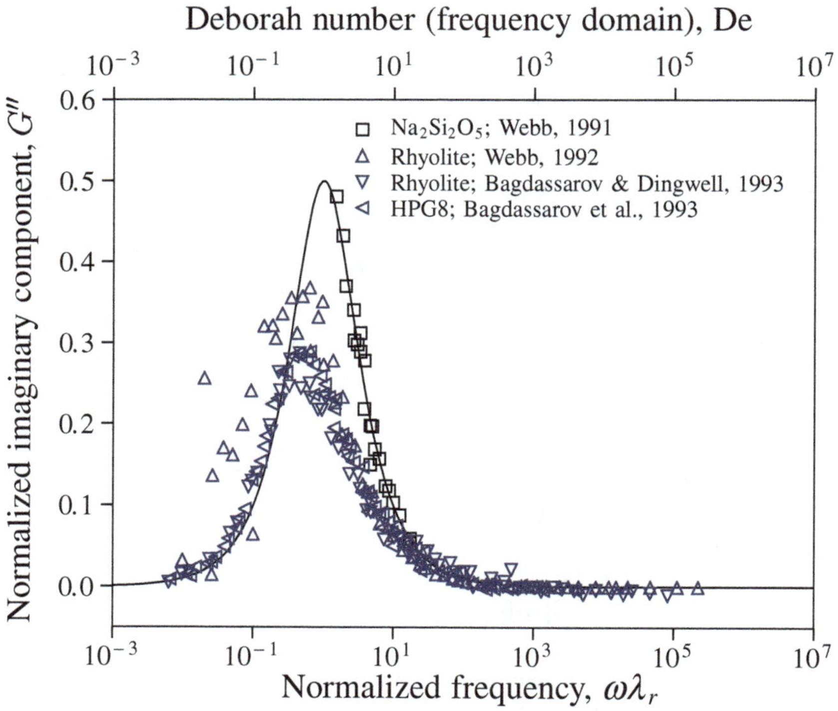

Fig. 4 The response of synthetic (Webb 1991; Bagdassarov and Dingwell 1993a) and natural (Webb 1992; Bagdassarov and Dingwell 1993b) silicate liquids to low strain, high frequency oscillatory deformation tests. This is shown as a normalized imaginary component of the shear modulus G''/G_∞ as a function of the frequency of applied oscillation normalized by the relaxation timescale $\omega\lambda_r$, which is equivalent to a Deborah number De in the frequency domain. We show the viscoelastic prediction for a Maxwell liquid for comparison (*solid curve*)

frequency-domain version of the Deborah number. The metric that we choose to track is the imaginary component of the complex elastic modulus G'' normalized by G_∞. When this value G''/G_∞ is close to zero in the low Deborah number limit, the system is dominated by liquid behaviour and when it rises from zero at increasing Deborah number, there is an increasing component of the response to the forced oscillation that is elastic. We present collated data for a sodium disilicate synthetic composition (Webb 1991), rhyolitic compositions (Webb 1992; Bagdassarov and Dingwell 1993b) and a synthetic composition used as an analogue for type calc-alkaline rhyolite systems (Bagdassarov and Dingwell 1993a). This imaginary component can be described by the generalized Debye model for viscoelasticity for systems with a single characteristic relaxation time as $G'' = \omega\lambda_r/\left[(\omega\lambda_r)^2 + 1\right]$. The data reproduce the broad shape of the Debye model, albeit with an overprediction of G'' around $\omega\lambda_r \sim 1$ for the rhyolitic liquids, which is poorly understood. Nevertheless, we point out that the first onset of a measurable elastic component to the response of the liquid to deformation occurs at $\mathrm{De} = \omega\lambda_r \approx 10^{-2}$. Similarly, the point above which the majority of the response to deformation is elastic (the peak of G'') occurs at $\mathrm{De} = \omega\lambda_r \approx 1$ (Fig. 3). We use this observation to validate the two thresholds found in Fig. 3. This implies that fractures can propagate in silicate liquids when there is even a small component of elastic behaviour ($\mathrm{De} \geq 10^{-2}$) and that those fractures can propagate vigorously when the elastic behaviour dominates over the viscous behaviour ($\mathrm{De} \geq 1$). An interpretation might also be that $10^{-2} \leq \mathrm{De} \leq 1$ is the range in which fracture propagation is competing with viscous relaxation of stress and therefore the fractures are unlikely to be long, sharp-tipped or pervasive. And that at $\mathrm{De} \geq 1$, stress dissipation by fracture propagation can be localized onto longer, sharp and pervasive fracture networks.

The fact that a viscous limit to the expansion of vesicular magma also scales with our Deborah number criterion is tantalizing. Microphysically, in this case it is not fracture of a deforming homogeneous liquid, but fractures propagating at bubble walls as the local rate of strain in the liquid induced by expansion of the bubble meets the Deborah number criterion for fracture propagation (c.f. Kameda and Kuribara 2008). For complete rupture of the vesicular material—which is a fragmentation event or *violent rupture* in Fig. 3—the fractures propagating from the bubbles must interact. Presumably in the range $10^{-2} \leq \mathrm{De} < 1$, fractures propagate but the fracture tips are blunted during competing viscous relaxation such that brittle behaviour can be observed but is not catastrophic to the system. Then, as the local strain rate at the bubble wall increases further and De approaches 1, the fractures propagate in a dominantly elastic medium and can span the inter-bubble distances, interacting to produce violent rupture. Therefore, it is clear that a scaling of the Deborah number concept to bubble wall dynamics would provide a fragmentation criterion for viscoelastic vesiculating magma (c.f. Namiki and Manga 2005; Koyaguchi et al. 2008; Namiki and Manga 2008).

4 Laboratory-Scale Unrest Signals

When magma breaks in a laboratory experiment, acoustic emissions—packets of acoustic energy—are released and can be recorded (Benson et al. 2008; Lavallee et al. 2008). These signals appear to represent large total released amounts of energy when λ is short (high $\dot{\gamma}$), compared to when λ is long (low $\dot{\gamma}$) (Lavallee et al. 2008). This supports our posit that fracture networks are more likely to be pervasive and large when De is large than when De is small. For single phase silicate liquids deformed at high De, Tuffen et al. (2008) showed that the experimental acoustic event frequency range and sample fracture length scale scaled with natural constraints of in-conduit volcanic fracture systems and natural frequencies of low-frequency volcano-seismicity, indicating that these viscoelastic fracturing events are indeed the likely source mechanism for low-frequency earthquakes at volcanoes. This confirmed the conclusion of Neuberg et al. (2006), who showed that these low-frequency events were most likely to be associated with repetitive fracturing of magma at a given depth in the conduit during ascent. Neuberg et al. (2006) further modelled magma ascent in confined geometry and showed that a $\mathrm{De} \sim 1$ is met at a depth of 830 m below the surface using parameters typical for recent eruptions at Soufriere Hills volcano and the magma thereof. Thomas and Neuberg (2012) predicted a deeper source of 1500 m using the same model approach but by invoking a conduit restriction (see Chap. 9), consistent with inversions for low-frequency sources using seismic data. Therefore, viscoelastic magma fracturing in the high Deborah number regime appears to be a consistent model for the source mechanics of low-frequency volcano seismicity, often used for eruption forecasting.

Vasseur et al. (2015) showed that the forecastability of full sample rupture scales with the heterogeneity of the system—cast most simply as a porosity (Fig. 5). The implication is that the less vesicular the magma undergoing deformation, the less likely that accurate forecasts based on accelerated precursory signals can be made, with up to ~120% error on the timing of the rupture event for single-phase homogeneous liquids in the high De limit. It may be that the magma vesicularity, pore-network structure (Vasseur et al. 2017), crystallinity and textural anisotropy, play key roles in determining forecasting success based on low-frequency earthquakes. More experimental work is clearly required in this area, along with more rigorous scaling between acoustic and seismic events that originate from magma failure.

An implication of the models explored here, encapsulated by Eq. 1, is that the peak stress supported by a liquid σ_m can be predicted as a

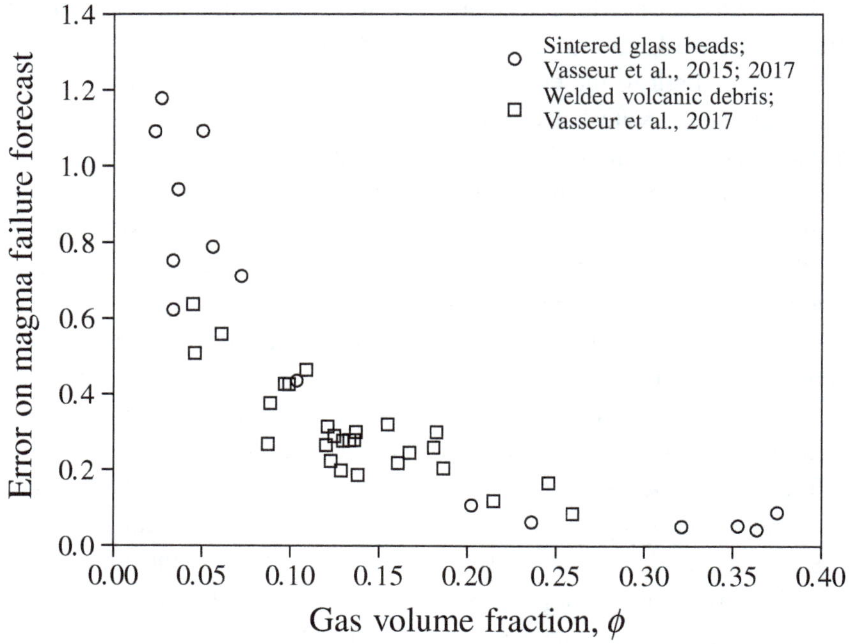

Fig. 5 In the high-De regime, the error on a prediction of failure times scales with the porosity of the material such that low porosity magmas are unpredictable and high porosity magmas are predictable (Vasseur et al. 2015). The failure times are recorded as a large stress drop during uniaxial loading, and the approach to failure is monitored using acoustic emissions generated by pre-failure micro-fracturing events

Fig. 6 The peak stress that can be supported by a shearing liquid in the low De limit is given by $\sigma_m = G_\infty \mathrm{De}$. And in the high De limit, this value appears to asymptotically approach $\sigma_m = 10^{-2} G_\infty$ which is equivalent to the strength of glassy materials in that same regime (Simmons et al. 1982)

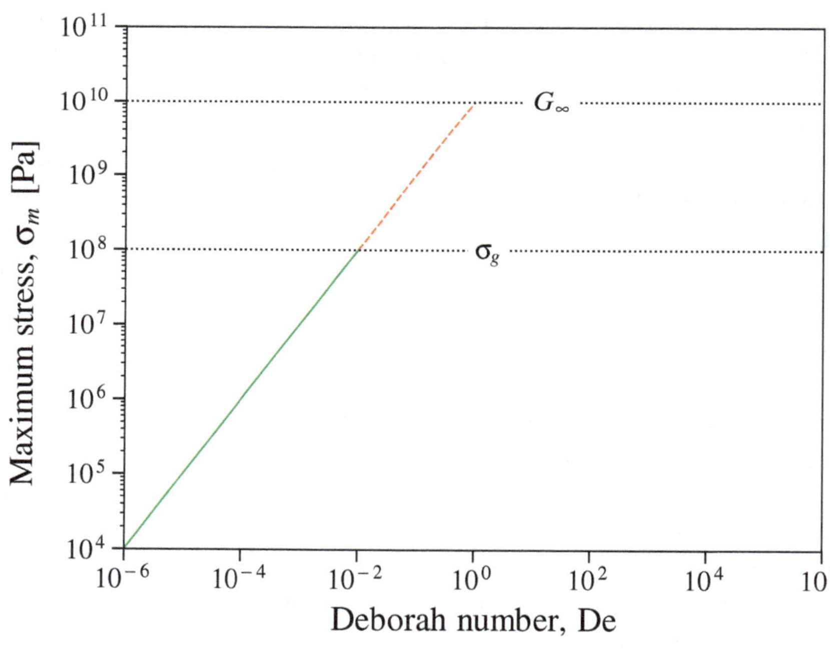

function of De by $\sigma_m = G_\infty \mathrm{De}$. Once $\mathrm{De} \geq 10^{-2}$, however, this linear relationship appears to be invalid as fractures can form in the liquid. It is perhaps significant that the average strength of glass in the high De regime is $\sim 10^8$ Pa (Simmons et al. 1982; Vasseur et al. 2013), consistent with a rupture threshold of $10^{-2}G_{\infty}$. These relationships are explored in Fig. 6.

Here we summarize just one example of how targeted laboratory experiments and dimensional analysis have demonstrated the most likely source mechanism of unrest signals commonly used to monitor magma movement and predict impending eruptions. Clearly, with scaling arguments for the applicability of an experimental set up—such as we show here with the Deborah number analysis—laboratory-based work can provide new insights into unrest mechanisms that facilitate more accurate forward-modelling of geophysical signals.

References

Alidibirov M, Dingwell DB (1996) Magma fragmentation by rapid decompression. Nature 380:146–148

Bagdassarov N, Dingwell DB, Webb SL (1993) Effect of boron, phosphorus and fluorine on shear stress relaxation in haplogranite melts. Eur J Miner 3:409–425

Bagdassarov NS, Dingwell DB (1993) Frequency dependent rheology of vesicular rhyolite. J Geophy Res: Solid Earth 98(B4):6477–6487

Benson P, Vinciguerra S, Meredith P, Young R (2008) Laboratory simulation of volcano seismicity. Science (80-) 322:249–252

Chouet BA (1996) Long-period volcano seismicity: its source and use in eruption forecasting. Nature 380:309–316. doi:10.1038/380309a0

Chouet BA, Page RA, Stephens CD et al (1994) Precursory swarms of long-period events at Redoubt Volcano (1989–1990), Alaska: their origin and use as a forecasting tool. J Volcanol Geotherm Res 62:95–135. doi:10.1016/0377-0273(94)90030-2

Cordonnier B, Caricchi L, Pistone M et al (2012a) The viscous-brittle transition of crystal-bearing silicic melt: direct observation of magma rupture and healing. Geology 40:611–614

Cordonnier B, Schmalholz SM, Hess K, Dingwell DB (2012b) Viscous heating in silicate melts: an experimental and numerical comparison. J Geophys Res 117:B02203

Costa A, Macedonio G (2005) Viscous heating effects in fluids with temperature-dependent viscosity: triggering of secondary flows. J Fluid Mech 540:21–38

Costa A, Melnik O, Vedeneeva E (2007) Thermal effects during magma ascent in conduits. J Geophys Res 112: B12205. doi:10.1029/2007JB004985

Deubelbeiss Y, Kaus BJ, Connolly JA, Caricchi L (2011) Potential causes for the non-Newtonian rheology of crystal-bearing magmas. Geochem Geophy Geosys 12(5)

Dingwell DB (1995) Relaxation in silicate melts; some applications. Rev Mineral Geochemistry 32:21–66

Dingwell DB (1996) Volcanic dilemma: flow or blow? Science (80-) 273:1054–1055. doi:10.1126/science.273.5278.1054

Dingwell DB, Webb SL (1989) Structural relaxation in silicate melts and non-Newtonian melt rheology in geologic processes. Phys Chem Miner 16:508–516

Giordano D, Russell JK, Dingwell DB (2008) Viscosity of magmatic liquids: a model. Earth Planet Sci Lett 271:123–134

Gonnermann HM, Manga M (2003) Explosive volcanism may not be an inevitable consequence of magma fragmentation. Nature 426:432–435

Hess KU, Dingwell DB (1996) Viscosities of hydrous leucogranitic melts: a non-Arrhenian model. Am Mineral 81:1297–1300

Hess K-U, Cordonnier B, Lavallee Y, Dingwell DB (2008) Viscous heating in rhyolite: an in situ experimental determination. Earth Planet Sci Lett 275:121–126. doi:10.1016/j.epsl.2008.08.014

Hornby AJ, Kendrick JE, Lamb OD, Hirose T, De Angelis S, Aulock FW, Umakoshi K, Miwa T, Henton De Angelis S, Wadsworth FB, Hess KU (2015) Spine growth and seismogenic faulting at Mt. Unzen, Japan. J Geophys Res: Solid Earth 120(6):4034–4054

Ichihara M, Rittel D, Sturtevant B (2002) Fragmentation of a porous viscoelastic material: implications to magma fragmentation. J Geophys Res: Solid Earth 107(B10)

Kameda M, Kuribara H, Ichihara M (2008) Dominant time scale for brittle fragmentation of vesicular magma by decompression. Geophy Res Lett 35(14)

Kameda M, Ichihara M, Shimanuki S, Okabe W, Shida T (2013) Delayed brittle-like fragmentation of vesicular magma analogue by decompression. J Volcanol Geoth Res 258:113–125

Kendrick JE, Lavallée, Y, Hirose T, Di Toro G, Hornby AJ, De Angelis S, Dingwell DB (2014) Volcanic drumbeat seismicity caused by stick-slip motion and magmatic frictional melting. Nat Geosci 7(6):438

Koyaguchi T, Scheu B, Mitani NK, Melnik O (2008) A fragmentation criterion for highly viscous bubbly magmas estimated from shock tube experiments. J Volcanol Geotherm Res 178:58–71

Lavallee Y, Meredith PG, Dingwell DB et al (2008) Seismogenic lavas and explosive eruption forecasting. Nature 453:507–510. doi:10.1038/nature06980

Lensky NG, Lyakhovsky V, Navon O (2001) Radial variations of melt viscosity around growing bubbles and gas overpressure in vesiculating magmas. Earth Planet Sci Lett 186:1–6

Mader HM, Llewellin EW, Mueller SP (2013) The rheology of two-phase magmas: a review and analysis. J Volcanol Geoth Res 257:135–158

Mastin LG (2005) The controlling effect of viscous dissipation on magma flow in silicic conduits. J Volcanol Geoth Res 143(1):17–28

Miller AD, Stewart RC, White R et al (1998) Seismicity associated wtih dome growth and collapse at the

Soufriere Hills Volcano, Montserrat. Geophys Res Lett 25:3401–3404

Mueller S, Llewellin EW, Mader HM (2010) The rheology of suspensions of solid particles. Proc R Soc A Math Phys Eng Sci 466:1201–1228

Namiki A, Manga M (2005) Response of a bubble bearing viscoelastic fluid to rapid decompression: implications for explosive volcanic eruptions. Earth Planet Sci Lett 236(1):269–284

Namiki A, Manga M (2008) Transition between fragmentation and permeable outgassing of low viscosity magmas. J Volcanol Geoth Res 169(1):48–60

Neuberg JW, Tuffen H, Collier L et al (2006) The trigger mechanism of low-frequency earthquakes on Montserrat. J Volcanol Geotherm Res 153:37–50. doi:10.1016/j.jvolgeores.2005.08.008

Prousevitch AA, Sahagian DL, Anderson AT (1993) Dynamics of diffusive bubble growth in magmas: isothermal case. J Geophys Res Solid Earth 98:22283–22307

Salvage R, Neuberg JW (2016) Using a cross correlation technique to refine the accuracy of the Failure Forecast Method: application to Soufrière Hills volcano, Montserrat. J Volcanol Geotherm Res 324:118–133. doi:10.1016/j.jvolgeores.2016.05.011

Sammis CG, Ashby MF (1986) The failure of brittle porous solids under compressive stress states. Acta Metall 34:511–526

Simmons JH, Mohr RK, Montrose CJ (1982) Non-Newtonian viscous flow in glass. J Appl Phys 53:4075–4080

Sparks RSJ (1978) The dynamics of bubble formation and growth in magmas: a review and analysis. J Volcanol Geotherm Res 3:1–37

Stevenson RJ, Dingwell DB, Webb SL, Bagdassarov NS (1995) The equivalence of enthalpy and shear stress relaxation in rhyolitic obsidians and quantification of the liquid-glass transition in volcanic processes. J Volcanol Geotherm Res 68:297–306

Thomas ME, Neuberg J (2012) What makes a volcano tick-a first explanation of deep multiple seismic sources in ascending magma. Geology 40:351–354. doi:10.1130/g32868.1

Tuffen H, Dingwell DB, Pinkerton H (2003) Repeated fracture and healing of silicic magma generate flow banding and earthquakes? Geology 31:1089–1092. doi:10.1130/g19777.1

Tuffen H, Smith R, Sammonds PR (2008) Evidence for seismogenic fracture of silicic magma. Nature 453:511–514

Vasseur J, Wadsworth FB, Lavallée Y et al (2013) Volcanic sintering: timescales of viscous densification and strength recovery. Geophys Res Lett 40:5658–5664

Vasseur J, Wadsworth FB, Lavallée Y, Bell AF, Main IG, Dingwell DB (2015) Heterogeneity: the key to failure forecasting. Sci Rep 5

Vasseur J, Wadsworth FB, Heap MJ, Main IG, Lavallée Y, Dingwell DB (2017) Does an inter-flaw length control the accuracy of rupture forecasting in geological materials? Earth Planet Sci Lett

Voight B (1988) A method for prediction of volcanic eruptions. Nature 332:125–130

Webb SL (1991) Shear and volume relaxation in $Na_2 Si_2 O_5$. Am Mineral 76(9–10):1449–1454

Webb SL (1992) Low-frequency shear and structural relaxation in rhyolite melt. Phys Chem Mineral 19 (4):240–245

Webb SL, Dingwell DB (1990a) The onset of non-Newtonian rheology of silicate melts. Phys Chem Miner 17:125–132

Webb SL, Dingwell DB (1990b) Non-Newtonian rheology of igneous melts at high stresses and strain rates: experimental results for rhyolite, andesite, basalt, and nephelinite. J Geophys Res Solid Earth 95:15695–15701

Zhang Y, Stolper EM, Wasserburg GJ (1991) Diffusion of a multi-species component and its role in oxygen and water transport in silicates. Earth Planet Sci Lett 103:228–240

Volcano Seismology: Detecting Unrest in Wiggly Lines

R.O. Salvage, S. Karl and J.W. Neuberg

Abstract

Seismology is a useful tool to gain a better understanding of volcanic unrest in real time as it unfolds. The generation of seismic signals in a volcanic environment has been linked to a number of different physical processes occurring at depth, including fracturing of the volcanic edifice (producing high frequency seismicity) and movement of magmatic fluids (producing low frequency seismicity). Further classification of seismic signals according to their waveform similarity, in addition to their frequency content, allows greater detail in temporal and spatial changes of seismicity to be detected. At Soufrière Hills volcano, Montserrat, one of the target volcanoes of the VUELCO project, families of similar waveforms provided valuable insight into evaluating the significance of ongoing unrest. In June 1997 over 6000 more events were able to be identified over a 5 day period of interest (22 to 25 June) by using families of seismic events, rather than a standard amplitude-based detection algorithm. In total, 11 families were identified, with the events clustering into a number of swarms, suggesting a repeating and non destructive cyclic source mechanism. Since each family is believed to represent a distinct source location and mechanism, identifying 11 coexisting families

R.O. Salvage (✉)
Observatorio Vulcanológico y Sismológico
de Costa Rica, Universidad Nacional,
Apartado Postal: 2386-3000, Heredia, Costa Rica
e-mail: beckysalvage@gmail.com

S. Karl
Shell, NAM offices, Schepersmaat 2,
9405 TA Assen, The Netherlands
e-mail: sandra.karl84@gmail.com

J.W. Neuberg
School of Earth and Environment, Institute of
Geophysics and Tectonics, University of Leeds,
Leeds LS2 9JT, United Kingdom
e-mail: j.neuberg@leeds.ac.uk

Advs in Volcanology (2019) 185–201
DOI 10.1007/11157_2017_11

Published Online: 02 August 2017

reflects the complex diversity of physical processes which act simultaneous at this volcano. In July 2003, conditions at the volcano had clearly changed since only one family of seismicity was identified. The source location of this family appeared to shift with time from 8 July (when no events from the family were identified) to 12 July (where most events had a cross correlation coefficient over 0.9). In addition, the use of families appears to greatly aid hindsight forecasting attempts for the large scale dome collapses of 1997 and 2003 using the Failure Forecast Method. Knowledge of the temporal and spatial extent of seismicity during periods of unrest, its source mechanism and its relationship to physical processes at depth is essential for decision and policy makers for risk mitigation. However, the source mechanisms of such volcanic seismicity is still much debated and appears to often be misinterpreted because of compromising assumptions used in the numerical modelling of inverting such sources. Use of a spatially extended source such as a ring fault structure, rather than a single point for determining the origin of low frequency seismicity, is now thought to be more realistic for the mechanism of such events since it more accurately represents the movement of magma through a conduit. However, use of this spatially extended source instead of a simple single point results in a large underestimation of slip from P-wave amplitudes, which may lead to an underestimation in magma ascent rates, with large consequences for eruption forecasting. Additionally, the P-wave radiation patterns exhibited by these two mechanisms are remarkably similar, and can only be distinguished if the small radial radiation lobes can be determined. In a volcanic environment this is extremely difficult due to large uncertainties in earthquake source depth locations, and the implementation of small aperture seismic networks.

Español

La sismología es una herramienta geofísica valiosa que brinda información en tiempo real, permitiendo una mejor comprensión del comportamiento de sistemas volcánicos que inician un proceso de reactivación o de intensificación de la actividad. La generación de señales sísmicas en ambientes volcánicos se ha relacionado con un número diverso de procesos geofísicos que ocurren en el interior de los volcanes, incluyendo fracturamiento del edificio volcánico (produciéndose sismicidad de alta frecuencia) y movimiento de fluidos magmáticos (produciéndose sismicidad de baja frecuencia). La clasificación de señales sísmicas basada en la similitud de las formas de onda, además del contenido de frecuencias, ha permitido detectar cambios temporales y espaciales de la sismicidad con mayor detalle.

En el volcán Soufriere Hills en Monserrat, uno de los volcanes investigados como parte del Proyecto VUELCO, el reconocimiento de familias de señales sísmicas con formas de onda similares proveyó un entendimiento valioso en la evaluación de la significancia de la reactivación de su actividad volcánica. En junio de 1997, se pudieron identificar más de 6000 eventos ssmicos dentro de unperiodo particular de 4 días (entre el 22 y el 25 de junio) mediante la determinación de familias de eventos sísmicos que fueron determinados por un algoritmo estándar de detección basado en la amplitud de la señal sísmica. En total, 11 familias de eventos sísmicos fueron identificadas, con grupos de eventos conformando enjambres sísmicos, lo que sugiere un mecanismo cíclico repetitivo de una fuente sísmica no destructiva. Ya que se considera que cada familia representa una fuente con ubicación espacial y mecanismo distinto, la identificación de 11 familias refleja la compleja diversidad de los procesos geofísicos que operan simultáneamente en este volcán en particular. En julio del 2003, el régimen del volcán cambió definitivamente ya que solamente 1 familia de eventos sísmicos fue identificada. La ubicación de la fuente sísmica de esta familia parece haber migrado con el tiempo entre el 8 de julio (cuando ningún evento sísmico de esta familia fue identificado) y el 12 de julio (cuando la mayoría de eventos tuvo un coeficiente de cros-correlación superior a 0,9). Además, el establecimiento de familias de sismos parece haber sido de gran ayuda en los intentos de pronosticar los colapsos del domo en gran escala de 1997 y el 2003 utilizando el método determinístico de pronóstico de rompimiento por fatiga (Failure Forecast Method, FFM). El conocimiento de la extensión temporal y espacial de la sismicidad durante periodos de reactivación volcánica, el mecanismo de la fuente, y su relación con los procesos geofísicos a niveles profundos son aspectos esenciales para los tomadores de decisiones y ejecutores de políticas relacionadas con la mitigación de riesgos.

En general, los mecanismos de la fuente generadora de la sismicidad volcánica continúan aún bajo gran debate y parecen ser con frecuencia mal interpretados debido a simplificaciones comprometedoras usadas en los modelos numéricos de inversión de dichas fuentes. El uso de una fuente espacial extendida tal como una falla estructural tipo anular, en lugar de un solo punto para determinar el origen de sismicidad de baja frecuencia, se piensa es más realista para visualizar el mecanismo de este tipo de eventos ya que representa con más exactitud el movimiento del magma a través de un conducto. Sin embargo, el uso de este modelo resulta en la subestimación significativa del deslizamiento en la falla determinada a través de las amplitudes de las ondas P, lo que podría llevar a una subestimación de las tasas de ascenso del magma, con consecuencias cruciales para el pronóstico de erupción inminente. Por otra parte, los patrones de radiación de la onda P mostrados por estos dos mecanismos son marcadamente similares, y pueden ser solamente distinguidos si los pequeños lóbulos radiales de radiación pueden ser determinados. En un contexto volcánico, esta es una tarea extremadamente difícil debido a la gran incertidumbre inherente a la localización de la profundidad de la fuente sísmica, y a la implementación de redes sísmicas de pequeña extensión.

Keywords

Volcano seismology · Families of similar seismic events · Magmatic source mechanism · Eruption forecasting

Index Terms

Volcano seismology · Soufrière Hills · Chiles-Cerro Negro · Low frequency seismicity · Families · Cross correlation · Forecasting · Failure forecast method · Source mechanisms · Point source · CLVD · Ring fault · Spatially extended source · P-wave radiation patterns

Volcanic Unrest

The monitoring of active volcanoes for the protection of society has excelled in recent decades fuelled by rapid technological advances, which have allowed the development and deployment of more cost-effective monitoring solutions. It is now possible to detect geophysical and geochemical signals from a volcano which previously would have been below the detection threshold. The routine monitoring of volcanoes during periods of quiescence is crucial, although not always feasible, in order to assess background levels of activity at volcanoes, and thus more rapidly detect the onset of future unrest. Most simply, volcanic unrest is defined as a deviation from background levels of activity towards a level which is a cause for concern over short time scales of hours to days. Volcanic unrest does not necessarily lead to eruption, although this is the most likely outcome, with 64% of 228 volcanoes that have experienced unrest since the year 2000 culminating in eruptive events (Phillipson et al. 2013).

Current monitoring efforts at volcanoes can be grouped into three categories: measurements of surface degassing; deformation; and seismic activity, which are all thought to result from the movement of magmatic fluids at depth, and therefore may provide key insights into an impending eruption. Unrest must be detected on short timescales which is appropriate for decision-making, and therefore seismicity has remained a primary monitoring tool, as it can be remotely analysed in real-time, often by an automated system, if a number of sensors are placed around the volcano. Deviation from the background level is also often easier to determine for seismicity than for other signals.

Seismic Event Characterisation

A wide variety of seismic signals exist within volcanic settings, associated with magmatic and hydrothermal fluid movement at depth, pressurization of the volcanic edifice, and/or the surface manifestation of the interaction of these processes. The variety of signals is a reflection of the number of different processes and the great structural heterogeneities found within this contex. The characterisation of seismicity can be based upon waveform similarities, but is traditionally based upon the signals' time and frequency characteristics: different bands of frequency relate to different active source processes at depth, which can then be distinguished from one another, although the frequency bands associated with each process may overlap (Lahr et al. 1994).

The identification of seismic events in volcanic settings can assist with the detection of the onset and cessation of unrest, while the spatial and temporal patterns of occurrence may be informative of magma movement and changes in stress at depth, as well as the spatial extent of concern. An understanding of the physical processes occurring at depth is essential if accurate

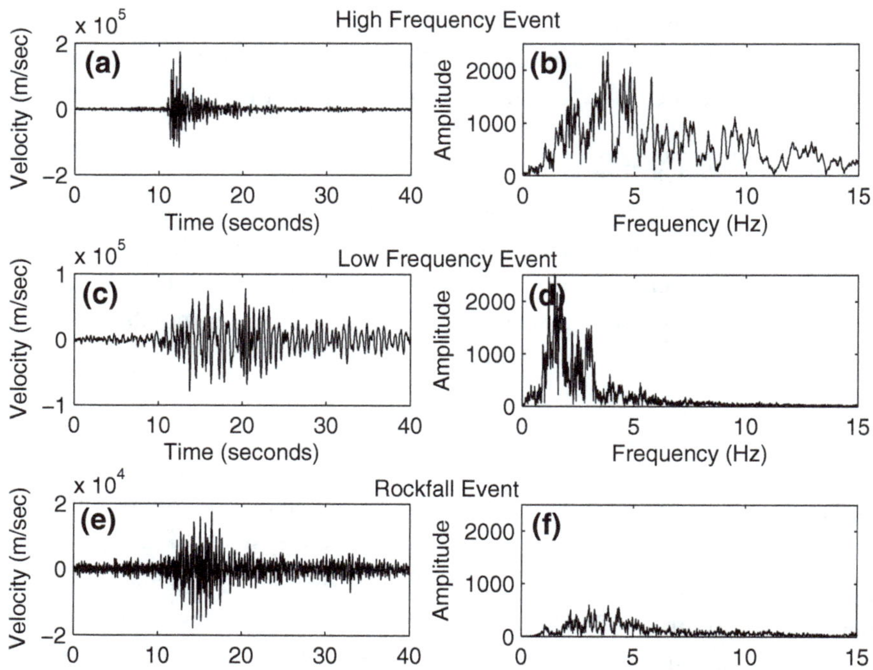

Fig. 1 Examples of waveforms and their frequency content seen in volcanic environments taken from Soufrière Hills Volcano, Montserrat in 1997. Soufrière Hills Volcano was a target volcano identified by the VUELCO project for investigation. **a**, **b** High frequency waveform with clear phase arrivals. **c**, **d** Low frequency waveforms with an emergent onset. Waveform filtered between 0.5 and 5 Hz. **e**, **f** Rockfall event with classic "cigar" shape

and timely forecasts of volcanic eruptions, and developing unrest scenarios, are to be made.

Classification by Frequency Content

Seismic signals originating from processes occurring at depth within the volcanic system are usually split into high- and low-frequency end-members, although in reality a continuum across the spectrum exists between the two (Chouet and Matoza 2013). High frequency seismic signals (Fig. 1a, b), also known as Volcano-Tectonic (VT) events, have energy concentrated in the frequency range of 1 to 20 Hz, generally peaking between 6 and 8 Hz (Lahr et al. 1994). They are characterized by clear, impulsive P- and S-wave arrivals, followed by a short coda. High frequency seismicity is usually attributed to brittle failure within the volcanic edifice, where magmatic processes create enough elastic strain to force the surrounding rocks into failure (Arciniega-Ceballos et al. 2003), similar to the generation of tectonic earthquakes.

Low frequency (LF) seismic signals (Fig. 1c, d) usually occupy the spectral range of 0.2 to 5 Hz (Chouet and Matoza 2013), and are frequently characterized by emergent P-wave onsets and lack of S-wave arrivals. It has been suggested that the occurrence of low frequency waveforms is linked to resonance of seismic energy trapped at a solid-fluid interface either within a crack (e.g. Chouet 1988), or a volcanic conduit (e.g. Neuberg et al. 2000). The trigger mechanism of such seismic energy is further disputed, with suggestions that it may be generated by: (1) a stick-slip motion along the conduit walls as magma ascends (e.g. Iverson et al. 2006); (2) the brittle failure of magma itself either through an increase in viscosity and strain rates (Lavallée et al. 2008), which may be due to an increase in the ascent rate of magma through the conduit (Neuberg et al. 2006), changes in the crystal and/or bubble concentration in the magma (Goto 1999), or through a change in the geometry of the conduit (Thomas and Neuberg 2012); (3) the interaction between the magmatic and hydrothermal system at depth (e.g. Nakano and Kumagai 2005); or (4) through slow rupture and failure of unconsolidated material on volcanic slopes (Bean et al. 2014).

Many volcanic seismic events fall between these two end-member categories and are termed "hybrid" events. Hybrid events are characterised by a high frequency onset with a long resonating low frequency coda, therefore distributing energy across a wider frequency spectrum (Chouet and Matoza 2013). Hybrid and LF events are often classified in the same group of volcanic seismicity, since source and path effects can result in a LF event recorded at one station being recorded as a hybrid event at another.

With the deployment of broadband sensors in many volcanic environments, it is now possible to detect seismicity within a much wider frequency band, up to 120 s periods (Chouet and Matoza 2013), known as Very Long Period (VLP) earthquakes. VLP events occupy the spectral range below 0.01 Hz. The generation of these waveforms is not yet fully understood, in particular how such long wavelengths can be generated in apparently small source volumes, although it has been linked to perturbations in the flow of fluid or gas within pressurized volcanic conduits or cracks (e.g. Dawson et al. 2011). Their large wavelength, sometimes over one hundred kilometers, means few path effects on the waveform and as such if identified, these waveforms provide an excellent choice for performing waveform inversion techniques to identify source characteristics (Chouet and Matoza 2013).

Furthermore, seismicity can be generated by surface processes, such as landslides, rockfall events, pyroclastic flows and lahars (Fig. 1e, f). These are particularly dominant during dome building eruptions and at volcanoes with glaciers during the spring and summer months due to partial melting of the ice (McNutt 2005). These signals can be exploited to determine the size and magnitude of such events, their location and their direction of travel (e.g. De Angelis et al. 2007). Typically rockfall events (small free falling rock events) form a "cigar shaped" waveform with an emergent onset (Fig. 1e, f), whereby there is an initial increasing amplitude of the waveform as the amount of material falling down slope increases. Pyroclastic flow signals are distinguishable from rockfalls since their waveforms are at least an order of magnitude larger and they often occur over a longer duration since larger amounts of material are involved moving down slope (De Angelis et al. 2007), however the two are likely to exist on a continuum.

Classification by Waveform Similarity

Seismic waveforms can also be classified according to their similarity with other detected seismic events. The frequency content of seismic waveforms is indicative of the active processes that may be occurring within the volcanic environment and the source mechanism involved in the generation of such seismicity. The further classification of seismic events into families which all have a similar waveform shape, as well as the same frequency content, allows the depiction of temporal and spatial changes in the source mechanism and the source location on a much smaller scale (e.g. Thelen et al. 2011; Salvage and Neuberg 2016). For example, the relative relocation of families of similar seismic events at Soufrière Hills volcano, Montserrat has produced very precise source locations (e.g. De Angelis and Henton 2011). By definition, families of seismic events must be generated by the same mechanism and at the same location in order for the detected waveforms to have the same shape at the seismometer, and therefore changes in either of these parameters affect the similarity of waveforms. In many instances it is assumed that families of seismic events are generated by the same mechanism and within a similar source location, estimated at between one quarter and one tenth of the wavelength (Geller and Mueller 1980; Neuberg et al. 2006).

Waveform similarity in terms of shape and duration can be evaluated by cross correlation. Identical signals will result in a cross correlation coefficient of 1 or −1, dependent upon their

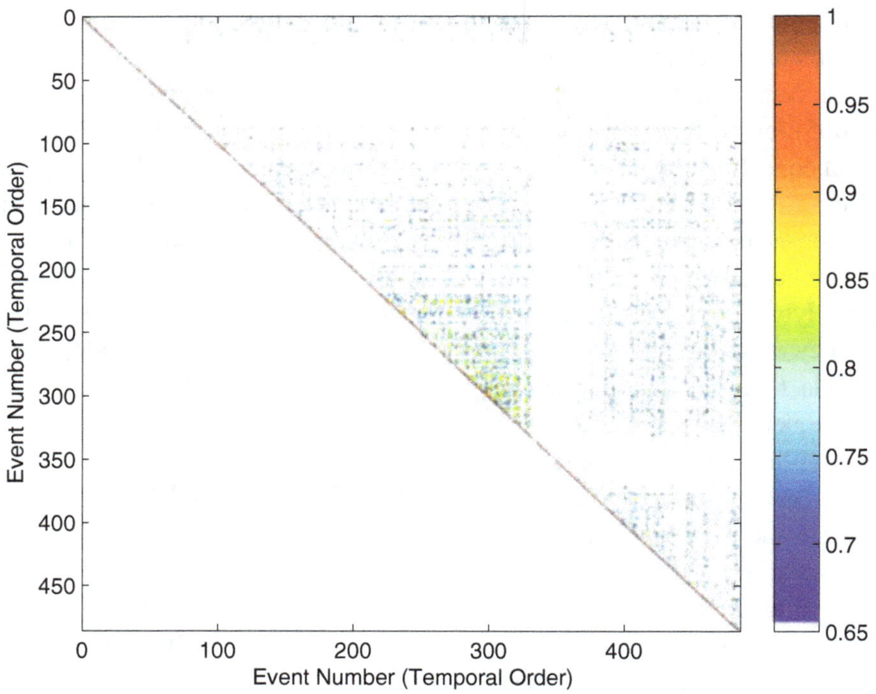

Fig. 2 Cross Correlation Matrix of events identified at Soufrière Hills volcano, Montserrat at a single station in June 1997. A total of 486 events were identified on 24 June 1997 and are shown in temporal order along the x and y axis. Events with a cross correlation coefficient of greater than 0.7 are shown on a colour scale, with those close to one being more similar. The autocorrelation of each event with itself is shown in *dark red* along the diagonal and is equal to a cross correlation coefficient of 1

relative polarity. Signals with no correlation result in a cross correlation coefficient of 0. A threshold must be chosen above which waveforms can be considered similar. Waveforms which are deemed similar can be grouped together into a family of events. The choice of similarity threshold is important: if it is too low there is a risk of placing events which are not similar into the same family; if it is too high similar events can be missed. Green and Neuberg (2006), Thelen et al. (2011) and Salvage and Neuberg (2016) suggest a cross correlation coefficient threshold of 0.7, since this is significantly above the correlation coefficient that can be produced from random correlations between noise and a waveform. Higher cross correlation coefficient thresholds can be used to identify families of almost identical waveforms, however Petersen (2007) suggests that this is probably not appropriate in volcanic settings due to additional noise in this environment.

The similarity between identified seismic events can be determined by cross correlating each individual seismic event with every other seismic event. The result are typically presented as a similarity matrix, as seen in Fig. 2, where events which are deemed to be similar are shown on the colour spectrum. However, such a matrix may include a number of families of similar events since it only determines whether each event shows similarity to any of the other earthquakes analysed. In order to identify families of similar events, events with a high cross correlation coefficient as decided by the user are grouped together and removed from the matrix. This procedure is repeated across the entire investigated time period until all events have been classified into a family, or have been removed from the matrix. A master event is then determined from each family of events as the average of the stack of similar waveforms. This is representative of the family in terms of waveform shape.

Families of similar seismicity have been identified at a number of active volcanoes around the world, including Redoubt volcano, Alaska

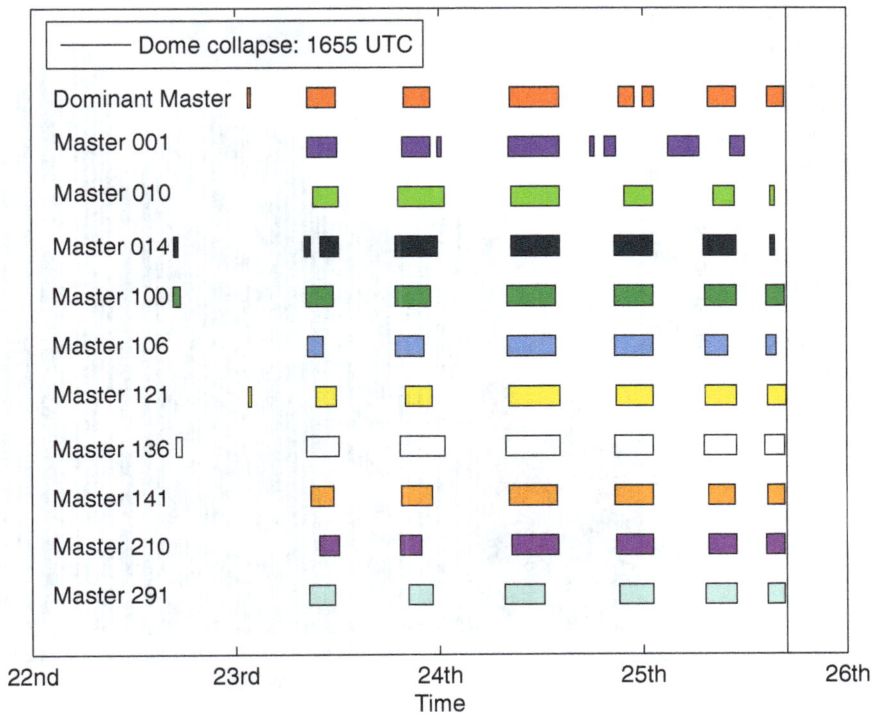

Fig. 3 Comparison of the timing and duration of swarms of families of events identified at a single station at Soufrière Hills volcano, Montserrat in June 1997. The timing of the dome collapse is represented by the vertical line on the 25 June 1997. The y axis is only an indication of each of the families present separated in space for the purpose of clarity on the plot and does not represent time or dominance; each master event is simply drawn below the last so that all can be compared. Each coloured *rectangular box* represents the times when the families were active during the 22–25 June analysis period

(e.g. Buurman et al. 2013); Mt. St. Helens, USA (Thelen et al. 2011); Colima, Mexico (Arámbula-Mendoza et al. 2011); Merapi, Indonesia (Budi-Santoso and Lesage 2016); Katla, Iceland (Sgattoni et al. 2016); and Soufrière Hills volcano, Montserrat (Green and Neuberg 2006; Ottemöller 2008; Salvage and Neuberg 2016). The identification of families rather than simply detecting seismic events and classifying them according to their frequency content is advantageous as subtle temporal and spatial patterns can be identified, allowing detailed source information to be uncovered. In addition, this technique dramatically increases the number of identified events from the continuous seismic record, since low amplitude events and closely spaced events can still be identified. For example, using a standard amplitude-based detection algorithm, 1435 seismic events were identified at Soufrière Hills volcano, Montserrat between 22 and 25 June 1997, a period of interest due to increased seismicity before a lava dome collapse. The cross correlation technique identified 7653 similar seismic events during the same time period, offering a five-fold increase in the number of detected earthquakes (Salvage and Neuberg 2016).

Low frequency families of seismicity identified during this unrest period at Soufrière Hills volcano were followed by a dome collapse on 25 June 1997. Soufrière Hills Volcano was chosen by the VUELCO project as a target volcano due to the longevity of its dome building and collapse cycles which have been ongoing since 1995, providing a wealth of associated geophysical data (Sparks and Young 2002; Wadge et al. 2014). In total, 11 distinct seismic sources (i.e. 11 families of seismicity) were identified during this period of unrest [Fig. 3; Green and Neuberg (2006); Salvage and Neuberg (2016)], which all broadly follow the same temporal pattern in the number of identified swarms present over this time period. However, the timing and duration of these

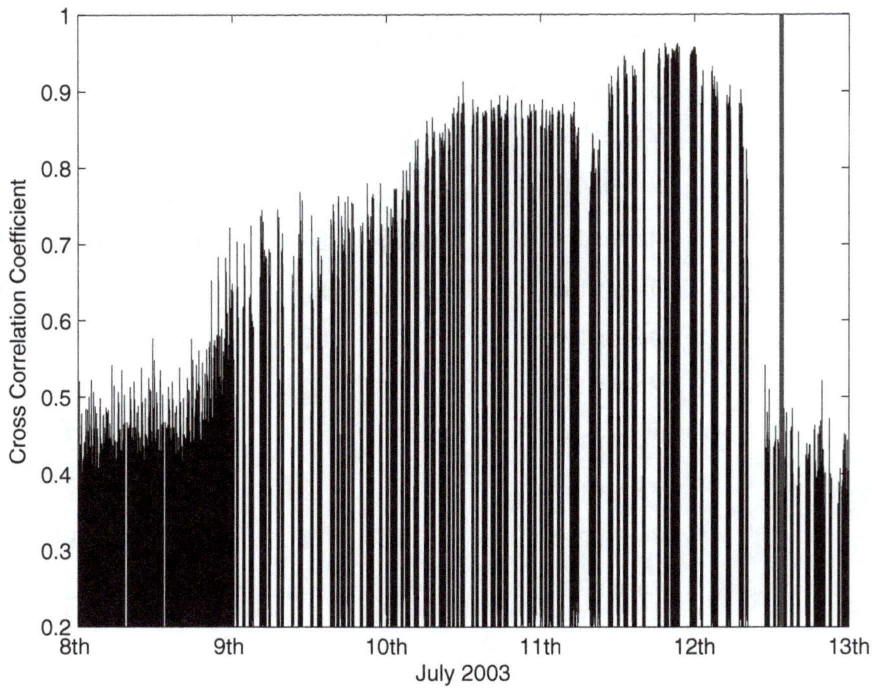

Fig. 4 The evolution of the dominant cross correlation coefficient with time at Soufrière Hills volcano, Montserrat in July 2003. A dome collapse event occurred at the time of the *vertical line* (13:30 on 12 July 2003). The temporal gaps in the data represent drops in the seismometer recordings rather than a change in the cross correlation coefficient, indicated by the white space on the x-axis

swarms can be seen to be different for each family of similar seismic events. Low frequency seismicity is associated with the movement of magmatic fluid at depth, and therefore in this case would suggest cyclic flow dynamics to generate such swarm-like behaviour. The source process for the generation of this seismicity must be stable and non-destructive in order to be repeatable (Green and Neuberg 2006; Petersen 2007), and must be able to occur at a number of different locations and/or by a number of different sources at the same time in order to generate a number of active families of events, reflecting the complex diversity of seismic sources and physical processes which act simultaneously at this volcano.

The identification of families can also be used to understand evolving seismicity with time. An evolving cross correlation coefficient with time, if not an artefact of data processing, may be indicative of a migrating source location or source mechanism. This was observed at Soufrière Hills volcano, Montserrat in July 2003 (Fig. 4; Salvage and Neuberg (2016)). The largest dome collapse to date observed at this volcano occurred on 12 July 2003, with removal of 210×10^6 m^3 of material (Herd et al. 2005), following a 4 day period from 8 to 12 July 2003 of heightened seismicity at Soufrière Hills. A migrating source mechanism can be identified from changing amplitudes in seismic events within the same family, however this characteristic cannot be identified from analysing cross correlation coefficients alone. The amplitudes of seismic events within the single family identified in July 2003 were relatively constant (Ottemöller 2008), suggesting the changing cross correlation coefficient is a consequence of a migrating source location at depth, rather than an evolving source mechanism. The generation of families ceased immediately prior to the dome collapse event, and no similar earthquakes were detected after the collapse (Fig. 4). This suggests that the physical conditions required for the generation of families were not met in the hours before, and after, the collapse event.

The analysis of families of seismicity in the time domain may also allow for the identification of spatial patterns in seismicity. Families detected at Chiles-Cerro Negro, a volcano within the Northern Andes on the border between Ecuador and Colombia in October 2014, suggests distinct temporal and spatial patterns of seismicity. The last eruption of the volcanic complex of Chiles-Cerro Negro is believed to have been

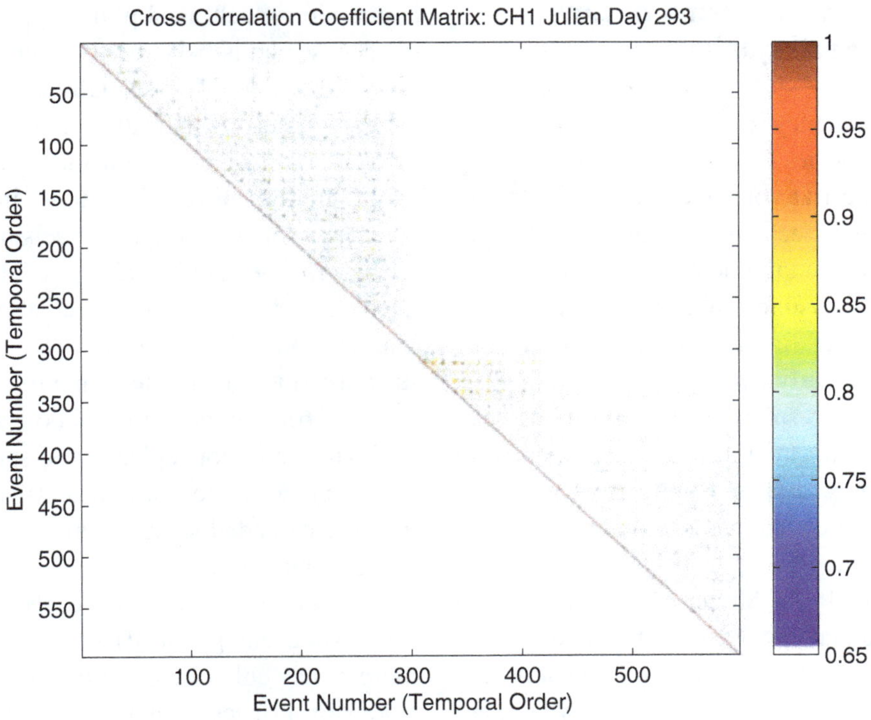

Fig. 5 Cross Correlation Matrix of events identified using a simple amplitude based detection algorithm at Chiles-Cerro Negro on 20 October 2014, at a single station. A total of 597 events were identified and are shown in temporal order along the x and y axis. Events with a cross correlation coefficient of greater than 0.7 are shown on a colour scale, with those close to one (*red*) being more similar. The autocorrelation of each event with itself is shown in dark red along the diagonal and is equal to a cross correlation coefficient of 1. Distinct clusters of similar events can be identified, thought to suggest a temporal evolution in the dominant similar seismicity

3400 years ago, although seismicity has since been detected in the area, thought to be related to an active hydrothermal system (Ruiz et al. 2013). The volcanic complex is dissected by a large fault system, believed to have been active as recently as 1868, when two large seismic events occurred (Mw 6.6 and 7.2) (Beauval et al. 2010). Seismic activity increased in 1991 and then again in July 2013. However, a dramatic increase in seismicity from less than 50 events a day to over 150 events occurred in October 2014, concentrated beneath the summit of Chiles volcano at depths of less than 10 km (Ruiz et al. 2013). Although originally not a target volcano for the VUELCO project, the volcanic complex of Chiles-Cerro Negro is an excellent example of a re-awakening volcano, having shown no signs of magmatic unrest in recent history until the events of 2014. Analysis of seismicity identified in October 2014 suggested not only the occurrence of families, but also their occurrence in distinct temporal patterns. The clustering of similar seismic events around the diagonal in a number of box-like formations within a similarity matrix suggests that a number of sources were active for discrete periods of time generating families of seismicity (Fig. 5). The distinct clusters of similar seismic waveforms may relate to a changing source location or mechanism at depth (Salvage 2015). Over this time period, no significant changes in the amplitude of events (indicative of a changing source mechanism) were identified. Since similar seismicity is thought to be generated through a similar source mechanism and a similar source location, distinct cluster of similar seismicity is most likely related to its own distinct spatial region, which generated seismicity during distinct periods of time.

The Source Mechanisms of Low Frequency Earthquakes

Since low frequency events, and in particular families, appear to be important in detecting changes in unrest at volcanoes, it is important to understand their mechanism of generation. Synthetic modelling and moment tensor inversions of low frequency seismic wavefields are powerful tools for gaining information on the source mechanisms underlying volcanic earthquakes. Once instrument response and path effects have been accounted for, real data can be compared to synthetic models, and on the basis of a best-fit approach the obtained model parameters allow insights into the nature and geometry of the source (Chouet 1996; Shuler et al. 2013). However, as the fundamental assumptions behind the commonly used moment tensor inversions are based on plane surface geometries which are believed to be too simple to explain the generation of low frequency events in a volcanic environment, the application to more complex seismic sources has so far been inconclusive. In the framework of the VUELCO project, slip along bent surfaces (a complex source) was thought of as the underlying physical motion responsible for generating seismic energy. This novel way of investigating low frequency earthquakes can explain several features of the earthquakes under investigation without introducing compromising assumptions such as slip along a single, unbent surface, which is believed to be unrealistic.

An example of a spatially extended source generating seismicity within a volcanic environment is a volcanic conduit through which magmatic fluids move. In these instances, the generation of low frequency seismicity may be related to the brittle failure of magma itself (Neuberg et al. 2006; Lavallée et al. 2008; Thomas and Neuberg 2012) or through a stick-slip motion at the conduit edge (Iverson et al. 2006). In either case, shallow source depths (1–2 km) and short epicentral distances to seismic receivers (a few kilometers) suggest that a spatially extended source is more realistic than a single point source.

The occurrence of slip (i.e. the generation of the seismic energy itself) of spatially extended sources may either be instantaneous along two or more slip surfaces, or may occur on different slip surfaces at different times, offset by a given time increment, delta *t*. A ring fault structure is a numerical description of seismogenic slip of magma along all of the conduit walls within a volcanic edifice, and can be numerically modelled by considering a cylinder representing the volcanic conduit with instantaneously slipping double couple (single point) sources bounding the circumference (Fig. 6). Upward movement inside the cylinder and downward movement outside represents the movement of magma

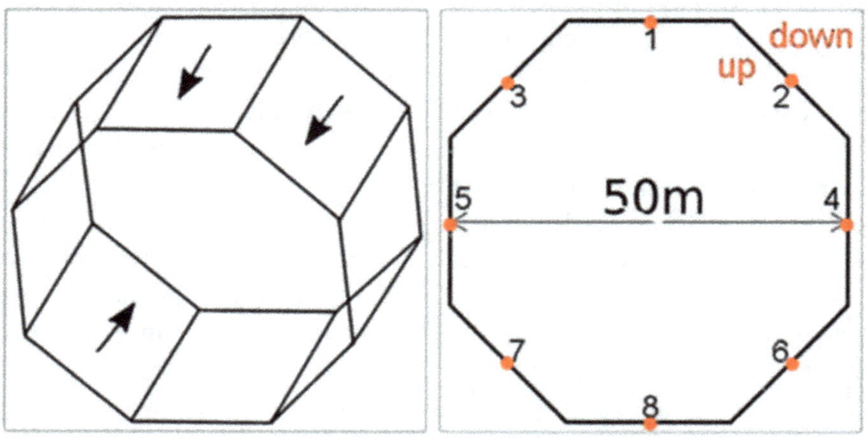

Fig. 6 Schematic representation of a ring fault structure with movement directed upwards within the cylinder to represent the flow of magma through a conduit. Each planar surface is host to a single double couple source i.e. a point source (labelled *1–8*)

through this channel. Other spatially extended sources which may evoke the generation of low frequency seismicity include: the upward movement of magma through a narrow dyke, numerically modelled by two oppositely directed double couple sources; slip along distinct segments of the volcanic conduit i.e. due to geometry changes (Thomas and Neuberg 2012), numerically modelled with double couple sources on distinct segments; or the generation of a number of seismic swarms of families of earthquakes occurring simultaneously (Salvage and Neuberg 2016), which can be numerically modelled by movement on two or more simultaneously acting ring fault structures.

Green and Neuberg (2006) suggested that accelerated magma movement at Soufrière Hills volcano can be linked to observed deformation cycles through low frequency seismic swarms, and that seismicity is only generated if significant magma movement takes place. Considering slip through extended sources brings us one step closer to estimating magma ascent rates. Once calibrated, the link between observed waveform amplitudes and the amount of seismogenic slip occurring during a seismic swarm will yield magma ascent rates and will ultimately contribute to forecasting volcanic eruptions more accurately. Point and extended source models yield great differences in observed P-wave amplitudes and waveforms, leading to remarkable differences when interpreting the amount of

(a) CLVD source (b) Ring fault structure

Fig. 7 P-wave radiation pattern generated for a CLVD source, and for an extended ring fault structure source. The *red* lobes are compressional, the *blue* lobes are dilatational. In both radiation patterns a large compressional lobe is found above the source, with dilatational lobes extending radially which can lead to confusion for interpretation of first motion polarity patterns

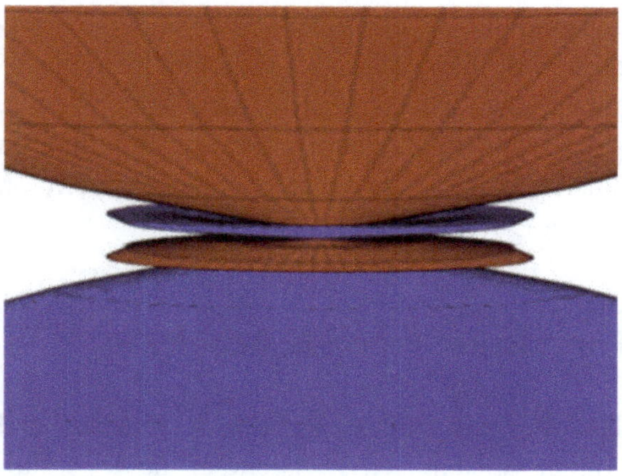

Fig. 8 Zoom of smaller radial lobes of P-wave radiation pattern for a spatially extended ring fault structure. Compressional lobes are *red*, dilatational lobes are *blue*

slip occurring and the slip rates. In the case of slip along a ring fault, P-wave amplitudes are greatly reduced due to destructive interference, in comparison to simple double couple (point) sources. As a result, observed amplitudes yield an underestimation of actual slip by more than a factor of 3 if interpreted as a point source, when in reality a spatially extended source acts at depth. This underestimation in seismic moment consequently may lead to an underestimation of magma flow rate at depth, which in turn has severe implications for eruption forecasting (Karl 2014).

Furthermore, the P-wave radiation pattern for a spatially extended ring fault structure (i.e. for modelling the movement of magma within the entire conduit made up of a number of double couple sources) shows remarkable similarity to the P-wave radiation pattern for a compensated linear vector dipole (CLVD) source, which instead is a conservation of energy (Fig. 7). The derived ring fault radiation pattern shows rotational symmetry around the depth axis, and consists of a large compressional lobe directly above the source and an inversely polarised, dilatational lobe with the same amplitude below it. Only if the small radial P-wave radiation lobes can be determined for the ring fault structure (Fig. 8) is it possible to distinguish between these two source mechanisms. Seismic networks with small apertures typical in volcanic settings and uncertainties in earthquake source depth locations will likely lead to difficulties in distinguishing between the two radiation patterns, since both can explain observed first motion polarity patterns of low frequency seismicity on volcanoes (Karl 2014).

Forecasting Eruptive Activity

The ability to forecast the timing, intensity and type of volcanic activity is one of the key issues facing volcanologists today. The most notable instances of successful volcanic forecasting use precursory activity at andesitic-dacitic volcanoes. The cataclysmic eruption of Mt. Pinatubo, Philippines on 15 June 1991 was preceded by at least two months of heightened seismicity (Harlow et al. 1996). With increases in seismicity and an alarming sudden drop in SO_2, scientists were able to successfully evacuate over 45,000 local people and 14,500 military personnel to safety by 14 June, such that less than 300 people were killed in the ensuing volcanic activity on 15 June. More recently, the 2010 eruption of Merapi, Indonesia, on 26 September was preceded by approximately 6 weeks of precursory activity (Budi-Santoso et al. 2013): rates of seismicity and SO_2 during this time were comparable to, or higher than, the highest rates observed during previous (smaller) Merapi eruptions (1992–2007), and rapid deformation was observed. Consequently, one day prior to the explosive eruption, several tens of thousands of people were evacuated from a radius extending 10 km from the volcano, resulting in a greatly lowered death toll of 35.

Volcanic eruptions are often preceded by accelerating geophysical signals, associated with the movement of magma or other fluid towards the surface. Of these precursors, seismicity is at the forefront of forecasting volcanic activity since it is frequently observed and the change from background level can be observed in real time. Since forecasting of volcanic eruptions relies on the ability to determine the timing of magma reaching the surface, low frequency seismicity may act as a forecasting tool due to its potential correlation with the movement of magmatic fluid at depth.

The Failure Forecast Method (FFM) is based on an empirical power-law relationship, which relates the acceleration of a precursor $(d^2\Omega/dt^2)$ to the rate of that precursor $(d\Omega/dt)$ (Voight 1988) method by:

$$\frac{d^2\Omega}{dt^2} = K\left(\frac{d\Omega}{dt}\right)^{\alpha} \tag{1}$$

where K and α are empirical constants. Ω can represent a number of different geophysical precursors, for example low frequency seismic event rate (Salvage and Neuberg 2016), event rate of all recorded seismicity (Kilburn and Voight 1998), or the amplitude of seismic events (Ortiz et al. 2003). The parameter α is thought to range between 1 and 2 in volcanic environments (Voight 1988), or may even evolve from 1 towards 2 as seismicity proceeds (Kilburn 2003). α has also been calculated in hindsight as high as 3.3 for accelerating seismicity in 1991 at Mt. Pinatubo, although this extreme value appears rare and was calculated with only a small amount of seismic data (Smith and Kilburn 2010). An infinite $d\Omega/dt$ suggests an uncontrolled rate of change (a singularity) and in this environment is associated with an impending eruption. The inverse form of $d\Omega/dt$ is linear if $\alpha = 2$, and therefore in this case the solution for the timing of failure is a linear regression of inverse rate against time, with the timing of failure relating to the point where the linear regression intersects the x-axis (Voight 1988).

Although assuming that $\alpha = 2$ is the simplest method to estimate the timing of an eruption through a linear regression and therefore the most common application of the FFM in hindsight analysis, some authors have suggested that it may not be an appropriate assumption for use with the FFM (e.g. Bell et al. 2011). Additionally, some authors have argued that α may evolve with time as precursory sequences develop, which is not detailed in the FFM (Kilburn 2003). As the FFM follows a least squares regression analysis when α is equal to 2, the residual error between the observed event rate and the mean event rate of seismicity should follow a typical Gaussian distribution (Bell et al. 2011). Greenhough and Main (2008) have suggested that since earthquake occurrence is a point process, the rate uncertainties are best described by a Poisson distribution. In this instance, a generalised linear model (GLM) where $\alpha = 1$, rather than a least squares regression model ($\alpha = 2$) may be more appropriate, since it can allow for a distribution of data that is non-Gaussian (Bell et al. 2011). At Soufrière Hills volcano, however, the use a GLM to forecast the timing of eruptive events in 1997 and 2003 failed to generate an appropriate forecast (Salvage and Neuberg 2016). Hammer and Ohrnberger (2012) suggested that this may be related to the fact that a Poisson process, and therefore the GLM, is a memoryless system, meaning that past events do not influence future patterns. A memoryless system is not consistent with the fundamental assumptions of the FFM, since previous geophysical observables form the basis of such a forecast.

One of the first instances of real time forecasting using the FFM was at Redoubt volcano, when the inverse average amplitude of seismic events followed a linear regression trend for 4 days prior to a dome collapse event on 2 January 1990. Due to this trend, and the fact that the seismic intensity was far above background levels, the Alaskan Volcano Observatory issued a "formal warning" of an impending eruptive event on the morning of the 2 January, a few hours before the eruption began, although the FFM calculations suggested an eruption was likely within 0.5–2 days. A similar, if not clearer trend, that supported the forecast was found using the same precursory sequence but only using seismic events within the spectral range of 1.3–1.9 Hz (Cornelius and Voight 1994), suggesting an increased accuracy in forecasts when focusing on a single source process at depth.

Swarms of seismic events, i.e. a number of similar events within a short period of time, with typical swarm durations of hours to day, are not observed at all volcanoes, but have been commonly observed at Soufrière Hills volcano (e.g. Green and Neuberg 2006) and Redoubt volcano (e.g. Buurman et al. 2013). Using precursory seismicity and the FFM, Salvage and Neuberg (2016) forecast in hindsight the timing of a dome collapse event on 25 June 1997 at Soufrière

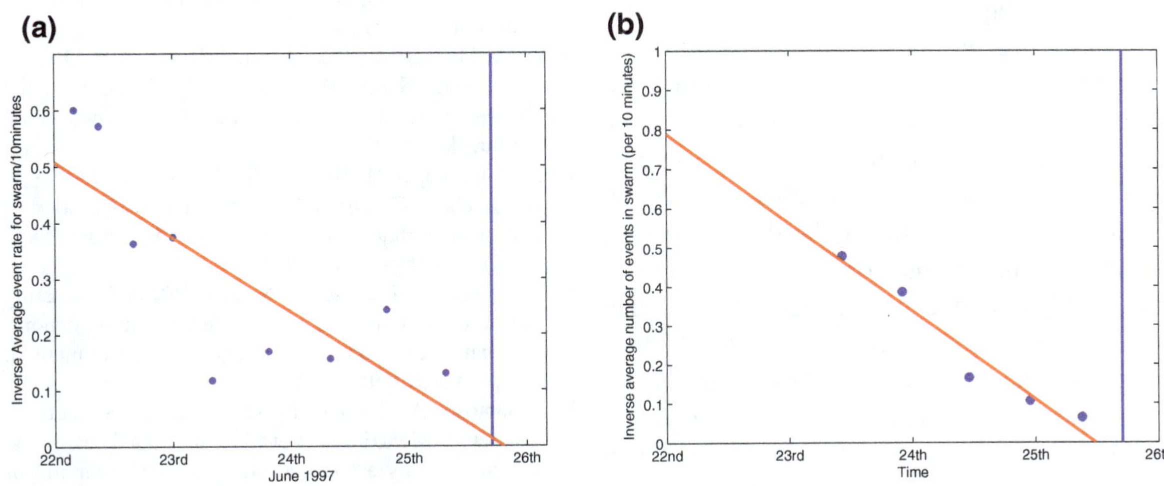

Fig. 9 Application of the FFM: the inverse average event rate per 10 min within swarms from 22 to 25 June 1997 at a single station at Soufrière Hills volcano. Each data point represents the inverse average event rate for each individual identified swarm of seismicity. The *vertical line* represents the known timing of dome collapse on the 25 June 1997 at 16:55 UTC. The graphical representation of the FFM is depicted by the linear regression (it is assumed that $\alpha = 2$ for simplicity) and the forecasted timing of failure can be read off the x-axis at the point where the linear regression crosses it. **a** All triggered low frequency seismicity. **b** Single family of similar seismicity

Hills, based upon accelerating rates of low frequency earthquakes which occurred in swarms, rather than simply the number of low frequency events over the precursory time period (Fig. 9). Using the average event rate per swarm showed a clearer accelerating pattern over the entire seismic sequence, rather than using the traditional method of binning data in units of time. More accurate forecasts were determined when using only one single family of similar events to forecast the dome collapse, rather than all low frequency seismicity mixed together. A dome collapse on 12 July 2003 at Soufrière Hills volcano was also more accurately forecast when using a single family of similar events, rather than all low frequency seismicity, which occurred during the period of unrest (Salvage and Neuberg 2016). Consequently, the use of families of seismicity, and therefore concentration upon a single active system at depth, may allow a more accurate forecast of the timing of an eruptive event in these instances.

Summary

Seismology is a powerful tool which can be used to understand processes occurring at depth, and their relationship to the surface at volcanoes, especially since seismic events are easily detected and the deviation from the background level is often notably pronounced. An increase in seismicity may be an indication of volcanic unrest, since it relates to a number of physical processes at depth including brittle failure and fracturing of the conduit or of the surrounding edifice (high frequency seismicity) and the movement of magmatic fluids at depth (low frequency seismicity). Using families of seismicity as an indicator of a single active system at depth, temporal and spatial patterns in the seismic events can be used to assess the potential migration of seismicity towards the surface, as well as to forecast the timing of volcanic eruptive events related to the acceleration of seismicity. However, a full understanding of the source mechanism of the generated seismic events is essential to ensure that the magma flow rate is estimated accurately and therefore an accurate forecast can be generated for the timing of eruption. Using seismicity in combination with other monitoring tools, we are now closer to gaining a better understanding of evolving magmatic systems at depth.

Acknowledgements The past and present staff at the Montserrat Volcano observatory are fully acknowledged for their ongoing support in the upkeep and maintenance of the seismic network, and the sharing of data. All staff at the Instituto-Geofisico in Ecuador are also fully acknowledged for providing data from Chiles-Cerro Negro volcano, for useful discussions regarding the volcanic complex, and for their continued monitoring efforts of all volcanoes in Ecuador. We thank two anonymous reviewers for their detailed comments and suggestions which greatly enhanced the quality of this manuscript. Muchas gracias también a Maria Martinez-Cruz y Javier Pacheco por su ayuda con la traducción al español.

References

Arámbula-Mendoza R, Lesage P, Valdés-González C, Varley N, Reyes-Dávila G, Navarro C (2011) Seismic activity that accompanied the effusive and explosive eruptions during the 2004–2005 period at Volcán de Colima, Mexico. J Volcanol Geoth Res 205(1):30–46

Arciniega-Ceballos A, Chouet B, Dawson P (2003) Long-period events and tremor at Popocatepetl volcano (1994–2000) and their broadband characteristics. Bull Volc 65(2):124–135

Bean CJ, De Barros L, Lokmer I, Métaxian J-P, O'Brien G, Murphy S (2014) Long-period seismicity in the shallow volcanic edifice formed from slow-rupture earthquakes. Nat Geosci 7(1):71–75

Beauval C, Yepes H, Bakun WH, Egred J, Alvarado A, Singaucho J-C (2010) Locations and magnitudes of historical earthquakes in the Sierra of Ecuador (1587–1996). Geophys J Int 181(3):1613–1633

Bell A, Naylor M, Heap M, Main I (2011) Forecasting volcanic eruptions and other material failure phenomena: an evaluation of the failure forecast method. Geophys Res Lett 38:L15304

Budi-Santoso A, Lesage P, Dwiyono S, Sumarti S, Jousset P, Metaxian J-P et al (2013) Analysis of the seismic activity associated with the 2010 eruption of Merapi Volcano, Java. J Volcanol Geoth Res 261:153–170

Budi-Santoso A, Lesage P (2016) Velocity variations associated with the large 2010 eruption of Merapi volcano, Java, retrieved from seismic multiplets and ambient noise cross-correlation. Geophys J Int 206 (1):221–240

Buurman H, West ME, Thompson G (2013) The seismicity of the 2009 Redoubt eruption. J Volcanol Geoth Res 259:16–30

Chouet B (1988) Resonance of a fluid-driven crack: radiation properties and implications for the source of long-period events and harmonic tremor. J Geophys Res Solid Earth (1978–2012) 93(B5), 4375–4400

Chouet BA (1996) New methods and future trends in seismological volcano monitoring. In: Monitoring and Mitigation of Volcano Hazards. Springer, pp. 23–97

Chouet BA, Matoza RS (2013) A multi-decadal view of seismic methods for detecting precursors of magma movement and eruption. J Volcanol Geoth Res 252:108–175

Cornelius R, Voight B (1994) Seismological aspects of the 1989–1990 eruption at Redoubt Volcano, Alaska: the materials Failure Forecast Method (FFM) with RSAM and SSAM seismic data. J Volcanol Geoth Res 62(1):469–498

Dawson PB, Chouet BA, Power J (2011) Determining the seismic source mechanism and location for an explosive eruption with limited observational data: Augustine Volcano, Alaska. Geophys Res Lett 38(3):L03302

De Angelis S, Henton S (2011) On the feasibility of magma fracture within volcanic conduits: constraints from earthquake data and empirical modelling of magma viscosity. Geophys Res Lett 38(19):L19310

De Angelis S, Bass V, Hards V, Ryan G (2007) Seismic characterization of pyroclastic flow activity at Soufrière Hills Volcano, Montserrat, 8 January 2007. Nat Hazards Earth Syst Sci 7:467–472

Geller R, Mueller C (1980) Four similar earthquakes in central California. Geophys Res Lett 7(10):821–824

Goto A (1999) A new model for volcanic earthquake at unzen volcano: melt rupture model. Geophys Res Lett 26(16):2541–2544

Green D, Neuberg J (2006) Waveform classification of volcanic low-frequency earthquake swarms and its implication at Soufrière Hills Volcano, Montserrat. J Volcanol Geoth Res 153(1):51–63

Greenhough J, Main I (2008) A poisson model for earthquake frequency uncertainties in seismic hazard analysis. Geophys Res Lett 35(19):L19313

Hammer C, Ohrnberger M (2012) Forecasting seismo-volcanic activity by using the dynamical behavior of volcanic earthquake rates. J Volcanol Geoth Res 229:34–43

Harlow DH, Power JA, Laguerta EP, Ambubuyog G, White RA, Hoblitt RP (1996) Precursory seismicity and forecasting of the June 15, 1991, eruption of Mount Pinatubo. Eruptions and lahars of Mount Pinatubo, Philippines, Fire and mud, pp 223–247

Herd RA, Edmonds M, Bass VA (2005) Catastrophic lava dome failure at Soufriere Hills volcano, Montserrat, 12–13 July 2003. J Volcanol Geoth Res 148(3):234–252

Iverson R, Dzurisin D, Gardner C, Gerlach T, LaHusen R, Lisowski M, Major J, Malone S, Messerich J, Moran S et al (2006) Dynamics of seismogenic volcanic extrusion at Mount St Helens in 2004–05. Nature 444(7118):439–443

Karl S (2014) The source mechanisms of low frequency seismic events on volcanoes. Ph.D. thesis, University of Leeds, UK, Online at: http://etheses.whiterose.ac.uk/id/eprint/8406

Kilburn C (2003) Multiscale fracturing as a key to forecasting volcanic eruptions. J Volcanol Geoth Res 125(3–4):271–289

Kilburn CR, Voight B (1998) Slow rock fracture as eruption precursor at Soufriere Hills volcano Montserrat. Geophys Res Lett 25(19):3665–3668

Lahr J, Chouet B, Stephens C, Power J, Page R (1994) Earthquake classification, location, and error analysis in a volcanic environment: implications for the magmatic system of the 1989–1990 eruptions at Redoubt Volcano, Alaska. J Volcanol Geoth Res 62 (1):137–151

Lavallée Y, Meredith P, Dingwell D, Hess K-U, Wassermann J, Cordonnier B, Gerik A, Kruhl J (2008) Seismogenic lavas and explosive eruption forecasting. Nature 453(7194):507–510

McNutt SR (2005) Volcanic seismology. Annu Rev Earth Planet Sci 32:461–491

Nakano M, Kumagai H (2005) Response of a hydrothermal system to magmatic heat inferred from temporal variations in the complex frequencies of long-period events at Kusatsu-Shirane Volcano, Japan. J Volcanol Geoth Res 147(3):233–244

Neuberg J, Luckett R, Baptie B, Olsen K (2000) Models of tremor and low-frequency earthquake swarms on Montserrat. J Volcanol Geoth Res 101(1–2):83–104

Neuberg J, Tuffen H, Collier L, Green D, Powell T, Dingwell D (2006) The trigger mechanism of low-frequency earthquakes on Montserrat. J Volcanol Geoth Res 153(1):37–50

Ortiz R, Moreno H, Garcì A, Fuentealba G, Astiz M, Peña P, Sánchez N, Tárraga M (2003) Villarrica volcano (Chile): characteristics of the volcanic tremor and forecasting of small explosions by means of a material failure method. J Volcanol Geoth Res 128(1):247–259

Ottemöller L (2008) Seismic hybrid swarm precursory to a major lava dome collapse: 9–12 July 2003, Soufriere Hills Volcano, Montserrat. J Volcanol Geoth Res 177 (4):903–910

Petersen T (2007) Swarms of repeating long-period earthquakes at Shishaldin Volcano, Alaska, 2001–2004. J Volcanol Geoth Res 166(3):177–192

Phillipson G, Sobradelo R, Gottsmann J (2013) Global volcanic unrest in the 21st century: an analysis of the first decade. J Volcanol Geoth Res 264:183–196

Ruiz G, Cordova A, Ruiz M, Alvarado A (2013) Informe Tecnico de los volcanes Cerro Negro y Chiles. Tech. Rep., IG-EPN, in Spanish

Salvage RO (2015) Using seismic signals to forecast volcanic processes. Ph.D. thesis, University of Leeds, UK, Online at: http://etheses.whiterose.ac.uk/12268/. Restricted until April 2018

Salvage R, Neuberg J (2016) Using a cross correlation technique to refine the accuracy of the failure forecast method: application to Soufrière Hills volcano, Montserrat. J Volcanol Geoth Res 324:118–133

Sgattoni G, Jeddi Z, Gudmundsson Ó, Einarsson P, Tryggvason A, Lund B, Lucchi F (2016) Long-period seismic events with strikingly regular temporal patterns on Katla volcano's south flank (Iceland). J Volcanol Geoth Res 324:28–40

Shuler A, Ekström G, Nettles M (2013) Physical mechanisms for vertical-CLVD earthquakes at active volcanoes. J Geophy Res Solid Earth 118(4):1569–1586

Smith R, Kilburn C (2010) Forecasting eruptions after long repose intervals from accelerating rates of rock fracture: the June 1991 eruption of Mount Pinatubo, Philippines. J Volcanol Geoth Res 191(1):129–136

Sparks R, Young S (2002) The eruption of Soufrière Hills Volcano, Montserrat (1995–1999): overview of scientific results. Geol Soc Lond Mem 21(1):45–69

Thelen W, Malone S, West M (2011) Multiplets: their behavior and utility at dacitic and andesitic volcanic centers. J Geophys Res Solid Earth (1978–2012) 116 (B8):B08210

Thomas ME, Neuberg J (2012) What makes a volcano tick—a first explanation of deep multiple seismic sources in ascending magma. Geology 40(4):351–354

Voight B (1988) A method for prediction of volcanic eruptions. Nature 332:125–130

Wadge G, Voight B, Sparks R, Cole P, Loughlin S, Robertson R (2014) An overview of the eruption of Soufriere Hills Volcano, Montserrat from 2000 to 2010. Geological Society, London, Memoirs 39(1):1–40

The Ups and Downs of Volcanic Unrest: Insights from Integrated Geodesy and Numerical Modelling

J. Hickey, J. Gottsmann, P. Mothes, H. Odbert, I. Prutkin and P. Vajda

Abstract

Volcanic eruptions are often preceded by small changes in the shape of the volcano. Such volcanic deformation may be measured using precise surveying techniques and analysed to better understand volcanic processes. Complicating the matter is the fact that deformation events (e.g., inflation or deflation) may result from magmatic, non-magmatic or mixed/hybrid sources. Using spatial and temporal patterns in volcanic deformation data and mathematical models it is possible to infer the location and strength of the subsurface driving mechanism. This can provide essential information to inform hazard assessment, risk mitigation and eruption forecasting. However, most generic models over-simplify their representation of the crustal conditions in which the deformation source resides. We present work from a selection of studies that employ advanced numerical models to interpret deformation and gravity data. These incorporate crustal heterogeneity, topography, viscoelastic rheology and the influence of temperature, to constrain unrest source parameters at Uturuncu (Bolivia), Cotopaxi (Ecuador), Soufrière Hills (Montserrat), and Teide (Tenerife) volcanoes. Such model complexities are justified by

J. Hickey and H. Odbert, previously at School of Earth Sciences, University of Bristol, UK.

J. Hickey (✉)
Camborne School of Mines, University of Exeter, Exeter, UK
e-mail: J.Hickey@exeter.ac.uk

J. Gottsmann
School of Earth Sciences, University of Bristol, Bristol, UK

J. Gottsmann
Cabot Institute, University of Bristol, Bristol, UK

P. Mothes
Instituto Geofísico, Escuela Politecnica Nacional, Quito, Ecuador

H. Odbert
Met Office, Exeter, UK

I. Prutkin
Institute of Geosciences, Jena University, Jena, Germany

P. Vajda
Earth Science Institute, Slovak Academy of Sciences, Bratislava, Slovakia

Advs in Volcanology (2019) 203–219
DOI 10.1007/11157_2017_13

Published Online: 15 July 2017

geophysical, geological, and petrological constraints. Results highlight how more realistic crustal mechanical conditions alter the way stress and strain are partitioned in the subsurface. This impacts inferred source locations and magmatic pressures, and demonstrates how generic models may produce misleading interpretations due to their simplified assumptions. Further model results are used to infer quantitative and qualitative estimates of magma supply rate and mechanism, respectively. The simultaneous inclusion of gravity data alongside deformation measurements may additionally allow the magmatic or non-magmatic nature of the source to be characterised. Together, these results highlight how models with more realistic, and geophysically consistent, components can improve our understanding of the mechanical processes affecting volcanic unrest and geodetic eruption precursors, to aid eruption forecasting, hazard assessment and risk mitigation.

Extended Spanish Abstract

La deformación volcánica, caracterizada por pequeños cambios medibles en la morfología del volcán, a menudo, pero no siempre, precede a una erupción volcánica. Esta cuestión, sin embargo, se complica por el hecho de que los eventos de deformación (por ejemplo, inflación o deflacción) pueden ser el resultado de una fuente magmática, no magmática o de fuentes mixtas/híbridas. Utilizando tanto la amplitud y patrones espacio-temporales de datos de deformación volcánica registrados, así como la utilización de modelos matemáticos, es posible inferir la ubicación y la fuerza de la fuente impulsora subyacente. Estos métodos pueden proporcionar información esencial para la evaluación y mitigación de riesgos, así como para el pronóstico de erupciones volcánicas. Sin embargo, la mayoría de los modelos genéricos son insatisfactorios en su representación de las condiciones de la corteza en las que reside la fuente. En este trabajo presentamos una selección de estudios que emplean modelos numéricos avanzados para la interpretación de datos de deformación y gravedad. Dichos datos incorporan la heterogeneidad de la corteza, la topografía, la reología inelástica y los efectos termo-mecánicos para constreñir los parámetros asociados a la fuente de perturbación en cuatro sistemas volcánicos. Las complejidades de estos modelos están justificadas por limitaciones geofísicas, geológicas y petrológicas. El estudio realizado en el volcán Uturuncu, localizado en Bolivia, destaca la importancia de la estructura sub-superficial y de los procesos dependientes de tiempo en la fuente para explicar los patrones de deformación espacial-temporal. La combinación de dichos resultados indica un ascenso del magma de tipo diapírico. En el volcán Cotopaxi, localizado en Ecuador, los nuevos modelos de inversión que emplean el Análisis por Elementos Finitos esclarecen la ubicación y el volumen de una intrusión magmática durante un episodio de actividad asísmico y no eruptivo con una baja tasa de suministro de magma. Estos modelos también proporcionan señales observables que podrían estar asociadas con futura actividad volcánica intrusiva o eruptiva. El análisis de la deformación intra-eruptiva en el volcán Soufrière Hills, en Montserrat, mostró cómo la inflación registrada podría deberse a una serie de reservorios magmáticos apilados o un depósito alargado verticalmente, debido a respuestas termo-mecánicas de la corteza similares. Las condiciones de falla derivadas para las tasas de suministro de magma y reservorio son consistentes con las restricciones térmicas y mecánicas independientes. Utilizando datos gravimétricos y la curiosa falta de deformación asociada, el estudio en el volcán Teide, en Tenerife, se identifican, de forma separada, las contribuciones de fuentes poco profundas como de aquellas con más profundidad. La interpretación de ambos elementos sugiere que una intrusión magmática profunda activó un sistema hidrotermal superficial localizado encima de ésta, y por lo tanto demuestra un efecto causal de un origen mixto. Estos casos de estudio representan una contribución significativa para la comprensión de los procesos volcánicos durante los periodos de actividad intra-eruptiva y no eruptiva. La combinación de estos resultados enfatiza cómo una mejor parametrización de condiciones mecánicas de la corteza puede alterar fundamentalmente la forma en que el estrés y la tensión se reparten en la sub-superficie. Del mismo modo, una topografía compleja, como es el caso en los estratovolcanes con laderas empinadas, puede afectar la partición de la deformación superficial. Estos dos efectos impactan la inferencia de la localización de las fuentes y presiones magmáticas previo al fallo y la erupción, e indican cómo los modelos genéricos pueden conducir a interpretaciones engañosas debido a la simplificación de inferencias acerca de la corteza y a espacios planos y a medias. Resultados adicionales de la modelización son utilizados para inferir estimaciones cuantitativas y cualitativas de la tasa de suministro de magma y el mecanismo, respectivamente, contribuyendo así al entendimiento de la dinámica del transporte del magma. Además, la inclusión simultánea de datos de gravedad junto a mediciones de deformación, permiten la caracterización de la naturaleza magmática o no magmática de la fuente. Juntos, estos estudios destacan cómo los modelos que cuentan con componentes más plausibles y geofísicamente consistentes, pueden mejorar nuestro entendimiento de los procesos mecánicos que afectan la reactivación volcánica y de precursores geodésicos de erupciones. Estos también proporcionan un marco para ayudarnos a avanzar en el pronóstico de una erupción, así como en la evaluación y mitigación de los riesgos, proporcionando datos cuantitativos derivados de la modelización de mecanismos físicos adecuados y robustos.

Keywords

Volcano deformation · Gravity · Modelling · Crustal mechanics · Geodesy

Palabras clave

deformación volcánica · gravedad · modelización · mecánica de la corteza · geodesia

Introduction

Volcano deformation is a key observable during periods of volcanic unrest, and one of the main tools used to monitor developing crises (Sparks et al. 2012). Non-magmatic causes of volcanic deformation during an unrest period do not involve movement of new magma and are most commonly related to active hydrothermal systems (e.g., Fournier and Chardot 2012; Rouwet et al. 2014, and references therein). Magmatic causes reflect the active migration and accumulation of new magma, with usually deeper origins and wider deformation footprints. Hybrid mechanisms have also been proposed, where contributions from magmatic and hydrothermal systems combine to produce a single complex deformation pattern (Gottsmann et al. 2006a).

A variety of both ground and satellite based geodetic monitoring techniques are used to assess the spatial and temporal evolution of volcanic related surface deformation. The two most common techniques employed today are the Global Positioning System (GPS) and Interferometric Synthetic Aperture Radar (InSAR), which track changes in position due to the ground deforming and complement each other with advantages in their temporal and spatial coverage, respectively. Volcano gravimetry studies, the monitoring and interpretation of continuous or time-lapse spatio-temporal gravity variations, can also provide additional information on any density changes associated with the driving mechanism (Battaglia et al. 2008), and thus enable an estimate of its nature (e.g., magmatic or hydrothermal), which is particularly important in cases where there is no significant (observable) surface deformation.

The magnitude, spatial pattern, and temporal evolution of volcanic deformation can be used to infer the location and 'strength' of a causative subsurface source. These source parameters have important implications for volcanic hazards and risk mitigation. Geodetic data are also a key component in eruption forecasting efforts (e.g., Sparks et al. 2012), and constraints on source parameters (e.g., pressure or volume changes) from previous deformation episodes can help to quantitatively improve these forecasts. However, making the transition from surface deformation observations to subsurface source processes requires the use of geodetic models, which demand assumptions about crustal mechanics.

Generic analytical models of volcanic deformation (e.g., Mogi 1958) are often oversimplified and do not capture the intrinsic complexities of subsurface systems, which are essential for reliable assessment of causative processes, and thus eruption forecasting and hazard assessment. They usually represent the Earth's crust as a homogeneous, isotropic, elastic, half-space with a flat, free surface. These assumptions can cause misleading interpretations when compared to numerical models that allow for more realistic crustal conditions (Masterlark 2007). Numerical models of volcanic deformation are thus becoming increasingly popular due to this ability to estimate source parameters in crustal conditions that are beyond the analytical realm. These improvements are synonymous with the recent advances in geodetic monitoring, which deserve a more in-depth analysis than generic analytical models can offer.

The extra complexities that numerical models can account for relate to both the source itself, as well as the crustal rocks in which it resides. The deformation source does not have to be represented as a point or cavity, but can contain its own material properties (e.g., Hickey et al. 2013; Gottsmann and Odbert 2014). Moreover, numerical models can incorporate inferences from other geophysical, geological and petrological observables and thus maintain a higher level of consistency by relaxing some of the assumptions that restrict generic analytical models. For example, this enables the inclusion of subsurface heterogeneity, topography, viscoelastic rheology, and temperature-dependent mechanics (e.g., Hickey et al. 2016, and references therein). Consequently, more robust source parameters and magma transport processes can

Table 1 Summary of geodetic modelling

Volcano	Unrest period	FWM	IVM	Data	Topo	CMX	TMX	Model	Ref
Uturuncu	1992–2006	✓	–	InSAR	–	✓	–	2D	Hickey et al. (2013)
Cotopaxi	2001–2002	✓	✓	EDM	✓	✓	✓	3D	Hickey et al. (2015)
Soufrière hills	2003–2005	✓	–	cGPS	✓	✓	✓	2D	Gottsmann and Odbert (2014)
Central volcanic complex	2004–2005	–	✓	GPS & gravity	✓	–	–	3D	Prutkin et al. (2014)

FWM forward modelling, *IVM* inverse modelling, *Topo* topography, *CMX* crustal mechanics, *TMX* thermomechanics, *2D* two-dimensional axisymmetric, *3D* three-dimensional, *Ref* reference

be inferred, thereby improving the understanding of the links between deformation and eruption.

In this chapter we summarise the key findings from investigations of a variety of unrest episodes at a selection of the VUELCO target volcanoes; Cotopaxi (Ecuador), Soufrière Hills (Montserrat, British West indies) and the Central Volcanic Complex on Tenerife (Spain), as well as at Uturuncu volcano in Bolivia (Table 1). These examples highlight not only different timescales of unrest and spatial patterns of volcano deformation, but also the influence of crustal mechanical heterogeneity, thermal effects, and topography, on observed signals. The unrest episodes are investigated using both forward and inverse numerical modelling procedures and demonstrate the process of interpreting different geodetic data sets to constrain realistic source parameters. Thus, the purpose of this chapter is not to provide a detailed treatment of the mathematical and numerical approaches. Instead, it is intended to provide the reader with a general overview of the use of advanced numerical models to infer volcanic unrest driving mechanisms with realistic source characteristics.

Implementing Complex Crustal Mechanics

One of the most important aspects of a deformation model is the representation of crustal mechanics. This has a fundamental control on the way in which stress and strain is distributed and transferred through the Earth's crust, from the source to the surface. In this regard, as briefly mentioned above, generic analytical models are limited by their necessary assumptions of homogeneous and elastic conditions throughout the entire model domain. In reality, the Earth's crust is known to be layered, and volcanic regions in particular can have wide-ranging regions of stiff (high Young's Modulus) and soft (low Young's Modulus) rocks relating to the type of volcanic deposit that formed them, e.g., lava flows (stiff) compared to tuffs (soft) (Gudmundsson 2011). This is what we call subsurface heterogeneity. Where stiff and soft regions are adjacent in the crust, complex subsurface partitioning alters the way stress and strain are transferred to the surface. The outcome is a different surface deformation pattern compared to one that would be seen if the crust was in fact homogeneous. Consequently, generic analytical models restricted to homogeneous crustal mechanics will not adequately represent likely subsurface conditions, and the inferred source parameters from these models can be misleading.

Seismic studies are capable of delineating areas of relatively high and low seismic velocities. They can be used to estimate the dynamic Young's Modulus, E_D, Poisson's Ratio, v, and density, ρ, of the crust (Brocher 2005):

$$v = 0.5 \times \left[\left(\frac{V_P}{V_S}\right)^2 - 2\right] \Big/ \left[\left(\frac{V_P}{V_S}\right)^2 - 1\right] \quad (1)$$

$$\rho = 1.6612V_P - 0.4721V_P^2 + 0.067V_P^3 - 0.0043V_P^4 + 0.000106V_P^5 \quad (2)$$

$$E_D = \frac{V_P^2\rho(1+\nu)(1-2\nu)}{(1-\nu)} \quad (3)$$

where V_P and V_S are the primary and shear seismic velocities, respectively. The static Young's Modulus, E, is usually a factor of 2–9 smaller than E_D, and is more appropriate for deformation studies as the dominant processes taking place are significantly slower than the propagation of seismic waves (e.g., Gudmundsson 2011, and references therein). Numerical approaches can incorporate these spatially-variable calculations of Young's Modulus and Poisson's Ratio by allowing the mechanical parameterisation of the crust to vary within the model (Hickey and Gottsmann 2014). This can be achieved in one-dimension, where the values are only changed with depth, or in three-dimensions, where the mechanical representation additionally changes in the two horizontal axes. The Finite Element Method is the most common technique used when incorporating subsurface heterogeneity. With this approach, a heterogeneous model domain can be built using multiple different sized homogeneous 'blocks' of varying material properties (Hickey et al. 2013), with continuous interpolations as a function of depth or distance (Gottsmann and Odbert 2014), or with a three-dimensional coordinate interpolation (Hickey et al. 2016).

A further advantage of a numerical approach using the Finite Element Method is the ability to include a spatially variable inelastic rheology that is dependent on an estimated temperature distribution. Rocks do not always behave in an elastic manner (Ranalli 1995). Instead, where crustal conditions are hotter than the brittle-ductile transition, or subject to long-term loading, rocks can deform like very high viscosity fluids, and so viscoelastic effects are more likely (Ranalli 1995). This is especially important when a hot magmatic component is assumed to be present, or where elevated geothermal gradients are observed. Also, when models are restricted to elastic mechanical behaviour it can be difficult to constrain realistic source processes, particularly if there is a complex temporal deformation pattern that can not be reproduced with the instantaneous elastic stress-strain relationship, and may be better represented by a non-instantaneous viscous stress response. A model that incorporates temperature-dependent mechanics can overcome these difficulties. This is achieved using the Finite Element Method with a two-step procedure (Hickey and Gottsmann 2014). First, a spatially-variable, stationary, temperature distribution is derived, primarily dependent on the local geothermal gradient(s) and an assumed (magmatic) source temperature. This temperature distribution, T, is then used to define a crustal viscosity, η, for example through an Arrhenius relationship, where:

$$\eta = A_d \exp\left(\frac{H}{RT}\right) \quad (4)$$

and A_d is the Dorn Parameter, H is the activation energy, and R is the universal gas constant. The viscosity is used in a viscoelastic material representation (Hickey and Gottsmann 2014), so where the viscosity is high the material response is dominated by elastic behaviour, and when the viscosity is lower the rocks show an increasingly higher tendency for viscous behaviour. To demonstrate the effect of temperature-dependent mechanics, and subsurface heterogeneity, the following case studies include examples of both.

Case Studies

Uturuncu

Uturuncu volcano is located in southern Bolivia, within the Altiplano-Puna Volcanic Complex. It has been steadily inflating since at least 1992 (Pritchard and Simons 2002), and combined with shallow seismicity and near-summit active fumaroles represents a volcanic system showing significant signs of unrest (Sparks et al. 2008).

Hickey et al. (2013) focused on the mechanism driving the 70 km wide region of ground uplift between 1992 and 2006. The aim was to constrain first-order source parameters that explain both the observed uplift rate of 1–2 cm/year and the large spatial deformation footprint (Pritchard and Simons 2002). Stress and strain from pressurised finite sources were solved numerically using Finite Element Analysis, accounting for both homogeneous and heterogeneous subsurface structure in elastic and viscoelastic rheologies. Crustal heterogeneity was constrained from seismic velocity data, which indicates a pervasive large low-velocity zone ~17 km below the surface. This is deduced to represent one of the world's largest known regions of partial-melt: the Altiplano-Puna Magma Body (APMB) (Sparks et al. 2008).

The comparison between crustal heterogeneity and homogeneity highlights the significant effect of a mechanically weak source-depth layer (Fig. 1). The weak layer, with a lower Young's Modulus, alters surface deformation patterns by accommodating more of the subsurface strain than its surrounding layers, thereby acting as a mechanical buffer. Continuous and regular time-dependent deformation, the long-lived nature of the source, and an anomalously high regional crustal heat-flux break the assumption of elastic conditions (e.g., Ranalli 1995), so a viscoelastic crustal rheology was tested, using the standard linear solid representation (e.g. Hickey and Gottsmann 2014). The elastic models could also only account for the spatial component of the observed uplift so their results were used solely to guide the parameters tested in the viscoelastic models. A range of possible source geometries were assessed, but spherical and oblate shapes were rejected on the grounds of their depth below the APMB and likely unsustainable pressurisation given the expected crustal mechanics. This left a prolate shaped source, whose minimum size was determined using maximum laboratory values for host-rock tensile strength. The final preferred model suggests that temporally-continuous pressurisation of a magma source protruding from the top of the APMB is causing the observed spatial and temporal surface uplift, whereas the previous models could only infer that a simple-geometry source was being pressurised somewhere within the APMB (Pritchard and Simons 2002). Hence, this also demonstrates how a pressure-time function plays a first-order role in explaining time-dependent deformation.

Simultaneous work on an extended InSAR data set shows how the central uplift region at Uturuncu is surrounded by a 'moat' of subsidence (Fialko and Pearse 2012). To explain this observation they also constrained a model where magma rises out and up from the APMB with a diapiric-type ascent mechanism. Further evidence for this magmatic process is available through a complementary gravimetric study (del Potro et al. 2013). Therefore, this highlights how combinations of geodetic data and numerical models can not only constrain more plausible deformation source parameters, but can also infer magma transport dynamics.

Cotopaxi

Cotopaxi is a large, glacier-clad stratovolcano situated in the Eastern Cordillera of the Ecuadorian Andes. The 1 km^3 glacier presents a substantial lahar risk to people in the surrounding areas, and particularly to the 100,000 inhabitants that reside in the path of the 1877 lahar which descended the Inter-Andean Valley (Pistolesi et al. 2013).

Unrest was detected at Cotopaxi in 2001 and 2002. There was a significant increase in the amount of volcanic seismicity in the NE quadrant of the volcano, and this was originally interpreted to represent a dyke intrusion with subsequent gas-release and resonance of the crack (Molina et al. 2008). Hickey et al. (2015) revisited this unrest period, but approached it from a volcano deformation viewpoint. Analysis of an electronic distance meter (EDM) network over the 2001–2002 period also indicated an asymmetric inflation of the edifice that accompanied the recorded seismicity (Fig. 2). However, the irregular

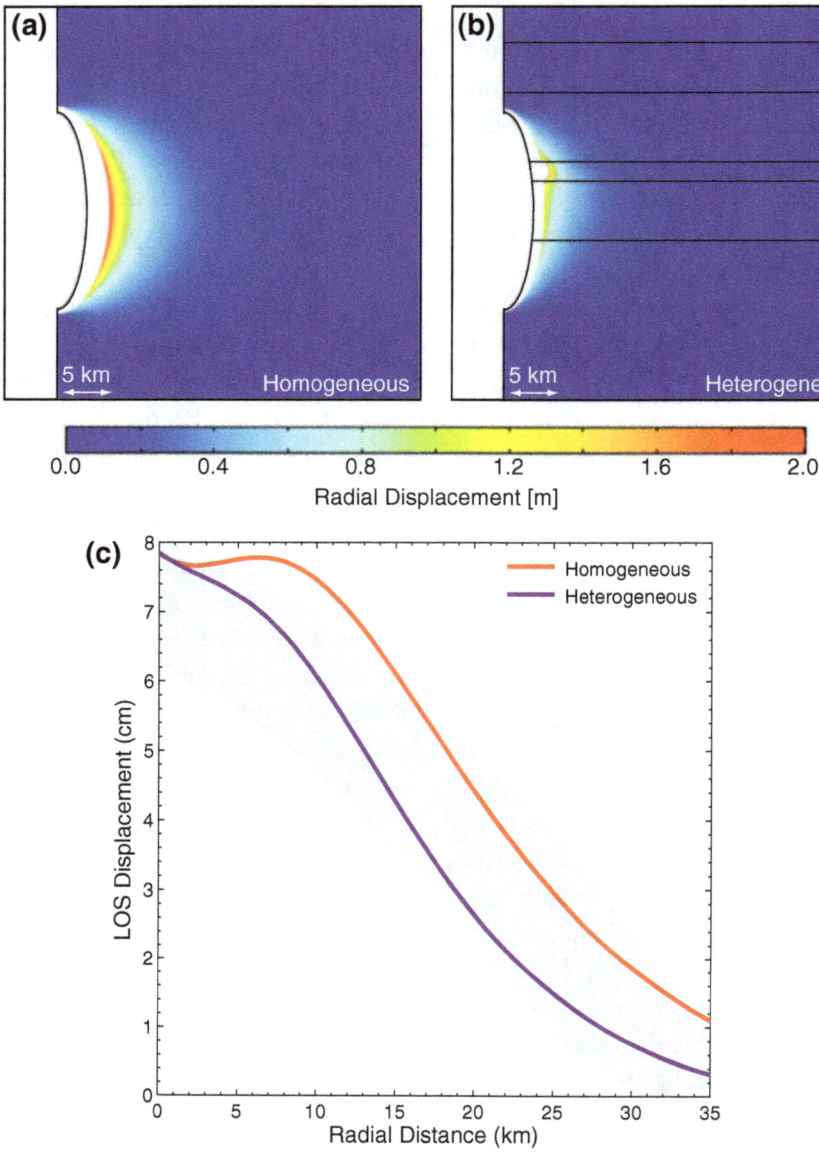

Fig. 1 A comparison of the effect crustal heterogeneity for the same prolate source geometry and depth. **a**, **b** The effect of a source depth soft layer on the subsurface deformation of a source. *Both panels* show the same source, embedded in a homogeneous (**a**) or heterogeneous domain (**b**). *Colours* relate to the radial displacement, and the *white shape* shows the exaggerated outline of the deformed source after the pressure is applied. In the heterogeneous model the source preferentially deforms into the softer layer (D), compared to the homogeneous medium, which exhibits a concentric deformation pattern. **c** Modelled surface displacement profiles from the homogeneous (**a**) and heterogeneous (**b**) models. The subsurface layering in **b** alters the displacement pattern produced at the surface as the soft layer modifies the subsurface strain partitioning. The *blue shaded area* represents the observed InSAR data and its estimated error bounds

acquisition in time of the EDM data prevented any systematic comparison between the two data sets. To solve for the optimum deformation source parameters, Hickey et al. (2015) implemented a novel numerical inversion procedure using Finite Element models. This is the first volcanic deformation inversion study to explicitly account for both subsurface heterogeneity and surface topography while searching for a best-fit solution with a range of source shapes. The method works by solving for the predicted EDM deformation with an initial model and source configuration. It then continually changes the source location and/or overpressure, within some predefined parameter limits, to minimise the misfit to the recorded EDM data (Fig. 2). After each inversion, the parameter limits (e.g., X and Y coordinates) are reduced around the previous solution to produce a set of decreasing size 'Russian-doll-like' parameter constraint grids that ensure a robust solution. Within this workflow, the Finite Element model geometry and mesh are automatically rebuilt, removing the need for repeated manual editing.

The inversion models converge on a shallow source beneath the SW flank. The individual best-fit model is inferred to represent a small oblate-shaped magmatic reservoir, approximately 4–5 km beneath the summit, with a volume increase of roughly 20×10^6 m^3. A deformation source location in the SW is substantially different to the NE location proposed by Molina et al. (2008) when explaining the recorded seismicity. Despite this, when the deformation source was restricted to the NE quadrant the predicted

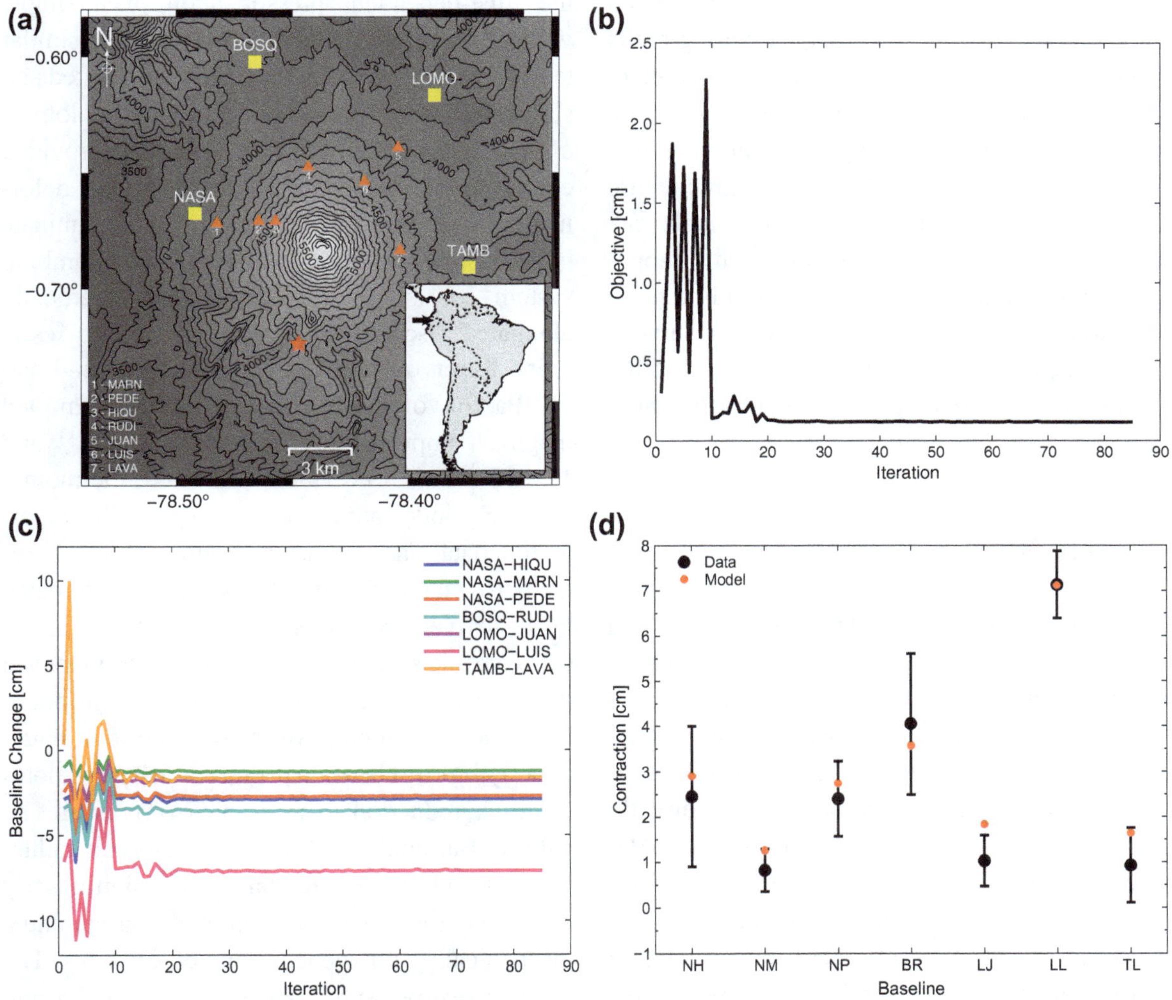

Fig. 2 Deformation and modelling results from Cotopaxi volcano. **a** Map of the EDM network operational between 2001 and 2002 around the summit of the volcano, with the best-fit source from the Finite Element inversions indicated by a *red star*. The *yellow squares* represent the EDM base-stations and the *orange triangles* are the reflecting prisms. The *inset* map and *arrow* shows the location of Cotopaxi within Ecuador, South America. **b** The variation, and eventual reduction, of the misfit objective function with each iteration from the final, best-fit Finite Element inversion model. **c** The modelled values from the final inversion for the EDM baseline changes converge to their near-absolute values after approximately 25 iterations. **d** The best-fitting model (*red circles*) fits six out of seven of the EDM observations (*black circles with error bounds*)

EDM measurements had a very poor fit to the observed data.

To clarify the difference in source location between the seismic and geodetic studies, Hickey et al. (2015) applied the best-fit source parameters from the elastic inversion models in a suite of temperature-dependent viscoelastic forward models to assess the rheology of the host-rock for a range of thermal parameters. The results indicated the most likely subsurface conditions would have promoted a large component of viscous deformation and that the deformation source in the SW would have consequently been pressurised aseismically. Fluid migration from the SW along existing NNE-SSW trending faults could have then caused the observed seismicity in the NE due to mass transport and excess pore pressures. The lack of eruption following this 2001–2002 unrest period and the aseismic nature of the event suggests that the magma supply rate for this period was low. A higher magma supply rate during a future unrest period would be more likely to produce seismicity around the reservoir, and could possibly indicate a level of unrest that signifies an increased likelihood of a forthcoming eruption.

Soufrière Hills

Starting its latest eruptive episode in 1995 and displaying a remarkable diversity of eruptive activity, Soufrière Hills volcano (SHV) on Montserrat (British West Indies) is one volcano where a correlation between observed deformation and subsequent eruption can be directly established. Ground deformation data at SHV indicate cyclic behaviour of the andesitic magmatic system from periods of several hours to a few years (Odbert et al. 2014). This section focuses on the analysis of intra-eruptive unrest associated with island-wide ground uplift observed via a network of continuous GPS receivers between 28/07/2003 and 01/08/2005 (Odbert et al. 2014). After a major lava dome collapse in 2003, which marked the end of the second phase of dome extrusion at SHV, this period preceded a restart of eruptive activity and renewed dome extrusion in August 2005.

Gottsmann and Odbert (2014) developed numerical models to test for the influence of temperature- and time-dependent stress evolution in a mechanically heterogeneous crust to explain the deformation data. Full details on the model setup and parameter derivation are given in Gottsmann and Odbert (2014) and not repeated here. They implemented two types of magma reservoir model. The first explored a series of pressurising, vertically-stacked reservoirs as proposed by Hautmann et al. (2010), while the second explored the time-dependent pressurisation of a single, vertically-elongated reservoir. Due to a similar temperature distribution from both the stacked and single reservoirs, and similar resultant rheological crustal properties, both suites of models provided equally good fits to the observed ground deformation. The study could hence not discriminate between pressurisation in a magmatic plumbing system consisting of either a single vertically elongated reservoir or a series of stacked reservoirs. Reservoir pressure changes between 4 and 7 MPa, for volumes between 60 and 100 km^3 and magma compressibility between 4×10^{-11} and 1×10^{-9} Pa^{-1}, provided plausible thermomechanical model parameters to explain the deformation data. The associated magma volume fluxes are between 0.015 and 0.021 km^3/year and match those derived from thermal modelling of active sub-volcanic systems (Annen 2009). Introducing a deep-crustal hot zone in the model, which modulates the partitioning of strain into the hotter underlying crust beneath the reservoir(s), promotes a further reduction in reservoir overpressures to values of around 1–2 MPa upon reservoir failure (Fig. 3). These pressure changes are significantly lower than those derived from models assuming a mechanically homogeneous and elastic crust. The deduced overpressures match those for sudden and rapid transcrustal reservoir activation prior to explosions at SHV from the analysis of volumetric strain data (Hautmann et al. 2014). The emerging eruption model at SHV hence involves the periodic failure of a compressible magma mush column beneath the volcano.

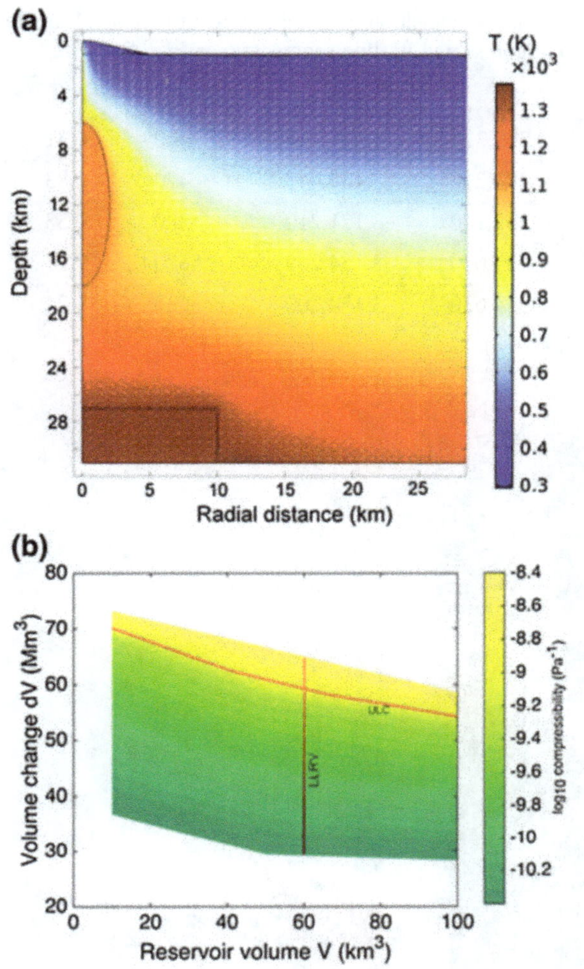

Fig. 3 **a** The temperature distribution (in K) of the 2-D axisymmetric model domains, caused by a combination of a basal heat flux, the heat from the plumbing system (magma reservoir plus feeder pipe), and a deep-crustal hot zone. Background thermal distribution in the far field is caused exclusively by a basal heat flux of 0.09 W/m^2. *Arrows* indicate deformation of the medium by reservoir pressurisation. Substantial deformation is accommodated by hot and ductile parts of the mid and lower crust. **b** The model parameter space for reservoir volume, volume change, and magma compressibility (shown in log units on the *colour bar*) from the best fit model solutions to the 2003–2005 intra-eruptive deformation time series at Soufrière Hills Volcano. The overall fit quality decreases substantially toward higher compressibility (order 10^{-9} Pa^{-1}) and toward smaller reservoir volumes (<50 km^3). The preferred parameter space is encompassed between the mapped upper limit of compressibility (*ULC*) and the lower limit of the magma reservoir volume (*LLRV*)

Las Cañadas

The Las Cañadas caldera (LCC) hosts the central volcanic complex (CVC) on Tenerife (Canary Islands), which includes the twin-volcanoes of Pico Viejo (PV) and Pico Teide (PT). Volcanic unrest on Tenerife started in April 2004 and ended a period of quiescence after the most recent magmatic eruption on the island in 1909. The unrest was geophysically characterised by anomalous seismic activity including a number of felt earthquakes and significant changes in the acceleration of gravity, yet, an absence of significant ground deformation (Gottsmann et al. 2006b). The initial interpretation of the gravity changes by Gottsmann et al. (2006b) pointed towards shallow (~2 km depth) fluid migration as the main source of the unrest, possibly related to a magma intrusion at greater depth. As part of the VUELCO project, Prutkin et al. (2014) revisited the microgravity data recorded between May 2004 and July 2005 and inverted the spatio-temporal residual gravity changes for the gravitational attraction of three-dimensional line segments.

The line segments are defined by 7 parameters: two end point coordinates (X, Y, Z) and a line strength (see Table 2). The latter is a proxy for the amplitude of sub-surface mass change, whereby a "weak" line segment represents a small amount of added mass compared to a "strong" line segment of greater added mass. An initial non-linear inversion yielded three line segments of different strengths at depths between 1 km a.s.l. and 2 km b.s.l. (i.e. between 1 and 4 km beneath the surface). Two of the segments were located to the NW of the PV-PT complex, while a third segment was located below the SW rim of the LCC (sources marked Sh 1–3 in Fig. 4). The locations of the segments correspond broadly to zones of heightened seismic activity and the unrest source location identified earlier by Gottsmann et al. (2006b).

Table 2 UTM coordinates X (Easting in m), Y (Northing in m), and Z (elevation in m with respect to sea level) of the end points of the modelled shallow (Sh) and deep (Dp) sources represented by line segments (see Fig. 4)

Line segment	X1	Y1	Z1	X2	Y2	Z2	ΔM
Sh1	327,500	3,137,930	1280	330,330	3,137,260	600	0.31
Sh2	336,560	3,132,050	590	334,210	3,134,040	420	0.83
Sh3	335,460	3,120,730	1380	332,100	3,121,210	380	0.53
Dp1	334,890	3,133,910	−5440	334,630	3,134,200	−5680	7.25
Dp2	334,310	3,134,590	−5840	334,630	3,134,200	−5680	8.24

Mass additions (ΔM) are given in 10^{10} kg. Data Prutkin et al. (2014)

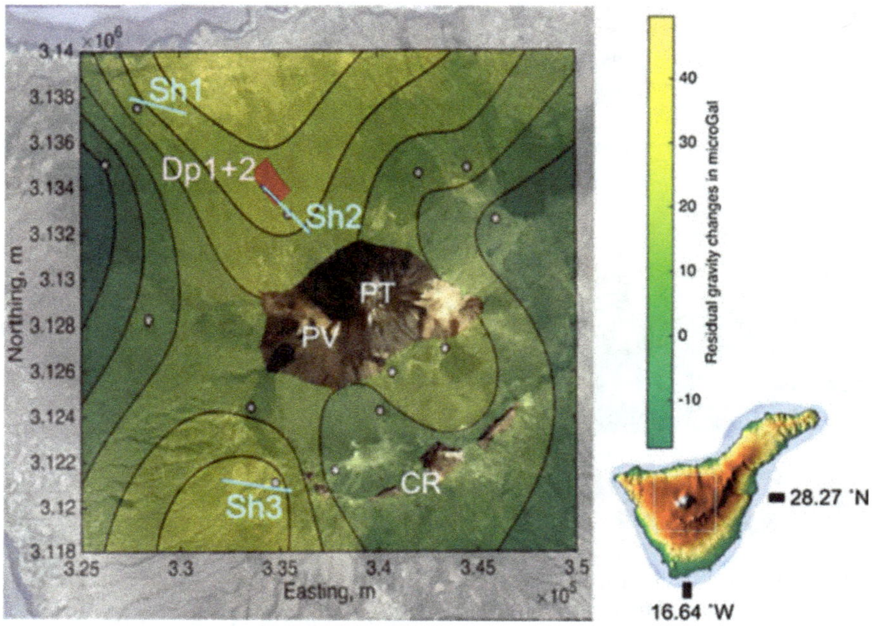

Fig. 4 Location of line segments from gravity data inversion. The figure shows the surface projections of three shallow sources (line segments Sh1–3 in *turquoise*) and the two deep sources (line segments Dp1 and 2 in *red*) superimposed over a Google Earth image of the Central Volcanic complex of Tenerife island (Spain). The source locations are derived by a nonlinear inversion of spatiotemporal residual gravity changes shown by coloured contours (in μGal) observed between 2004 and 2005 (see Prutkin et al. (2014) for details on inversion routine). The thickness of a line segment is indicative of the "strength" of the deduced mass change; i.e., the amount of added mass. The mass added to the deep sources is more than 10 times higher than to the shallow sources. Line segments Sh1–3 are interpreted to represent near-surface sources in the NW and SW part of the Pico Teide (PT; 3718 m a.s.l.) and Pico Viejo (PV; 3135 m a.s.l.) volcanic complex, associated with fluid migration as a result of an intrusion of magma at around 5.8 km b.s.l. (Dp sources). See Table 2 for details on segments and associated mass changes. The southern caldera rim (*CR*) is shown for reference and the *inset* shows a digital elevation model of Tenerife, with the study area identified by a *white rectangle*

However, closer inspection of the initial inversion results revealed that the line segments represent the superposition of deep and shallow seated sources. The decomposition of the gravity change data into shallow and deep fields (see Prutkin et al. 2014 for details) provided the basis for the separate inversions of the two fields using the three-dimensional line segment approximation. The deep field inversion constrained two connected and strong line segments at a depth of about 5.8 km b.s.l. (marked Dp1 and Dp2 in Fig. 4), while the inversion of the shallow field identified three similarly weak line segments situated at near-surface depths (<2 km from ground surface; see Fig. 4 and Table 2).

The most plausible interpretation of the inversion results is that the weak line segments represent sources dominated by hydrothermal fluids. In contrast, the deeper-seated sources can be interpreted as parts of an intrusion of new magma. This intrusion may have released fluids which consequently migrated towards shallower depths where they excited a shallow hydrothermal system. The geophysical signals resulting from these coupled magmatic-hydrothermal processes point towards a hybrid source nature for the unrest on Tenerife in 2004–2005. The identified link between deep and shallow unrest sources suggests the presence of permeable pathways for shallow fluid migration at the CVC.

Discussion

The Effect of Crustal Mechanics on Stress, Strain and Pressure

The presented investigations at Uturuncu, Cotopaxi and SHV all incorporate subsurface heterogeneity, however the effects are somewhat different. At Cotopaxi the final inferred deformation source is shallow, at a level where the model does not incorporate substantial heterogeneity due to the limited amount of available seismic data. Hence, with this model configuration, the heterogeneity has not played a significant role in altering the location of the source. It is likely, however, that in reality the volcanic edifice and shallow subsurface does have a certain level of heterogeneity. It is therefore possible that had this been taken into account the inferred deformation source might be located more centrally beneath the edifice. Conversely, at Uturuncu and SHV subsurface heterogeneity played a crucial role in determining the deformation source locations. In both cases, the inclusion of vertical layering in the Young's Modulus distribution altered the inferred depth of the source; the heterogeneous models predict deeper sources than the generic homogeneous analytical models. From this, it also follows that a horizontal variation in mechanical properties would influence the horizontal location of a deformation source (e.g., Hickey et al. 2016).

A second effect of subsurface heterogeneity relates to source pressure requirements. It is common with homogeneous crustal mechanics to require unrealistically high source pressurisation (>100's MPa) when attempting to fit an observed deformation signal, yet there is no petrological or mechanical evidence for such conditions. Theoretical work from a mechanical viewpoint suggests that the maximum overpressure a source can sustain without failing is equal to the tensile strength of the host rock (e.g., Gudmundsson 2011). Above this limit, a magmatic reservoir would trigger a dyke intrusion. Work at Uturuncu and SHV shows that elastic heterogeneous deformation models bring source over-pressure requirements more in line with both in situ and laboratory values of tensile strength. This is due to a relative reduction in the average Young's Modulus above the source, compared to a higher homogeneous Young's Modulus through the entirety of the crust, and thus a greater amount of deformation for a given pressure increment. Furthermore, with a maximum value for the source over-pressure, constraints can be placed on the minimum size of a deformation source, given the two are directly linked. This allows, for example, an estimate to be placed on the volume of a magma reservoir. When multiple magmatic sources might be present numerical models can additionally account for stress interactions between the two, something not considered by generic analytical models (Pascal et al. 2013).

As a next step to subsurface heterogeneity, the case studies at Cotopaxi and SHV incorporated temperature-dependent mechanics to evaluate the effect of a viscoelastic rheology. At Uturuncu, investigations were also carried out using a viscoelastic rheology without temperature-dependence. In all three cases, a viscoelastic medium reduced the over-pressure requirements further when compared to the purely elastic models. This is due to the viscous expansion that follows an initial elastic inflation. The effect of thermomechanics is greater than just modifying source over-pressure requirements, however. Prolonged magma emplacement over thousands to millions of years in active volcanic areas builds up a significant thermal legacy within the crust. This results in elevated geothermal gradients, and in the case of continued active magmatism, deep crustal hot-zones (Annen 2009). These thermal perturbations are significant for the transfer of stress and strain. For example, at SHV, the combined thermomechanical effects of a deep-crustal hot zone and hot encasing rocks around a mid-crustal andesitic reservoir fundamentally alter the time-dependent subsurface stress and strain partitioning upon priming of the magma reservoir. These effects substantially influence surface strains recorded by volcano geodetic monitoring.

Hybrid Unrest and Source Characterisation

An intrusion of magma commonly leads to the exsolution of fluids upon decompression, and fluid migration (and accumulation) in itself can produce measurable geodetic surface signals (e.g., Fournier and Chardot 2012; Rouwet et al. 2014, and references therein). A deep intrusion of magma, such as the 2004–2005 unrest on Tenerife, may not necessarily lead to observable surface deformation, meaning deformation data on its own can not provide any meaningful insights on the source process(es). In this case, the combination of deformation and gravity surveys allowed the characterisation of the unrest sources in much greater detail owing to their density contrasts, relative depths and mass additions. This highlights that, especially for the case of hybrid mechanisms where both magmatic and hydrothermal components are present, multi-disciplinary geodetic surveys can provide valuable information on source characterisation to help distinguish the processes driving unrest. This identification and discrimination of sources driving volcanic unrest via mathematical modelling of surface data is of vital importance for hazard and risk characterisation, given the different connotations associated with magmatic and non-magmatic unrest, and plays a major role in probabilistic eruption forecasting (Rouwet et al. 2014).

Application to Eruption Forecasting

The primary objective of the case studies was to develop advanced geodetic models to interpret spatial and temporal deformation monitoring signals, and provide better constraints on the subsurface processes causing volcanic unrest. This has been achieved by incorporating more plausible model components, such as subsurface heterogeneity, topography and temperature-dependent mechanics, to relax the assumptions that hinder analytical models and maintain consistency with inferences from geophysics, geology and petrology. Crucially, this highlighted the importance of pressure-time functions and inelastic rheology in deciphering temporal deformation patterns. In turn, a more thorough understanding of how a volcanic system behaves through time will benefit eruption forecasting, as quantitative estimates of key parameters such as magma supply rate and mechanism can be deduced.

When only considering the spatial deformation pattern, this work has further demonstrated some of the pitfalls associated with models assuming homogeneous, elastic, half-spaces (e.g., Mogi 1958). Crustal heterogeneity significantly effects the horizontal and vertical location of a deformation source by altering the subsurface strain distribution. Surface strain partitioning by complex topography is equally important, such as at steep-sided stratovolcanoes. The Cotopaxi case highlighted complex partitioning behaviour along deep ravines and adjacent lava flow ridges. If ground-based geodetic monitoring sites are positioned at localities with steep topographic gradients, surface strain partitioning plays a first-order role and needs to be accounted for during data modelling. Consequently, estimates of source parameters from generic, analytical models which cannot account for heterogeneities or complex topography may skew eruption forecasting and risk mitigation efforts. On the other hand, more reliable deformation source locations from numerical models can be used to better estimate where and when an eruption might occur, and the associated hazard. It should also be pointed out that data limitations at some volcanoes may prevent complex numerical models being used, in which case simpler models may be the only option and their results should be carefully scrutinised and only applied cautiously to further analyses.

Conclusions

The case studies provide new interpretations of volcanic processes during intra-eruptive and non-eruptive unrest. They also provide observations for distinct magmatic settings, thus contributing to the ongoing global comparisons of deformation and unrest, and the process of pattern recognition for identification of eruption precursors. The incorporation of multi-disciplinary data into integrated geodetic models has closed the gap between observations and interpretations of volcanic deformation and gravimetric changes. This will help to improve eruption forecasting as it moves away from a qualitative approach towards incorporation of more quantitative data derived from well-constrained physical mechanisms.

Acknowledgements Work presented herein has received funding by the European Commission (FP7; Theme: ENV.2011.1.3.3-1; Grant 282759: VUELCO). We are very grateful to Alvaro Guevara and Irving Munguía Gonzalez for their help translating Spanish, and to two anonymous reviewers for their constructive comments.

Glossary Terms

Deformation the action of changing the shape of a physical object (e.g., a volcano) under the influence of some stress.

Elastic rheological behaviour in which an applied stress causes an immediate strain that is 100% recoverable when the stress is removed. See also rheology and viscoelastic.

Finite Element Method/Analysis a numerical modelling technique that subdivides an entire problem into a set of smaller 'elements'. Mathematical problems are then solved in each of the elements and combined together to calculate the response of the whole object.

Geodesy applied mathematics with the aim of measuring the geometric spatial representation of the Earth and its gravitational field in time-varying three-dimensions, as well as their orientation in space. Volcano geodesy specifically deals with recording the spatial and temporal patterns of crustal deformation and gravimetric changes in volcanic settings.

Gravimetry field of geophysics devoted to observing, processing and interpreting minute changes in the Earth's gravitational field.

Hydrothermal system an area in which heat and fluids from a partially molten magmatic body interact with a multi-phase groundwater system, causing chemical and thermal (heating) perturbations to the water.

Mechanics (crustal, thermal) the branch of physics that studies how stress can effect a physical object. Crustal mechanics relates to the values of the material parameters that describe the mechanical behaviour of the Earth's crust. Thermal mechanics relates to the variation in mechanical material properties due to changes in temperature.

Model (inverse, forward) a simulation of a process. Inverse models solve 'backwards' to determine optimal model parameters that fit a set of data. Forward models use predefined constant input parameters to calculate the expected model response.

Rheology (crustal) the behavioural response of the Earth's crust to forces that act upon or within it. See also elastic and viscoelastic.

Strain the change in dimension of an object (e.g., ΔX) relative to the original dimension of the object (e.g., X). It has no units, as the units cancel: $\Delta X/X$.

Stress a measure of force per unit area, with the unit of pascals, Pa. 1 Pa = 1 N/m^2.

Viscoelastic rheological behaviour in which an applied stress causes an elastic response, followed by a delayed (viscous) response. See also rheology and elastic.

Index Terms

Magma reservoir, deformation, gravity, GPS, InSAR, Cotopaxi, Soufrière Hills, Las Cañadas, Uturuncu, unrest, thermomechanics, modelling, finite element, inversion/inverse model, crustal mechanics, hybrid unrest, Young's Modulus, Poisson's Ratio, viscosity.

References

Annen C (2009) From plutons to magma chambers: thermal constraints on the accumulation of eruptible silicic magma in the upper crust. Earth Planet Sci Lett 284(3–4):409–416

Battaglia M, Gottsmann J, Carbone D, Fernández J (2008) 4D volcano gravimetry. Geophysics 73(6):WA3–WA18

Brocher TM (2005) Empirical relations between elastic wavespeeds and density in the Earth's crust. Bull Seismol Soc Am 95(6):2081–2092

del Potro R, Dez M, Blundy J, Camacho AG, Gottsmann J (2013) Diapiric ascent of silicic magma beneath the Bolivian Altiplano. Geophys Res Lett 40(10):2044–2048

Fialko Y, Pearse J (2012) Sombrero uplift above the Altiplano-Puna magma body: evidence of a ballooning mid-crustal diapir. Science 338(6104):250–252

Fournier N, Chardot L (2012) Understanding volcano hydrothermal unrest from geodetic observations: insights from numerical modeling and application to White Island volcano, New Zealand. J Geophys Res 117(B11):B11208

Gottsmann J, Odbert HM (2014) The effects of thermo-mechanical heterogeneities in island-arc crust on time-dependent pre-eruptive stresses and the failure of an andesitic reservoir. J Geophys Res. doi:10.1002/2014JB011079

Gottsmann J, Rymer H, Berrino G (2006a) Unrest at the Campi Flegrei caldera (Italy): a critical evaluation of source parameters from geodetic data inversion. J Volcanol Geoth Res 150(1–3):132–145

Gottsmann J, Wooller L, Marti J, Fernandez J, Camacho AG, Gonzalez PJ, Garcia A, Rymer H (2006b) New evidence for the reawakening of Teide volcano. Geophys Res Lett 33(20)

Gudmundsson A (2011) Rock fractures in geological processes. Cambridge University Press, New York

Hautmann S, Gottsmann J, Sparks RSJ, Mattioli GS, Sacks IS, Strutt MH (2010) Effect of mechanical heterogeneity in arc crust on volcano deformation with application to Soufrière Hills volcano, Montserrat, West Indies. J Geophys Res 115(B9):B09203

Hautmann S, Witham F, Christopher T, Cole P, Linde AT, Sacks IS, Sparks RSJ (2014) Strain field analysis on Montserrat (W.I.) as tool for assessing permeable flow paths in the magmatic system of Soufriere Hills Volcano. Geochem Geophys Geosyst 15:2013G. doi:10.1002/C005087

Hickey J, Gottsmann J (2014) Benchmarking and developing numerical finite element models of volcanic deformation. J Volcanol Geoth Res 280:126–130

Hickey J, Gottsmann J, del Potro R (2013) The large-scale surface uplift in the Altiplano-Puna region of Bolivia: a parametric study of source characteristics and crustal rheology using finite element analysis. Geochem Geophys Geosyst 14(3):540–555

Hickey J, Gottsmann J, Mothes P (2015) Estimating volcanic deformation source parameters with a finite element inversion: the 2001–2002 unrest at Cotopaxi volcano, Ecuador. J Geophys Res Solid Earth 120(3):1473–1486

Hickey J, Gottsmann J, Nakamichi H, Iguchi M (2016) Thermomechanical controls on magma supply and volcanic deformation: application to Aira caldera, Japan. Sci Rep 6(August):32691

Masterlark T (2007) Magma intrusion and deformation predictions: sensitivities to the Mogi assumptions. J Geophys Res 112(B06419)

Mogi K (1958) Relations between the eruptions of various volcanoes and the deformations of the ground surfaces around them. Bull Earthq Res Inst 36:99–134

Molina I, Kumagai H, Garca-Aristizábal A, Nakano M, Mothes P (2008) Source process of very-long-period events accompanying long-period signals at Cotopaxi Volcano, Ecuador. J Volcanol Geothermal Res 176(1):119–133

Odbert HM, Stewart RC, Wadge G (2014) Cyclic Phenomena at the Soufrière Hills Volcano, Montserrat, The Eruption of Soufrière Hills Volcano, Montserrat from 2000 to 2010, vol 39. The Geological Society, London, pp 41–60. doi:10.1144/M39.2

Pascal K, Neuberg J, Rivalta E (2013) On precisely modelling surface deformation due to interacting magma chambers and dykes. Geophys J Int 196(1):253–278

Pistolesi M, Cioni R, Rosi M, Cashman KV, Rossotti A, Aguilera E (2013) Evidence for lahar-triggering mechanisms in complex stratigraphic sequences: the post-twelfth century eruptive activity of Cotopaxi Volcano, Ecuador. Bull Volcanol 75(3):698

Pritchard ME, Simons M (2002) A satellite geodetic survey of large-scale deformation of volcanic centres in the central Andes. Nature 418(6894):167–171

Prutkin I, Vajda P, Gottsmann J (2014) The gravimetric picture of magmatic and hydrothermal sources driving hybrid unrest on Tenerife in 2004/5. J Volcanol Geoth Res 282:9–18

Ranalli G (1995) Rheology of the Earth. Chapman and Hall, London

Rouwet D, Sandri L, Marzocchi W, Gottsmann J, Selva J, Tonini R, Papale P (2014) Recognizing and tracking volcanic hazards related to non-magmatic unrest: a review. J Appl Volcanol 3:1–17

Sparks RSJ, Folkes CB, Humphreys MC, Barfod DN, Clavero J, Sunagua MC, McNutt SR, Pritchard ME (2008) Uturuncu volcano, Bolivia: volcanic unrest due to mid-crustal magma intrusion. Am J Sci 308(6):727–769

Sparks RSJ, Biggs J, Neuberg JW (2012) Monitoring volcanoes. Science 335(6074):1310–1311

Fluid Geochemistry and Volcanic Unrest: Dissolving the Haze in Time and Space

Dmitri Rouwet, Silvana Hidalgo, Erouscilla P. Joseph and Gino González-Ilama

Abstract

The heat and gas released by a degassing magma affects the overlying predominantly meteoric aquifers to form magmatic-hydrothermal systems inside the solid body of a volcano. This chapter reviews how fluid geochemical signals help to track the evolution throughout the various stages of volcanic unrest. A direct view into a degassing magma is possible at open-conduit degassing volcanoes. Nevertheless, in most cases gas is trapped (i.e. scrubbed) by abundant water, leading to the loss of the pure signal the magma ideally provides. Deciphering how magmatic gas rises through, reacts, and re-equilibrates with the liquids in the magmatic-hydrothermal system in time and space is the only way to trace back to the pure signal. The most indicative magmatic gas species (CO_2, SO_2–H_2S, HCl and HF) are released as a function of their solubility in magma. The less soluble gas species are released early from a magma at higher pressure conditions (CO_2) (deeper), whereas the more soluble species are released later, at lower pressures (SO_2, HCl and HF) (shallower depth). When these gases hit the water during their rise towards the surface, they will be more or less scrubbed. Depending on the chemical equilibria inside the magmatic-hydrothermal system (e.g. SO_2–H_2S conversion, acidity), the gas that eventually reaches the surface will carry the history of its rise from bottom to top. Tracking volcanic unrest implies a time frame; the kinetics of magma degassing throughout the liquid cocktail inside the volcano impose the maximum resolution the volcano provides and hence the monitoring time window to be adopted for

D. Rouwet (✉)
Istituto Nazionale di Geofisica e Vulcanologia, Sezione di Bologna, Via Donato Creti 12, 40128 Bologna, Italy
e-mail: dmitri.rouwet@ingv.it

S. Hidalgo
Instituto Geofisico, Escuela Politécnica Nacional, Quito, Ecuador

E.P. Joseph
Seismic Research Centre, University of the West Indies, St. Augustine, Trinidad & Tobago

G. González-Ilama
Centro de Investigaciones en Ciencias Geológicas, Universidad de Costa Rica, San José, Costa Rica

Advs in Volcanology (2019) 221–239
DOI 10.1007/11157_2017_12

Published Online: 19 July 2017

each volcano. Gas-dominated systems are "faster" and require a higher monitoring frequency, water-dominated systems are slower and require a lower monitoring frequency.

Resumen

El calor y gas liberados por la desgasificación del magma afecta los acuíferos de origen predominantemente meteórico para formar sistemas magmático-hidrotermales dentro el cuerpo sólido del volcán. Este capítulo revisa como la geoquímica de fluidos puede ayudar a trazar la evolución a través de las varias etapas de "unrest" volcánico. Una visión directa dentro de un magma en desgasificación es solo posible para volcanes de conducto abierto. Sin embargo, en la mayoría de las situaciones el gas queda atrapado (i.e. "scrubbing") en el agua, que conduce a la pérdida de la señal pura que el magma idealmente puede proporcionar. Descifrar como el magma sube a través de los líquidos, y reacciona y re-equilibra con ellos dentro el sistema magmático-hidrotermal, en un marco de tiempo y espacio, es la única manera para rastrear el origen de la señal del magma. La desgasificación de magma se da por cuatro procesos: (1) durante la subida de magma, (2) por la descompresión debido al eliminar una porción del edificio volcánico, (3) debido a la convección interna dentro la cámara magmática, o (4) después de "ebullición secundaria" siguiendo el enfriamiento y consecuente cristalización. Las especies gaseosas magmáticas más indicativas (CO_2, SO_2–H_2S, HCl y HF) se liberan en función de su solubilidad en el magma. Las especies menos solubles se liberan antes del magma, bajo regímenes de presiones más altas (CO_2), mientras que las especies más solubles se liberan después, bajo regímenes de presiones más bajas (SO_2, HCl y HF). En términos espaciales, CO_2 se libera a lo largo de una área espacial más amplia (desde lo más profundo). La presencia de SO_2 es una indicación clara de un magma que sube hacia un ambiente más somero. La llegada de HCl en la superficie generalmente indica la presencia de una remesa de magma somera (cientos de metros hasta pocos kilómetros). Especialmente un aumento en la proporción CO_2/SO_2 es indicativo para elucidar un estado de "inquietud" ("unrest"). Una disminución consecutiva en el CO_2/SO_2, después de un aumento, es una indicación de que el magma está cerca de la superficie y es propenso a una erupción. Cuando estos gases alcanzan el agua durante su ascenso hacia la superficie, serán más, o menos, absorbidos. Dependiendo de los equilibrios químicos dentro el sistema magmático-hidrotermal (e.g. conversión de SO_2–H_2S, acidez), el gas que al final llega a la superficie lleva consigo la historia de su ascenso desde el fondo hasta la superficie. La desgasificación magmática es un proceso más rápido, mientras que la dinámica hidrotermal en el sistema rocoso FeO–$FeO_{1.5}$ es más lenta. Por eso, el H_2S se suele llamar un "gas hidrotermal", y el SO_2 un "gas magmático". Trazar "unrest volcánico" implica un encuadramiento de tiempo más especifico. Si la ventana de tiempo de monitoreo es más largo que el tiempo definido por la cinética de la migración de gas, los detalles de la dinámica de desgasificación se perderán inevitablemente. Contrariamente, si la ventana de tiempo de monitoreo es más corto que la ventana definida por la cinética de la migración de gas, en tal caso, resultará en una visión demasiado detallada de lo que el sistema magmático-hidrotermal puede proveer. Estudios recientes han demostrado que acuíferos extremadamente ácidos pueden "desacelerar" la señal dejada por el sistema magmático-hidrotermal dominado por el gas (e.g. fumarolas), pero pueden "acelerar" la señal dejada por el sistema magmático-hidrotermal dominado por agua (e.g. lagos cratéricos ácidos). Estos hallazgos tienen implicaciones significativas para el encuadramiento de tiempo en reconocer la desgasificación magmática, y por tanto, para la frecuencia del monitoreo.

Keywords
Fluid geochemistry · Magmatic-hydrothermal systems · Volcanic unrest · Volcano monitoring

Palabras clave
Geoquímica de fluidos · Sistemas magmático-hidrotermales · Inquietud volcánica · Monitoreo volcánico

Introduction

Fluid geochemical monitoring tracks variations in gaseous species and fluid phases released in various manners from a volcano. Fluids infiltrate, move, migrate, rise, react and re-equilibrate in the water- and vapor-filled solid body of the volcano. These processes are invisible as they occur in the subsurface, and can only be deduced from measurements at the surface. Nevertheless, gas and water are more mobile than rock and, when a volcano shifts into the gear of unrest, a change in degassing is often the first sign to be detected. As such, fluid geochemistry offers a crucial means to recognize unrest in a timely matter.

Most volcanic edifices store a large volume of water, as cold or thermal aquifers, with various hydrogeological architectures. When gas hits water in its rise towards the surface, the original signature of the gas is mostly lost. This disadvantage can however be overcome. Understanding how gas absorbs and solutes react in the liquid phase, and how it is eventually released to the atmosphere, is key to linking surface manifestations to magma dynamics.

Volcano monitoring largely focuses on how, where and when magma migrates towards the surface, as, intrinsically, every magmatic eruption is anticipated by magma rise. Nevertheless, volcanoes can become hazardous even without an eruption. A volcano is in a state of magmatic unrest if we recognize signals of a magma-on-the-move; in any other situation, given volcanic unrest, the volcano is in a state of non-magmatic unrest (hydrothermal or tectonic) (Rouwet et al. 2014a). To recognize volcanic unrest, the background behavior of a volcanic system should be tracked for a sufficiently long period, in order to know when a deviation from this background becomes a cause for concern (i.e. unrest, Phillipson et al. 2013). This background behavior and deviations from it are volcano-specific and can be monitored in several ways, besides the geochemical approach reviewed here.

The aim of this chapter is to scan through magmatic-hydrothermal systems during the process of magmatic degassing, from bottom to top, and describe how fluids behave in time and space. What are the lessons learned from fluid geochemistry throughout the evolution of volcanic quiescence, re-awakening, volcanic unrest, magmatic unrest and non-magmatic/hydrothermal unrest?

Magmatic-Hydrothermal Manifestations

One of the first signs of re-awakening after prolonged volcanic quiescence to a state that eventually causes concern is often the appearance of fumarolic exhalations from a crater. This happened, for instance, in 1994 during the reawakening of Popocatépetl (Mexico), and recently, at Cotopaxi (Ecuador), both VUELCO target volcanoes (De la Cruz-Reyna and Tilling 2008; Hall and Mothes 2008) (Fig. 1a). For open-conduit volcanoes the presence of a plume (i.e. a visible gas-vapor cloud originating from an open volcano crater) can become the prominent manifestation of degassing.

Some volcanoes are characterized by decade to century long high-temperature fumarolic degassing in a closed-conduit setting (Fig. 1b), suggesting the presence of a stable, but shallow magma chamber. This constantly high-temperature background degassing is generally no sign of unrest (e.g. Momotombo's >700 °C, Satsuma-Iwojima's >900 °C and Kudryavy's >700 °C; Menyailov et al. 1986; Shinohara et al. 1993; Taran et al. 1995), whereas the increase of fumarolic temperatures from low (boiling point of water at a given altitude, hence atmospheric pressure) to high (above boiling to magmatic temperature) can be a sign of resumed unrest (e.g. the 1980–1990s crisis at Vulcano, Italy, Capasso et al. 1997).

Magmatic-hydrothermal systems are aquifers inside a volcano or beneath a volcanic area, heated by a magma, at an unspecified depth. The origin of the water is generally meteoric (i.e. rain, snow and its melt water). How hot, and how gas-rich such magmatic-hydrothermal systems are depend on the proportion of the water volume with respect to the heat and gas provided by the magma. The latter depends on the residing depth of the magma. Boiling-temperature fumaroles are a common manifestation at magmatic-hydrothermal systems (Fig. 1c). When the water table of such systems intersects the surface, in craters or volcano flanks, boiling pools appear (Fig. 1d). Such pools can manifest bubbling degassing, and are nothing less than a water-rich fumarole (Fig. 1 e). Depending on the dominant gas they exhale, paired with water vapor, the manifestations are called solfataras (S-rich gases) or mofettes (CO_2-rich gases); depending on the temperature and vapor/water proportion they emit they are called fumaroles (boiling or above boiling steam vents), thermal springs (liquid water emission) or geysers (water + vapor jets with a cyclic behavior). Thermal springs can discharge inside active craters or on volcano flanks in a degassed state, without bubbling or boiling (Fig. 1f). Heated water can fill (parts of) craters and form volcanic lakes (Fig. 1g). Depending on the degassing state and depth of the underlying magma, degassing features (bubbling or diffuse degassing) and evaporation can occur at the lake surface (Fig. 1h).

The pictures of the degassing manifestations in Fig. 1 show a trend from gas-dominated, active plume degassing in an open-conduit setting towards more water-dominated, hydrothermal, passive degassing in a closed-conduit setting. These visual observations only give a first glance of magmatic-hydrothermal activity, and do not reflect the state of unrest of a volcano. Throughout the life-time of magmatic-hydrothermal systems (centuries to millenia) volcanoes can evolve from gas-dominated to water-dominated, and vice versa. The next sections present what we know on the theoretical level, following the laws of chemistry. Despite these classic rules, it will become clear that the range of manifestations and variations in fluid signatures is wide, and volcano-dependent.

Fig. 1 Degassing manifestations at magmatic-hydrothermal systems. **a** Open-conduit degassing at Popocatépetl. **b** High-temperature fumarolic degassing at Vulcano. **c** Boiling-temperature fumarolic degassing at the Arbol Quemado fracture in the Turrialba crater area. **d** Boiling pools inside the El Chichón crater. **e** Bubbling thermal spring at the SE flank of El Chichón. **f** Thermal flank spring at El Chichón. **g** The El Chichón crater lake. **h** Evaporative degassing from Laguna Caliente, Poás. The *cyan arrow* points towards water-dominance; the *yellow arrow* towards more gas-dominance

Magma Degassing from Bottom to Top

Magma Degassing

The degassing of a magma increases when the confining pressure in the magma decreases. Magma decompression can occur in four ways (Fig. 2): (1) magma rise, (2) decompression by uncovering a portion of the volcanic edifice, (3) internal convection in a magma chamber, or (4) secondary boiling upon cooling and consequent crystallization. The first process is often induced by the input of a deeper magma, into a shallower magma chamber. Magma rise towards the lower pressure regime results from the buoyancy difference between the stagnant magma, the rising new melt and the surrounding rocks. The second degassing process can be triggered by the mass removal from part of the volcanic edifice. This superficial process can even trigger eruptions (e.g. 1980 Mt St. Helens). Internal magma convection causes degassing of a less dense, gas-rich magma batch. Once degassed, the now denser magma batch sinks (e.g. Stromboli, Aiuppa et al. 2009). The fourth process increases the gas/melt ratio in the magma, due to the loss of crystals from the cooling magma, hence favoring degassing.

Based on gas geochemistry only it is hard to rule out which process actually occurs at depth, and thus distinguish between magmatic and non-magmatic unrest. Geophysical signals are needed. Following the definition of magmatic unrest by Rouwet et al. (2014a) (i.e. the recognition of a magma-on-the-move), only the first process is initially consistent with the requisite of magmatic unrest.

Variations in magma degassing can be detected both qualitatively and quantitatively. Detailed insights in the degassing state of a magma can only be obtained if both are measured contemporaneously. Which gas species are released, how much and when? When a magma starts to degas, by any of the above processes, the less soluble species is released first (i.e. at higher confining pressure in the magma chamber). The order in solubility of indicative magmatic gas species is $CO_2 < SO_2 < HCl < HF$; the order of release when a magma progressively degasses is "CO_2-first till HF-last" (Giggenbach 1987). Hence, tracking variations in ratios between these species gives qualitative insights into the degassing state of a magma. A consecutive increase with time in first CO_2/SO_2, then SO_2/HCl, then HCl/HF ratios reflects the evolution in degassing state from a magma moving from depth towards the surface. Especially an increase in the CO_2/SO_2 is indicative of the state of unrest, pointing to an input of poorly degassed magma at great depths. A consecutive decrease in CO_2/SO_2 ratio, after the increase, is an indication of magma moving towards the surface. The latter two ratios come into play when eruption of magma is imminent, or even ongoing: the highly soluble species HCl and HF are released from a highly degassed magma, a situation that reflects near-surface degassing (Aiuppa et al. 2002). The arrival of HCl at the surface (e.g. in

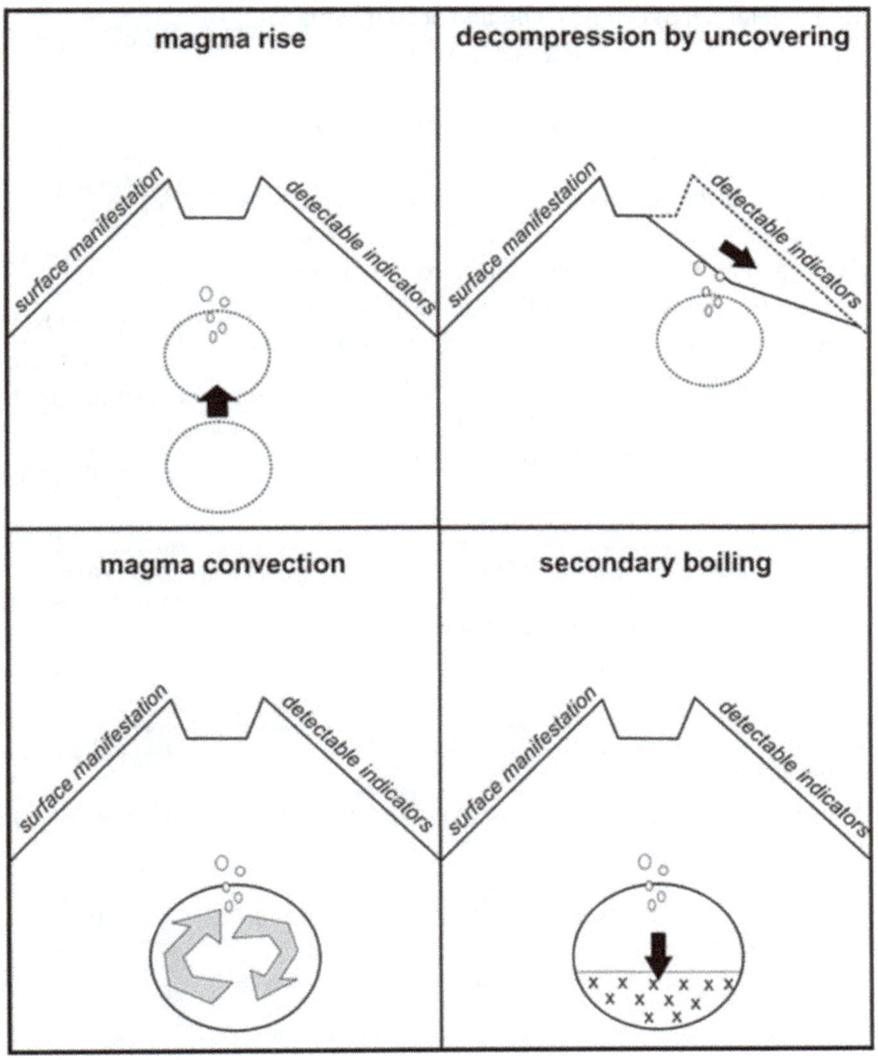

Fig. 2 Sketch of the four mechanisms that instigate magma degassing (not to scale)

fumaroles) generally indicates the presence of a shallow magma batch (hundreds of meters, or less).

The "purest" proxy of the magmatic gas is provided by direct sampling of a fumarole with a near-magmatic temperature, followed by the analysis of its chemical composition. During the past five-six decades fumarolic gases have been extensively sampled and analyzed representing a wide range of temperatures and states of volcanic activity. Table 1 presents a compilation of chemical compositions of fumarolic gases for the high- and low-temperature ranges.

Before dealing with the abundant water vapor in fumaroles, we tackle the other gas species in the "dry-gas phase". Regardless of the fumarole temperature, CO_2 is the most abundant gas species (Table 1). CO_2 is often released at the surface across a wider spatial extent, as it degasses from greater depth (Fig. 3). Old and deep magma bodies can continue to release CO_2 for tens to hundreds of millenia.

The second most abundant "dry-gas" species are the sulphur species SO_2 and H_2S. SO_2 is more soluble in magma than CO_2, and will thus be released at lower pressure. As SO_2 degasses at lower depth, the degassing tends to be more centralized along the central conduit (open or closed) of the volcano (Fig. 3). As such, SO_2 is often measured in volcanic plumes. An increase of SO_2 has often been interpreted as an indicator of magma rise into the shallower environment.

At high-temperature magmatic conditions the following reaction applies (Giggenbach 1987; Delmelle and Bernard 2015):

$$3SO_2 + 7H_2 = H_2S + 2S^\circ + 6H_2O \quad (1)$$

This explains the presence of both SO_2 and H_2S in high-temperature fumaroles. The SO_2/H_2S ratio is sensitive to temperature and the oxidation state (i.e. the role of H_2 in Eq. 1 reflects the oxidation vs reduction state, Giggenbach 1987).

On the other hand, it is noted that the low-temperature fumaroles lack SO_2, instead, the dominant sulphur species is H_2S (Table 1). Low-temperature hydrothermal conditions ($T < 300$ °C) favor the reduced S-species H_2S (Table 1), equilibrated by the rock phase, following the reaction:

$$SO_2 + 6(FeO)_{rock} + H_2O = H_2S + 6(FeO_{1.5})_{rock} \quad (2)$$

or simply through reduction of SO_2 in the gaseous environment:

$$SO_2 + H_2 = H_2S + O_2 \quad (3)$$

To convert SO_2 into H_2S in a rock matrix (FeO–$FeO_{1.5}$-system), following reaction (2), a major constraint is "sufficient time". Magmatic degassing (Eq. 1) is a faster process, while hydrothermal dynamics (Eq. 2) are slower. This

Table 1 Chemical composition of fumarolic gases (concentrations in micromol/mol), for high (magmatic, 280°–1130 °C) and low (hydrothermal, 83°–160 °C) temperature conditions, expressed as the minimum and maximum measured concentrations for 12 and 59 samples, respectively

	T(°C)	H_2O	CO_2	SO_2	H_2S	HCl	HF	H_2	N_2	CH_4	CO
High-T minimum (#12)	280	311000	1200	320	4	275	21	3	40	0.1	0.2
High-T maximum (#12)	1130	993000	672000	67800	21460	14200	2500	14900	1800	7.1	4600
Low-T minimum (#59)	83	638200	2655	0	0	0	0	16	32	0.4	0.011
Low-T maximum (#59)	160	997200	355000	0	3700	0	0	1220	6800	5330	1.6

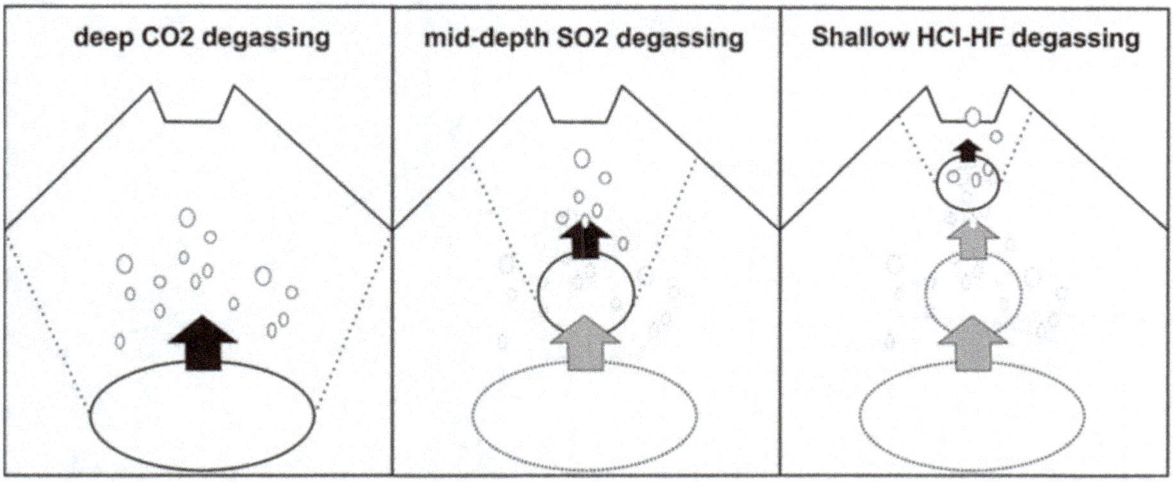

Fig. 3 Sketch of the passing "gas train" during magma degassing, resulting from the difference in solubility of the various gas species (not to scale)

is why H_2S is often called a "hydrothermal gas" and SO_2 a "magmatic gas" (Table 1).

As noted in Table 1, the major gas species in fumaroles is often water, regardless of their temperature. Even high-temperature fumaroles are water-dominated.

Knowing that water boils at 100 °C (at atmospheric pressure, at sea level), in theory, the temperature of a fumarole is buffered at 100 °C and cannot rise until water is exhausted in the underlying plumbing system. Moreover, the critical temperature of water is 374 °C (i.e. the temperature at which vapor and liquid water cannot coexist anymore), imposing a second temperature buffer. This implies that (1) water directly originates from a high-temperature magma under super-critical conditions ("andesitic water" with T > 374 °C, Taran et al. 1989), and/or (2) water is excessively present in the fumarole plumbing system with respect to the gas, and will hardly ever exhaust.

When the Gas Hits the Water

The shallow subsurface environment of the different sections of the Earth's crust hosts numerous aquifers at various depths originating from the infiltration and storage of meteoric water, or seawater in the case of low-lying islands; volcanic edifices are no different, being "small dots" at the Earth's surface. As described before, a magma degasses and heats the space between the magma and the surface, and will inevitably heat and modify the volcanic aquifers. As the magma heats the overlying aquifers, between the magma and the surface, from bottom to top, a gas-only, vapor + gas zone, a vapor + liquid zone and a liquid-only zone can be found (Fig. 4). When a magma rises, or the heat input from the magma increases, the vapor + liquid zone or vapor + gas zone will be pushed upwards until intersecting the surface (e.g. in a volcano crater), manifested at the surface as boiling or bubbling pools and fumaroles (Fig. 4). If the distance between the magma and surface is larger, the thermal aquifer will intersect the surface and create thermal springs (Fig. 4). Facing unrest, a liquid to vapor transition at a surface manifestation reflects heating of the hydrothermal system.

When the above "gas train" (Fig. 3) consecutively reaches the liquid-only zone, gas species will be absorbed and react depending on their specific chemical properties in water. The capacity to absorb magmatic gases in the liquid phase is called "scrubbing" (Symonds et al. 2001). The CO_2 that reaches the water from greater depths during magma degassing (Fig. 3), will create CO_2-dominated bubbling thermal springs, and HCO_3-rich slightly acidic springs (pH 5–7) (Fig. 4). The second least soluble gas species that hits the water is SO_2 that will be

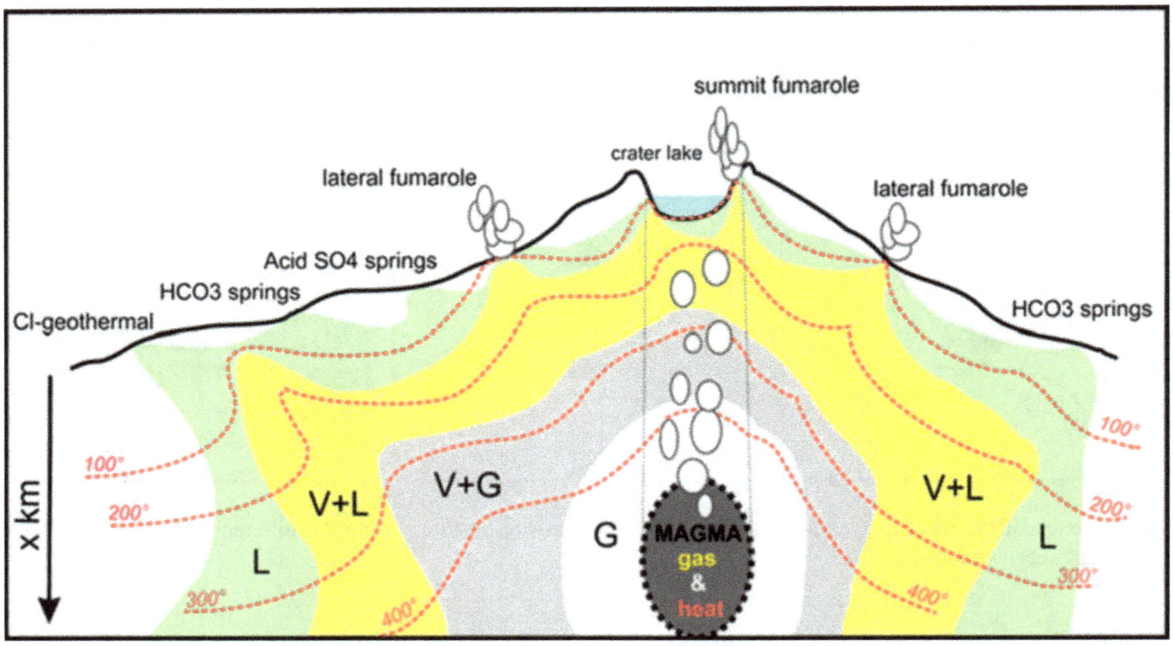

Fig. 4 Sketch of the gas that hits the water in an hypothetical "wet volcano" (not to scale). L = liquid only zone (turquoise area), V + L = vapor + liquid zone, (*yellow area*) V + G = vapor + gas zone (*grey area*), G = gas only zone (*white area* around the degassing magma, *dark grey*). *Red dotted lines* are isotherms. Cl-geothermal water are deep remnant waters, not of interest for geochemical monitoring of volcanoes for being "old and slow"

hydrolyzed as sulphuric acid (H_2SO_4) and dissociate into HSO_4^- or SO_4^{2-}, depending on the pH of the water. The dissolved H^+ creates the high acidity or low pH, following $pH = -\log[H^+]$. Volcanic environments are renowned for being acidic; the acid is generated by the scrubbing of acidic gases into the water.

The following SO_2 disproportionation reactions occur (Kusakabe et al. 2000):

$$3SO_2 + 3H_2O = 2HSO_{4^-} + S^\circ + 2H^+ \quad (4)$$

or

$$4SO_2 + 4H_2O = 3HSO_{4^-} + H_2S + 3H^+ \quad (5)$$

Reaction (4) occurs under relatively oxidizing conditions, low temperature and high total sulphur concentrations, whereas reaction (5) occurs under relatively reducing (i.e. oxygen poor) conditions, high temperature and low total sulphur concentrations. "To disproportionate SO_2" means that for each three or four moles of SO_2 that hit the water, one mole of S°, or one mole of H_2S is given in return (reaction 4 and 5, respectively).

As SO_2 degasses at shallower depth in a magma system than CO_2, the resulting acid SO_4-rich springs are found near the central degassing conduit (e.g. inside active craters). For SO_2-dominated magmatic systems a pH near 2 or less is common; for H_2S-dominated hydrothermal systems a pH of 2–2.5 is the most acidic water can get (i.e. hydrothermal or "steam heated" waters).

The next gas species to be added to the "liquid cocktail" are HCl and HF (Fig. 3). HCl is highly volatile, but also highly hydrophile. This *contradictio in terminis* means that when HCl_{gas} reaches water it will be trapped in the liquid phase as Cl^- and H^+. Generally, Cl^- is considered "conservative" in the liquid phase, and is therefore often used as a tracer in the hydrothermal environment (see Section "Rock Leaching Upon Weathering"). Conservative means that Cl^- can be hardly lost from the solution as Cl-salts are highly soluble in acidic and hot environments, and HCl should not degas from "high pH water" (>2). Sulphate minerals in their turn can be lost from solutions by precipitation, demonstrating their non-conservative character (Rouwet and Ohba 2015). For extremely acidic environments (pH 0 or <0) the reaction

$$HCl = H^+ + Cl^- \quad (6)$$

moves to the left due to an H^+ excess with respect to Cl^- (i.e. HCl degassing). The same counts for HF. Bearing in mind the abundance of the acid SO_2 in the magmatic-hydrothermal environment, providing a large part of the acidity, this "secondary HCl degassing from the liquid phase" is less unexpected than previously thought. This implies that HCl can degas from a low-temperature aquifer, as long as the aquifer is extremely acid (Capaccioni et al. 2016). Moreover, as hot water releases vapor, this loss enriches the remnant liquid in solutes, including H^+ (i.e. salinity and acidity increase), leading to the fact that even SO_2 tends to degas from the liquid, instead of remaining in the water phase as SO_4^{2-} or HSO_4^-.

Acid water (pH < 3.8) is completely transparent for the omni-present CO_2. This implies that CO_2, one of the "deepest signals" available for a degassing volcano, will completely outgas from bottom to top. In active craters, often underlain by acidic thermal aquifers, the release of CO_2 thus behaves as though there is no water present. This is a great advantage to monitoring volcanoes and tracking unrest, especially to detect the onset of unrest.

In conclusion, the above insights demythologize two generally accepted facts: (1) high-temperature fumaroles cannot contain water vapor (Table 1), and (2) low-temperature fumaroles cannot release acidic gases.

Within the scope of this book, tracking unrest using fluid geochemistry requires the introduction of a time frame, or a monitoring time window and frequency. Does the fumarole reflect the exact moment of degassing, or is it rather an average degassing for the longer previous period stored and steadily released from the magmatic-hydrothermal plumbing system? What is the time delay between the moment the gas hits the water and the eventual release at the surface? The kinetics (i.e. "speed") of the gas migration from a magma towards the surface are still poorly constrained. If we can estimate the residence time of gas and water in the magmatic-hydrothermal system, we are able to define a monitoring time window, and hence adopt an adequate monitoring frequency. As explained, the acidity of the feeding aquifer plays a role. For extreme acidic conditions, less gas scrubbing occurs and the fumarolic system will react faster, and hence, shorter monitoring time windows can be adopted. If the monitoring time window is longer than the window defined by the kinetics of gas migration, details in degassing dynamics will be lost. On the contrary, the monitoring time window should not be a lot shorter than the time window defined by the kinetics of gas migration, if so, it will provide a too detailed view of what the magmatic-hydrothermal system can maximally provide.

The Other Liquid: Elemental Sulphur

Whereas water melts at 0 °C and boils at 100 °C, sulphur melts at 119 °C and boils at 444 °C (Oppenheimer and Stevenson 1989, and references therein). This physical fact on phase transitions implies that in the hydrothermal environment (boiling water) elemental sulphur is solid, and that during the initial phase of transition towards a more magmatic, high-temperature regime sulphur will become liquid. Molten sulphur in the hydrothermal plumbing system can be remobilized, clear vugs and vents and eventually be expelled as a liquid sulphur flow from fumarole mouths, or "fill-and-freeze" pores in the shallower hydrothermal system. The first process opens up degassing pathways; the second process can decrease rock porosity and permeability near the surface, thus sealing a hydrothermal system. During the evolution from low-temperature (>119 °C, unrest) towards high-temperature (occasionally magmatic unrest), the viscosity of the liquid sulphur increases

2000-fold (>160 °C) to become an extremely efficient sealer of a magmatic-hydrothermal system. Pressure build-up beneath seals in hydrothermal systems can lead to phreatic eruptions (Rouwet and Morrissey 2015; Rouwet et al. 2016). Monitoring fumarolic temperatures is thus essential, and probably one the simplest methods to apply.

Tracking Hydrothermal Unrest and Related Hazards: Methods from Case-Studies

From Quiescence to Unrest, to Phreatic Eruptions, to Magmatic Eruptions

Turrialba, Costa Rica (2001–2016)

A transition from the stable, passively degassing hydrothermal system of Turrialba volcano (Costa Rica), to hydrothermal unrest, to phreatic eruptions, to magmatic eruptions, is a recent example of an evolution from volcanic quiescence heading towards eruption.

In 2001, increased fumarolic activity (appearance of SO_2 in late 2001) was paired to seismic swarms and ground deformation (Martini et al. 2010; Vaselli et al. 2009) (i.e. unrest). In 2007, the increased SO_2/H_2S molar ratio in fumaroles (>100), combined with an increase in exhalation temperature up to 282 °C (in early 2008, Martini et al. 2010; Vaselli et al. 2009), point to more oxidized and magmatic conditions (i.e. magmatic unrest). Clear plume degassing resumed in early 2007 (Fig. 5a), and SO_2 fluxes reached 740 t/d in January 2008 (Martini et al. 2010). In late 2009, fumarolic degassing was vigorous and extended into the Arbol Quemado fracture, newly formed in 2002. The first phreatic eruption occurred on 5 January 2010, in the inner crater wall of the actively degassing SW crater. Strong jet-like degassing occurred afterwards from this new vent (Fig. 5b), while diffuse fumarolic degassing diminished in the SW crater. A second phreatic eruption occurred on 12 January 2012, from a vent inside the Arbol Quemado fracture. The day before, this eruption was preceded by liquid sulphur flowing out of the Arbol Quemado fumaroles (González et al. 2015). A third phreatic eruption episode involved the 2010 and 2012 vents simultaneously (21 May 2013). During this 3.5-year long phreatic cycle the SO_2 flux from Turrialba's plume was high: from 2500 to 4300 t/d (Campion et al. 2012; Moussallam et al. 2014). CO_2/SO_2 molar ratios in March 2013 were relatively low (2.6), hinting at a CO_2-depleted and SO_2-rich magma (Moussallam et al. 2014). From 2001 to 2013, Moussallam et al. (2014) suggest the progressive "drying-out" of the underlying hydrothermal system.

The first magmatic eruption at Turrialba since the 1864–1866 phreatomagmatic activity occurred during the night of 29 and 30 October 2014 (Mora-Amador et al. 2015). At the time of writing, the last magmatic eruptions took place in September 2016.

Despite the well-monitored and tracked evolution from volcanic quiescence to magmatic eruption it remains unclear why some volcanoes quickly evolve from quiescence to eruption, while at Turrialba it took 14 years from quiescence to magmatic eruption, passing the complete range of unrest manifestations during this relatively long time span.

Cotopaxi, Ecuador (2015–2016)

Another example of volcanic quiescence to magmatic unrest is the one of Cotopaxi volcano in 2015. As Cotopaxi is a dangerous volcano whose activity would potentially affect densely populated areas its monitoring network has been continuously improved since the late 70s. After 73 years of quiescence, the first sign of unrest at Cotopaxi was a progressive increase in the amplitude of transient seismic events in April 2015. SO_2 is measured at Cotopaxi by DOAS stations installed on the flanks of the volcano since 2008. The permanently measured SO_2 emissions showed an increase on May 20 from almost non detectable up to ~3000 t/d. The fumaroles showed increased activity and a gas plume from the crater was usually observed on clear days. By early June SO_2 emissions yielded up to 5000 t/d. On July 20 a green lake was observed filling the crater of the volcano,

Fig. 5 Hydrothermal unrest at Turrialba, Poás and Irazú volcanoes. **a** Vegetation die-back due to resumed plume degassing. **b** The 2010 phreatic eruption vent (picture by S. Calabrese). **c** Lack of vegetation at Cerro Pelón (Poás) due to plume degassing and acid rain fall downwind. **d** Efficient rock mass removal through Río Sucio, 30 km downstream Irazú volcano

nevertheless no significant changes in the SO_2 emission was observed, suggesting that the lake was of high acidity and/or too small to be an efficient scrubber. The first phreato-magmatic explosions occurred on August 14 and produced ash and gas columns reaching up to 9 km above the crater. The satellite-borne instruments such as OMI and OMPS reported 16,400 and 12,500 t/d of SO_2 released to the atmosphere on that day (http://so2.gsfc.nasa.gov/pix/daily/0815/ecuador_0815z.html). Continuous ash emissions followed the initial explosions producing a permanent gas and ash plume drifting westward. SO_2 measured in this permanent plume, by mobile-DOAS traverses or by the permanent stations, reached 24,000 t/d and decreased progressively until the end of the activity in late November 2015.

Since June and more consistently since 14 August 2015, BrO was also detected in the plume (Dinger et al. 2016). Airborne Multi-GAS measurements showed that the plume had a low CO_2/SO_2 ratio, and that SO_2 was >99% of total sulfur (SO_2 + H_2S), indicating a shallow magmatic origin for the gas. At the time of writing (September 2016), SO_2 emissions decreased to background levels. SO_2 permanent monitoring proved to be a useful tool at Cotopaxi providing real time data contributing, together with other geophysical methods, to better evaluate volcanic unrest scenarios.

Gas Impact and Acid Rain

Persistent high-temperature fumarolic and plume degassing impact volcano flanks down-wind (Fig. 5c). This can happen when it rains through a plume, generating acid rain. Many volcanoes are subject to such long-term, non-eruptive hazard, making the surrounding ground harsh living environments. Other volcanoes suffer vegetation die-back, when activity resumes after prolonged quiescence (Fig. 5a). This was clearly visible during the increased activity at Turrialba volcano (Section "Turrialba, Costa Rica (2001–2016)", Fig. 5a, b; González et al. 2015) and downwind Poás' western flank (Cerro Pelón, Fig. 5c). To assess such hazards, a meteorological station (wind direction, speed, air humidity and rainfall), and a DOAS device to measure SO_2 fluxes are valuable tools.

Rock Leaching upon Weathering

Absorption of magmatic gases into aquifers creates acidic magmatic-hydrothermal systems that, sooner or later, will exit the volcano. Acidic water attacks the wall rock and becomes loaded with solutes (Delmelle et al. 2015). If (1) meteoric recharge is high, (2) acid input is high, (3) wall rock is fresh, and (4) (thermal) spring water discharge is high, rock leaching capacity can reach thresholds to even mechanically destabilize volcano flanks. Enhanced chemical leaching for long periods favors physical rock removal, causing rock fall, landslides, or even flank and sector collapses of volcanic edifices. Even if magmatic degassing seems absent and a volcano may be long dormant, hazards loom due to the scrubbing capacity of deeper aquifers (Delmelle et al. 2015, and references therein).

The best suited method to quantify and track rock mass removal from a volcanic edifice is by monitoring the discharge rates from thermal springs, and their dissolved solutes and solids. The "Cl-inventory" uses Cl as the conservative tracer. As mentioned earlier, this is true for less acidic magmatic-hydrothermal systems (Ingebritsen et al. 2001; Taran and Peiffer 2009; Chiodini et al. 2014; Collard et al. 2014). Measuring the Cl-release from rivers draining thermal springs, and knowing the Cl-content and Cl/solute ratios in thermal spring waters, the rock mass removal rate can be estimated by:

$$Q_r * C_r = Q_s * C_s \tag{7}$$

where Q_r and Q_s are the discharges of rivers and thermal springs, respectively, in L/s; C_r and C_s are the concentrations of Cl of rivers and thermal springs, respectively, in mg/L. Measuring the river discharge (Q_r) and analyzing the river and spring waters for its Cl content (C_r and C_s, respectively) thus enables to estimate the spring discharge (Q_s), which would otherwise be impossible to directly measure in the field (e.g. numerous spring discharges, too irregular spring mouths). This method combines gas-water-rock interaction and hydrology of magmatic-hydrothermal systems in order to assess indirect hazard. Volcanoes with high rock mass removal rates are e.g. Irazú (Costa Rica, Fig. 5d); extremely acidic magmatic-hydrothermal rock removers are Kawah Ijen (Java), Poás (Costa Rica), and Copahue (Argentina-Chile). In the most extreme cases, the acidity and toxic metal load affects agricultural activities and human health.

Moreover, through the same Cl-inventory approach, the geothermal potential (i.e. heat output) from springs can be estimated, by multiplying the enthalpy of discharged spring waters, often based on geothermometric temperatures of the deep system, with the spring discharge rates. Such estimates were obtained for the active magmatic-hydrothermal systems of El Chichón and Tacaná (both in Chiapas, Mexico; Taran and Peiffer 2009; Collard et al. 2014), and Domuyo (Argentina; Chiodini et al. 2014), and originally of the Cascades Volcanic Range by Ingebritsen et al. (2001). Understanding the state of unrest on the long term of a specific volcano is needed to rule out if the volcano would be a feasible target for geothermal exploitation, or not.

Volcanic Lakes

Acid Peak-Activity Lakes in a State of Unrest

Volcanic lakes are the intersection of the crater surface and the underlying aquifer (Fig. 1g, h). Hence, they become "windows" into the depths of magmatic-hydrothermal systems. While fumaroles directly lose their signal from depth to the atmosphere as a "snapshot" (but see Section "When the Gas Hits the Water"), volcanic lakes preserve a past gas marker for a certain period. The time we can track back by studying the water chemistry depends on the duration the water resides in the lake (i.e. residence time). The residence time (RT) is estimated by dividing the lake volume (V in m^3) by the input or output rate (Q in m^3/s) of fluids, assuming steady-state conditions (Varekamp 2003; Rouwet et al. 2014b):

$$RT = V/Q \quad (8)$$

Small lakes with high rates of fluid flushing offer a better time-resolution than large lakes with slow rates of fluid in- and output. The monitoring frequency (e.g. lake water sampling) should be tuned to this residence time; a higher monitoring frequency will oversample the lake chemistry, while a lower monitoring frequency will lead to the loss of information the lake potentially provides (Rouwet et al. 2014b).

Volcanic lakes are excellent gas scrubbers. Nevertheless, recent studies quantify the gas release from the lake surface of the most acidic lakes (e.g. Aso, Copahue, Poás, Kawah Ijen; Shinohara et al. 2015; Tamburello et al. 2015; Capaccioni et al. 2016; de Moor et al. 2016; Gunawan et al. 2016). Under the most extreme pH conditions (<0) HF, HCl and even SO_2 can degas freely from the lake. This means that acidic lakes are more sensitive than thought before, as acid gas flashes through the water body with only minor scrubbing. Monitoring frequency can thus increase and, hence, spectroscopic or electrochemical sensor tools become extremely useful (DOAS to measure SO_2 fluxes from volcanic plumes, Multi-GAS to measure ratios between gas species, e.g. CO_2/SO_2).

Considering acidic lakes as "open-air" fumaroles (and fumaroles as "buried" acidic lakes) has changed the monitoring time frame, which might lead to better chances to forecast phreatic eruptions (Rouwet et al. 2014a, b; Rouwet and Morrissey 2015).

Volcanic Lake Response to External Triggers in the Absence of Magmatic Unrest

The Boiling Lake, in Dominica of the Lesser Antilles (a VUELCO target volcano), is a high-temperature volcanic crater lake, that is believed to be formed as a result of phreatic or phreatomagmatic activity (Lindsay et al. 2005). It is approximately 50 m × 60 m in size and ca. 12–15 m deep, with an estimated volume of $\sim 1.2 \times 10^4$ m^3 when filled (Fournier et al. 2009). Over the last 150 years, temperatures taken at the edge of the lake have generally ranged between 80 and 90 °C, and the pH from 4 to 6 (Joseph et al. 2011), however, the lake experiences periods of instability where water level, temperature and state of hydrothermal activity fluctuate suddenly. The long term stability of the Boiling Lake is attributed to the hypothesis that the lake is suspended above the water table by the buoyancy of rising steam bubbles from the underlying hydrothermal system (Fournier et al. 2009). Perturbations of this interaction as a result of landslides into the lake or regional seismicity is attributed to the cause of the sudden periods of instability (i.e. lake drainage and refill cycles), rather than changes in underlying magmatic activity. This is mainly because no coincident anomalous hydrothermal activity was observed in the Valley of Desolation geothermal area, which is located nearby the Boiling Lake.

The most recent episode of instability occurred during December 2004 to April 2005 (Fig. 6), when the lake water level dropped by ~8–10 m, and the temperature at the water's edge decreased to <30 °C. Water acidity went from the usual acidic pH of 4–6 to neutral, while

Fig. 6 The refill of an empty Boiling Lake during a period of unusual activity where water levels and geothermal activity were rapidly fluctuating. **a** An empty Boiling Lake on 7 April 2005. **b** A full Boiling Lake on 13 April 2005. Pictures by Arlington James, Forestry Officer, Forestry and Wildlife Division, Dominica (used with permission)

Cl concentration dropped from the typical 2000–6000 to 29–50 mg/L, and SO_4 concentration dropped from 1500–4000 to 100–270 mg/L, indicating a drastic decrease of hydrothermal fluid input into the lake (Joseph et al. 2011). Additionally, total dissolved solid content decreased from 13,400 to 4500 mg/L, suggesting strong dilution by fresh water. Measurements of temperature, pH and chemical composition taken in August 2006 indicate that the lake had returned to its normal steady-state of activity (Joseph et al. 2011). This episode is reported to have been triggered as a result of the extensional strain induced by a regional Mw 6.3 earthquake that occurred on November 21, 2004 offshore Les Saintes (Guadeloupe) (Feuillet et al. 2011), which may have contributed to diminished water inflow.

It should be noted, however, that a phreatic explosion and gas release occurred at an "empty"

Table 2 Geochemical signals and what they indicate with respect to volcanic unrest

Type of unrest	Geochemical signal	Indication
Unrest	CO_2 flux above background	Changes in deep degassing dynamics
	Increase in T of hot springs and/or fumaroles	Increased heat input
	Changes in H_2O/CO_2 ratios in fumaroles	Changes in water/gas ratio
	Appearance of new fumaroles and/or hot springs	Aerial extension of activity
Magmatic unrest	Appearance of acidic gases (SO_2, HCl, HF)	Changes in mid- to shallow magma dynamics
	T fumarole >119 °C	Remobilisation of sulphur
	SO_2 flux > X t/d	SO_2 flux above background, volcano-dependent
	Increase in CO_2/SO_2 ratio	Arrival of an undegassed magma at depth
	Extreme increase in T fumaroles (>300 °C)	Towards magmatic T
Magmatic eruption	Decreasing CO_2/SO_2 ratios after increase	More superficial magma degassing
	Increase in Cl, Br, F concentrations in hot springs/pools	Input of highly soluble acidic gases
	Decrease in H_2O/CO_2 and/or H_2S/SO_2 and/or SO_2/HCl ratios	More gas with a more magmatic signature
Hydrothermal unrest	New fumaroles	Aerial extension of activity
	Anomalous glacier defrosting	Sudden removal of water mass… lahars
	Water to vapour transition	Pushing vapour front from below
	Changes in hydrothermal features	Variations or aerial extension of activity
	Increase in B and/or NH_4 in waters	Input of vapour
	Increase in CH_4/CO_2 in fumaroles	A more hydrothermal signature in fumaroles
	Variations in phreatic level in aquifers	Pushing vapour front from below
Hydrothermal eruption	120 °C < T fumarole <200 °C	Self-sealing by a change in S viscosity
	Extension of alteration areas or fumarolic fields	Aerial extension of activity
	Appearance of muddy pools	Clearing bugs and vents, unplugging
	Boiling/bubbling of pools that previously didn't	Rising vapour front and/or extra heating and degassing

Boiling Lake on 10 December 1901 that resulted in the deaths of two individuals (Elliot 1938; Bell 1946). This suggests that hazards related to volcanic lakes such as the Boiling Lake, may occur without magmatic input.

Take-Home Ideas: Implications for Geochemical Monitoring

Over the past five-six decades, gas geochemistry at magmatic-hydrothermal systems has mainly focussed on chemical equilibria and kinetics in the subsurface environment. Over the last 10–15 years more attention has been paid to remote sensing of volcanic gas plumes (DOAS, Multi-GAS) with the obvious advantage of increased safety and frequency of data gathering. Nevertheless, the relationship between the fumarole and plume has yet to be better constrained. The best proxy of a magmatic gas remains a direct sample of a high-temperature fumarole, although such target fumaroles are often inaccessible, especially during eruptive phases. Compromises between data fidelity and safety of the operators, and the frequency of data gathering should be framed in terms of what we want and maximally can unravel. The advantage of fluid geochemistry in volcano monitoring arrises from the fact that volatiles are mobile and thus reach the surface often before physical changes manifest. Timely recognition of unrest and especially hydrothermal unrest is often possible. Table 2 summarizes geochemical signals and how they relate to the various states of volcanic unrest, useful for monitoring based on deterministic research and probabilistic modeling.

Future research should focus on better constraining degassing dynamics at the surface-atmosphere boundary. Recent studies have demonstrated that extremely acidic aquifers can "slow down" the signal released from gas-dominated magmatic-hydrothermal systems (fumaroles), but "speed up" the signal released from water-dominated systems (e.g. acidic crater lakes). These findings have strong implications for the time frame of magma degassing, and hence for the monitoring frequency.

References

Aiuppa A, Federico C, Paonita G, Valenza M (2002) S, Cl and F degassing as an indicator of volcanic dynamics: the 2001 eruption of Mount Etna. Geophys Res Lett 29(11):1559. doi:10.1029/2002GL015032

Aiuppa A, Federico C, Giudice G, Giuffrida G, Guida R, Gurrieri S, Liuzzo M, Moretti R, Papale P (2009) The 2007 eruption of Stromboli volcano: insights from real-time measurements of the volcanic gas plume CO_2/SO_2 ratio. J Volcanol Geoth Res 182(221):230. doi:10.1016/j.jvolgeores.2008.09.013

Baldoni E, Rouwet D, Mora-Amador R, Ramírez C, González-Ilama G, Lucchi F, Capaccioni B, Tranne CA, Pecoraino G (submitted) Hydrogeochemical model of the Irazú-Turrialba volcanic complex (Costa Rica) and implications for hazard assessment and volcanic surveillance. In: Caudron C, Capaccioni B, Ohba T (eds) GSL special volume geochemistry and geophysics of volcanic lakes

Bell H (1946) Glimpses of a governor's life—Dominica. Samson Low, London

Campion R, Martínez-Cruz M, Lecocq T, Caudron C, Pacheco J, Pianrdi G, Hermans C, Carn S, Bernard A (2012) Space- and ground-based measurements of sulphur dioxide emissions from Turrialba Volcano (Costa Rica). Bull Volcanol. doi:10.1007/s00445-012-0631-z

Capaccioni B, Rouwet D, Tassi F (2016) HCl degassing from extremely acidic crater lakes: empirical results from experimental determinations and implications for geochemical monitoring. In: Caudron C, Capaccioni B, Ohba T (eds) GSL special publications 437 geochemistry and geophysics of volcanic lakes. doi:10.1144/SP437.12

Capasso G, Favara R, Inguaggiato S (1997) Chemical features and isotopic composition of gaseous manifestations on Vulcano Island, Aeolian Islands, Italy: an interpretative model of fluid circulation. Geochim Cosmochim Acta 61(16):3425–3440

Chiodini G, Liccioli C, Vaselli O, Calabrese S, Tassi F, Caliro S, Caselli A, Agusto M, D'Alessandro W (2014) The Domuyo volcanic system: an enormous geothermal resource in Argentine Patagonia. J Volcanol Geoth Res 272:71–77. doi:10.1016/j.jvolgeores.2014.02.006

Dinger F, Arellano S, Battaglia J, Bobrowski N, Galle B, Hernández S, Hidalgo S, Hörmann C, Lübcke P, Platt U, Ruíz M, Warnach S, Wagner T (2016) Variations of the BrO/SO_2 molar ratios during the 2015 Cotopaxi eruption. EGU 2016-1001

Collard N, Taran Y, Peiffer L, Campion R, Jacome Paz MP (2014) Solute fluxes and geothermal potential of Tacana volcano-hydrothermal system, Mexico-Guatemala. J Volcanol Geoth Res 288:123–131. doi:10.1016/j.jvolgeores.2014.10.012

De la Cruz-Reyna S, Tilling RI (2008) Scientific and public responses to the ongoing volcanic crisis at Popocatépetl Volcano, Mexico: importance of an effective hazards-warning system. J Volcanol Geoth Res 170:121–134

Delmelle P, Bernard A (2015) The remarkable chemistry of sulfur in hyper-acid crater lakes: a scientific tribute to Bokuichiro Takano and Minoru Kusakabe. In: Rouwet D, Christenson B, Tassi F, Vandemeulebrouck J (eds) Book chapter in volcanic lakes, Springer, Heidelberg, pp 239–260. doi:10.1007/978-3-642-36833-2_10

Delmelle P, Henley RW, Opfergelt S, Detienne M (2015) Summit acid crater lakes and flank instability in composite volcanoes. In: Rouwet D, Christenson B, Tassi F, Vandemeulebrouck J (eds) Book chapter in volcanic lakes. Springer, Heidelberg, pp 289–306. doi:10.1007/978-3-642-36833-2_12

de Moor JM, Aiuppa A, Pacheco J, Avard G, Kern C, Liuzzo M, Martinez M, Giudice G, Fischer TP (2016) Short-period volcanic gas precursors to phreatic eruptions: insights from Poás Volcano, Costa Rica. Earth Planet Sci Lett 442:218–227. doi:10.1016/j.epsl.2016.02.056

Elliot EC (1938) Boiling lake—the 1900 story. In: Broken Atoms. Unpublished report presented to the Government of Dominica

Feuillet N, Beauducel F, Jacques E, Tapponnier P, Delouis B, Bazin S, Vallée M, King GCP (2011) The Mw = 6.3, November 21, 2004, Les Saintes earthquake (Guadeloupe): Tectonic setting, slip model and static stress changes. J Geophys Res 116(B10):1–25. doi:10.1029/2011JB008310

Fournier N, Withal F, Moreau-Fournier M, Bardou L (2009) The Boiling Lake of Dominica, West Indies: high temperature volcanic crater lake dynamics. J Geophys Res 114:B02203. doi:10.1029/2008JB005773

Giggenbach WF (1987) Redox processes governing the chemistry of fumarolic gas discharges from White Island, New Zealand. Appl Geochem 2:143–161

González G, Mora-Amador R, Ramirez C, Rouwet D, Alpizar Y, Picado C, Mora R (2015) Actividad historica y analysis de la amenaza del volcano Turrialba, Costa Rica. Rev Geol Am Centr 52:129–149. doi:10.15517/rgac.v0i52.19033

Gunawan H et al (2016) New insights into Kawah Ijen's volcanic system from the wet volcano workshop experiment. In: Caudron C, Capaccioni B, Ohba T (eds) GSL special volume geochemistry and geophysics of volcanic lakes

Hall M, Mothes P (2008) The rhyolitic-andesitic eruptive history of Cotopaxi volcano, Ecuador. Bull Volcanol 70:675–702. doi:10.1007/s00445-007-0161-2

Ingebritsen SE, Galloway DL, Colvard EM, Sorey ML, Mariner RH (2001) Time variation of hydrothermal discharge at selected sites in the western United States: implications for monitoring. J Volcanol Geoth Res 111:1–23

Joseph EP, Fournier N, Lindsay J, Fischer T (2011) Gas and water geochemistry of geothermal systems in Dominica, Lesser Antilles island arc. J Volcanol Geoth Res 206(1–2):1–14. doi:10.1016/j.volgeores.2011.06.007

Kusakabe M, Komoda Y, Takano B, Abiko T (2000) Sulfur isotopic effects in the disproportionation reaction of sulfur dioxide in hydrothermal fluids: implications for the d34S variations of dissolved bisulphate and elemental sulfur from active crater lakes. J Volcanol Geoth Res 97:287–307

Lindsay J, Robertson R, Shepherd J, Ali S (2005) Volcanic hazard atlas of the Lesser Antilles. In: 279. St. Augustine: Seismic Research Unit, University of the West Indies

Martini F, Tassi F, Vaselli O, Del Potro R, Martínez M, Van der Laat R, Fernandez E (2010) Geophysical, geochemical and geodetical signals of reawakening at Turrialba volcano (Costa Rica) after 150 years of quiescence. J Volcanol Geoth Res 198:416–432. doi:10.1016/j.jvolgeores.2010.09.021

Menyailov IA, Nikitina LP, Shapar VN, Pilipenko VP (1986) Temperature increase and chemical changes of fumarolic gases at Momotombo volcano, Nicaragua, in 1982–1985: are these indicators of a possible eruption? J Geophys Res 187:12199–12214

Mora-Amador R, Ramírez-Umaña C, González G, Rouwet D, Lucchi F, Forni F, Sulpizio R, Baldoni E, Alpízar-Segura Y, Tranne CA (2015) First documentation of the ongoing phreatic-strombolian eruptions of Turrialba volcano (Costa Rica). IUGG General Assembly, Prague 3746

Moussallam Y, Peters N, Ramirez C, Oppenheimer C, Aiuppa A, Giudice G (2014) Characterisation of the magmatic signature in gas emissions from Turrialba Volcano, Costa Rica. Solid Earth 5:1341–1350. doi:10.5194/se-5-1341-2014

Oppenheimer C, Stevenson D (1989) Liquid sulphur lakes at Poás volcano. Nature 342:790–793

Phillipson G, Sobradelo R, Gottsmann J (2013) Global volcanic unrest in the 21st century: an analysis of the first decade. J Volcanol Geoth Res 264:183–196

Rouwet D, Morrissey M (2015) Mechanisms of crater lake eruptions: physical and numerical modeling. In: Rouwet D, Christenson B, Tassi F, Vandemeulebrouck J (eds) Book chapter volcanic lakes. Springer, Heidelberg, pp 73–92. doi:10.1007/978-3-642-36833-2_3

Rouwet D, Ohba T (2015) Iotope fractionation and HCl partitioning during evaporative degassing from active crater lakes. In: Rouwet D, Christenson B, Tassi F, Vandemeulebrouck J (eds) Book chapter volcanic lakes. Springer, Heidelberg, pp 179–200. doi:10.1007/978-3-642-36833-2_7

Rouwet D, Sandri L, Marzocchi W, Gottsmann J, Selva J, Tonini R, Papale P (2014a) Recognizing and tracking hazards related to non-magmatic unrest: a review. J Appl Volcanol 3:17. doi:10.1186/s13617-014-0017-3

Rouwet D, Tassi F, Mora-Amador R, Sandri L, Chiarini V (2014b) Past, present and future of volcanic lake monitoring. J Volcanol Geoth Res 272:78–97. doi:10.1016/j.jvolgeores.2013.12.009

Rouwet D, Mora-Amador R, Ramírez-Umaña C, González G, Inguaggiato S (2016) Dynamic fluid recycling at Laguna Caliente (Poás, Costa Rica) before and during the 2006-ongoing phreatic eruption cycle (2005–2010). In: Caudron C, Capaccioni B, Ohba T (eds) Geochemistry and geophysics of volcanic lakes. Geological Society London Special Publications. doi:10.1144/SP437.11

Shinohara H, Giggenbach WF, Kazahaya K, Hedenquist JW (1993) Geochemistry of volcanic gases and hot springs of Satsuma-Iwojima, Japan: following Matsuo. Geochem J 27:271–285

Shinohara H, Yoshikawa S, Miyabuchi Y (2015) Degassing activity of a volcanic crater lake: volcanic plume measurements at the Yudamari crater lake, Aso volcano, Japan. In: Rouwet D, Christenson B, Tassi F, Vandemeulebrouck J (eds) Book chapter volcanic lakes. Springer, Heidelberg, pp 73–92. doi:10.1007/978-3-642-36833-2_8

Symonds RB, Gerlach TM, Reed MH (2001) Magmatic gas scrubbing: implications for volcano monitoring. J Volcanol Geoth Res 108:303–341

Tamburello G, Agusto M, Caselli A, Tassi F, Vaselli O, Calabrese S, Rouwet D, Capaccioni B, Di Napoli R, Cardellino C, Chiodini G, Bitetto M, Brusca L, Bellomo S, Aiuppa A (2015) Intense magmatic degassing through te lake of Copahue volcano, 2013–2014. J Geophys Res doi:10.1002/2015JB012160

Taran YA, Hedenquist JW, Korzhinsky M, Tkachenko SI, Shmulovich KI (1995) Geochemistry of magmatic gases from Kudryavy volcano, Iturup, Kuril Islands. Geochim Cosmochim Acta 59:1749–1761

Taran YA, Peiffer L (2009) Hydrology, hydrochemistry and geothermal potential of El Chichón volcano-hydrothermal system, Mexico. Geothermics 38:370–378

Taran YA, Pokrovsky BG, Dubik YM (1989) Isotopic composition and origin of water from andesitic magmas. Dokl Acad Sci 304:440–443

Varekamp JC (2003) Lake contamination models for evolution towards steady state. J Limnol 62(1):67–72

Vaselli O, Tassi F, Duarte E, Fernández E, Poreda RJ, Delgado Huertas A (2009) Evolution of fluid geochemistry at the Turrialba volcano (Costa Rica) from 1998 to 2008. Bull Volcanol. doi:10.1007/s00445-009-0332-4

Geophysical Footprints of Cotopaxi's Unrest and Minor Eruptions in 2015: An Opportunity to Test Scientific and Community Preparedness

Patricia A. Mothes, Mario C. Ruiz, Edwin G. Viracucha, Patricio A. Ramón, Stephen Hernández, Silvana Hidalgo, Benjamin Bernard, Elizabeth H. Gaunt, Paul Jarrín, Marco A. Yépez and Pedro A. Espín

Abstract

Cotopaxi volcano, Ecuador, experienced notable restlessness in 2015 that was a major deviation from its normal background activity. Starting in April and continuing through November 2015 strong seismic activity, infrasound registry, hikes in SO_2 degassing and flank deformation with small displacements were some of the geophysical anomalies that were registered. Obvious superficial changes, such as small hydromagmatic eruptions, emission of vapor and ash columns, thermal hotspots around the crater and in nearby orifices and exacerbated glacier melting were also observed. Our contribution provides an overview of the 2015 Cotopaxi unrest by presenting the patterns of geophysical data and the sequence of events produced by the volcano. Cotopaxi's last important VEI 4 eruption was in 1877. Then it had devastating effects because of the transit of huge lahars down 3 major drainages. Comparatively, the 2015 activity never surpassed a magnitude VEI 2 and principally produced limited hydromagmatic explosions and semi-continuous low energy emissions and light ashfalls. Given the potential of major destruction from a large Cotopaxi eruption it is important to understand the geophysical fingerprints that characterized the 2015 episode with an eye to identifying onset of future restless periods. Overall, the monitoring activities, the data interpretation, formulation of reasonable eruptive scenarios, and finally, the preparation of a stream of constant information being relayed to concerned authorities and the public, was a real test of the IGEPN's capacity to deal with a complicated eruption situation whose outcome was not apparent at the beginning, but which concluded in a very small eruptive episode.

P.A. Mothes (✉) · M.C. Ruiz · E.G. Viracucha · P.A. Ramón · S. Hernández · S. Hidalgo · B. Bernard · E.H. Gaunt · P. Jarrín · M.A. Yépez · P.A. Espín
Instituto Geofísico, Escuela Politécnica Nacional, Quito, Ecuador
e-mail: pmothes@igepn.edu.ec

Advs in Volcanology (2019) 241–270
DOI 10.1007/11157_2017_10

Published Online: 19 July 2017

Resumen

En 2015 el volcán Cotopaxi, Ecuador experimento un notable cambio, que fue una desviación importante de su actividad normal de base. A partir de abril y hasta noviembre de 2015 fuerte actividad sísmica, registros de infrasonido, incremento en la desgasificación de SO_2 y pequeños cambios en la deformación de los flancos fueron algunas de las anomalías geofísicas registradas. Evidentes cambios superficiales también fueron observados como pequeñas erupciones hidromagmaticas, emisión de vapor, columnas de cenizas, puntos calientes alrededor del cráter y el deshielo de los glaciares. Nuestra contribución proporciona una visión general de las anomalías del Cotopaxi en el 2015, mediante la presentación de patrones de los datos geofísicos y la secuencia de eventos producidos por el volcán. La última erupción importante del Cotopaxi fue un VEI 4 en 1877. Esta tuvo efectos devastadores debido al descenso de enormes lahares por sus tres drenajes mayores. Comparativamente, la actividad del año 2015 nunca superó una magnitud VEI 2, principalmente produciendo explosiones hidromagmaticas, escasas emisiones y leves caídas de ceniza. Debido a la potencial destrucción por una eventual erupción grande del Cotopaxi es importante entender los registros geofísicos que caracterizó el episodio de 2015 para poder identificar el inicio de futuros períodos eruptivos. En general, las actividades de vigilancia, la interpretación de datos, formulación de escenarios eruptivos razonables y por último, la preparación de un flujo de información constante que llegue a las autoridades interesadas y el público, fue una verdadera prueba de la capacidad del IGEPN para hacer frente a una situación de erupción cuyo resultado no era evidente al principio, pero que finalizó como una erupción pequeña.

Keywords

Volcanic unrest • Precursory geophysical patterns • Precursory LPs and VLPs • Volcano monitoring • Cotopaxi volcano-Ecuador • State of preparedness

Introduction

Long dormant volcanoes that begin to awaken may have start and stop activity that has to be evaluated with respect to the volcano's known past. A volcano's past activity is known by study of its stratigraphy, other physical evidence and possibly historical chronicles (Tilling 1989). Many uncertainties preclude knowing the final outcome of a restless volcano (Newhall 2000; Sparks and Aspinall 2004), since a volcano may awaken for short term low-level activity, then resume repose until additional magma inputs herald an episode of further unrest (Phillipson et al. 2013). Eruptions that barely bring magma to the surface may be classified as "failed", since so little juvenile magma erupts (Moran et al. 2011).

At Cotopaxi ample geological and historical information exists with regard to its past activity (Hall and Mothes 2008; Garrison et al. 2011; Pistolesi et al. 2012). Formulation of eruptive scenarios with respect to the 2015 unrest period were based on our collective knowledge of the volcano's geology and eruptive history and published information as well as interpretation of the abundant geophysical data streams available through instrumental and observational networks operated by the Instituto Geofísico of the Escuela Politécnica Nacional (IGEPN)-Quito, Ecuador, the entity in charge of volcano and tectonic monitoring in Ecuador. The combination of these inputs allowed scientists at the IGEPN to transmit a coherent image of the evolving unrest presented during 2015 and to indicate the most likely eruption/activity scenarios. Two earlier unrest periods are known: 1975–1976, when the IGEPN had limited seismic equipment operating on the volcano and then in 2001–2002. Both periods were comprised of increased fumarolic activity both inside and outside of the crater and a hike in seismicity for the 2001 period (Molina et al. 2008). The 2015 unrest displayed important changes in seismic and deformation patterns, gas output and superficial activity, compared to background, whose level was established since around 1990. In sum, Cotopaxi's 2015 unrest displayed a progressive crescendo of geophysical signals, then minor hydromagmatic explosions, followed by overall seismic energy decrease at the end of 2015, which was accompanied by fewer superficial manifestations. Like many other volcanoes that have displayed unrest, this recent episode did not culminate in a full-fledged eruption with large volumes of juvenile pyroclastics (Phillipson et al. 2013). Moran et al. (2011) maintain that an eruption is "failed" when magma reaches but does not pass the "shallow intrusion" stage, i.e., the magma gets close to, but does not reach the surface. In the actual case, the amount of erupted material was minor, and had a dense rock equivalent volume of $\sim$0.5 Mm^3 (Bernard et al. 2016).

Cotopaxi Volcano

Cotopaxi volcano, located in central Ecuador atop the Eastern Cordillera, is a large, symmetrical stratocone with a basal diameter of 18 km and an altitude of 5897 m (Hall and Mothes 2008). Its actual glacier cap of 10.49 km^2 is rapidly diminishing due to climatic change Cáceres et al. (2016) (Fig. 1). The volcano's last important VEI = 4 eruption was on 26 June 1877. Then it generated highly erosive pyroclastic flows that melted glaciers and triggered voluminous lahars ($\sim$100 Mm^3 per drainage). Each lahar traveled hundreds of kilometers down several drainages enroute to the Amazon basin, Pacific Ocean and to the Atlantic, respectively (Mothes et al. 2004; Mothes and Vallance 2015). These past lahar routes now host sprawling suburbia, important economic activities and vital infrastructure. Ecuador's second most visited national park (Parque Nacional Cotopaxi-PNC) is centered on the volcano and draws some 200,000 tourists a year.

The volcano's five most important eruptive episodes during the historical period (since 1532) have been of andesitic composition and ranged between VEI = 3 and 4 (Pistolesi et al. 2012). Nonetheless, the volcano is bi-modal and produces VEI = 5 rhyolitic eruptions about every 2000 years (Hall and Mothes 2008). The last important rhyolitic eruption, the Peñas Blancas event, occurred about 2800 years BP (Mothes et al. 2015a). The youngest andesitic eruptive products contain intergrowths of plagioclase and pyroxene and four different populations of plagioclase crystals which indicate pervasive magma mixing (Garrison et al. 2011).

Given the high probability for the generation of long-distance lahars, wide dispersal of pyroclastic fall, and the consequential negative impact on many economic activities and the compromise of critical infrastructure should an eruptive period last for months to years, Cotopaxi is considered a "National" volcano, located in the center of Ecuador, near to Quito, the capital and other populated areas. Even a short-lived VEI 4 eruption (VEI = Volcano Explosivity Index) (Newhall and Self 1982) has the potential to gravely affect Ecuador's overall productivity and major transport lines. Lastly, the volcano is considered one of the most dangerous in Ecuador, given the possible exposure of >300,000 residents to primary lahars and ashfalls during

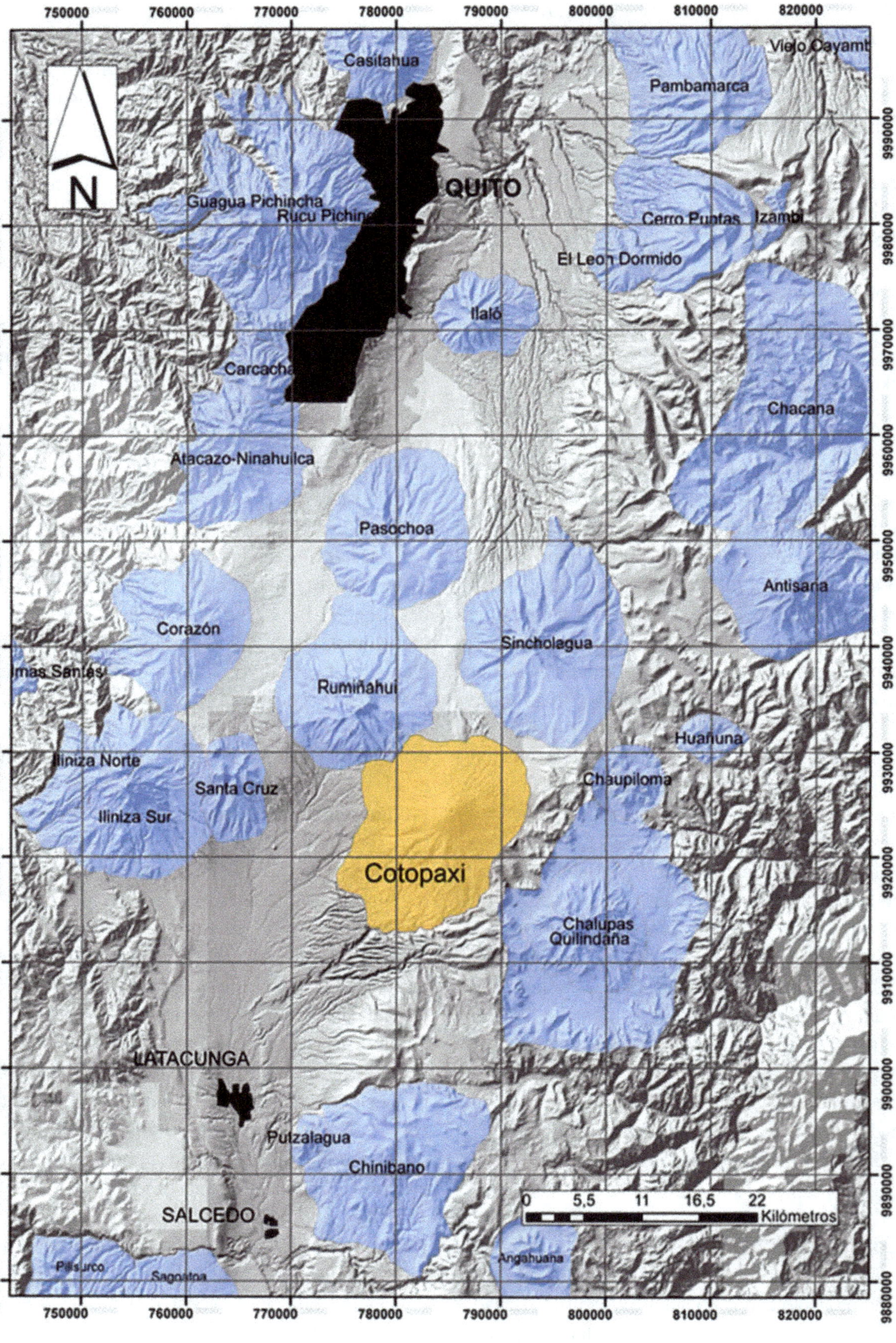

Fig. 1 Location of Cotopaxi volcano in Ecuador's Inter–Andean Valley and major cities

future VEI = 3 or greater magnitude eruptions (Mothes et al. 2016a).

Monitoring Cotopaxi

Cotopaxi has long been a producer of swarms of LP's, and Ruiz et al. (1998) hypothesized that they were related to the interaction of the hydrothermal system with heat ascending from depth. During unrest in 2001–2002 thousands of long period, volcano-tectonic and distal volcano-tectonic seismic events were registered and magma was hypothesized to have ascended to shallow levels in the center of the edifice (Molina et al. 2008). Its flanks also experienced deformation from magma input estimated at ~ 20 Mm^3 from modeling of EDM data (Hickey et al. 2015). But, observable superficial manifestations were meager, no magma erupted and the volcano returned to a relatively quiet state with only short-lived LP swarms (Lyons et al. 2012) and sporadic VLP events being registered until April, 2015 (Márquez 2012; Arias et al. 2015).

The IGEPN has monitored Cotopaxi volcano since 1986. Subsequently, over the years the monitoring coverage has become denser and more robust (Fig. 2) (Kumagai et al. 2007, 2009, 2010). Presently there are approximately 60 telemetered geophysical sensors operating on its flanks. Cotopaxi hazard maps have been published in several versions since 1976, with the newest version published in late 2016 (Mothes et al. 2016a). The IGEPN has carried out a long-term program of community education for areas that are at highest risk, although as noted by Christie et al. (2015), the attention over such a vast area ($\approx$2000 km^2), was uneven.

Cotopaxi was a VUELCO target volcano from 2013 to 2015. As part of the VUELCO project, in November 2014 a simulation exercise was carried out with the purpose of presenting a timeline of potential unrest and expected events and to test the communication between scientists, decision-makers and the public (www.vuelco.net). This present contribution is written in the spirit of holistically documenting this recent and most serious unrest of Cotopaxi to date—since 1942, when slight ash emissions and mild explosive activity were then reported (Siebert et al. 2010). Here we present the macro patterns of seismic, gas, deformation and visual observation monitoring data associated with the awakening volcano. The data are provided by the monitoring networks operated by the IGEPN. We avoid dwelling on details, as forthcoming contributions dedicated to exploiting specific datasets are in preparation. We also provide comment on selected actions in which IGEPN scientists participated to make the overall societal outcome more favorable in case Cotopaxi produced a major eruption. We impart with the philosophy, stated in Marzocchi et al. (2012) that "sound scientific management of volcanic crises is the primary tool to reduce significantly volcanic risk in the short-term". We also maintain that a constant and rapid analysis of the monitoring data is key to giving forecasts that include reasonable scenarios. Some of the IGEPN actions were guided by experiences gained in the VUELCO project, since one of the scenarios in the simulated eruptions was that the volcano would wake up, be active then return to repose.

The 14th of August, 2015 explosions and subsequent emissions pushed the first evidence of new magma to the surface, although in a limited way (Gaunt et al. 2016). Documentation of the geophysical signals and observations that we registered through late 2015 leads to the depiction of what transpired—mainly of an intrusion, which stayed deep, although the signs of intrusions that stall at depth may be very similar to those produced by intrusions that finally do erupt (e.g. Moran et al. 2011).

Synthesis of the Geophysical Fingerprints of the Unrest

Having passed 13 years with low levels of activity since cessation in 2002 of its last reactivation, in mid-April 2015 Cotopaxi began departing from background levels: higher seismic energy release, gas outputs and superficial manifestations transpired. The height of activity was

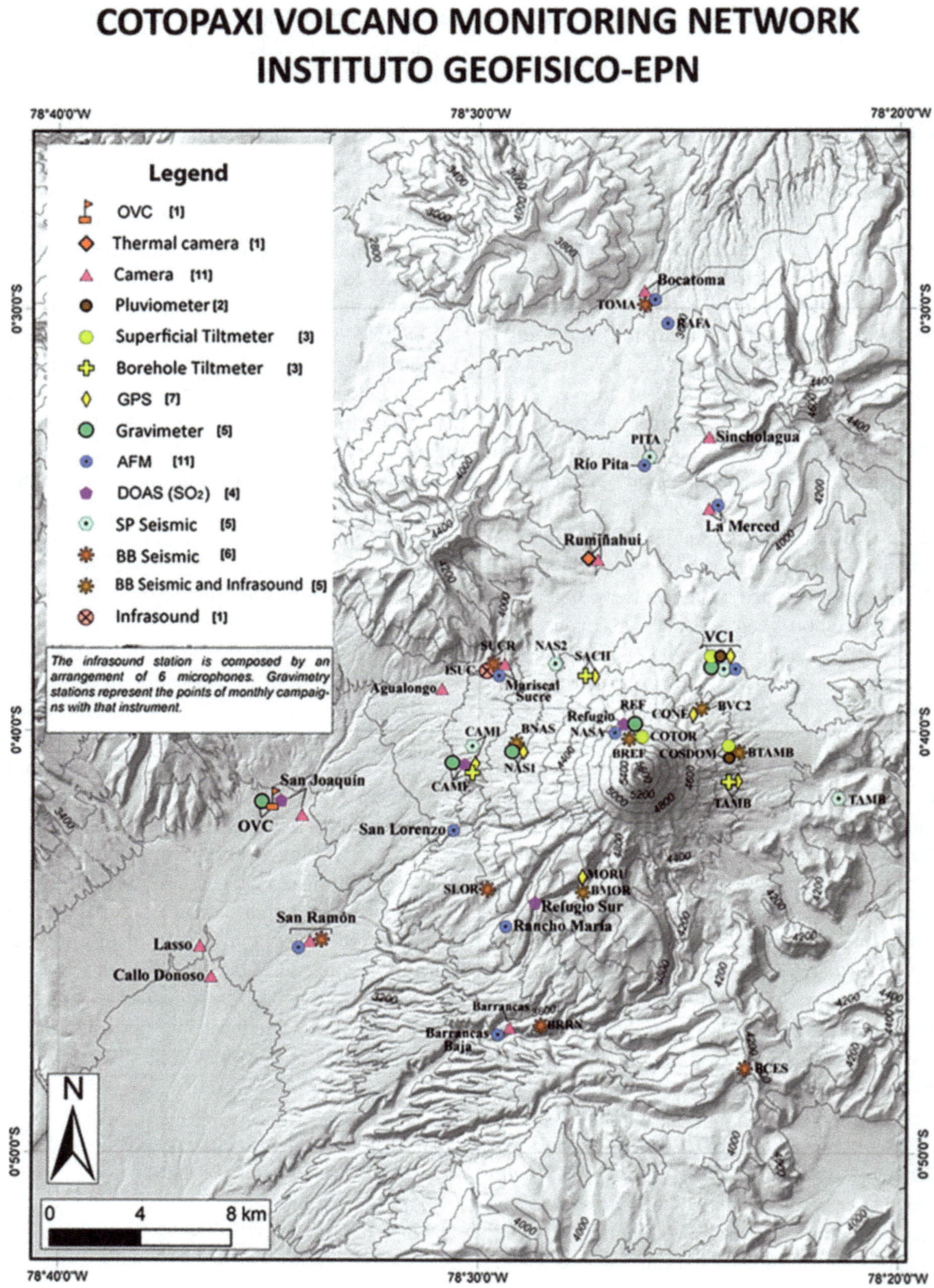

Fig. 2 Map of the instrumental monitoring network around Cotopaxi volcano, April 2016

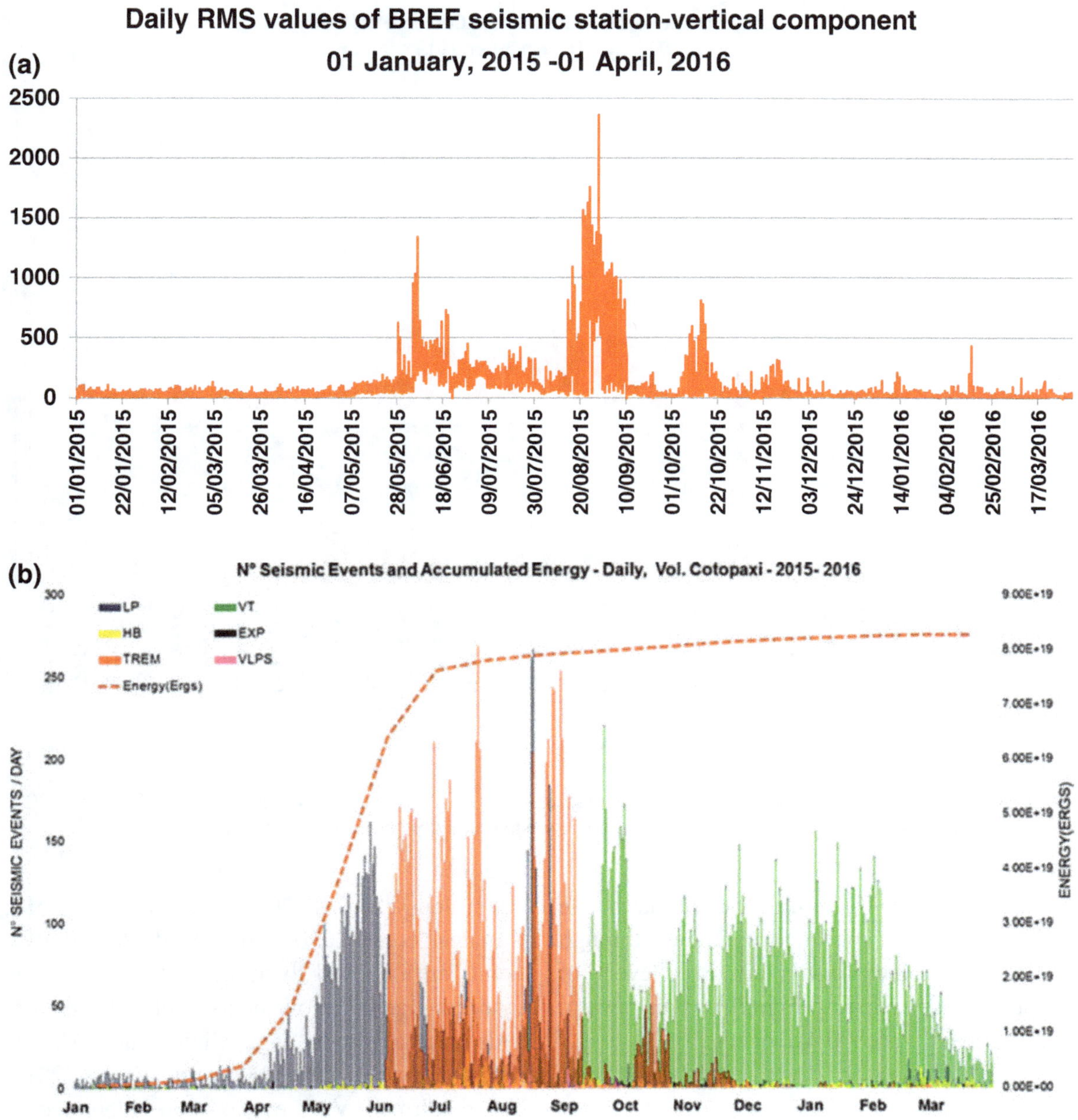

Fig. 3 **a** Plot showing RMS values of seismic data that has been segmented into 1 min windows and smoothed with a 31-point median filter. Of note is the calm period from January to April 2015, than a slight increase in seismic energy release in May. The increase in June is associated with greater emissions and strong tremor. A sharp decline in early August and later the notable increase in August and September represents the 14th of August explosions/emissions and subsequent ash emissions and emission tremor. Accumulated seismic energy values through the end of 2015 show a marked decline. **b** Number of daily seismic events versus accumulated seismic energy of these events. The acronyms for different seismic events are: *LP* Long Period; *HB* Hybrid; *VT* Volcano-tectonic; *TREM* High Frequency Emission Tremor; *EXP* Explosion; *VLPs* Very Long Period

a series of 5 explosions/energetic emissions on the 14th of August, which expulsed preexisting conduit plug material, ash and gases, but whose size did not surpass VEI = 1. By late September 2015, activity mostly died back and RSAM values showed a decline except for brief hikes in October and in November, when light ashfalls occurred. By December 2015 nearly all monitoring parameters were down to background levels (Fig. 3a), except for a protracted

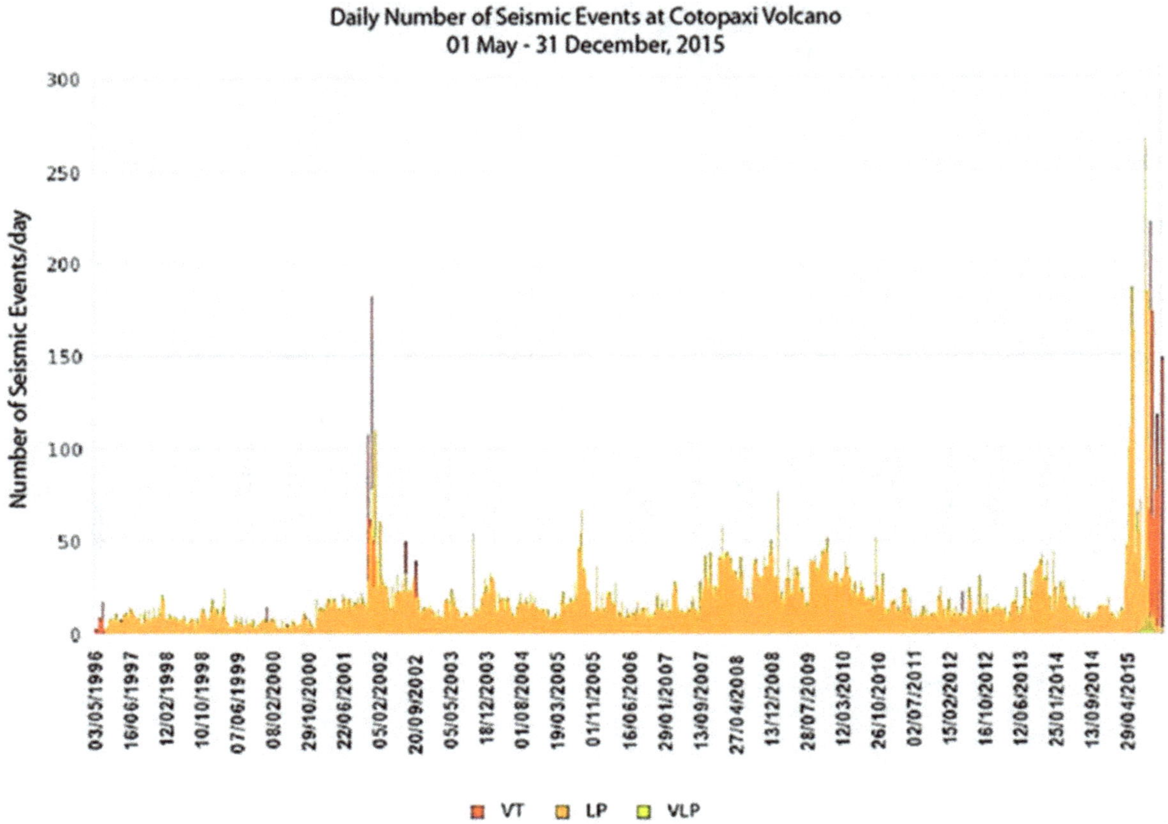

Fig. 4 Registry of VT and LP seismic events at Cotopaxi since May 1996 to 31 December, 2015. The 2001–2002 period was the only other period with a marked jump in seismic activity before the 2015 episode

volcano-tectonic (VT) seismic swarm that began on 10 September, 2015 and continued through March 2016, albeit displaying low levels of seismic energy release (Fig. 3b). This swarm produced nearly 15,000 VT events.

Geophysical Registry of Cotopaxi's Restlessness in 2015

From 2002 to April, 2015, seismic registry of mostly long period (LP) seismic events averaged around 10 events/day. In April 2015 the monthly tally was about 630 earthquakes, then rose to 3000 events in May (Fig. 4), with a jump to about 180 events/day registered on 23 May (Fig. 4).

Of significance also was the notable increase in very long period seismic events (VLPs) recorded since late May 2015. VLPs are often interpreted to signify magma movement (Zobin 2012; Jousset et al. 2013; Maeda et al. 2015; Kumagai et al. 2010; Arias et al. 2015). VLP events are believed to be generated by volume changes and movements of magmatic-hydrothermal fluids (e.g., Chouet and Matoza 2013). Between June 2006 and October 2014, 106 confirmed VLP events were identified at Cotopaxi (Márquez 2012; Arias 2015). In 2015 Cotopaxi, VLPs rarely passed 11 events/day (Fig. 5), but commonly had magnitudes of 2–3. The recent VLP events that were located under the Cotopaxi's edifice, occurred in sectors of the volcano where VLP's had been previously located by Molina et al. (2008) (Fig. 6). The greatest number of VLPs, of the 114 located events, were registered during the third week of July up to the explosions on the 14th of August. While most were between 1 and 2 magnitude, some were greater than 2.5 (Fig. 5a). The relationship between the great number of LPs which started the awakening process at Cotopaxi and

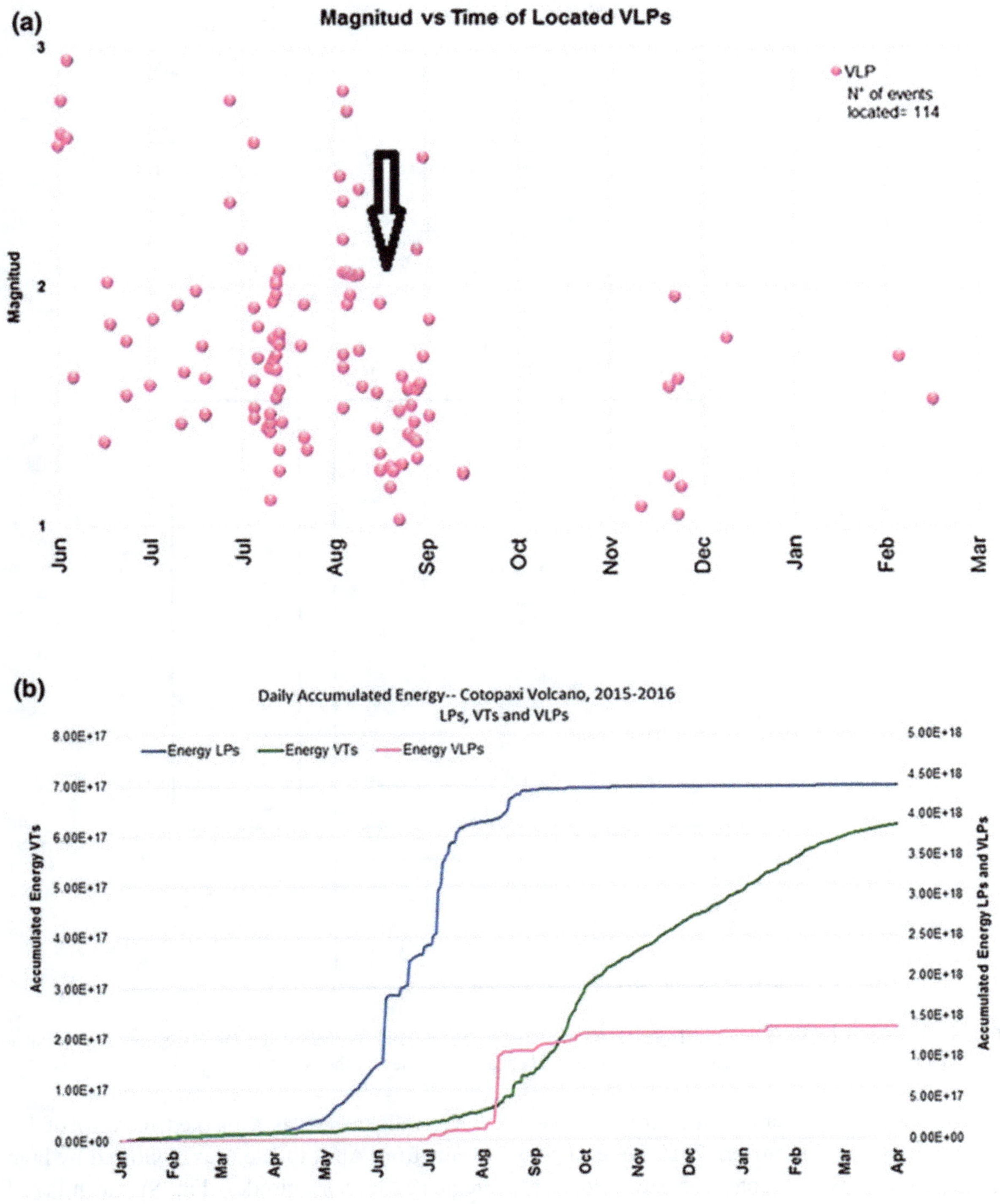

Fig. 5 a Occurrence of VLP events in 2015 at Cotopaxi. *Black arrow* represents 14 August hydromagmatic explosions. **b** Accumulated seismic energy from VTs, LPs and VLPs in 2015–2016 at Cotopaxi

afterwards the stalling out of these events to be followed immediately by the strong VLPs is another possible indicator of the precursory nature of this type of volcanic earthquake before the discrete eruptions on 14 August (Fig. 5b).

Most VLP events had frequencies between 0.1 and 1 Hz and had strong P and S waves, such as the example given for 04 August, 2015 which was located 3 km below summit on the NE flank of the volcano (Fig. 6).

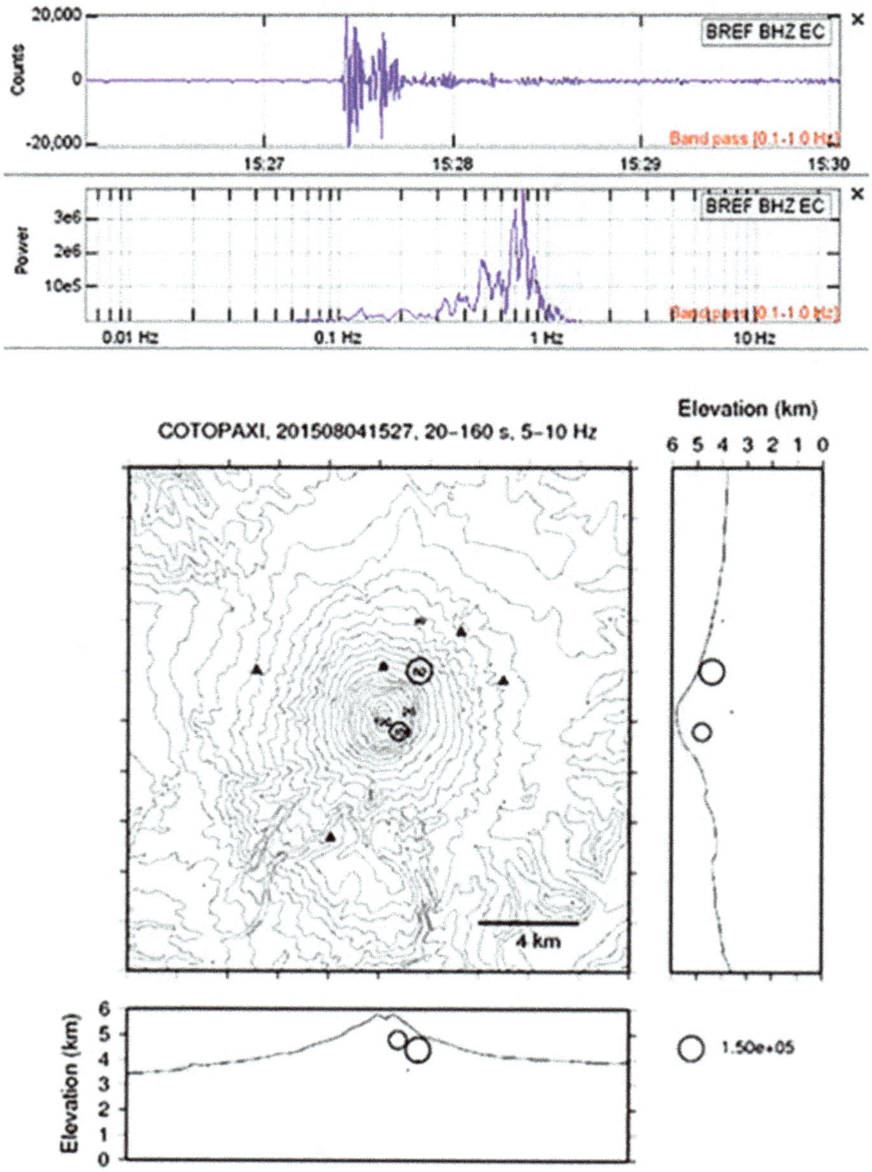

Fig. 6 Waveform of VLP registered on 04 August, 2015 15h27 GMT is one of the largest events registered at Cotopaxi (local magnitude 2.5) and was located 3 km beneath the crater on the NE flank

The locations of earthquakes (of all types) from January to December, 2015 were at two levels: at depths of about 3–5 km below the crater (Fig. 7a) and at a deeper level of 7–15 km below the crater. Most events were aligned with the conduit. However some distal VTs were registered about 15 km due north of the volcano (Fig. 7b) and were interpreted as fault slips due to stress transfer from the volcano (White and McCausland 2016). Distal VTs were also important in the reactivation of Pinatubo volcano (Harlow et al. 1996).

Overall, there was a marked increase of LP events from April to late May, followed by high frequency tremor episodes (Fig. 8) which lasted until the onset of high frequency tremor related to gas emissions and which became prominent from 04 June and lasted to the second week of August (represented by black bars), and could have been related to the boiling of the volcano's hydrothermal system, and coincided with the high water vapor and SO_2 flux then emitting from the crater (Bernard et al. 2016). In Fig. 8, the VLPs that were important especially in July

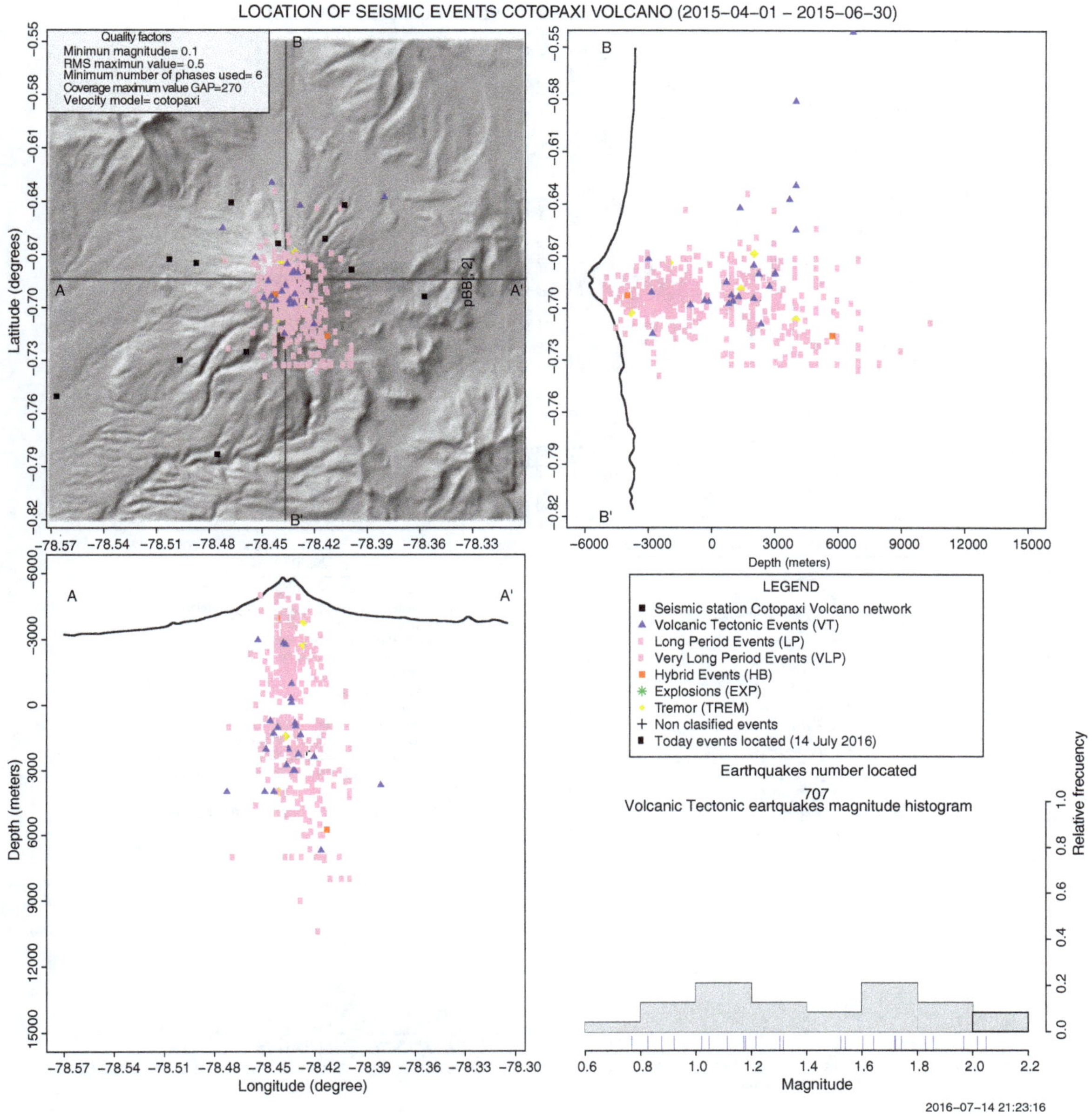

Fig. 7 Top-Locations of seismic events from a–01 April to 30 June, 2015 and Base—01 July to 31 August, 2015. Most are aligned with the conduit, however the SE flank is favored for harboring event locations

to mid-August are masked by this tremor signal, but can be observed in Fig. 3 and their accumulative energy levels are shown in Fig. 5.

The increase in SO_2 gas emissions, rose to 3000 ton/day by the end of May with a clear SO_2 signal progressively more notable through late May into June (Fig. 9) (Hidalgo et al. 2016). For example, on the 22nd–23rd of May odors of sulphur were very evident above the 5700 m level on the volcano's northern flank, as reported by Cotopaxi Park personnel.

GPS stations on the W and S flanks showed horizontal displacements of almost 16 ± 0.5 mm toward the W and SW. GPS stations on the NE and E flanks showed displacements to the N at a reduced velocity (Fig. 10). The vertical component registered a maximum uplift of 15 ± 2.3 mm. The movement to the west could have been accentuated by the volcano's morphology, as the W flank is poorly buttressed and sits upon Inter-Andean Valley volcaniclastic fill. In comparison the east and northern part of the cone

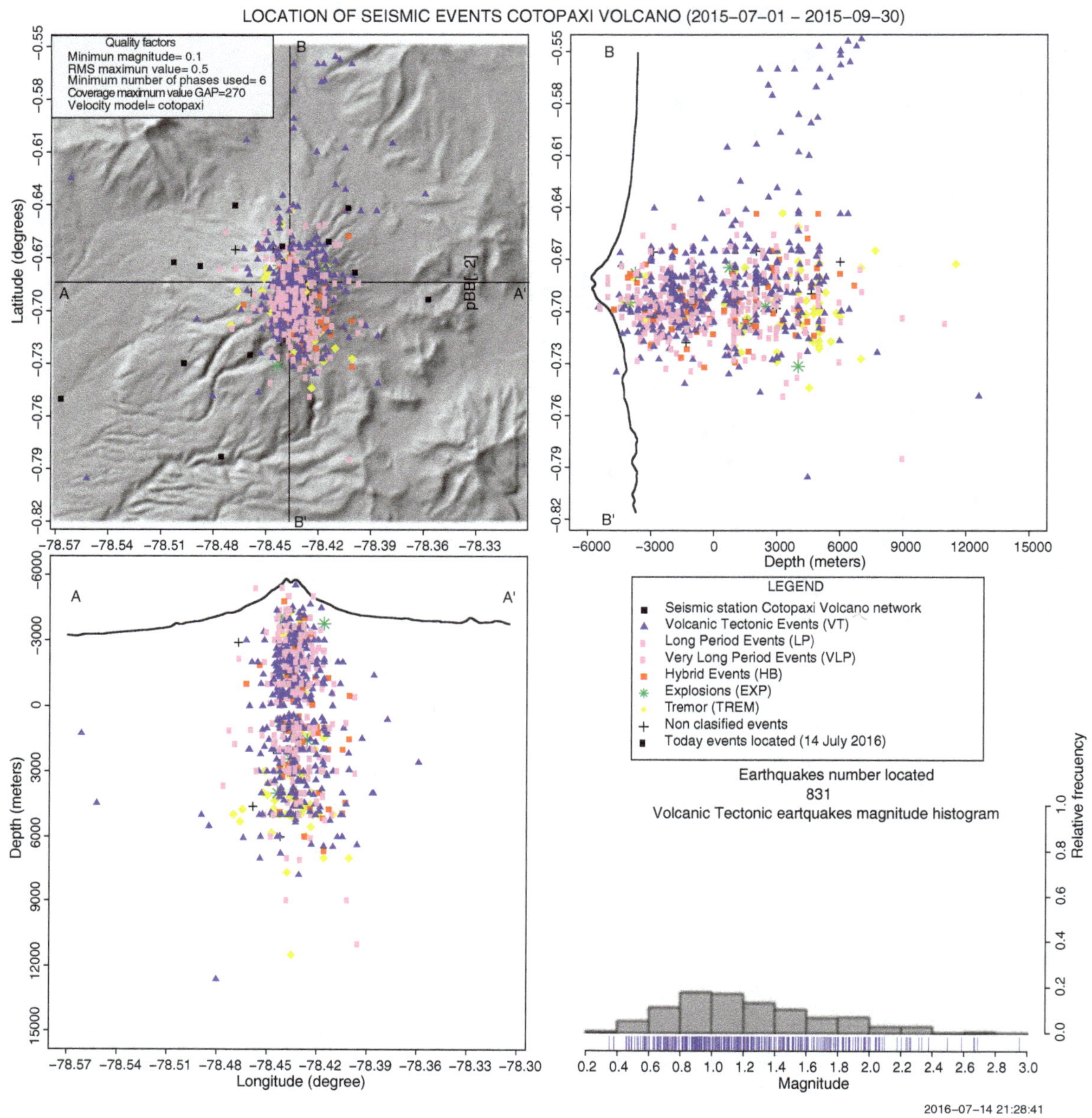

Fig. 7 (continued)

Fig. 8 a Comparative graph of LP seismic events (*orange bars*) and high frequency tremor (*black bars*) possibly related to the boiling of the hydrothermal system and gas movements from 01 April to 14 August, 2015. Ash and gas emission-related tremor (*pink bars*) abruptly began the second week of August, 2015, after the hydromagmatic explosions on the 14th. **b** Seismogram (11 June, 2015) of BREF station showing registry of spasmodic tremor related to internal fluid movements in the upper part of the edifice

lies upon a thick lava package and basement crystalline metamorphic rock and may be more resistant to lateral movement. Data processing employed the program GAMIT/GLOBK (Herring et al. 2015) and used a local reference frame with respect to fixed South America (Nocquet et al. 2014. We also defined a long-term displacement model for each GPS site by estimating a trend and annual and semi-annual components using all available data between 2008 and 2015. The transient displacements identified during the 2015 unrest period are with respect to this model. In a second step, we applied a common-mode filtering estimated from the average time series residuals for sites ~50 km away from the volcano. Short-term repeatabilities are of the order of 1–2 mm on the

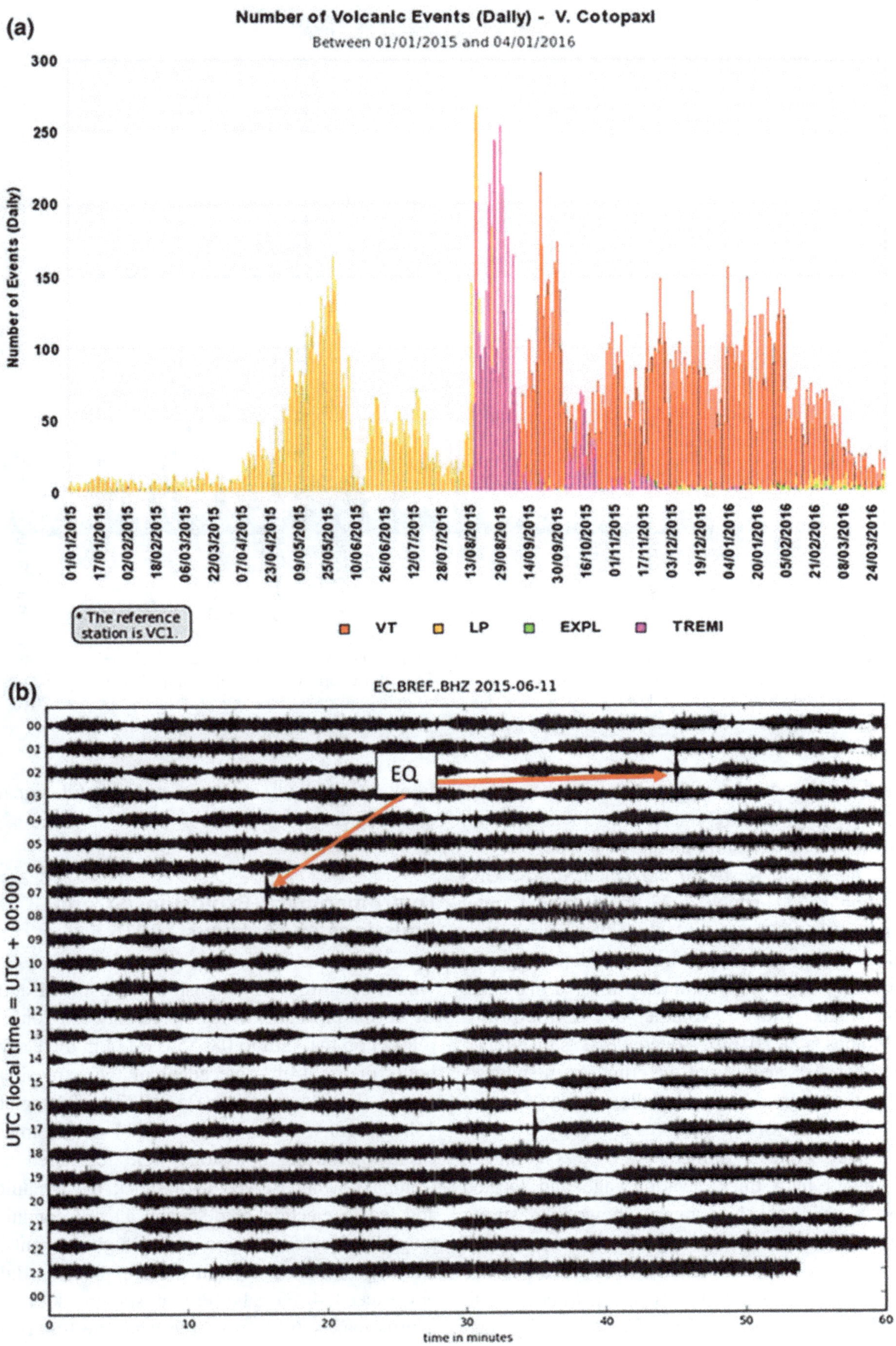
(a)
Number of Volcanic Events (Daily) - V. Cotopaxi
Between 01/01/2015 and 04/01/2016
Number of Events (Daily)
300
250
200
150
100
50
0
01/01/2015
17/01/2015
02/02/2015
18/02/2015
06/03/2015
22/03/2015
07/04/2015
23/04/2015
09/05/2015
25/05/2015
10/06/2015
26/06/2015
12/07/2015
28/07/2015
13/08/2015
29/08/2015
14/09/2015
30/09/2015
16/10/2015
01/11/2015
17/11/2015
03/12/2015
19/12/2015
04/01/2016
20/01/2016
05/02/2016
21/02/2016
08/03/2016
24/03/2016
* The reference station is VC1.
VT
LP
EXPL
TREMI
(b)
EC.BREF..BHZ 2015-06-11
EQ
UTC (local time = UTC + 00:00)
00
01
02
03
04
05
06
07
08
09
10
11
12
13
14
15
16
17
18
19
20
21
22
23
00
0
10
20
30
40
50
60
time in minutes

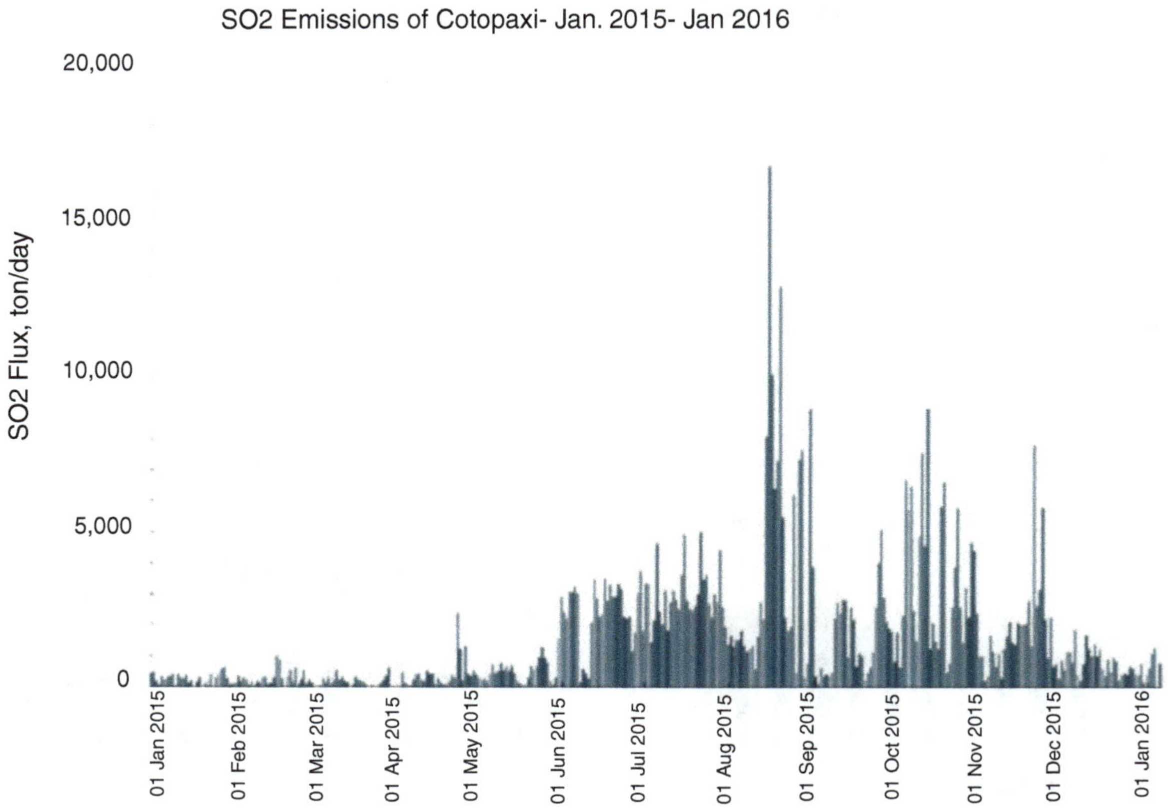

Fig. 9 Registry of SO_2 values for Cotopaxi, January 2015 until 05 January 2016, with a significant increase in SO_2 observed since early May. Data was processed daily using a single wind speed and direction obtained from the NOAA and VAAC alerts. Graph included in Cotopaxi Special Report, No. 1, 2016: http://www.igepn.edu.ec/cotopaxi/informes-cotopaxi/coto-especiales/coto-e-2016/14074-informe-especial-cotopaxi-n-01/file

horizontal components and 3 mm on the vertical component, enabling us to extract the small GPS signal observed during the unrest period.

Due to the westward movement on the GPS station PSTO, which is 22 km W of the crater (Fig. 10), we surmised that the source was deep. Subsequent modeling of the data suggested a source of about 24 km deep located under the SE flank with a volume of 42 ± 26 Mm^3 (Mothes et al. 2016b). Nonetheless, as mentioned in Sect. 3.1, analysis of the erupted ash suggests that the magma source is shallow, as least for the initial small volume that was emitted.

Data from a tiltmeter (VC1G on Fig. 2) installed in a thick lava package and located 6 km NE of the crater, showed a strong inflationary pattern that had started in April, 2015 on both axis. This tilt anomaly coincided with the notable increase in seismicity (Mothes et al. 2016b). Generally, when LP seismicity and tremor were both strong, a positive tilt signal predominated.

Hydromagmatic Explosions/Strong Emissions of 14 August, 2015

On the evening of 13 August, a swarm of VT and LP seismic events was registered between 20h03 (GMT) on the 13th to 08h55 (GMT) in the early hours of the 14th, antecedent of the explosion events (Fig. 11). At 09h02, 09h07 and later at 15h25, 18h45 and 19h29 (UT) five small explosions/strong emissions were registered at Cotopaxi which served to unblock the conduit and led to ejection of degassed altered conduit plug material and scarce juvenile components. Although infrasound from these explosions did not exceed 4 Pascals (Pa) at stations located approximately 6 km from the vent, the first two

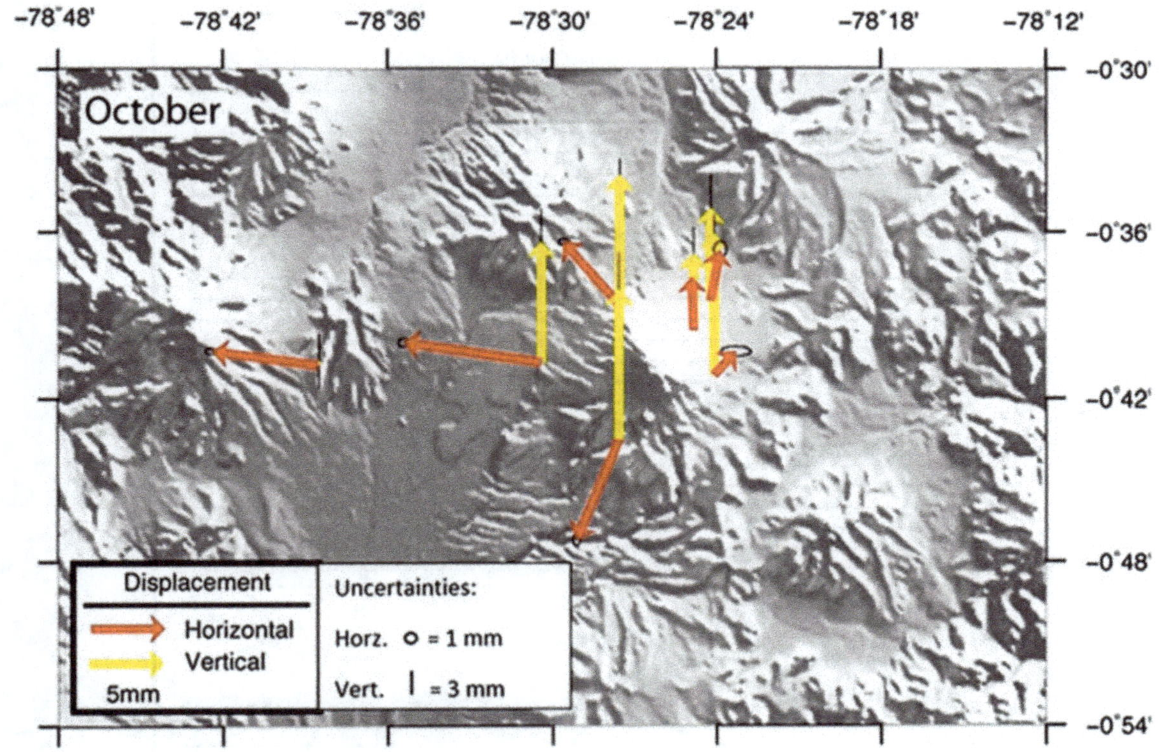

Fig. 10 GPS vectors for stations around Cotopaxi's cone and one to the west for October, 2015. Displacements are expressed with respect to the North Andean Silver and represent the comparison of GPS data collected from 01 January to 01 October, 2015 at the 7 station Cotopaxi CGPS network

explosions were heard by climbers in the Refuge on Cotopaxi's north flank, where lapilli-size fallout reached the Cotopaxi Refuge.

Two months earlier public and authorities had been forewarned in the special IGEPN reports (No. 3 and 4) that phreatic explosions would be a likely phenomenon in precursory eruptive activity (http://www.igepn.edu.ec/cotopaxi/informes-cotopaxi/coto-especiales/coto-e-2015/12990-informe-especial-cotopaxi-11-06-2015/file).

With these explosions the eruption column at 15h25 rose to 9 km above the crater rim and was clearly visible from the SW (Fig. 12a, b). Infrasound values of the explosions were less than 10 Pa at station BNAS (5 km from the crater), but the seismic source amplitudes of the tremor associated with the first two explosions were greater than those of most Cotopaxi LP events and also of some explosions registered at Tungurahua volcano (Kumagai et al. 2015). The initial explosions had evidence of water involvement. In previous weeks a small lake was observed in the crater's bottom; this was totally evacuated by the explosions. Observers also reported that the fallout had a "wet aspect" and many of the fragments were agglutinated by a fine clay-size patina. The eruption is categorized as hydromagmatic, since the rapid interaction with water caused overpressures beneath the plug, raising lithostatic pressures that overcame the capacity of the altered conduit plug rock. After these main vent-opening events the presence of hydrothermally altered material gradually waned and possible juvenile material became more prevalent (Gaunt et al. 2016).

The ash emissions from this first activity covered agricultural lands to the NW and W of the volcano with a ≤ 1 mm thick dusting of altered silt to sand lithic grit and crystals (Fig. 13a) and caused poor visibility along major highways that enter Quito from the south. This ash emission mantled over 500 km^2 with more than 80 gr/m^2 and amounted to a volume of 118,000 m^3, keeping it within the range of a VEI = 1 (Bernard et al. 2016) (Fig. 13b).

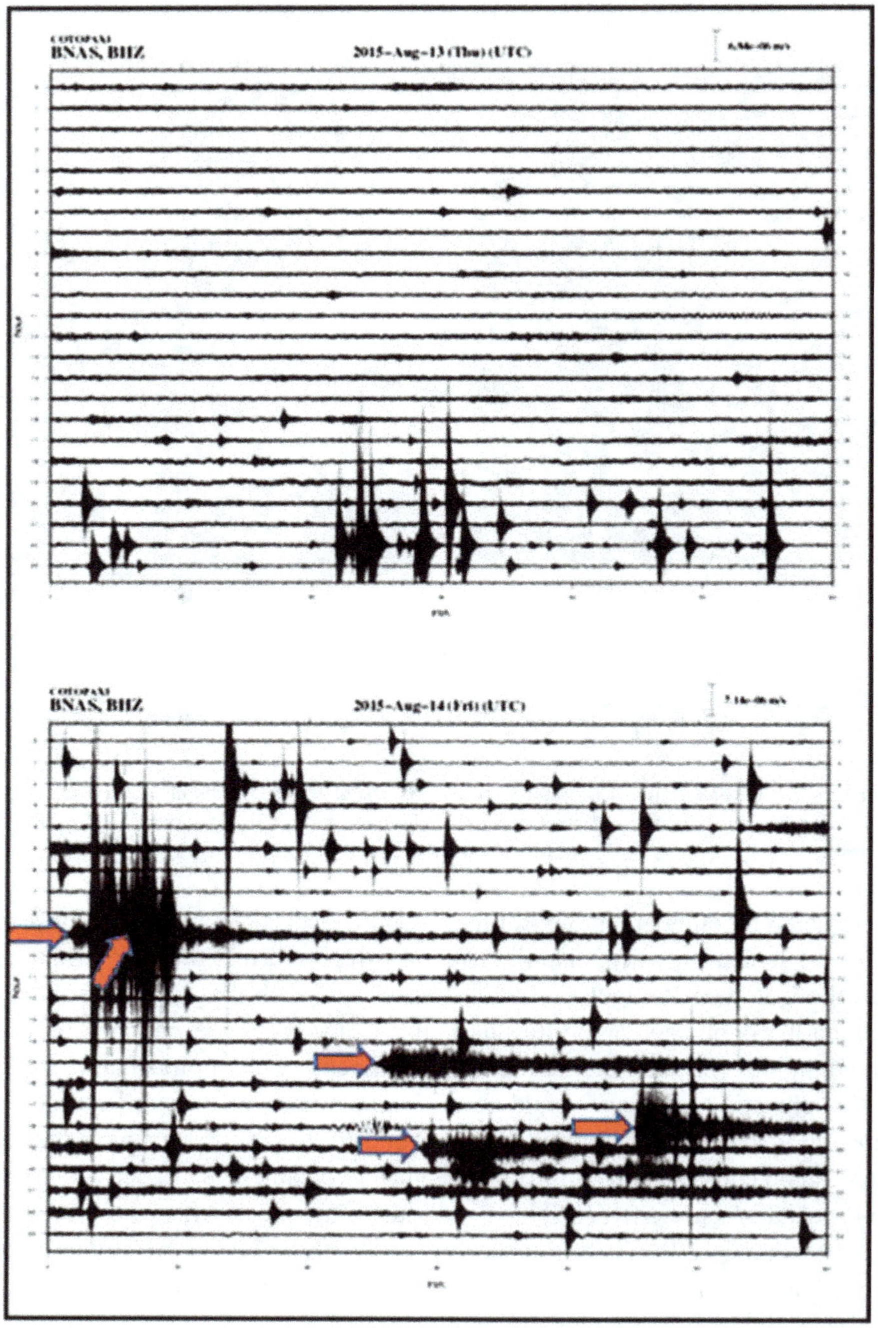

Fig. 11 Registry of VT—LP seismic swarm which begin late 13 August, 2015 and was followed by the 5 small explosions/strong emissions on 14 August, all indicated with *red arrows*. Seismograms are of the IGEPN's monitoring network

Post 14 August, 2015: Open Conduit Degassing and Ash Emissions

Ashfalls were prevalent towards the N and NW after 14 August into October and became scarce in late November (Fig. 14). A common scene was that of the ash and gas plume cascading down the W flank, with only the initial pulse rising to <1 km upon emitting from the crater (Fig. 14).

Emission tremor of varying amplitudes accompanied the ash emissions and permitted IGEPN monitoring scientists to forecast if the

Fig. 12 *Left* Cotopaxi's 14 August, continuing emission—view of the volcano from SW at 14h10UT. *Photo* E. Pinajota, IGEPN. Right- The 15h25UT strong emission produced a column that ascended 6–8 km above the summit. *Photo* by Santiago Tapia, at the Novacero company grounds, 20 SW of the volcano

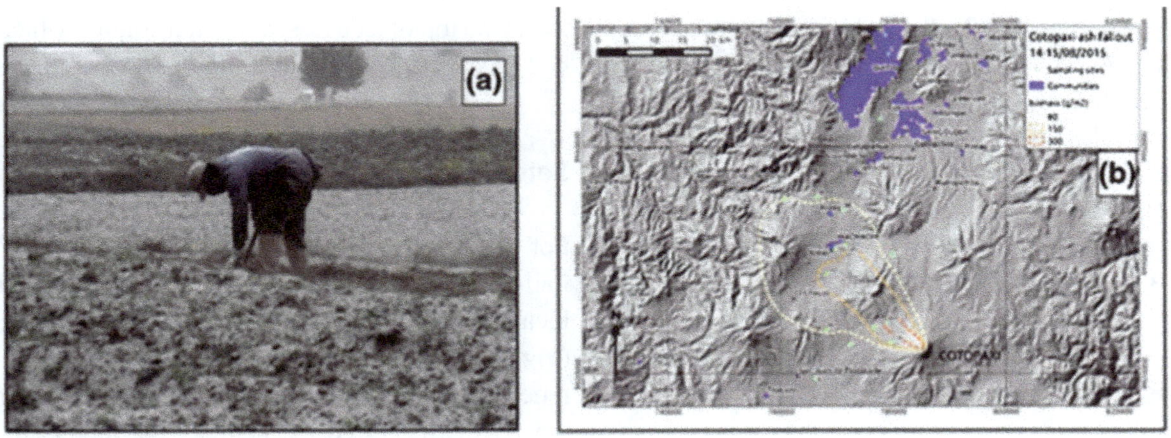

Fig. 13 **a** Ashfall from 01 September emissions accumulated in cultivated fields near El Chaupi town. **b** Ash fallout map associated with the eruptive activity of Cotopaxi on the 14th–15th of August, 2015. Map projection WGS 84, coordinates in UTMs. Values expressed in isomass of grams/m^2. *Source* Bernard et al. (2016)

rate of ash falls was increasing during frequent foggy, overcast conditions (Bernard et al. 2016). On the 14th of August, after the explosions, SO_2 levels reached 16,400 ton/day as registered by the satellite sensor OMI (http://so2.gsfc.nasa.gov/pix/daily/ixxxza/loopall3.php?yr=15&mo=08&dy=15&bn=ecuador) (Fig. 15). Subsequently, SO_2 levels were particularly high on the 15th and 24th of August, when OMI measurements gave readings of 6500 and 6600 ton/day, respectively.

In early September ash columns still rose to over three kilometers height above the vent and carried fine ash particles to cities on the piedmont of the coastal plain, such as Santo Domingo de los Colorados, located 120 km W of the volcano,

Fig. 14 *Left* Photo with view toward south taken from Autopista Ruminahui (SE of Quito) on 20 August. *Photo*- C. Zapata- EPN. *Right* Photo taken on the 23rd of August, 2015, from the north side of Cotopaxi. A low gas and ash column trending to the NW is observed. *Photo* P. Mothes, IGEPN

essentially situated under the red swath trending W in Fig. 15.

Ashfall was still prevalent in mid-October, but had all but terminated the third week of November where it was seen W-NW of the volcano. VAAC Washington again reported suspended fine ash above Santo Domingo as well as in Los Rios province to the SW. The ash column generally rose to only 1 and 3 km above the summit and had a velocity between 6 and 10 m/s and lasted about one week. Fieldwork permitted the estimation of a mass and volume total of 3.49×10^7 kg (22,100 m^3) for this late, waning period (Bernard et al. 2016). The total ashfall dense rock equivalent (DRE) volume for the entire eruption was calculated in 0.5 Mm^3 (Bernard et al. 2016).

Ash Componentry

Analysis of ash beneath both binocular and scanning electron microscope showed clearly that there was an evolution in ash componentry from the eruption's beginning on 14 August and later. The first ash from 14 August had more hydrothermal lithics (pyrite, scoria with vesicles filled with altered material and hydrothermal quartz). As the eruptions progressed we saw an increase in more fresh magmatic components, such as free crystals, glass particles with low vesicularity and a high percentage of microlites, which implied low magma ascent rates and stiffening of magma in the upper part of the column (Gaunt et al. 2016).

Ashes collected on the 20th of October, had a high concentration of dense microcrystalline material. Although there is evidence of few vesiculated clasts (diktytaxitic texture); about 65% of the ash is considered possibly juvenile. Gaunt et al. (2016) suggest that the origin of the ash is the top of a degassed magma column which had ascended from about 3 km below the crater.

Seismicity

For most of the post explosive period after mid-August, seismic hypocenters still remained located at the two depths mentioned above (Fig. 7). Most relevant was the sporadic occurrence of VT events with magnitudes of 3 or greater that occurred. Sometimes spasmodic tremor was registered and continued for hours, as for example, that registered on 02 September, 2015.

Starting on 10 September, a swarm of VT seismic events kicked in with a rate of approximately 100 events/day and a daily registry of coeval small internal explosions which has associated infrasound signatures, (shown in green color in Fig. 16). This swarm lasted past the New Year, but the overall seismic energy release was low (Figs. 3b and 5).

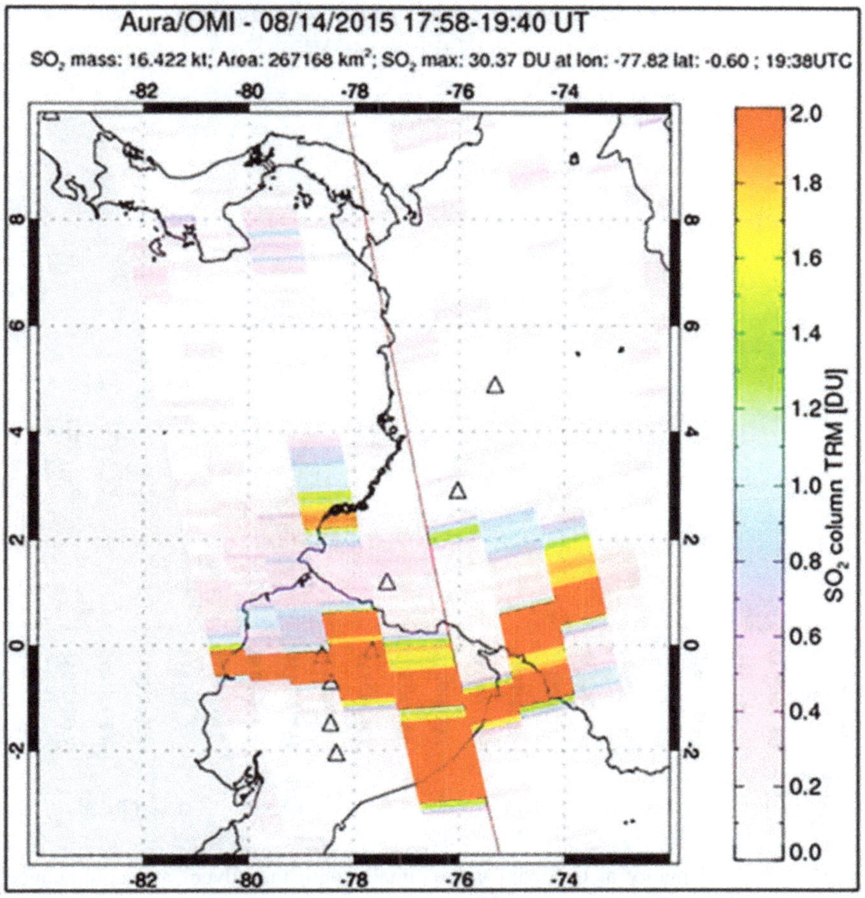

Fig. 15 Registry of SO_2 (16,400 ton/day) from satellite as detected by the OMI sensor on the 14th of August, 2015

Visual Observations and Secondary Effects

Thermal images (August/2015) showed the presence of new thermal anomalies (~15 °C) inside the crevices on the N side glaciers, at the same time fumarolic gases were observed coming out from those fractures. The highest temperature obtained was about 200 °C (Fig. 17) from gases ascending the crater.

On September 3, water emerging from the basal fronts on the northern glaciers was clearly observed, and countless new crevices in the majority of glacier ends and on the upper flanks were evident. All this led to the conclusion that an abnormal process was producing increased melting of the glaciers. Starting in mid September it was possible to observe the presence of small secondary lahars descending several streams and we estimated that many of them were due to increased glacier melting.

Orthophotos made on August 18 and then again on October 8, show a decrease of about 0.49 km^2 of the area covered by glaciers. This represents a very high rate of glacier melting, not explained exclusively by climate change (Cáceres et al. 2016).

We estimate that small volumes of magma reached surface levels in the volcano conduits causing increased circulation of hot fluids inside the edifice, apparently reaching the basal area of the glaciers and producing increased melting. It is necessary to further investigate the hazard due to instability in the melting glaciers and their eventual collapse that could lead to greater secondary lahars. Numerical modeling by

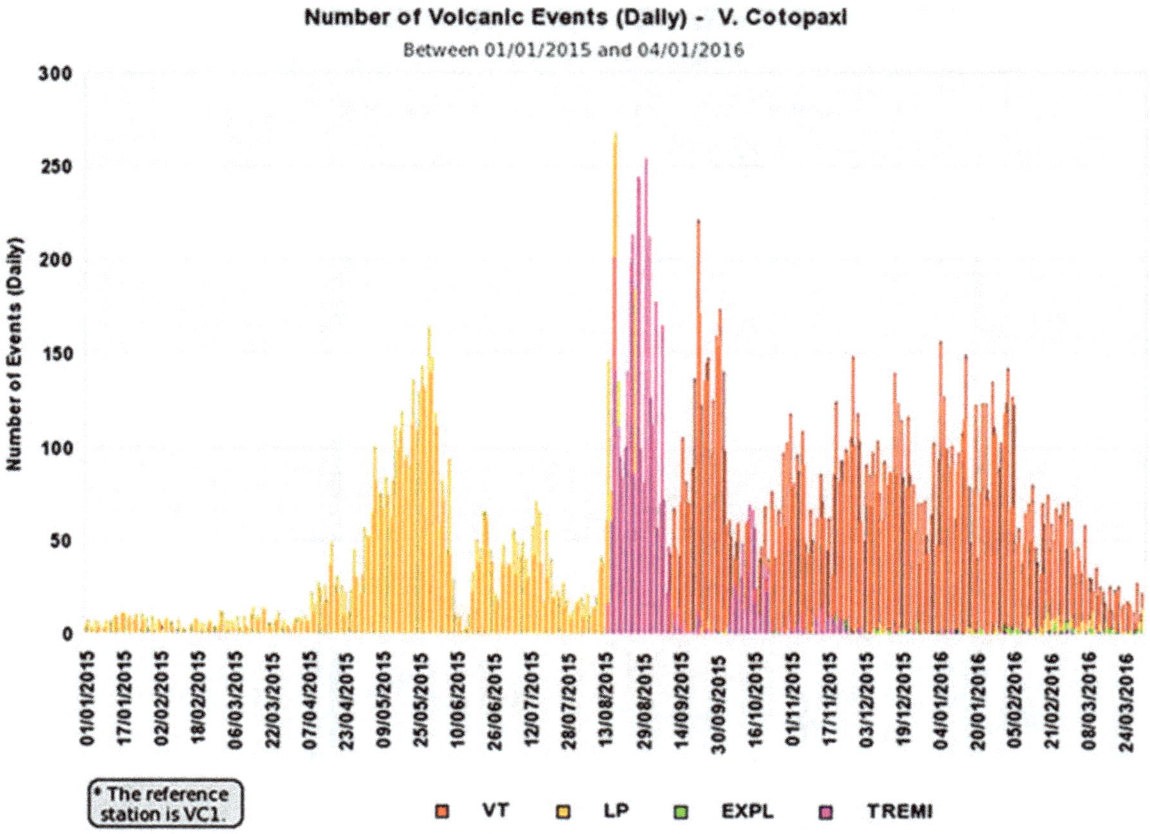

Fig. 16 Registry of overall seismicity at Cotopaxi volcano from 01 January 2015 to 01 April, 2016. Most notable is the presence of LP earthquakes in the first semester of 2015, later followed by emission tremor and finally with the advent and continuance of frequent VT seismic events with accompanying internal explosions that had no superficial manifestation, except infrasound registry

Hemmings et al. (2016) has shown the importance of hyrdrothermal perturbations at Cotopaxi in generating watery flows.

Incandescence was also occasionally observed on cold still nights with a thermal camera or by simple vision. These events were considered to have been caused by pulses of hot gases.

The glacier around the crater, on the W and N flanks, became partially covered by ash. This coating of dark ash decreased the glacier albedo and consequently increased the absorption of solar rays. Therefore expedite melting of the glacier tongues increased, leaving obvious melt water channels issuing from the glaciers' base.

As a result of the afternoon melting by insolation and perhaps also by higher temperatures of the rock beneath the glaciers, runoff increased, especially off the W flank glaciers and there were frequent small secondary lahars. Those lahars that have been especially associated with rain storms obtained the highest discharges—on the order of 10–30 m^3/sec (D. Andrade-IGEPN, Pers, Comm, 2015). The Agualongo channel, on Cotopaxi's W side was frequently flooded by lahars and on three occasions they covered partially the main road giving access to the volcano.

Interpretation and Model

As shown in Fig. 16, LP events gradually increased starting in April, 2015, beginning with small magnitudes and low energy levels (Fig. 3). We interpret the LPs to imply fluid movement occurring at 10–12 km below the crater and then up to shallower levels (Fig. 7). There were only

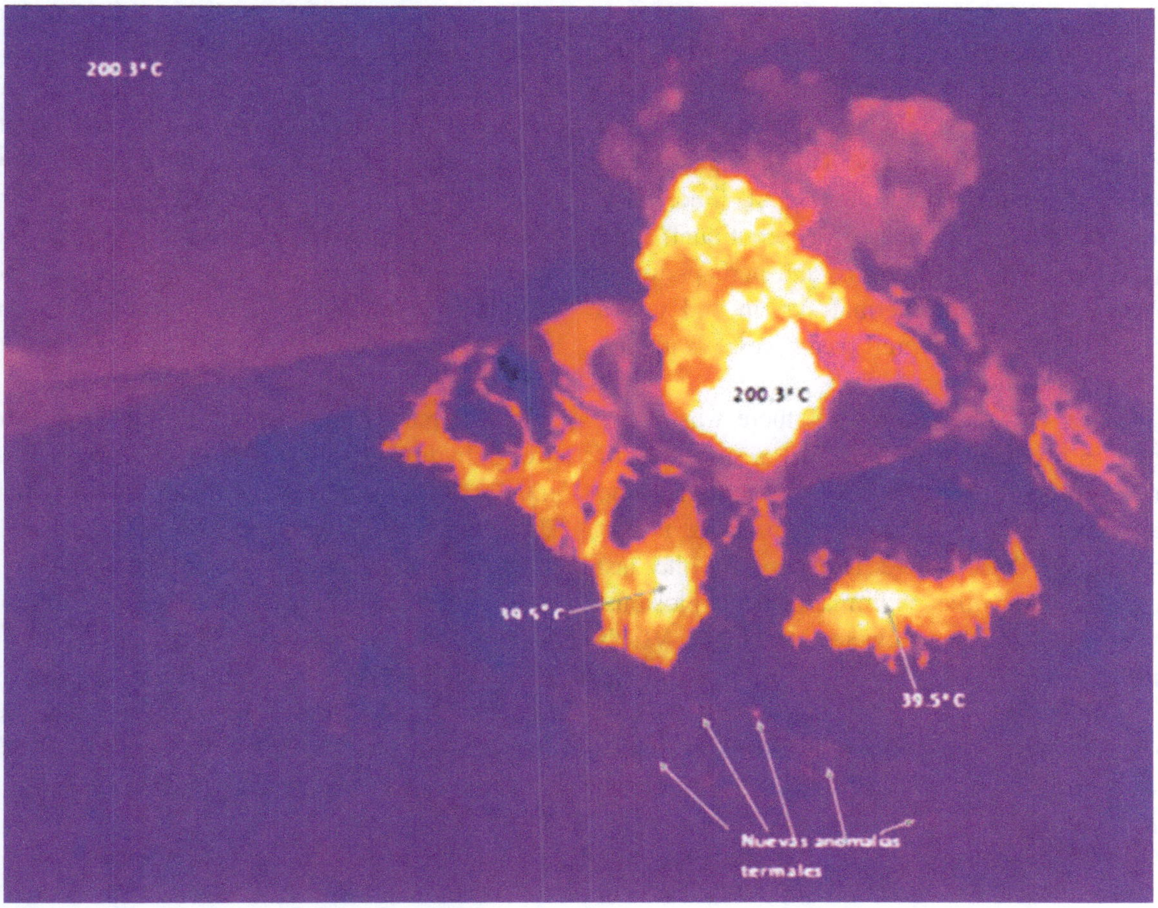

Fig. 17 Thermal image taken on 03 September, 2015 looking toward the SE sector of the upper cone. The infrared camera detected temperatures of 200 °C associated with emission from the crater and lower temperatures from fractures below the summit rim. *Photo* P. Ramón, IGEPN. *Source* http://www.igepn.edu.ec/cotopaxi/informes-cotopaxi/coto-especiales/coto-e-2015/13529-informe-especial-no-14/file

scarce VTs that occurred in concert with the LPs. In late May VLPs were registered (Fig. 5) and were interpreted to be possible mass magma transport processes as reported by Arias et al. (2015). Often VLPs are identified as eruption precursors, e.g. at Redoubt (Power et al. 2012), for example. In hindsight, vigorous VLPs were also registered in 2009–2010 at Cotopaxi and correlated with recoverable deformation patterns at borehole tilt stations (Mothes et al. 2010), but did not result in an extended seismic crisis or magmatic activity on the surface.

Nonetheless, LP and VLP events registered in April–August, 2015 were apparently responding to a slow ascent of a small magma slug and associated fluids and in April 2015 deformation recorded by tilt and GPS stations began almost synchronically with the jump in LP seismicity, implying that there may already have been magma ascent from a deeper depth to a shallower reservoir in order to show changes on the most proximal tiltmeters. The GPS stations begin to show minor displacements at the same time, particularly on the W-SW flank, where no strong evidence was detected in seismicity, but to the contrary seismicity was concentrated more on the E-SE flanks (Fig. 7). In sum, we registered both shallow and deep seismic activity. A leading hypothesis for this pattern likely was the interaction between fluids being released by a deep-seated source, say at 24 km depth as proposed in our geodetic model. These fluids ascended and perturbed a preexisting shallow-seated source, which may be the magmatic remnant that drove the 2001–2002 unrest reported by Hickey et al. (2015).

With more new magma in Cotopaxi's system, SO_2 output became prevalent in mid-May, one month after the hike in seismicity. Background SO_2 levels of <500 ton/day were surpassed and rose to over 3000 ton/day. The strong onset of bands of tremor about the 1st of June, were conjectured to be related to continual fluid movement within the edifice and perhaps to the boiling of the hydrothermal system and was a signal that more overall heat was circulating within the edifice.

Along with the rise in SO_2 there was also a trend in production of more VLP's since 114 of these events were registered between May and mid-August, 2015. Of great significance is that the largest VLPs were registered in the last 3 weeks before the hydromagmatic explosions. Afterwards too there were infrequent VLP's (Fig. 5).

With the highest energy levels of the VLPs being logged before the explosions, these events seem to have been one of the detonators of the explosions (Fig. 5b). They seemed to herald that magma/fluids were ascending. Of particular note was the VT/LP swarm of the 13th–14th of August that began 12 h before the hydromagmatic eruptions on the morning of 14 August (Fig. 11). The swarm comprised of some 40 VTs and >50 LPs was the most energetic of any seismic swarm registered at Cotopaxi since 2002 and was a warning in hindsight which presaged the subsequent explosions/strong emissions some hours later. These seismic trends and the higher SO_2 flux, would indicate that magma was working in the upper part of the system—at least in the 0 to 6 km level below the crater.

The initial explosions had a phreatic component since water was available in the small pond at the crater's base, sub-glacial melting and from pore water within the hydrothermal system. Nonetheless, Gaunt et al. (2016) argued that the most likely driving force of the initial explosions was magmatic heat interacting with the hydrothermal system providing energy to trigger hydromagmatic eruptions at Cotopaxi. Textural evidence for this process was only preserved in the deposits of the initial eruptions, but not subsequent ones. Later emissions were likely the result of the repeated formation and destruction of a shallow magmatic plug by brittle fragmentation through mechanical stresses and decompression. Gas overpressure must have been accumulating beneath the conduit plug and may have contributed to the flank deformation, particularly as registered by the tiltmeters. In the succeeding post-explosion days SO_2 output rose to 16,000–18,000 ton/day (Hidalgo et al. 2016), a likely testimony to the accumulated gas that had been trapped in the plumbing system.

The expulsed material showed evidence of strong hydrothermal alteration and there was initially little evidence of juvenile components. There was also a low pH (3.6–5.1) and high sulfate- SO_4 concentrations (up to 13,000 mg/kg) in the expulsed ashes of the 14th to 25th of August, as detected by leachate analysis (P. Delmelle, Pers. Comm, 2015). Later, as described above, the percentage of juvenile components increased through time, to the last erupted material collected in late November, 2015.

After the explosions and strong emissions of the 14th of August, the conduit lost its retaining plug and remained open and continual fluid movement was facilitated, although pulsatile superficial activity continued, and few shallow explosions occurred. After the explosions LP events were initially high (>200 events/day), but dropped to <20 events/day by the first week of September, where they remain at this writing.

While the LPs diminished, to the contrary the VT events rose notably. Around the 1st of September 15–20 LP events/day were registered. By the third week of September these increased to >200 events/day (Fig. 16), and subsequently this value decreased to 50–100 events/day, but in all totaled nearly 15,000 events. In the last 3 months of 2015 persistent VT daily activity was registered as well as waning tilt and GPS offsets. Even with the VT swarm overall seismic energy levels decreased compared to the levels registered in May to August, which may be due to the overall successful degassing of the system, but could also be explained by closing of the conduit by a degassed magma column, thus impeding freer liberation of gases. Between

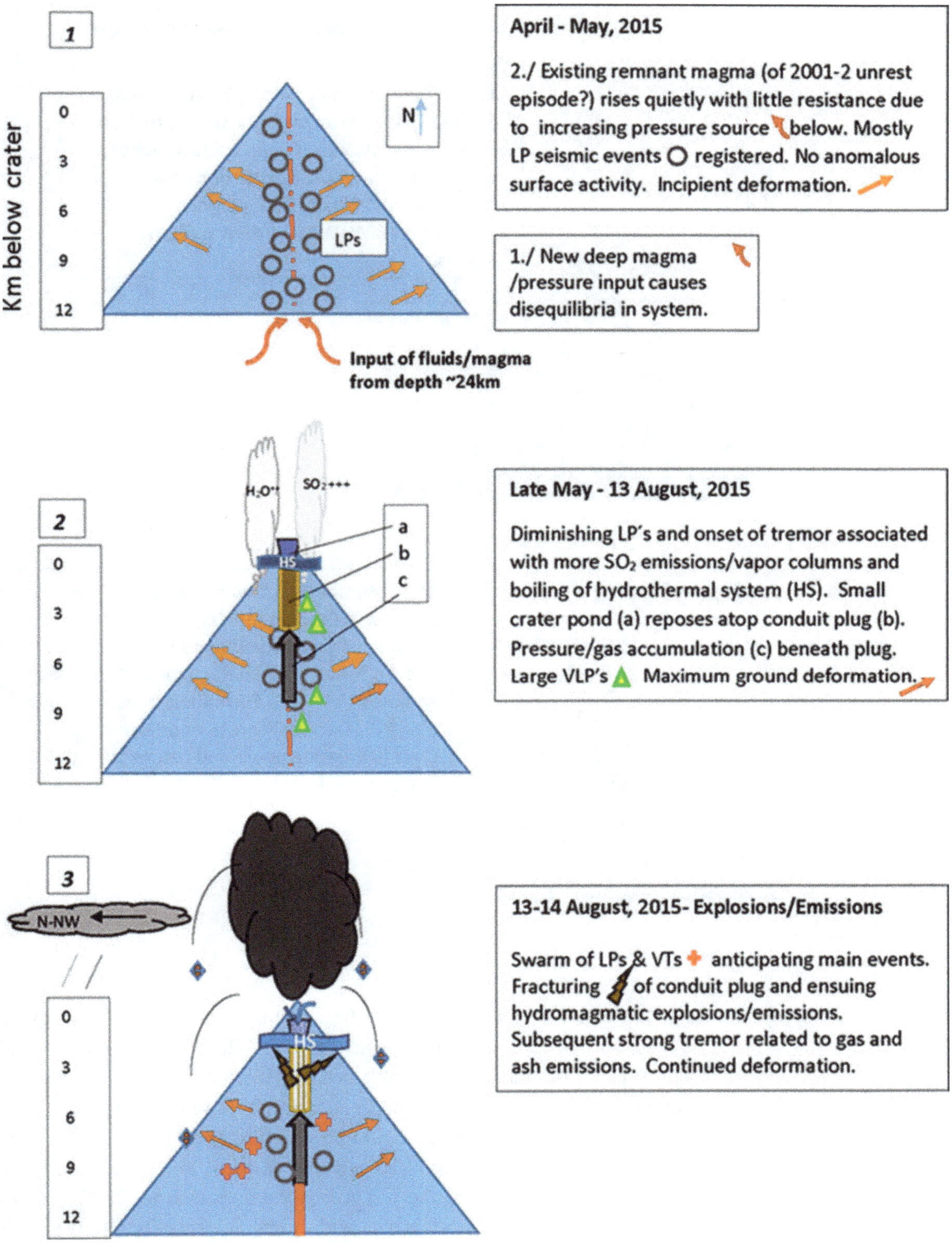

Fig. 18 Cartoons synthesizing the internal and superficial processes observed from April to December, 2015

October 2015 through April, 2016 internal explosions of deep providence were registered at a rate of 20–30 such events/month. These explosions could be interpreted as gas passing through restrictive areas within the conduit. While displaying only minimal infrasound and no detectable superficial vestiges, these explosions may be occurring due to pressurization

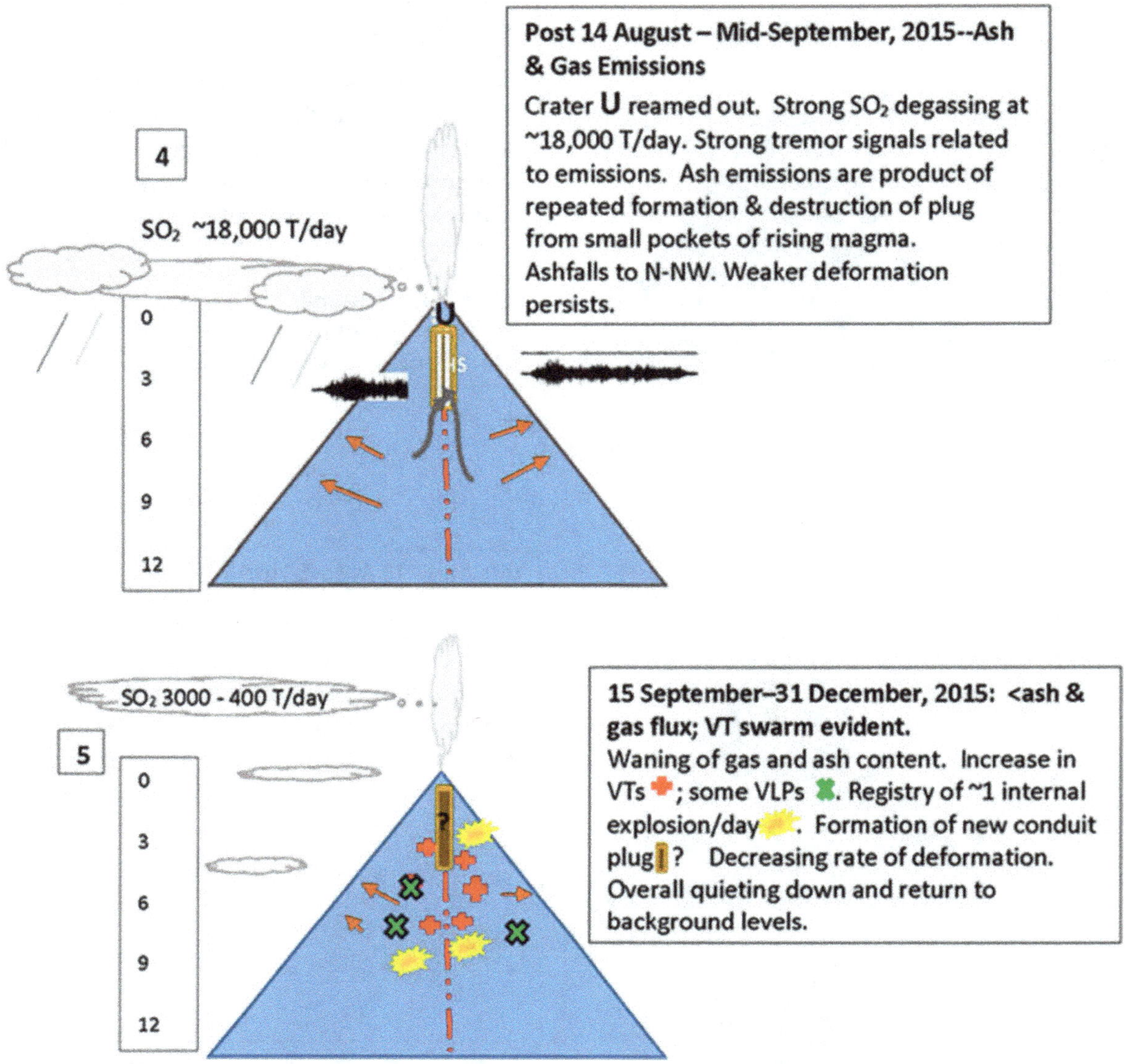

Fig. 18 (continued)

deep within the system (e.g. Valentine et al. 2014).

A synthesis of the 2015 unrest at Cotopaxi, the hypothesized driving forces and possible paths taken are shown in the following schematic cartoons (Fig. 18).

The Science-Society Interface

Strong evidence points to a magmatic component in Cotopaxi's 2015 reactivation and that this may be a precursor for future episodes, given that the bulk of the new magma that perturbed the system remained at depth. As a comparison, the 2001–2002 restless period had essentially been internal with only the weakest of superficial manifestations. For the 2015 reactivation IG scientists took the warnings very seriously and put all their collective experience to the test to make interpretations of the monitoring data and to manage successfully the great expectations of the public and authorities during the crisis period described above. Overall IGEPN volcanologists wrote 23 special reports that were disseminated via multiple media (www.igepn.edu.ec). Given the low levels of acceleration of seismic energy, small ground displacements and visual observations, eruption scenarios that the IGEPN formulated stayed within the realm of VEI 1–2 levels and clearly stated that the least likely scenarios was the generation of a paroxysmal eruption in which PDC's, voluminous ashfalls and giant lahars would be formed.

With each convincing sign that Cotopaxi was displaying stronger activity, IG scientists were proactive in improving and strengthening all monitoring systems while simultaneously helping to prepare the populations and authorities for what a major eruption of Cotopaxi could mean. Work by Christie et al. (2015) (an earlier VUELCO contribution) had shown that residents of the Chillos valley, to the east of Quito, were particularly ill-prepared to confront lahar hazards due to their recency of living in that valley. On the other hand, residents of the Latacunga valley had a clearer memory of lahar hazards, since many of their distant relatives had lived through Cotopaxi eruptions and their collective memory is better preserved. Nonetheless, social media, both beneficial and alarming, steered perceptions and actions of residents. The area of influence by the volcano, especially with respect with lahars, includes four important provinces, several counties and Quito's jurisdiction. It was particularly difficult to meet the demands and expectations and provide the personal attention of monitoring scientists to the authorities in each of these different municipalities as well as to meet with other community groups and respond to their uncertainties. There were also the constant attacks on social media of several particularly meanly-intentioned individuals who constantly tried to steer attention away from the IGEPNs scientific work by saying that it was operating with poor instrumentation or that the monitoring work at IGEPN was "carried out by amateurs".

All told, IGEPN scientists provided abundant custom guidance to local and national officials and residents with regards to volcano hazards and the proposed scenarios. A total of about 125 talks were given by IGEPN personal during the unrest period. Additionally, there was broad coordination with Ecuador's Secretary for Risk Management at all levels and participation in guiding eruption simulations. Some of the discussions with them were based on what had been reviewed in the VUELCO workshop-simulated eruption exercises carried out in late 2014 in Quito. The IGEPN also greatly benefitted by the strong collaboration and presence of members of the USGS/USAID Volcano Disaster Assistance Team who led informed discussions on the trends of the geophysical precursors and also helped to reinforce the lahar-detection network. Personal from Chalmers University of Technology (Sweden), JICA (Japan), IRD (France) and NASA (USA), DEMEX-EPN with help in SEM, LMU-Germany with grain size x-ray diffraction and UCL (Belgium) with leachates, also collaborated during the crisis.

As the eruption process waned, it was obvious that high-risk populations were tired of being constantly alert and didn't want to be perpetually attentive to volcanic processes that could threaten their livelihoods and families. This issue will have to be acknowledged and dealt with in future reactivations.

Conclusions

Seismic activity and its evolution in event types, energy release, shallowing depths and locations, elevated degassing and ash emissions and flank deformation typified the restlessness of Cotopaxi during 2015. The important accumulative energy release first of LPs than followed by registry of an important suite of large VLP's was a significant geophysical pattern indicating fluid movement, followed then by a more convincing transfer up conduit of small slugs of magma and gases to beneath the conduit plug. In the late hours of the 13th of August this plug fractured and ruptured, evidenced by the vigorous swarm of VTs and LPs before the hydromagmatic explosions that occurred early on the 14th of August 2015. Minor ground deformation, the small, limited explosions in August and subsequent ash emission suggest that the ascended magma volume was small, and indeed as calculated by Bernard et al. (2016), was only about

0.5 Mm^3 DRE. This value is far inferior to the possible volume of 42 ± 26 Mm^3, which is hypothesized to be at depth based on modeling of the observed GPS displacements (Mothes et al. 2016b). A second energetic magmatic pulse did not arise, and certainly not one with a sufficient volume to produce a VEI 3 or 4 eruption, which was one of the least likely scenarios, but nonetheless dreaded by the society and scientists.

Following the 14 August explosions and subsequent ash emissions we did not observe a new phase of outward GPS displacement trends in the deformation data, which could have implied a new magma input to cause another phase of deformation. The post-explosion VT seismic swarm which lasted 5 months was indicative of persistent internal perturbation but did not transpire in a new phase of deformation, thus we assumed that we were dealing with a small magma volume. The magma that tipped off the 2015 unrest may have been a remnant of that which provoked the 2001–2002 episode and was reported by Hickey et al. (2015). This residual magma could have been disturbed by the ascending heat and fluids from the new magma input at depth ($\sim$24 km) whose source was possibly under the SE flank, and which provoked the recorded ground deformation and the LP and VLP seismicity.

The volcano was benevolent and had awakened to only a VEI 2 level. No major damage was imparted upon the population or on livelihoods, except for temporary local economic depression, increased anxiety of the population, mild crop losses and premature selling of livestock due to fears of future losses. Overall, the volcano's manifestations served as a warning to everyone to keep attentive of Cotopaxi's capacity to cause destruction and possible severe ruin by lahar transit down major drainages which are heavily populated and host important strategic infrastructure.

An eruption process can last months to decades, and we need only to look at Tungurahua, an andesitic stratocone also in Ecuador's Cordillera Real, with ongoing eruptions for 17 years (Mothes et al. 2015a, b), or Soufrière Hills on Montserrat (Sparks and Young 2002) to suggest that the next round of Cotopaxi eruptions could last more than just several months. In the case of Tungurahua, activity started gradually in 1999 and displayed oscillating low-level behavior over the years to finally generate a rapid-onset VEI 3 eruption in 2006 (Hall et al. 2013). Such long waits test the population's resilience, but is also a time for monitoring scientists to become acquainted with the volcano's eruption style. During a reactivation period of a long dormant volcano there are many uncertainties and this demands stringent work and continual mindfulness by monitoring scientists and frequent ongoing interactive and personal communication with local communities and authorities.

During the 2015 unrest period at Cotopaxi, people living in high-risk zones (i.e. Latacunga and Valle de los Chillos) were swayed by speculation, rumors and lies concerning the status of the volcano. Some people also tended to weigh-in toward imprecise information posted on Facebook or Twitter and heed pseudo volcanologists and detractors, rather than rely on information from official channels. It was not uncommon to receive telephone calls from hysterical residents in either of these population centers inquiring if a Cotopaxi eruption was *imminent?* All told the IGEPN put out 3 reports each day about the volcano's activity and more than 24 special reports, all which are available on the IG website (www.igepn.edu.ec). The number of followers on the IGEPN's Facebook page grew to >1 million. To help stem the flow of bad information at the community level Ecuador's 911 system, in coordination with IGEPN personal, formed a pan-volcano *vigía* network comprised of volunteer observers who report via radio several times a day about their visual and audible observations of Cotopaxi or the rivers that are borne on it. This network with 55 volunteers, is in many ways a replica of the successful community-based *vigía* system that has functioned at Tungurahua volcano since 2001 (Stone et al. 2014; Mothes et al. 2015b). The information provided by the *vigía* volunteers compliments the ongoing geophysical monitoring and also serves to strengthen their capacities as community leaders and guides during volcano crisis (Espín Bedón et al. 2016).

A hypothesis for a future trend in activity weighs heavily towards hydromagmatic to Vulcanian explosions which may have a rapid onset, similar to the 14th of August episode, then evolve to sub-Plinian to Plinian eruptions of VEI 3–4 magnitude, if enough magma has accumulated at a relatively shallow depth (maybe 0–7 km below the crater) as shown in Fig. 7 for the upper level seismicity, and can make it to the surface before degassing. Vulcanian eruptions have been prominent in the volcano's historical activity (Gaunt et al. 2015). Unraveling the story will be difficult.

As the volcano is well-monitored 24 h/day/365, we anticipate that the IGEPN will provide early warnings to the public and officials before onset of important eruptive activity. This 2015 "dry run" allowed for diversification and hardening of Cotopaxi's monitoring network, frequent preparation and reappraisal of eruption scenarios and for the creation of a society-wide discussion of the possible consequences of a large Cotopaxi eruption. Some of these steps were facilitated by previous work in a VUELCO workshop. Essentially, attending to the 2015 activity was an opportunity to test the level of preparedness of the scientists and of the Ecuadorian society. All IGEPN scientists strived hard to be ready to "call it right", had the occasion arisen and a large eruption was in preparation. Since so little new magma erupted and there was no detectable subsequent shallow magmatic recharge, we consider the eruption as extremely small, and that the residual magma is in repose until a future time. Overall, the crisis was an important opportunity for learning about Cotopaxi's restlessness, with particular recognition of the increase in the VLP events and their energy levels just weeks before the mid-August explosions and the synchronous but progressive ground deformation signals, albeit small, that coincided with the increased seismicity. These two patterns more than any other geophysical signals, announced the ascent of Cotopaxi's magma, although finally only a small quantity breached the surface.

In all likelihood little or no evidence of the 2015 restless period will be preserved in the geological record. We know from written chronicles (1534–1877) that Cotopaxi often had weeks to months of ramping up before unleashing VEI 3 or 4 eruptions, i.e. there were probably several poorly preserved 2015-sized like events, and therefore unrest has been poorly documented. In this recent case, scientists had the benefit of observing and analyzing the geophysical monitoring output during the entire episode and knowing what level of activity the volcano was at. But, monitoring scientists, just like the citizens of Ecuador, experienced the anxiety of pondering what could be the volcano's next steps, i.e., the possible rapid intrusion of a new batch of volatile-rich magma or returning to calm. Fortunately, in this time around, the first scenario did not transpire. However, this training opportunity that we experienced could prove invaluable for when the next scenario is played out.

Acknowledgements This work was supported by a grant from European Commission FP 7 program (ENV.2011.1.3.3-1; grant no: 282759; VUELCO). We would like to thank the IGEPN staff for keeping all monitoring operations at Cotopaxi volcano optimally functioning. Members of the IGEPN's Volcanology group gave most of the talks and explanations of eruptions scenarios to authorities, community groups and wherever was necessary during the crisis. We acknowledge the VUELCO project for opportunities to share opinions about future eruptive crisis scenarios. Also, we are grateful for the support of the VDAP/USAID team during the crisis. The EPN, SENESCYT and SENPLADES provided funding for much of the instrumentation and daily operations. Support by JICA instrumentation and guidance is also noted. The Ecuadorian military provided overflights of Cotopaxi, while IGM provided ortophotos. The LMI project was instrumental in carrying out near real time petrologic monitoring. Thanks to Viviana Valverde for preparation of several figures. Recognition is given to the expressions of interest by STREVA project members on the societal implications of eruptive activity at Cotopaxi.

References

Arias G (2015) Estudio de las Señales Sísmicas de Muy Largo Periodo del volcán Cotopaxi. Proyecto de Titulación, Facultad de Ciencias, Escuela Politécnica Nacional

Arias G, Molina I, Ruiz M, Hernandez S, Alvarado A, Plain M, Mothes P, M Yépez, Hidalgo S, Barrington, C (2015) very long period seismicity accompanying increasing shallower activity at Cotopaxi Volcano. In Conference Paper No. S51D-2724, AGU Fall Meeting, San Francisco, 14–18 December

Bernard B, Battaglia J, Proaño A, Hidalgo S, Vásconez F, Hernandez S, Ruiz M (2016) Relationship between volcanic ash fallouts and seismic tremor: quantitative assessment of the 2015 eruptive period at Cotopaxi volcano, Ecuador. Bull Volcanol. doi:10.1007/s00445-016-1077-5

Cáceres B (2016) Dramatic reduction of Cotopaxi glaciers during the last volcano awakening 2015–2016. Conference: AGU Fall Meeting, San Francisco-Moscone Center. doi:10.13140/RG.2.2.17483.80168

Chouet BA, Matoza RS (2013) A multi-decadal view of seismic methods for detecting precursors of magma movement and eruption. J Volcanol Geoth Res 252:108–175

Christie R, Cooke O, Gottsmann J (2015) Fearing the knock on the door: critical security studies insights into limited cooperation with disaster management regimes. J Appl Volcanol 4:19. doi:10.1186/s13617-015-0037-7

Espín Bedón P, Mothes P, Montesdeoca M, Ruiz M, Alvarado A, Córdova M, Santamaria S (2016) Communication network of "Vigias" implemented at Cotopaxi Volcano, Ecuador, Abstract, Cities on Volcanoes-9, Puerto Varas, Chile, November, 2016

Garrison JM, Davidson JP, Hall M, Mothes P (2011) Geochemistry and petrology of the most recent deposits from Cotopaxi Volcano, Northern Volcanic Zone, Ecuador. J Petrol. doi:10.1093/petrology/egr023

Gaunt HE, Mothes P, Chadderton A, Lavallee Y (2015) Physical characteristics of conduit plug rocks during Vulcanian eruptions at Cotopaxi and Tungurahua volcanoes, Ecuador. Abstract, IUGG meeting, Prague-Czech Republic, June 2015.

Gaunt EH, Bernard B, Hidalgo S, Proaño A, Wright H, Mothes P, Criollo E, Kueppers U (2016) Juvenile magma recognition and eruptive dynamics inferred from the analysis of ash time series: the 2015 reawakening of Cotopaxi volcano 328 October 2016. doi:10.1016/j.jvolgeores.2016.10.013

Hall M, Mothes P (2008) The rhyolitic–andesitic eruptive history of Cotopaxi volcano Ecuador. Bull Volcanol 70(6):675–702

Hall ML, Steele A., Mothes P, Ruiz M (2013) Pyroclastic density currents (PDC) of the 16–17 August 2006 eruptions of Tungurahua volcano, Ecuador: geophysical registry and characteristics. J Volcanol Geotherm Res 265(2013):78–93. https://doi.org/10.1016/j.jvolgeores.2013.08.011

Harlow DH, Power JA, Laguerta EP, Ambubuyog G, White RA, Hoblitt RP (1996) Precursory seismicity and forecasting of the June 15, 1991, eruption of Mount Pinatubo. Fire and Mud: Eruptions and Lahars of Mount Pinatubo, Philippines. University of Washington Press, Seattle, pp 285–306

Hemmings B, Whitaker F, Gottsmann J, Hawes MC (2016) Non-eruptive ice-melt driven by internal heat at glaciated stratovolcanoes. Submitted to J Vol Geotherm Res

Herring TA, King RW, Floyd MA, McClusky SC (2015) Introduction to GAMIT/GLOBK, Release 10.6. Department of Earth, Atmospheric, and Planetary Sciences, Massachusetts Institute of Technology, PSF: http://www-gpsg.mit.edu/~simon/gtgk/Intro_GG.pdf, 50p

Hickey J, Gottsmann J, Mothes P (2015) Estimating volcanic deformation source parameters with a finite element inversion: The 2001–2002 unrest at Cotopaxi volcano, Ecuador. Journal of Geophysical Research: Solid Earth 120(3):1473–1486

Hidalgo S, Bernard B, Battaglia B, Gaunt E, et al. (2016) Cotopaxi volcano's unrest and eruptive activity in 2015: mild awakening after 73 years of quiescence. Geophys Res Abstracts Vol. 18, EGU2016-5043-1, 2016 EGU General Assembly, Viena-Austria

Jousset P, Budi-Santoso A, Jolly AD, Boichu M, Dwiyono S, Sumarti S, Hidayati S, Thierry P (2013) Signs of magma ascent in LP and VLP seismic events and link to degassing: an example from the 2010 explosive eruption at Merapi volcano, Indonesia. J Volcanol Geoth Res 261:171–192. doi:10.1016/j.jvolgeores.2013.03.014

Kumagai H, Yepes H, Vaca M, Caceres V, Nagai T, Yokoe K, Vasconez F (2007) Enhancing volcano-monitoring capabilities in Ecuador. Eos 88 (23):245–252

Kumagai H, Palacios P, Maeda T, Castillo DB, Nakano M (2009) Seismic tracking of lahars using tremor signals. J Volcanol Geoth Res 183(1):112–121

Kumagai H, Nakano M, Maeda T, Yepes H, Palacios P, Ruiz M, Yamashima T (2010) Broadband seismic monitoring of active volcanoes using deterministic and stochastic approaches. J Geophys Res: Solid Earth 115 (B8)

Kumagai H, Mothes P, Ruiz M, Maeda Y (2015) An approach to source characterization of tremor signals associated with eruptions and lahars. Earth, Planets, and Space. 67:178, 67:178, doi:10.1186/s40623-015-0349-1

Lyons J, Segovia M, Ruiz M (2012) Reporte del Enjambre Sísmico en la Zona de Volcán de Cotopaxi—Mayo 2012. Internal report. Not published. IGEPN, Quito.

Maeda Y, Kumagai H, Lacson RJ, Figueroa MS, Yamashina T, Ohkura T, Baloloy AV (2015) A phreatic explosion model inferred from a very long period seismic event at Mayon Volcano, Philippines. J Geophys Res Solid Earth 120:226–242. doi:10.1002/2014JB011440

Márquez M (2012) Caracterización de los sismos de muy largo periodo en el volcán Cotopaxi y sus implicaciones. Tesis de Grado, Carrera de Geofísica, Facultad de Ciencias Astronómicas y Geofísicas, Universidad Nacional de La Plata. Argentina

Marzocchi W, Newhall CG, Woo G (2012) The scientific management of volcanic crises. J Volcanol Geoth Res 247–248:181–189

Molina I, Kumagai H, García-Aristizábal A, Nakano M, Mothes P (2008) Source process of very-long-period events accompanying long-period signals at Cotopaxi Volcano, Ecuador. J Volcanol Geoth Res 176(1):119–133

Moran SC, Newhall C, Roman DC (2011) Failed magmatic eruptions: Late-stage cessation of magma ascent. Bull Volc 73(2):115–122. doi:10.1007/s00445-010-0444-x

Mothes P, Hall ML, Espín, PA, Vasconez F, Sierra D, Córdova M, Santamaría S, Andrade D (2016a) Mapa Regional de las Amenazas del Volcán Cotopaxi- Norte and Sur, scale 1:50,000, Instituto Geofísico and Instituto Geografico Militar, Quito http://www.igepn.edu.ec/mapas/mapas-volcan-cotopaxi.html

Mothes PA, Nocquet JM, Jarrín Tamayo P, Morales Rivera AM, Lundgren PR, Yépez YM, Viracucha EG, Gaunt EH (2016b) Geodetic signature associated with unrest at Cotopaxi in 2015–2016: modeling of GPS data, a deep magma source, synchronous seismic swarms and petrologic constraints. Abstract, Cities on Volcanoes-9, Puerto Varas, Chile, November, 2016

Mothes P, Hall ML, Andrade D, Yepes H, Pierson TC, Ruiz AG, Samaniego P (2004) Character, stratigraphy and magnitude of historical lahars of Cotopaxi volcano (Ecuador). Acta Vulcanol 16(1/2):1000–1023

Mothes P, Lisowski M, Ruiz M, Ruiz G, Palacios P (2010) Borehole Tiltmeter and CGPS Response to VLP Seismic Events under Cotopaxi Volcano. Ecuador, Fall AGU, San Francisco Abstract

Mothes PA, Vallance JW, Hall ML, Garrison JM (2015a) Cotopaxi's most recent rhyolitic eruptions, 2800 y BP. In Conference Abstract, 26th IUGG General Assembly, Prague

Mothes PA, Yepes HA, Hall ML, Ramón PA, Steele AL, Ruiz MC (2015b) The scientific–community interface over the fifteen-year eruptive episode of Tungurahua Volcano Ecuador. J Appl Volcanol 4(1):1–15

Mothes PA, Vallance JW (2015) Lahars at Cotopaxi and Tungurahua Volcanoes, Ecuador: highlights from stratigraphy and observational records and related downstream hazards. In Volcanic Hazards, Risks and Disasters. doi:10.1016/B978-0-12-396453-3.00006-X, chapter 6

Newhall CG (2000) Volcano Warnings. In: Sigurdsson et al. (Eds) Encyclopedia of Volcanoes. San Diego, CA, pp 1185–1198

Newhall, Christopher G, Self Stephen (1982) The Volcanic Explosivity Index (VEI): An Estimate of Explosive Magnitude for Historical Volcanism (PDF). J Geophys Res. 87 (C2): 1231–1238. Bibcode:1982JGR....87.1231 N. doi:10.1029/JC087iC02p01231

Nocquet JM, Villegas-Lanza JC, Chlieh M, Mothes PA, Rolandone F, Jarrin P, Martin X (2014) Motion of continental slivers and creeping subduction in the northern Andes. Nat Geosci 7(4):287–291

Phillipson G, Sobredelo R, Gottsmann J (2013) Global volcanic unrest in the 21st century: an analysis of the first decade. J Volcanol Geother Res 264:183–196. doi:10.1016/j.jvolgeores.2013.08.004

Pistolesi M, Rosi M, Cioni R, Cashman KV, Rossotti A, Aguilera E (2012) Physical volcanology of the post–twelfth-century activity at Cotopaxi volcano, Ecuador: behavior of an andesitic central volcano. Geol Soc Am Bull 123(5–6):1193–1215

Power JA, Stihler SD, Chouet BA, Haney MM, Ketner DM (2012) Seismic observations of Redoubt Volcano, Alaska—1989–2010 and a conceptual model of the Redoubt magmatic system. J Volcanol Geotherm Res 259:31–44

Ruiz M, Guillier B, Chatelain JL, Yepes H, Hall M, Ramon P (1998) Possible causes for the seismic activity observed in Cotopaxi volcano Ecuador. Geophys Res Lett 25(13):2305–2308

Siebert L, Simkin T, Kimberly P (2010) Volcanoes of the World. Univ. of California Press and Smithsonian Institute. 551 p

Sparks RSJ, Aspinall WP (2004) Volcanic activity: frontiers and challenges in forecasting, prediction and risk assessment. The State of the Planet: Frontiers and Challenges in Geophysics, Geophysical Monograph 150, IUGG Volume 19

Sparks RSJ, Young SR (2002) The eruption of Soufrière Hills Volcano, Montserrat (1995–1999): overview of scientific results Geological Society, London, Memoirs 21:45–69. doi:10.1144/GSL.MEM.2002.021.01.03

Stone J, Barclay J, Simmons P, Cole PD, Loughlin SC, Ramón P, Mothes P (2014) Risk reduction through community-based monitoring: the *vigías* of Tungurahua Ecuador. J Appl Volcanol Soc Volcanoes 3:11. doi:10.1186/s13617-014-0011-9

Tilling RI (1989) Volcanic hazards and their mitigation: progress and problems. U.S. Geological Survey, Menlo Park, California, p p33

Valentine G, Graettinger AH, Sonder I (2014) Explosion depths for phreatomagmatic eruptions. Geoph Res Lett doi:10.1002/2014GL060096

White R, McCausland W (2016) Volcano-tectonic earthquakes: A new tool for estimating intrusive volumes and forecasting eruptions. J Volcanol Geoth Res 309:139–155

Zobin, VM (2012) Long-Period and Very Long-Period Seismic Signals at Volcanoes. In: Zobin (Ed), VM Introduction to Volcanic Seismology (2nd ed). pp 327–354

Volcanic Unrest Simulation Exercises: Checklists and Guidance Notes

R. J. Bretton, S. Ciolli, C. Cristiani, J. Gottsmann, R. Christie and W. Aspinall

Abstract

When a volcano emerges from dormancy into a phase of unrest, the civil protection authorities charged with managing societal risks have the unenviable responsibility of making difficult decisions balancing numerous competing societal, political and economic considerations. A volcano that is threatening to erupt requires sound risk assessments incorporating trusted hazard assessments that are timely, relevant and comprehensible. Foreseeable challenges arise when the inevitable uncertainties of hazard assessment and communication meet societal and political demands for certitude. In some regions that host volcanic hazards, it would be both realistic and prudent to adopt three working assumptions. The complex legal and administrative infrastructures of risk governance will be largely untested and possibly inadequate. Many volcano observatory scientists, and probably even more risk managers and at-risk individuals/communities, will have inadequate recent experience of the challenges of hazard communication during a period of unrest. And lastly, the scientists may also have inadequate practical experience of the needs and management capacities of the risk-mitigation decision makers with whom they must communicate. "Practice doesn't make perfect. Practice reduces the imperfection." (Beta 2011). If this statement is correct, volcanic unrest simulation exercises (VUSE) have a vital role to play within the complex processes of volcanic risk governance. Consistent with the broad approach of the Sendai Framework for Risk Reduction 2015–30, this chapter argues that practical knowledge of VUSE can and should be analysed and recorded so that key lessons can be shared for the widest possible benefit. This chapter investigates five recent simulation exercises and presents six complementary checklists based upon data, insights and practice pointers derived from those exercises. The use of checklists, supported by guidance notes, is commended as a pragmatic way to create, test and develop acceptable standards of governance practice.

R. J. Bretton (✉) · J. Gottsmann · W. Aspinall
School of Earth Sciences, University of Bristol, Bristol, UK
e-mail: Richard.Bretton@bristol.ac.uk

R. J. Bretton · J. Gottsmann · R. Christie
W. Aspinall
The Cabot Institute, University of Bristol, Bristol, UK

S. Ciolli · C. Cristiani
Dipartimento della Protezione Civile (DPC), Rome, Italy

R. Christie
School of Sociology, Politics and International Studies, University of Bristol, Bristol, UK

Advs in Volcanology (2019) 271–298
DOI 10.1007/11157_2018_34

Published Online: 14 July 2018

It is argued here that well planned and executed simulation exercises are capable of informing and motivating a wide range of risk governance stakeholders. They can identify process and individual shortcomings that can be mitigated. Simulation exercises can and should play a vital role in reducing volcanic risks.

Keywords

Volcanic unrest · Risk governance · Simulations · Exercises · Training · Communication

1 Introduction

1.1 Simulation Exercises and the Sendai Framework

This chapter is about reconstructions of the evolution of past, and realistic simulations of hypothetical future, volcanic unrest events. They will be referred to as Volcanic Unrest Simulation Exercises (VUSE). A simulation is "a learning experience that occurs within an imaginary or virtual system or world" and involves role-play, which has been defined as "the importance and interactivity of roles in pre-defined scenarios" (van Ments 1999; Errington 1997, 2011; Dohaney et al. 2015). A fuller working description is set out in Sect. 3.1.

The Sendai Framework prioritises mitigation of risks before response and recovery. It also recognises the importance of "building the knowledge of, inter alia, government officials… through sharing experiences, lessons learned, good practices, and training and education on disaster risk reduction" (UN/ISDR 2015, 15). Mutual learning, dialogues and cooperation between risk governance stakeholders are encouraged. The Sendai Framework also identifies a need for quality standards.

The principal purpose of this chapter is to accept the challenge of the goals of the Sendai Framework and to share knowledge derived from several recent simulation exercises with a wide audience.

1.2 The Managerial and Scrutiny Dimensions of Risk

1.2.1 Managerial Dimension

When a volcano emerges from dormancy into a phase of unrest, the civil authorities in charge of managing volcanic risks have to make challenging decisions (Fiske 1984; Dohaney et al. 2015). Decisions balancing safety and cost typically must made with limited information (Sparks et al. 2012a, b; Marzocchi et al. 2012; Jenkins et al. 2012) and in real time and under uncertainty, in a context of intense pressure (Marrero et al. 2015, 2). While the primary objective is to minimise the loss and damage from any volcanic event, the socio-economic losses resulting from false alarms and evacuations must also be considered" (Woo 2008; Hincks et al. 2014, 2; Donovan and Oppenheimer 2014).

Poorly handled unrest periods cause social, economic and political problems, even without an eruption. Ill-considered responses may facilitate the release of inappropriate advice and emergency declarations, and may lead to unwarranted media speculation and the premature cessation of economic activity and community services (Johnston et al. 2002, 228).

Recent crises, including the 2010 Icelandic Eyjafjallajökull eruption, have highlighted the difficulty of co-ordinating and synthesising scientific inputs from many different disciplines and institutions, and translating these into useful policy advice at very short notice (Harris et al. 2012; Dohaney et al. 2015). Effective communication, collaboration and cooperation are necessary between many expert and technical advisors, emergency management agencies and lifeline organisations (Doyle et al. 2015).

Jordan et al. (2011) have emphasised the need for scientists to have a clear role, a clear and authoritative voice and effective communication skills. The Organisation for Economic

Co-operation and Development (OECD) (2015) has identified the need for scientific advisers to have: (1) permanent authoritative structures; (2) a central contact point; (3) clear reporting procedures; (4) a pre-defined public communication strategy; and, when necessary, (5) ways to coordinate their actions internationally.

Although civil protection "authorities may have theoretical knowledge of volcanos, few have any practical experience of eruptions" (Solana et al. 2008, 312). Furthermore, the timescales of periods of volcanic unrest, especially bigger ones, do not correlate well with the tenures of political and senior management appointments (Donovan and Oppenheimer 2012; Mothes et al. 2015).

The term 'standard equivocality' relates to the "absence of commonly recognised standards capable of guiding, measuring and evolving acceptable practice" (Hood 1986; Bretton 2014; Bretton et al. 2015; Rothstein 2002). It is suggested here that there are no readily accessible standards regarding how hazard communication should be conducted during a period of volcanic unrest.

1.2.2 Scrutiny Dimension

The International Federation of Red Cross and Red Crescent Societies (IFRC) and United Nations Development Programme (2015) (IFRC/UNDP) legal checklist requires national laws to establish and promote the training of public officials and relevant professionals.

There is an emerging international law duty upon sovereign states to have substantive regulatory systems to ensure that risks from natural hazards are mitigated so that they do not endanger human lives. States must inform at-risk communities of the potential of unmitigated risks and establish sufficient co-ordination and cooperation between administrative authorities. This is an onerous duty and it is argued here that it would be prudent to assume that the general duty includes several more specific subsidiary duties. One would be to consider the merits of simulation exercises as a way of satisfying: (1) the specific requirements of national laws; (2) the general expectations of international law; (3) the education, training and knowledge-sharing goals of the Sendai Framework; and (4) the requirements of the IFRC/UNDP legal checklist.

1.3 Academic Support for Training and Simulation Exercises

The experience and levels of expertise of observatory scientists are critical to making accurate forecasts and training is important (McGuire and Kilburn 1997). In the context of volcanic risks, Doyle et al. (2015) undertook a review of the literature on emergency management team response, decision making, mental models and situation awareness and exercising. They argue that science agencies and science advisory groups must embark on a suite of training activities to enhance their response during a disaster. These should include exercise and simulation programmes within their own organisations rather than participation solely as external players in emergency management activities. Structures, resources and time must be provided for these programmes.

It is further argued by Doyle et al. (2015) that training will enhance the future response capabilities of both scientists and risk-mitigation agencies in several ways.

2 Methodology

This chapter reviews five simulation exercises (four of which were conducted as part of the VUELCO project), which for ease of reference are summarised in Table 1. It investigates the ways in which these exercises were planned and undertaken. By design, it does not address in any detail: (1) the features of the various and varying volcanic settings in which the exercises were conducted; or (2) the nature, scope or analysis of the monitoring data that were painstakingly created for the volcanic hazard scenarios that underpinned them.

This chapter draws from ethnographic observational data (recorded in hand written field notes) collected during five simulation exercises from the perspectives of overt non-participant

Table 1 Brief details of the analysed five volcanic unrest simulation exercises

Name	Hosting organisations country/when	Volcanic hazard scenarios and interpretation of observables
Colima	VUELCO Mexico (Colima) 17–23/11/2012	Volcan de Colima—evidence of moderate effusive activity alternating with moderate explosive events; escalating unrest, dome growth, some explosions, a 2 km + eruptive column, ash-fall, possible evolutions included: effusive dome growth alone, a large explosive eruption after dome collapse, partial flank collapse, a mixed 'Merapi type' event and cessation of dome growth; culmination in risk sufficient to require the evacuation of a single small town—La Becerrera
Campi Flegrei	VUELCO Italy (Rome) 11–12/02/2014	Campi Flegrei—conflicting evidence of either a rise of magma from about 7 km or an increase in shallow hydrothermal activity; further evolution involving a possible 3 km sill or continued shallow fluid migration; further rapid evolution raised the possibility of phreatic explosions and small volume magmatic eruptions in the eastern sector; further dramatic evolution made explosions and eruptions likely within days or weeks in the Bagnoli-Solfatara area and small offshore eruptions were not excluded; culmination in eruption between Bagnoli and Monte Spina and a sustained eruptive column, ash fall and pyroclastic flows south and east
Mount Teide	Presidencia del Gobierno de Canarias and Instituto Geográfico Nacional (IGN), Ministerio de Fomento, Gobierno de España Tenerife, Spain (Santa Cruz de Tenerife) 25/04/2014	Teide-Pico Viejo—evidence of rumbling noises, small debris avalanches and no seismic/deformation changes; further evolution with more rumbling noises, small swarm of seismic events; further evolution with swarm of small high-frequency VT, an explosive event, new thermal events and ash emissions with LFs and tremors; further evolution with pronounced SO_2 decrease interpreted as near-surface sealing of rising magma with high probability of eruption; culmination in eruption
Cotopaxi	VUELCO Ecuador (Quito) 13/11/2014	Cotopaxi—evidence of anomalous seismic activity with returns to background levels and no physical-chemical changes; possible local tectonic activity, fluctuating gas emissions, some deformation, further evolution with many long-period events, increasing released energy, sub-vertical fault in NE flank, increasing deformation, high gas emissions, increased fumarolic activity, increased thaw in upper parts of the edifice; further evolution to sporadic vulcanian explosions, eruptive columns <5/8 km plus tephra; further evolution with escalating seismic activity, inflation and degassing changes; culmination in eruption
Dominica	VUELCO Dominica (Roseau) 14–15/05/2015	Southern volcanic region including Trois Piton—evidence of elevated seismicity, landslides and small hydrothermal changes; further evolution with many VT events and landslides, a return to hydrothermal background levels plus Boiling Lake drainage; further evolution with more VT events and hybrids followed by VT decrease with hybrids dominating, continuing geochemical background levels, no recent deformation and contraction in some areas, indications of a deeper source than in the early unrest phases, possible pressurisation of deep seated magmatic system or hydrothermal activity; culminated in the consideration of a lowering of the Alert level

observers and, to a very limited but recorded extent, of participant-observers. Four exercises relate to VUELCO volcano unrest simulations in Mexico, Italy, Ecuador and Dominica. Detailed analysis of the documents prepared before, during and after these exercises was undertaken. The challenges of avoiding researcher bias and unintended observer effects were recognised.

Additional observational and documentary data were acquired by the lead author during the Tenerife (Spain) exercise as an invited external non-participant observer commissioned to prepare a post-exercise evaluation at the request of the Presidencia del Gobierno de Canarias. No ethical agreement was signed before or during that exercise. Data from that exercise and extracts from the post-exercise report are included by kind written permission of the Presidencia del Gobierno de Canarias.

Within the main body of this chapter, no attempt is made to provide complete details of the five exercises listed in Table 1. Essential background information and additional reading sources are to be found in additional files A–E which, for ease of reference, have deposited at the 'Collaborative volcano research and risk mitigation' (VHUB) website under reference "Volcanic Unrest Simulation Exercises: checklists and guidance notes—Additional files A-E". The additional reading sources are also listed at the end of the references for this chapter.

3 Background

3.1 VUELCO Themes and Goals

As VUSE are purpose-driven learning activities, each will have tailored goals and an overall design based upon the needs of its participants. Consistent with VUELCO's stated goals, the principal purpose of each VUELCO exercise was to present a realistic simulation of the evolution of past and hypothetical future volcanic unrest events associated with a host volcano or wider volcanic setting. In the manner of a film set, a simulated volcanic event is the dynamic backdrop against which a selection of risk governance infrastructures, policies, procedures and people is tested.

The VUELCO exercises involved the provision of raw or partially analysed monitoring data and information to geo-scientists. Their role was to undertake a scientific analysis of the data with the purpose of passing on characterisations of likely volcanic hazard scenarios, by means of hazard communications, to risk mitigation decision-makers (civil protection authorities and the representatives of at-risk individuals) and the mass media. The civil protection authorities, having considered not only societal issues but also political, economic and other values, made and communicated risk mitigation decisions. In some exercises, representatives of at-risk communities, emergency services and relevant government entities played active roles in testing emergency, law and order, rescue, medical and evacuation procedures.

3.2 Checklists and Guidance Notes

3.2.1 *Checklists*

The checklists within this chapter are presented in response to the perceived importance of such documents as sources of good practice in the eyes of recent commentators (e.g. Gawande 2010; Newhall 2010; IAVCEI 2015). For example, within the 'Safe Surgery Saves Lives' project, which started in 2007, the World Health Organisation (WHO) introduced a Surgical Safety Checklist developed by Dr. Atul Gawande based, inter alia, upon the success of pre-flight checklists in enhancing safety within the aviation industry. "A systematic review and meta-analysis of the effect of the WHO checklist" strongly suggests a related "reduction in post-operative complications and mortality" (Bergs et al. 2014, 150; Treadwell et al. 2014).

Drawing upon the research of Gawande (2010), this chapter's checklists aim to incorporate several critical features that are output and outcome-focussed. As far as reasonably practicable, the authors' aim has been to ensure that each checklist is: (1) concise and preferably short as well; (2) simple, precise and unambiguous;

(3) targeted by addressing only evidence-based priorities that are considered either critical or significant to risk governance; and (4) non-prescriptive and non-comprehensive. Checklists may encourage rational, systematic (i.e. both consistent and complete), routine and transparent practices whilst recognising the importance of, and encouraging careful attention to, a wide variety of constraints and expectations. However, it is not envisaged that they should be used as a regulatory device, an enforceable legal requirement, or part of a blame-avoidance strategy (Hood 2011; OECD 2015).

The checklists presented in this chapter address the general planning, funding and execution of purposeful simulation exercises. The exercises also highlighted the many challenges of hazard communication and the difficulties that may result if the needs of other stakeholders are not identified and responded to.

This chapter presents six checklists namely: (1) Planning; (2) Logistics; (3) The Volcano team; (4) The Scientific Advisory Committee (SAC); (5) The Risk Managers - Civil Protection Authorities (CPA); and (6) The Observers/ Auditors. The first two relate to exercise 'phases' and the other four relate to 'roles'. It is not suggested that these checklists will be relevant to all simulation exercises, however, they will provide an indication of some of the critical phases and major roles that must be considered.

The checklists inevitably reflect the fact that VUELCO's exercises placed particular emphasis on the scientific analysis of monitoring data, the communication of hazard characterisations to risk managers, and interactions between scientists, risk managers, the media and the general public. It is accepted that future exercises may have other goals for which more or different checklists may be helpful.

The 'Planning' checklist encourages the identification of goals and objectives, and the need for clear leadership, careful design and realistic financing.

The 'Logistics' checklist is probably the most important. The actual task of arranging the 'planned' exercise may have many benefits in its own right as it will require the careful identification of key laws, policies, procedures and people.

The 'Volcano team' has the very difficult obligation to craft and deliver monitoring data consistent with realistic volcanic hazard scenarios. The data must be suitable to meet the hazard analysis challenges confronted by the experts within the 'Scientific Advisory Committee' (SAC). It is noted here that, in some situations, the roles of monitoring and analysis may be undertaken by the same team or with substantial personnel overlaps. This chapter retains the separation to cover circumstances where there are distinct group remits, functions and responsibilities.

The SAC receives and handles the data provided utilising, as appropriate, a variety of soft skills (including those of analysis, deliberation and communication), tools (including expert elicitation and probabilistic models) and established procedures and protocols.

The 'Risk Managers - Civil Protection Authorities' (CPA) make risk mitigation decisions based in part upon scientific communications received from the SAC or, where the existence of a SAC is not foreseen within the relevant governance system, the Volcano team. An important part of the CPA role is usually the challenging duty, not only to advise individuals and entities driven by political values, but also to interact with members of the public and respond to mass and social media demands. The press and social media "can play an important role in the dissemination of information, true or false" and social media can rapidly ferment public, anxiety, distrust or dissent (OECD 2015, 37).

Simulation exercises are learning/training exercises and, accordingly, the importance of 'Observers' should not be underestimated. If properly briefed, observers and auditors will not only contribute their own candid views about all aspects of the exercise but also encourage other participants to be reflective about their own contributions and actions.

3.2.2 Guidance Notes

To support and supplement the checklists, it is suggested here that guidance notes should be issued from time to time to provide a dynamic and helpful knowledge and innovation resource. Their aim should be: (1) to gather together, record and share the accumulated experience of other practitioners in relevant fields of expertise; and (2) to suggest ways to find optimal solutions to the most critical issues.

It is readily accepted that the observations within the guidance notes are inevitably subjective. They are not intended to be in any respect either comprehensive or prescriptive. They are presented as options to be considered along with other issues that will be found within best practice guidelines such as those identified in Doyle et al. (2015).

4 Checklists (in bold italics) and Guidance Notes (in normal font)

4.1 Planning

Leadership

Who (individual and entity) has overall responsibility for planning and delivering the exercise?

The choice of exercise leaders should be dictated by the planned goals and activities of the exercise. If a significant involvement of civil protection authorities is planned (for example, where risk mitigation decisions are to be made and/or related mitigation measures are to be tested) those authorities should probably take the lead.

The VUELCO exercises have suggested that overall responsibility for a VUSE should be assigned to just one person within the risk governance stakeholder (e.g. the civil protection authority) likely to gain most from it. That person, who will need suitable and sufficient support from a working team or steering group, should have the gravitas, personality, authority, experience and resources (both human and financial) needed to plan and handle a complex high profile project that will inevitably attract political, societal and media attention.

The MIAVITA Handbook (2012, 118) suggest that it is good practice for "a steering group to be in charge of co-ordination and leadership of the various preparatory activities".

After one VUELCO exercise it was suggested that at least "a full-time scientist" should be dedicated to preparing any exercise that focusses on scientific analysis.

Purpose and Goals

What is the overall purpose of the VUSE?

What are the short and long term goals of the exercise and its players?

A VUSE is a learning activity and the principal reasons for it must be identified and stated. It should respond to the perceived core needs of the participating stakeholders and, like any learning activity, be carefully planned with stated assumptions, aims, objectives and themes.

Before the exercise, each and every participant should know what they will learn from the exercise. Thought should also be given to how and by whom the success of the exercise will be evaluated and what will be done by whom to build upon the exercise.

The assumptions, aims, objectives and themes should be set out in the pre-exercise briefing note referred to in Sect. 4.2 of this checklist.

Scope

Which parts of the risk governance process are going to be tested?

A VUSE should be focussed, purpose-driven and planned accordingly.

No VUSE can realistically attempt to replicate all aspects and phases of, and all stakeholders interested in, a societal risk governance regime. It is possible that more regular exercises, which concentrate upon very carefully defined aspects of the system (e.g. communicating in real time with the general public and the mass media), may be more cost-effective and beneficial.

Consideration might be given to those issues and functions that require the most inter-stakeholder planning, cooperation and collaboration and accordingly excellent means of communication. A VUSE can and should be a learning exercise in communication between many key players such as:

- Geoscientists and other geoscientists.
- Geoscientists and risk governance advisers (e.g. weather forecasters, aviation and marine space managers, communication specialists etc.).
- Geoscientists and other stakeholders (e.g. individuals within civil protection authorities, interested and affected members of the public, representatives of community, religious, public utility and commercial interests, and representatives of international, national, regional, municipal and other levels of government).
- Local geoscientists (e.g. volcano observatory staff) and visiting scientists (e.g. academics contributing to a scientific advisory committee) and visiting researchers.
- Hazard analysts (e.g. local and visiting geoscientists) and experts in risk assessment and management.
- All scientific analysts/risk decision makers and (1) the general public and (2) the mass media.

In 1999 several of these interactions were addressed by IAVCEI's Subcommittee for Crisis Protocols, which issued a report entitled "Professional conduct of scientists during volcanic crises" (Newhall et al. 1999). In the context of the governance of the risks of volcanic hazards, this paper represents a rare, if not unique, example of an attempt to offer authoritative practice guidance based on past events and an extensive literature review. It was concerned principally with personal and institutional interactions during a volcanic crisis.

Consideration should also be given to whole or part of an exercise being in 'degraded mode'. The MIAVITA Handbook (2012, 118) suggests that exercises should be planned to test system level capabilities of response when some parts of the system are not fully operative. By this means, depending upon which aspects are being tested, allowance can be made for, inter alia, holidays, the malfunctioning or destruction of monitoring or telecommunication equipment, blocked escape or rescue routes, and weather conditions.

By way of illustration of needs that an exercise might seek to target, Exercise Capital Quake identified eleven functions—"Public Information Management, Governance, National Financial System, Logistics and Other Support Co-ordination, International Assistance and Liaison, Rescue, Health, Welfare, Building Safety, Restoration of Access and Restoration of Lifelines" (NZ/MCDEM 2008).

Another possible example comes from the "Metodo Augustus", the organisation of emergency management, used at the Italian Department of Civil Protection. That organisation provided several support functions including technical-scientific, health and veterinary assistance, mass-media and information, volunteers, means and materials, transportations and mobility, telecommunications, essential services, damage assessment, operational structures, local administrations, dangerous materials, people assistance, cultural heritage and coordination (Galanti et al. 2006).

VUELCO chose a science-orientated focus for its four exercises concentrating upon the interfaces between:

- Hazard monitoring and hazard assessment:
  - Long-term monitoring
  - The host volcano's main precursors of volcanic unrest.
  - The host volcano's short-term monitoring resources (e.g. equipment, employed staff and volunteers etc.), the nature, adequacy and timing of the data output, and their capacity to respond to changing demands.
- Hazard assessment and risk assessment
  - As the period of unrest evolves: (1) real time characterisations of the host volcano's possible and most likely hazards and their temporal, physical and spatial parameters; and (2) other advice e.g. the merits/safety of further/different short-term monitoring.
- The communication of scientific analysis (with its inherent assumptions, limitations, complexities and uncertainties) to a variety of stakeholders each possibly having different requirements and expectations and related communication challenges.

If an exercise focusses predominantly upon scientific issues, it may be very difficult to engage participants from civil protections agencies and more 'risk' related functions.

A VUSE can be a good opportunity to test the processes by which scientific analysis is integrated into risk assessment agencies, such as civil protection authorities, that do not have embedded scientists.

What is the proposed active participant scope?

In a VUSE, if time and resources permit, a wide range of stakeholders can be represented and actively involved including, but not limited to:

- the host volcano as the source of scientific and visual data, which are provided sequentially in discrete pre-planned "phase" briefings/reports (hereinafter called the Volcano team);
- host volcano observatory scientists (the VOS);
- local and external scientists (collectively called the scientific advisory committee or SAC);
- risk assessors and managers possibly from a range of national, federal, state, regional and municipal tiers of government (hereinafter called civil protection authorities or CPA);
- the media;
- interested and affected members of the public.

In a more sophisticated exercise, it may be worthwhile ensuring that a "dissenting/minority view" and/or "maverick" scientists are represented to test robustly relevant democratic and communication processes.

The MIAVITA Handbook (2012, 114–116) details the benefits of informing, sharing and training involving: (1) national, regional and local authorities; (2) scientists; (3) volunteers; (4) the media; (5) pupils and students; and (6) the public.

The VUELCO Colima exercise incorporated the planned evacuation of an at-risk community. A VUSE can usefully focus upon the real time implementation of the risk mitigation strategies that result from the dynamic characterisation of an evolving unrest. However, the time and resources required to identify and engage multiple stakeholders, including members of the public, should not be underestimated.

Care must be exercised if one of the hazard scenarios will result in the need for the evacuation of representatives of at least one vulnerable community. Of course, this will have to be organised and announced well in advance of the start of the exercise. This may affect (i.e. make rather implausible) other parts of the exercise which may depend upon realistic and continuing uncertainty as to the future evolution of the initial hazard.

What is the proposed passive participant (e.g. observers and monitors) scope?

In the VUELCO exercises, very valuable contributions were made by monitors and observers. They can be from the participating organisations or entirely 'independent'. Monitors and observers (particularly with experience of previous exercises and the practices of countries other than the host country) can play a critical role in providing both hot and cold feedback whilst also gaining invaluable personal experience to assist in the planning of future exercises.

VUSE are invariably observed by local and visiting students and early career academics. With careful planning, it may be possible for less experienced scientists and civil protection authority decision makers to be involved directly (as happened in the Campi Flegrei VUSE) or indirectly in the 'shadow' SAC and CPA teams.

Reference should be made to the separate Observers/Auditors checklist in Sect. 4.6.

What is the proposed geographical scope?

The geographic scope of a VUSE should be considered carefully. A balance must be achieved between the spatial parameters that produce a risk exposure area that is realistic in the context of the host volcano and those that create an area that is too large in terms of the duration of the exercise, data production and/or the roles of the active participants.

What is the proposed administrative scope?

National laws tend to identify, authorise and fund risk governance bodies (e.g. government departments and agencies, public corporations) and public officials (e.g. individuals such as governors, mayors, prefects and village heads) within a coherent legal and administrative framework—a risk governance infrastructure.

These laws often use and build upon existing entities with existing administrative frameworks that have multi-level national, regional, district, municipal etc. political divisions and subdivisions (Bretton et al. 2015).

The MIAVITA Handbook (2012, 118) suggests that it is good practice for an exercise to "be based on the regulations and laws of the [host] country".

The administrative scope of a VUSE should be considered carefully and it is likely this scope will be related to those parts of the risk governance process selected for review and testing. It may be very difficult to involve actively all levels of governance and therefore a decision will have to be made as which levels will participate and which will merely observe.

It is likely that a good starting point will be a complete flow diagram of the existing societal risk governance infrastructure for the host volcanic region. This can then be annotated to indicate which national, regional, local bodies and individuals will participate and their respective roles (both active and passive) in the exercise.

For the SAC and CPA at least, consider setting out clearly its respective:

- legal status;
- legal remit and reporting lines; and
- the rights, responsibilities and liabilities of its members.

Duration and type

How long will the exercise last and why?

This is a very important issue with funding, resources, logistical and many other implications.

VUSE vary greatly in length as shown by the data in Table 1. Much will depend upon the critical decisions that must be made about scoping—process, participants, geographical area and administrative levels.

What type of exercise will be undertaken and why?

There are many types of management exercises including: (1) full-scale; (2) reduced; (3) orientation; (4) drill; (5) table-top/discussion; and (6) functional (see Doyle et al. 2015).

Exercises can be also announced or unannounced depending upon their objectives. It has been suggested that the use of unannounced exercises is necessary to verify the real strength of systems and levels of preparedness, when people do not expect them (MIAVITA 2012, 116–117)

How frequently should exercises be held?

The MIAVITA Handbook (2012, 116) notes that exercises are fundamental for testing existing procedures, plans, and preparedness and for maintaining the attention of stakeholders "on the spot". It also argues that exercises should be scheduled frequently with the frequency depending upon several factors including: (1) the behaviour of the host volcano; (2) the social-economic context; (3) the levels of risk perception; and (4) the democratic trend.

Finance

What are the planned budgets for all phases of the exercise including:

- ***Before?***
  Consider planning, field trips, printed field guides, data and scenario creation, documentation etc.

- ***During?***
  Consider venue, catering, data/scenario communication etc.

- ***After?***
  Consider feedback (hot and cold) and follow up meetings, reports, presentations, interviews, actions.

Who is funding:

- ***The exercise?***
- ***The participants, including invited guests, experts and invited observers/auditors (including travel, accommodation, field trips, other expenses etc.)?***

Consider whether there should be an underwriter of last resort for any unplanned deficiencies.

Do the funders have any demands, expectations etc. as to any aspects of the exercise?

If yes, these must be understood by the organiser, planned and budgeted for.

4.2 Logistics

How long will the planning stage take?

It would be easy to underestimate the time that will be needed for, and the complexity of, the planning stage upon which the whole success of the exercise will depend. It may be prudent to speak to the organisers of other VUSE to discuss this and other financial/practical issues. Reference could also be made to Dohaney et al. (2015).

The MIAVITA Handbook (2012, 118) notes that planning an announced exercise takes time and states "six to twelve months are necessary to prepare a full-scale exercise. If the promoted exercises are repeated on a fixed schedule, three can be sufficient". We consider these to be the minimum planning periods.

Will every participant/observer receive in advance a full briefing note covering:

- ***Purpose***

  Refer to the Sect. 4.1 regarding the overall purpose of the training exercise.
- ***Scope***

  Refer to the Sect. 4.1 regarding scoping. Issues covered may include processes, participants, geographical, administrative etc.
- ***Themes***

  If some important themes have been identified for particular attention (e.g. expressions (numerical and narrative) of likelihood, expressions of analytical/diagnostic confidence, dealing with communication network failures, using social media etc.), it might make sense to mention these in advance to encourage pre-exercise thought and preparation.

- ***Models, methods and protocols to be used/made available for use***

The VUELCO exercises have proved that VUSE can be used successfully to test in simulated real time the merits of probabilistic methods and tools such as BET_EF (Bayesian Event Tree for Eruption Forecasting), HASSET (Hazard Assessment Event Tree) and QVAST (Quantifying Volcanic Susceptibility) (Sandri et al. 2009; Sobradelo and Bartolini 2014; Bartolini et al. 2013).

In advance of the exercise, there must be communication between the modellers and the Volcano team. This should ensure that all appropriate monitoring data (perhaps deliberately not all the data) is available for all relevant phases of the exercise scenario including periods between phases.

Feedback from the VUELCO exercises suggests that input from the modellers during each step of an exercise may be welcomed by other participants. Testing simultaneous integration was a specific VUELCO goal. However, if this is to be done, it must be carefully planned and sufficient time allocated within the timetable.

It is important to state in advance of the exercise, for example in the briefing documents, what role the models, tools etc. will play during the exercise. Consideration should be given to whether the results will be available in real time to the participating SACs and CPAs. If yes, how and when will the results be communicated? If no, when will feed back regarding the utility of the items tested be communicated?

VUSE can also be used to give SAC's the opportunity to use formal expert elicitation (EE) methods. EE may provide formality and direction to the deliberation of monitoring data, assist the framing of likely hazard scenarios, and facilitate the drafting of expressions of temporal certainty and analytical/diagnostic confidence.

EE were conducted in the VUELCO exercises in Colima and Campi Flegrei.

The Somma Vesuvio MESIMEX exercise undertaken in 2006 was used to conduct a before and after volcanic risk perception survey (Ricci et al. 2013).

- ***Players (participants, observers etc.):***
  - ***by name and organisation***
  - ***by role in the exercise***

It is very likely that a large number of people will be involved from a wide range of organisations and, perhaps, several countries. It would be prudent to assume that nobody will be known to all of the other participants. Consider the use of name badges stating Name, Organisation, and Exercise Role. As found during the Campi Flegrei exercise, colour coding for teams and active/passive roles might also be helpful. The scheme adopted should be explained carefully to save time during the start of exercise when valuable time can be lost easily.

By way of illustration, more than 100 participants from over 10 European and Latin American countries attended the VUELCO Campi Flegrei exercise. The respective figures for the VUELCO Cotopaxi exercise were about 50 participants from 13 countries.

After the Dominica exercise it was suggested that, at the very start of the exercise, introductions would have been helpful "to establish who each person was and what their role was…It was challenging accepting information from a person whom we did not know or have any clue as to their background or capabilities. Also the scientist [s] must be aware of who the[y] are working with in order to frame their advice appropriately".

For numerous purposes (before, during and after the exercise) a comprehensive, up-to-date and accurate email list containing all participants will be necessary. Inadequate email lists caused difficulties during several VUELCO exercises.

- ***Teams***

Consider how many teams will be required and how and when their members will be allocated. For the sake of simplicity, it may be necessary to ask representatives of several different organisations to work together. By way of illustration, this chapter refers to four teams (Volcano, SAC, CPA and Observers) but others can and should be used as dictated by the specific goals of the planned exercise.

Consideration should also be given to the use of a technical team to support the SAC's deliberations in respect to issues such as real time GPS mapping, model trials and expert elicitations. The time needed for these complex matters must be considered particularly during short exercises. In VUELCO's Campi Flegrei simulation, a technical team composed of civil protection personnel operated continuously with the aim of supporting the SAC's simulated 'real-time' deliberations. The team's results were only presented at the end of the exercise because they lagged behind the 'real time' evolution of the phases of the exercise's hazard scenarios.

- ***Venue (full address, plan, transport and contact details)***
  - ***Sub-venues for all aspects of the exercise (including any evening events) with a plan***
  - ***Layout for the plenary sessions***
  - ***Layout for any breakout sessions***

These critical issues should not be overlooked. The exercise organiser should decide in advance, in respect for each phase of the exercise, which 'team' will sit where and why, probably by reference to which other participants they will have to communicate with. For the sake of intended realism, consider whether they should be near or far apart and by what means they will communicate.

Consider whether, in a real emergency, participants would speak to each other face to face. Consider whether it would be more realistic to separate some teams physically in order to force the use of video conferences, emails and phone calls in team interactions. At the planning stage it will be necessary to allow more time for this nuance, which may highlight the challenges of long-distance communication, data sharing, data analysis, collaboration, consensus building and decision-making.

Feedback from the VUELCO exercises specifically mentioned the problems associated with having too many people in the same room and failing to plan in advance who should be where and near to whom.

It is worth considering that each refreshment break will take longer than expected unless refreshments are either provided in all rooms or available at all times. Will working lunches simulate the difficulties of a tense period of emerging unrest? For a short period of intense training, is a degree of inconvenience and/or discomfort in fact to be encouraged?

- ***Timetable***
- ***Dates for each part of the exercise and timings for each day addressing:***
- ***Registration (including perhaps the handing out of name badges in accordance with the stated scheme)***
  This will need very careful planning and resourcing to avoid a late start on Day 1.
- ***Start, finish***
  Start and finish promptly.
- ***Food/drink breaks (if needed!)***

Refer to the above guidance about the need for these in an exercise that attempts to simulate real time challenges.

- ***Evening events, banquets etc.***

Consider how long it will take participants to return to their accommodation before evening events. Give clear advice about the dress code, if any, and other cultural expectations (e.g. the use of cameras or the need for modest clothes and head coverings).

- ***Venue tours***

It is likely that many of the participants will expect/request a tour if the exercise takes place in a major risk management/communication centre. Allow plenty of time for this and consider making these visits part of the pre-exercise field trip to avoid wasting time when all of the participants (and probably the media) are present.

- ***Greetings, Introductions, Reviews, Thank you's etc.***
- ***Presentations to or by representatives of central, local government, etc.***

Predicting accurate timings for these two components is notoriously difficult and careful planning in advance and strict control on the day will be necessary. Consider carefully whether these are really necessary. Consider whether it would be better for relevant entities to be given formal active or passive roles within the exercise itself and be appropriately involved in that way.

- ***Dissemination of new data***

Consider when, by whom, to whom and how. During the VUELCO exercises, difficulties arose due to format incompatibilities and internet access issues.

- ***Deliberation of new data***

The time necessary for deliberation may depend upon the amount/ format of the data, the need to communicate with other participants, expectations about output deliverables (e.g. written reports) etc. As a general rule, allow plenty of time within the context of the realism that the exercise seeks to simulate.

- ***Dissemination of 'injects'***

Consideration should be given to the timing and drafting of 'inputs' (for example, requests for local, national and international media interviews or briefings with concerned public officials, unwelcome social media exchanges, etc.), which will be distributed to the participants at planned stages during the exercise. Participants might be asked to think about how they would respond and who would be delegated to meet these urgent requests.

- ***Outputs from the SAC and other teams***

As a general rule, allow plenty of time for the dissemination of outputs within the context of the realism that the exercise seeks to simulate.

Will information about current CPA mitigation actions, risk alert levels etc. be provided at any time during the exercise?

If yes, why, what, when and to whom?

Feedback from the VUELCO exercises suggests that scientists benefit greatly from improved knowledge of civil protection authority systems and in particular better appreciation of those hazard parameters that are most relevant to risk mitigation decisions.

- ***Factual background and related resources***

Consider whether a field trip is necessary before Day 1 of the exercise so that all of the participants can gain a basic grasp of the topography, geography, geology (including major structures, faults, tectonic plates, existing volcanic and other hazards, and aquifers), geo-history (a time-line of major events is often helpful), infrastructures and other essential information.

All the VUELCO exercises incorporated well-planned and informative pre-exercise field trips that were supported by printed, carefully researched field guides with relevant histories, maps, reading lists etc.

A field guide can be supplemented by more detailed essential information about the monitoring history, thresholds etc. Feedback from VUELCO exercises suggests that the briefing pack should include maps showing the positions of all relevant observatories, monitoring equipment and stations, GPS positions, cameras, sample sources etc.

In addition to a field trip, it may be helpful to have a day/half-day of short presentations about the host volcano with careful oblique references to features that will be relevant during the exercise.

- ***Real/assumed legal/regulatory framework and related duty holders***

Each participant must have a clear and comprehensive understanding of their role, the roles of all other participants and all planned lines and methods of communication. To ensure a higher degree of realism, these roles should accurately reflect the governance infrastructures, roles and duties required by:

- the national laws of the host country as encouraged by the UN Sendai Framework and the IFRC legal checklist;
- any regional emergency management or response arrangements, such as those in the Caribbean involving the Caribbean Disaster Emergency Agency (CDEMA), which are set out in inter-country agreements, memoranda of understanding and protocols; and
- any applicable international law standards.

- ***Information Technology (IT) including disseminating/sharing data****

This critical issue should not be underestimated based upon difficulties encountered during VUELCO exercises.

It is inevitable that most of the participants will wish to have easy access to the internet. Ensure that the Wi-Fi network has sufficient capacity and range to allow easy access.

Consideration should be given to the format requirements of the computer models and probabilistic tools that will be tested during the exercise.

- ***Language arrangements****

Consider the dominant language for the exercise and its documents (particularly the pre-exercise briefing pack) and what arrangements can be made, if any, for translations/translators.

Feedback from the VUELCO exercises suggests that some participants (e.g. locally based risk managers and members of the emergency services) may become bored and disengaged if they cannot understand fully what is going on and contribute.

- ***Other equipment****

Consider whether it will be necessary, for the purposes of training, publicity or planning future exercises, to use other equipment including:

- Display screens, smart boards and white boards

- Fixed and roving microphones
  A shortage of roving microphones caused irritation at several VUELCO exercises.
- Visual and audio recording equipment
- A projector for PowerPoint presentations
  Feedback sessions are enhanced by such presentations, if time allows for their preparation.
- Laser pointers
- Video conference links

- ***Presentations of data, results etc.****

How, why and when will the data, results etc. be presented?

Consider imposing a strict timetable to avoid timing problems.

- ***Requirements regarding pre-Day 1 reading, preparation, queries etc.***

Consider stating clearly that there will no lengthy introductions or briefings at the start of Day 1 because the participants are required and expected: (1) to read in advance the pre-exercise briefing pack; and (2) to raise any queries before Day 1 with the exercise organiser.

- ***Procedures for daily and end of exercise feedback.***

If daily feedback or announcements are necessary, consider in advance what issues are likely to be covered, who will give them and allow adequate time within the timetable.

*Note

It is accepted that during a real emergency these facilities may not be readily available. Accordingly, any difficulties encountered during an exercise may provide helpful insights into the challenges that would be encountered during a fast evolving crisis.

Will role/team leaders receive in advance?

- ***A full pre-exercise briefing note about the overall legal/regulatory infrastructure for the exercise?***
- ***Checklists for the main aspects of the exercise?***

Day 1 (Start of exercise)

Ensure that registration does not delay a prompt start, which will thereby set the standard for all other timings during the exercise.

Who will lead the exercise and ensure that the timetable is adhered to strictly?

Consider giving this role to a person (probably supported by an assistant) who does not have any other role in the exercise and therefore can move easily from room to room and deal promptly with any difficulties that arise.

Is there a need for a short introduction? If yes, why, who will give it and how long will it last?

Is there a need for a short end of day summary? If yes, why, who will give it and how long will it last?

Day 2 etc. (Continuation of exercise)

Is there a need for a short introduction? If yes, why, who will give it and how long will it last?

Is there a need for a short end of day summary? If yes, why, who will give it and how long will it last?

Day (End of exercise)

Is there a need for a short introduction? If yes, why, who will give it and how long will it last?

Is there a need for a short end of exercise summary? If yes, why, who will give it and how long will it last?

4.3 The Volcano Team

Membership

Who will lead the team and why?

Who will be the other Volcano members?

VUELCO's exercises indicated that considerable time, exceptional skills and infinite patience are required to write any coherent and challenging hazard scenario.

Members of the Volcano team must: (1) be selected and appointed at a very early stage in the planning process; and (2) be aware of, understand and accept the needs and goals of the exercise.

The overall hazard scenario and its planned phases must be tailored to fit within the agreed timetable for the exercise. The Volcano team will need to work very closely with the exercise leader and the steering group as soon as it is formed.

Where the Volcano team is the host volcano's actual monitoring team, a chance to test the monitoring team in 'real time' action will be lost.

Preparation

How long will be needed to prepare:

- ***the scenario/s?***
- ***all of the briefing documents?***
- ***the pre-exercise field trips?***

The exercise

What will be the time scale in months/years covered by the exercise?

VUELCO's exercise scenarios varied greatly in length—Colima (2 months), Campi Flegrei (7 years), Cotopaxi (5 years) and Dominica (2 years).

The period of time covered by the simulation, the number of phases and the length of the intervals between the phases, must take into account the needs and goals of the exercise. For example, consideration must be given to the time that will be required for data to be delivered, entered and assimilated, and for probabilistic models to be run.

How many exercise phases will be needed to cover the period of years chosen?

Each phase will involve, and must be allocated sufficient time for, the delivery, entry, analysis and modelling of the monitoring data and the drafting and delivery of any expected outputs (e.g. reports, press releases etc.).

VUELCO's exercise scenarios had between 3 and 6 phases - Colima (2 months in 4 phases over 10 h spread over 5 days), Campi Flegrei (7 years in 4 phases over 3 days), Cotopaxi (5 years in 6 phases in 1 day) and Dominica (2 years in 3 phases over 2 days).

As already stated, the precise specifications for the phases must be agreed with the exercise leader taking into account the needs of the modellers and other technical teams.

What data will be provided?

The monitoring data must be suitable and sufficient to test the whole range of geo-scientific disciplines represented in the SAC. Feedback from the VUELCO exercises identified the temptation to favour the overprovision of seismic data at the expense of adequate geochemical, geodetic, petrological and other data.

Although this might reflect real situations or even the architecture of existing monitoring networks, careful consideration should be given to the consequences of having experts from some disciplines having insufficient data to occupy them and therefore becoming disengaged from the exercise.

It may also be necessary to provide in the briefing pack related historical data and background information (such as historical thresholds) to allow the data used during an exercise to be considered in context.

The data must remain coherent, albeit deliberately unclear and uncertain, over the full duration of the scenario including periods of inactivity. Feedback after the Dominica exercise praised the quality of the data and the design of the scenario. The following expressions illustrate some of the features of a successful scenario—very realistic unsure signals creating a state of limbo, non-linearity, phreatic evidence that was not a precursor of magmatic activity, complex data and realistic missing data.

During some phases of the exercise scenario it is possible that the data will be incomplete (some critical data may be held back deliberately to await a request for more data) and/or difficult to analyse because it may be a stated objective of the exercise to test the SAC's ability to handle uncertainty, disagreement, monitoring inadequacies etc. and/or to request data from additional or different monitoring.

If data is provided for the periods between scenario phases, it is likely that the longer the inter-phase period the more complete the data will be. Time will have to be planned for data delivery and analysis. A large amount of data may also create very great difficulties for the utility of probabilistic tools such as BET.

The Volcano team should communicate with the exercise leader and modellers well before the exercise to avoid unnecessary surprises regarding the format and quantum of the monitoring data and the timing of their release. In the Colima exercise, three BET nodes were passed before Phase 1 of the exercise scenario and this degraded the utility of the BET trial.

How will the years between the periods that are covered by the data be described?

This difficult issue should not be overlooked. In the absence of data, what will the participants assume in absence of guidance? An absence of data may also create very great difficulties for the utility of probabilistic tools such as BET.

Consider issuing, along with the monitoring data for the next phase, a brief summary of the 'missing periods' in terms of volcanic activity, precursor evidence and/or the risk mitigation decisions and actions of civil protection authorities.

Will the exercise scenarios be (1) entirely fictitious; (2) based on suitably disguised real events; or (3) a mix of the two?

During the VUELCO exercises feedback suggested that:

- it may be very difficult to create realistic scenarios that are entirely fictitious because the data, under the very close expert analysis that they will receive, may appear to be inconsistent and improbable;
- a coherent and plausible chronology for the scenario is essential;
- after a period of sudden emerging unrest, a period of reducing unrest or quiescence may form the basis of a challenging scenario as proved during the Dominica exercise;
- in order to simulate reality, the Volcano team should provide monitoring data to the SAC without any form of analysis or interpretation, however the provision of tables comparing data sets may be helpful;
- the amount, relevance and format of the monitoring data provided should be considered carefully.

Will the exercise introduce secondary hazards (e.g. forest fires, contaminated aquifers, etc.)?

During the Mount Teide exercise, considerable attention was paid to secondary hazards. If secondary hazards are to be included, what data need to be provided before, at and after the start of the exercise?

Will plans be needed showing geographical, geological or societal features?

Will the exercise use fixed/dynamic data about weather, ground conditions etc.?

If yes, what data need to be provided before, at and after the start of the exercise? Will plans be needed showing geographical, geological or societal features?

Relevant weather/ground water data will be essential if ash dispersion and fall-out simulations are to be included within the exercise.

When, how and to whom will the monitoring data be disseminated:

- ***Before the start of the exercise?***
- ***At the start of the exercise?***
- ***During the exercise?***

This is a critical issue and important lessons were learned during the VUELCO exercises. During a VUSE, reliance on a limited Wi-Fi link may be very problematical.

In the Dominica exercise, monitoring information was provided within separate 'phase-specific' documents in password-protected pdf format and emailed before the start of exercise to all of the participants. At the beginning of each scenario phase, the relevant password was released. This solution worked very well.

During the Cotopaxi exercise, Wi-Fi and other communication difficulties were identified by the civil protection authorities and, as a result, they were better placed to improve their existing systems and resources.

Consideration should be given to the format requirements of the computer models and probabilistic tools that will be tested during the exercise.

Feedback from the VUELCO exercises suggests that when data are provided they should not duplicate data that have already been provided.

Will the SAC be able to ask for further monitoring data?

If yes, when and how can it be requested? If yes, when and how will it be provided?

Will any player other than the SAC be able to ask for monitoring data?

If yes, by whom, why, when and how can it be requested? If yes, when and how will it be provided?

Will anything other than monitoring data be provided?

Consider whether the SAC and CPA should be given, or have access to when they request it, fixed/dynamic data about weather, ground conditions etc.?

If yes, why, what, when and to whom?

4.4 The Scientific Advisory Committee (SAC)

Is a SAC necessary in every exercise?

Careful consideration should be given to this important question. It is appreciated that advice to civil protection authorities may only be provided by the head of the local volcano observatory, who is not, and would never be supported by a SAC.

Feedback from VUELCO exercises suggests that it makes good sense for exercises to simulate existing risk governance structures as far as reasonably practicable. It follows that a SAC team should be included in an exercise only: (1) where one exists, or could exist, in a period of evolving unrest or (2) where the civil protection authorities wish to test the utility of one.

If an exercise SAC is simulating the actions of a real SAC, the brief to the SAC and the briefing note for the exercise should set out clearly the SAC's constitution, mandate, responsibilities and powers.

Legal status and duties

A VUSE can be a good opportunity to consider the legal status of scientific committees and their full-time, part-time and seconded members.

Consideration should be given to the taking of appropriate steps to avoid or reduce managerial risks by the inclusion of disclaimers and exclusion statements in hazard assessments. A VUSE may serve as, at least, a prompt for committees to seek legal advice from a competent local lawyer.

During the Dominica exercise, a member of the Scientific Advisory Committee sought legal advice from the lead author about: (1) the wording of a liability disclaimer for the benefit of visiting scientists; and (2) the way in which 'risk-related' advice could be passed to civil protection authorities without liability for subsequent risk mitigation decisions. In respect of the first, it is likely that a visiting scientist will have little or no control in respect of either the absence or adequacy of monitoring data and the selection and competence of other committee members. Based upon this advice, a disclaimer was included in the Phase 1 hazard assessment report and the Phase 2 and 3 reports were more carefully worded.

Membership

Who will be leading the SAC and why?

The SAC will require a leader who should be identified in advance.

Feedback from the VUELCO exercises suggests the role of Chairperson of the SAC may be critical to the success of an exercise.

Consider whether a deputy Chairperson should be identified in advance of the exercise.

In several of the VUELCO exercises, it became evident that the Chairperson should not be given, or retain, prime responsibility for the initial drafting of the hazard assessment report.

Who will be the other SAC members?

Consider whether it is desirable to have a range of experts to simulate the issues that might arise during a real period of emerging unrest. Consider the inclusion of: (1) local volcano observatory scientists; (2) other host country scientists; (3) host country academics (4) non-national observatory scientists; (5) non-national academics; and (6) young researchers.

The involvement of non-local and foreign scientists may represent an opportunity to share knowledge and to have access to a wider range of opinions uninfluenced by local social contexts, non-scientific values or entrenched scientific assumptions or preconceptions.

Consideration should be given to the need for rules regarding the nature and extent of the interactions between local and non-local/foreign scientists.

During the Campi Flegrei exercise a number of young researchers were co-opted in rotation onto the SAC. This worked well and gave them a unique learning experience.

Which geoscience disciplines (e.g. geophysics, geochemistry, geodetics, geo-history, and petrology) will be covered within the SAC?

If the monitoring data resources of the host volcano permit, ensure that the Volcano team provides sufficient monitoring data to keep each discipline engaged and, if possible, challenged throughout the exercise.

Will any geoscience disciplines not be covered? Why?

Consider whether this is a realistic representation of what might happen in a real period of emerging unrest.

Will the SAC have local members who are usually based near or at the host volcano?

Consider whether the briefing pack should refer to the IAVCEI protocol mentioned above.

Will the SAC have international members who are not based in the host country?

Consider whether the briefing pack should refer to Newhall et al. (1999) mentioned above.

Will the SAC be realistic in size, too big or too small?

Will any attempt be made to deal with the issue of 'maverick' scientists?

If yes, how? If no, why?

A VUSE may present a good opportunity to consider and test procedures for dealing very difficult issues such as this.

Inputs

When and in what format will the SAC get the monitoring data?

This is a critical issue. During a VUSE, reliance on a limited Wi-Fi link may be very problematical.

Will the SAC be able to ask for further monitoring data?

If yes, when and how can it be requested? If yes, when and how will it be provided?

Will the SAC get anything other than monitoring data?

Consider whether the SAC should be given, or have access to on request, fixed/dynamic data about weather, ground conditions etc.?

If yes, what data need to be provided before, at and after the start of the exercise? Will plans be needed showing geographical, geological or societal features?

Deliberations

Will the SAC's meetings/deliberations follow standardised operating procedures (SOPs) and/or use a standardised reporting template?

Feedback following the Dominica exercise advocated the use of SOPs and the keeping of a log recording and timing, inter alia, main data inputs and analytical decisions, assumptions, outputs and communications.

Careful consideration should be given to the SAC adopting working assumptions that:

- All hazard communications, including all hazard assessments, in their format, content and delivery must be focussed upon: (1) facilitating informed risk mitigation decisions that may not only be difficult but also based upon many sources of knowledge and social, political, economic and other influences; and (2) responding to the identified needs and expectations of their makers.
- Before acting as hazard communicators, hazard analysts must (by means of active and careful two-way dialogue) canvass, note and respond to the needs and expectations of the risk decision makers and the legitimate foreseeable demands of mass and social media and allow sufficient time to do this.
- The utility of hazard communications (i.e. the outputs of hazard analysis) must be judged by empirical evidence of the sentiments and actions of risk decision-makers (i.e. the outcomes of hazard analysis within wider risk mitigation processes).
- No assumptions will be made as to what risk decision-makers know about: (1) the science of volcanic hazards and in particular its complexities, uncertainties and limitations;

(2) the role of scientists and the many temporal, financial, legal and other constraints under which they operate particularly during periods of emerging volcanic unrest; and (3) the role, benefits and limitations of long-term and short-term monitoring.

- No assumptions will be made about what risk decision-makers either need or want and, accordingly, the widest possible range of reasonably practicable options for not only hazard communication but also hazard analysis should be offered and discussed.
- No assumptions will be made that risk decision-makers know what general and/or bespoke analysis, information, data, advice and guidance scientists may be able to provide if asked and given adequate resources.

Will any guidance be given as to how the data should be considered (e.g. expert elicitation, probabilistic tools etc.)?

If an expert elicitation is a possibility or is actively encouraged/required, timetable enough time for this as it is likely that a briefing will be necessary before the elicitation starts to ensure that everyone is fully aware of what will be involved. Do not underestimate the amount of time that a worthwhile elicitation will take.

Consider providing essential information in the form of an elicitation toolkit in advance. Who will prepare this toolkit?

Can the SAC take into consideration any societal factors such as high societal exposure in a particular area? If yes, what 'exposure' and 'vulnerability' data will be provided, when and how? Do not underestimate the amount of time that this will take.

Consider how and when the results of probabilistic models will be delivered to the SAC and thereafter considered. Do not underestimate the amount of time that this will take.

Feedback after VUELCO's Dominica exercise showed that scientists involved in exercises need to be briefed about: (1) the claimed benefits offered by the probabilistic models being used; (2) their limitations; (3) when and in what form their results will be presented; and (4) how it is envisaged their results will be integrated within the overall analysis of relevant hazard scenarios.

Consider including in the pre-exercise briefing pack an additional file containing essential background information about expert elicitations and modelling.

Will the SAC have to liaise with any non-SAC players?

If yes, when and how will this be done?

During the Dominica exercise, sub-team leaders of the SAC were asked to attend early parts of the CPA's deliberations. In the post-exercise feedback, it was suggested that all of the SAC scientists would have benefitted greatly from hearing the CPA's deliberations which were often very critical of the scientific analysis upon which risk decisions had to be made.

Outputs—Hazard communications

What outputs (e.g. oral and written reports, written minutes etc.) will be required from the SAC?

The quality of the dialogue between risk management stakeholders, and the provision of information and advice, depends upon a mutual understanding by those stakeholders of their respective needs, responsibilities, functions demands and roles, and their capacity to anticipate other stakeholders' demands and decision needs (Salas et al. 1994; Ronan et al. 2008; Lipshitz et al. 2001; Paton and Jackson 2002; Doyle and Johnston 2011a, b; Doyle et al. 2014).

"Without knowing the concerns" and understandings "of the targeted audience, communication will not succeed" (Renn 2008, 147).

Careful consideration should be given to this critical issue. Based upon the VUELCO exercises, it may be very unwise to provide any guidance whatsoever as to how the outputs of the hazard analysis must be communicated to the CPA and other exercise participants. This may encourage the type of two-way dialogue referred to under the above heading "Deliberations". By providing guidance of any sort, a unique learning opportunity may be lost.

Consider briefing the observers in advance about this important issue, so that they can make an early invention if unwarranted assumptions are made about what risk decision makers know, need or want.

Will the SAC be required to communicate with teams representing:

- ***The CPA?***
- ***The public?***
- ***The media?***
- ***Representatives of local/central government?***

If yes, when and where?

A VUSE is a good opportunity to practise hazard communications and to consider the needs and expectation of risk decisions and their makers. Reference should be made to Doyle et al. (2015).

Consider how, when and by whom the needs and expectations of risk decision-makers will be sought.

4.5 The Risk Managers—Civil Protection Authorities (CPA)

Role

It is critical for the role of the CPA to be considered very carefully during the planning stage.

Feedback from the VUELCO exercises suggests that the CPA's role must be:

- clearly defined and integrated within the design of the hazard scenario;
- sufficiently interesting and demanding to keep the participants engaged; and
- known to all other participants.

The geoscientists will be engaged predominantly with the prompt analysis of monitoring data and this will lead subsequently to the delivery of hazard assessments in some form to the CPA. Depending upon the goals of the exercise, it is likely that, whilst this is being done, the CPA team must have something worthwhile to do.

Consideration might be given to a requirement within both the briefing pack and the timetable that representatives of the SAC discuss with the CPA:

- the timing, format, content etc. of hazard reports, communication channels and protocols (the preparation of a template was suggested after the Campi Flegrei exercise), and/or
- the arrangements for a possible mock joint press conference later in the exercise. A joint press conference might involve another press/media team (possibly made up of individuals from the CPA's public relations department) and/or invited observers from local and national press organisations.

Membership

Will the CPA represent a real/fictional entity?

Who will be leading the CPA and why?

Who will be the other CPA members?

Inputs

What inputs will the CPA get?

Will the CPA get monitoring data?

If yes, why?

Deliberations

Will any guidance be given as to how the CPA inputs should be considered?

Will the CPA have to liaise with any non-CPA players?

If yes, when and how will this be done?

Outputs

What outputs will be required from the CPA?

Will the CPA be required to communicate with teams representing:

- ***The SAC?***
- ***The public?***
- ***The media?***
- ***Representatives of local/central government?***

A simulated press conference, which involves both scientists and CPA officials within the same panel, can serve a number of useful purposes. In a potentially hostile environment, scientists will have to communicate effectively complex geo-scientific concepts and explain issues such as uncertainty and analytical/diagnostic confidence. They may also have to avoid being drawn into sensitive areas involving societal issues and risk decisions. Risk decision makers will have to demonstrate an understanding of complex hazard scenarios and to explain what risk mitigation decisions they have taken and the reasons for them.

Open and convincing displays of collaboration and agreement between different authorities and key individuals during real crises can build trust with representatives of the media and general public. Panel question and answers sessions involving key personnel, who will have 'communication' roles, can and should be practised during VUSE because it is likely that related training and resource needs will be identified.

The VUELCO Dominica exercise included three very realistic but successful simulated press conferences.

Feedback

The processes of hazard analysis and hazard communication should be outcome focussed and driven by the expectations and needs of risk decisions and risk decision-makers.

Consideration should be given to measuring the extent to which the exercise and its participants addressed: (1) the needs and expectations of the CPA; (2) the steps, if any, the SAC took to identify those needs and expectations; and (3) the extent to which the SAC satisfied those needs; and identified actions that might lead to improvements.

The VUELCO exercises identified the value of civil protection authorities having an internal technical-scientific structure: (1) to support the SAC in technical analysis (mapping, elicitation, models); and: (2) to improve interactions with the SAC.

4.6 Observers/Auditors

Is it necessary to have observers and/or auditors?

In the VUELCO exercises a very valuable contribution was made by monitors and observers. They can be from the participating organisations or entirely 'independent'. Monitors and observers (particularly with experience of previous exercises and the practices of countries other than the host country) can play a critical role in providing both hot and cold feedback whilst also gaining invaluable personal experience to assist in the planning of future exercises.

VUSE are invariably observed by local and visiting students and early-career geoscience academics. With careful planning, it may be possible for less experienced scientists and civil protection authority decision makers to be involved directly (as happened in the Campi Flegrei VUSE) or indirectly in 'shadow' SAC and CPA teams. The participation of young researchers in this kind of exercise represents a great chance to train future generations of advisory scientists.

What role will be played by nominated observers and auditors?

Consider whether it would be worthwhile differentiating between: (1) observers who might have responsibility for capturing and presenting immediate unstructured feedback; and (2) auditors who would compile a more formal report possibly in a format and using audit criteria agreed by the main participants in advance.

Would something be gained by having an 'independent' observer/auditor who is not part of any organisation taking part in the exercise?

Independent observers played a very useful role in the VUELCO exercises in Campi Flegrei, Cotopaxi and Dominica and the author was an invited external auditor of the Tenerife exercise.

During the Campi Flegrei exercise, it was recorded that undertaking the role of observer is particularly helpful for senior civil protection authority managers who are thinking of planning an exercise in their own locality.

Who will be nominated and/or invited to be observers/auditors?

This will depend largely upon the roles which have been chosen and whether some of the observers/auditors should be 'independent'.

Format

At the end of the exercise will there be a 'hot' feedback session?

The MIAVITA Handbook (2012, 118, 179) recommends an "on the spot debriefing straight after the exercises with the participants and another one approximately one month later with the exercise organisers, evaluators, observers, and a representative of the general public".

If yes, why, who will lead it and how long will it last?

When and how will the results be recorded, promulgated and acted upon?

Will there be 'cold' feedback meetings, surveys or questionnaires?

If yes, why and who will be responsible for their design and implementation?

When and how will the results be recorded, promulgated and acted upon?

Will a more formal audit of the exercise be undertaken?

If yes, consider whether it might address issues such as the identification of the need for better or additional:

- structures of risk governance
- information sources and resources
- long-term monitoring resources
- consideration of the role of tectonics and resulting faults and features
- geo-history data
- training
- financial, personnel and other resources
- documented procedures
- communication plans and equipment
- media response plans
- easily accessible database catalogues, which may be important and needed urgently during a crisis, of:
  - existing data (e.g. seismic, geodetic, gas, water geochemistry and geospatial) and common archiving procedures for newly acquired datasets; and
  - information (e.g. geographical information systems 'GIS' digital bases, overlays, metadata, geological maps, remote-sensing images, background levels of unrest indicators, such as seismic energy release rates, normal rates on inflation/deflation, aqueous and geochemistry and fluxes).
- post-exercise actions to review and revise existing procedures in particular those for any future exercises.

Will the observers give real-time feedback during the exercise?

Consideration should be given to this difficult issue. A balance must be achieved between, on the one hand, unhelpful and intrusive interventions and, on the other, allowing the exercise to proceed in ways that may waste valuable time and/or miss or diminish valuable learning opportunities.

During the Dominica exercise, the three 'independent' observers gave brief feedback to the SAC at the end of Phases 1 and 2 of the exercise. They also gave spontaneous suggestions to the CPA. The participants indicated that they found this helpful and there was clear evidence that suggestions for change were understood and put into practice immediately.

Preparation

Will instructions be given to the observers/auditors in advance?

Consider giving the observers/auditors a clear brief before the start of the exercise.

Experience from the VUELCO exercises suggests that providing observers with a list of more specific considerations, to include within their more general observations, may result in more focussed outputs.

Will instructions be given to the participants in advance of the exercise to facilitate richer feedback?

Consider asking each participant and/or each team (1) to list their main roles, responsibilities and tasks and (2) to keep a written list of needs that would help them accomplish their tasks covering for example better/different information, training, guidance, software, communication equipment, other equipment, personnel, and so on. These lists can then be used during the hot and cold feedback sessions.

Will the participants be told in advance about the planned arrangements for observing/auditing?

If yes, when? Consider whether the pre-exercise briefing pack is the best place to set out the planned arrangements and time table and what might be required from the participants in this regard.

5 Discussion

The checklists presented in this chapter are dedicated to the organisers and participants of the five exercises that this chapter has investigated. The checklists record and build upon the successes, experiences, occasional oversights, misjudgements and mistakes of a few in order to provide readily accessible sources of knowledge, learning and inspiration for others.

This chapter has identified a number of themes:

- VUSE are purpose-driven learning/training activities. Vital knowledge is acquired not only during but also before and after each exercise (see Doyle et al. 2015).
- VUSE must have clear recorded goals based upon the needs and expectations of their participants.
- VUSE are complex and require clear and firm leadership, and very careful planning, funding and execution. This finding is supported by the work of Dohaney et al. (2015), which contains very helpful guidance on design and evaluation methods.
- During the 'planning phase', the critical issues of leadership, purpose, scope, duration, type and financing should be considered.
- The mere planning of a VUSE will provide invaluable knowledge of the wider legal and administrative infrastructures in which it is framed.
- The 'logistics' phase will ensure that the execution of the exercise is not undermined by avoidable technical, communication and other related difficulties. The importance of a comprehensive and comprehensible pre-exercise briefing note and exercise plan cannot be overstated.
- A pre-VUSE field trip will serve numerous purposes including those of introducing relevant geographical, geological, cultural and governance histories and providing an opportunity for relationships to develop between the exercise participants.
- VUSE require and depend upon numerous acts of communication between a wide range of stakeholders with a variety of knowledge, expertise, experience, needs and expectations within many types of formal/informal relationships/associations (see Doyle et al. 2015).
- VUSE are not judged in terms of 'success' or 'failure' but rather in terms of whether relevant knowledge has been generated, recorded, considered and utilised.
- Whenever possible, VUSE simulate:
  - 'existing' risk governance arrangements, or those being considered for the future, with real policies, processes and people rather than contrived unrealistic governance scenarios and false role-play; and
  - volcanic hazard scenarios based upon, and significant within, the context of the host volcano region, which may experience other relevant or related natural hazards such as tectonic and weather-related hazards.
- VUSE provide unique opportunities:
  - to address many known challenges including those of inter-scientist deliberation, scientific uncertainty, analytical/diagnostic confidence, unorthodox/maverick science sources, hazard communication, and mass media and social networking relations and communications (see Doyle et al. 2015);

  - to test new/prototype risk arrangements and tools, models and protocols for analysis/communication.
- Great care should be taken to ensure that comprehensive feedback and goal-related knowledge can be captured during and after each exercise and disseminated as widely as possible.

6 Conclusions

"Practice doesn't make perfect. Practice reduces the imperfection". None of the five exercises considered in this chapter was completely perfect. They involved not only well-intentioned yet imperfect policies and procedures capable of improvement but also dedicated yet not infallible people keen to seek further knowledge, training and experience. Careful analysis of these exercises shows that, as suggested by Doyle et al. (2015), with very careful planning, execution and review, worrisome periods of emerging volcanic unrest and the dynamics of real-time hazard assessments and risk decisions can be simulated for a wide variety of worthwhile purposes.

However this chapter does not attempt to dictate how future exercises must be organised. Based upon a review of the published records of four exercises and the authors' personal experiences of the five further exercises listed in Table 1, it is believed that these lists will assist the organisers of future exercises to meet the specific challenges of the volcanic hazards they face and the societal, political, economic and legal contexts in which difficult and timely risk decisions have to be made.

The authors hope that the checklists will be considered and used, and, above all, improved. Candid feedback from future exercises will ensure that the guidance notes evolve and will be supplemented by the addition of further detailed exercise studies. Will these checklists facilitate more effective future exercises and improve volcanic risk mitigation? This question can be answered only if they are used and relevant evidence is generated within future post-exercise empirical studies.

The Sendai Framework sets ambitious goals for quality standards, learning, knowledge exchange, education and training. The paramount goal of this chapter is to ensure that the pooled experiences of the few, who have had the advantages and privilege of being exercise participants, will be accessible to the widest possible audience, to encourage future exercises and thereby to improve the governance of volcanic risks.

Disclaimer

The content of this paper reflects the authors' views and not necessarily the opinion of the organisations to which they belong. The authors and their organisations are not liable for any use that may be made of the information contained therein.

References

Bartolini S, Cappello A, Martin J, Del Negro C (2013) QVAST: a new quantum GIS plug for estimating volcanic susceptibility. Nat Hazards Earth Syst Sci 13, 3013-3-34

Bergs J, Hellings J, Cleemput I, Zurel Ő, De Troyer V, Van Hiel M, Demeere J-L, Claeys D, Vandijck D (2014) Systematic review and meta-analysis of the effect of the World Health Organization surgical safety checklist on postoperative complications. BJS 101:150–158. https://doi.org/10.1002/bjs.9381

Beta T (2011) Master of Stupidity. http://www.goodreads.com/book/show/8542012-master-of-stupidity

Bretton RJ (2014) The role of science in the rule of law. http://expertsinuncertainty.net

Bretton RJ, Gottsmann J, Aspinall WP, Christie R (2015) Implications of legal scrutiny processes (including the L'Aquila trial and other recent court cases) for future volcanic risk governance. J Appl Volcanol 4:18. https://doi.org/10.1186/s13617-015-0034-x

Dohaney J, Brogt E, Kennedy B, Wilson TM, Lindsay JM (2015) Training in crisis communication and volcanic unrest forecasting: design and evaluation of an authentic role-play simulation. J Appl Volcanol 4:12. https://doi.org/10.1186/s13617-015-0030-1

Donovan A, Oppenheimer C (2012) Governing the lithosphere: insights from Eyjafjallajökull concerning the role of scientists in supporting decision-making on active volcanoes. J Geophys Res 117:B03214. https://doi.org/10.1029/2011JB0090802012

Donovan A, Oppenheimer C (2014) Science, policy and place in volcanic disasters: insights from Montserrat. Environ Sci Policy 38, 150–161

Doyle EEH, Johnston DM (2011) Science advice for critical decision-making. In: Paton D, Violanti J (eds) Working in high risk environments: developing sustained resilience. Charles C Thomas, Springfield, Illinois, USA, pp 69–92

Doyle EEH, Johnston DM, McClure J, Paton M (2011) The communication of uncertain scientific advice during natural hazard events. N Z J Psychol 40(4):39–50

Doyle EEH, McClure J, Johnston DM, Paton D (2014) Communicating likelihoods and probabilities in forecasts of volcanic eruptions. J Volcanol Geoth Res 272 (2014):1–15

Doyle EEH, Paton D, Johnston DM (2015) Enhancing scientific response in a crisis: evidence-based approaches from emergency management in New Zealand. J Appl Volcanol 4:1. https://doi.org/10.1186/s.13617-014-0020-8

Errington EP (1997) Role-play. Higher education research and development society of Australasia, Jamison Centre, ACT

Errington EP (2011) Mission possible: using near-world scenarios to prepare graduates for the professions. Int J Teach Learn High Educ 23:84–91

Fiske R (1984) Volcanologists, Journalists, and the Concerned Local Public: a tale of two crises in the Eastern Caribbean. In: Explosive volcanism: inception, evolution and hazards. Geophys. Study Committee, National Research Council, National Academy Press, Washington, pp 170–176

Galanti E, Goretti A, Foster B, Di Pasquali G (2006) Geotech Geol Earthquake Eng 369–384

Gawande A (2010) The checklist manifesto—how to get things right. Profile Books Ltd, London

Harris AJL, Gurioli L, Hughes EE, Lagreulet S (2012) Impact of Eyjafjallajökull ash cloud: a newspaper perspective. J Gepphys Res 117:35. https://doi.org/10.1029/2011JB008735

Hincks TK, Jean-Christophe Komorowski J-C, Sparks SR, Aspinall WP (2014) Retrospective analysis of uncertain eruption precursors at La Soufrière volcano, Guadeloupe, 1975–77: volcanic hazard assessment using a Bayesian Belief Network approach. J Appl Volcanol 2014(3):3. https://doi.org/10.1186/2191-5040-3-3

Hood C (1986) Administrative analysis: an introduction to rules, enforcement and organisations. Wheatsheaf Books, Sussex

Hood C (2011) The blame game—Spin, bureaucracy and self-preservation in government. Princeton University Press, Princeton, New Jersey, USA and Woodstock, Oxfordshire, UK

IAVCEI (2015) News no. 1–3. http://www.iavcei.org/

International Federation of Red Cross and Red Crescent Societies (IFRC) and United Nations Development Programme (2015) The checklist on law and disaster risk reduction, Pilot version, March 2015. IFRC, Geneva, Switzerland

Jenkins S, Spence R, Baxter P, Komorowski J-C, Sara Barsotti S, Esposti-Ongaro T, Neri A (2012) Preparation of disaster risk scenarios at La Soufriere, Guadeloupe poster at volcanic and magmatic studies group annual meeting 2013

Johnston D, Scott B, Houghton B, Paton D, Dowrick D, Villamor P, Savage J (2002) Social and economic consequences of historic caldera unrest at the Taupo volcano, New Zealand and the management of future episodes of unrest. Bull NZ Soc Earthquake Eng 35 (4):215–229

Jordan TH, Chen Y.-T, Gasparini P, Madariaga R, Main I, Marzocchi W, Papadopoulos G, Sobolev G, Yamaoka K, Zschau J (2011) Operational earthquake forecasting: state of knowledge and guidelines for implementation, findings and recommendations of the international commission on earthquake forecasting for civil protection, submitted to the Department of Civil Protection, Rome, Italy. Ann Geophys **54**, 4. https://doi.org/10.4401/ag-5350

Lipshitz R, Klein G, Orasanu J, Salas E (2001) Focus article: taking stock of naturalistic decision making. J Behav Decis Mak 14(5):331–352

Marrero JM, García A, Llinares Á, Berrocoso M, Ortiz R (2015) Legal framework and scientific responsibilities during volcanic crises: the case of the El Hierro eruption (2011–2014). J Appl Volcanol 4:13. https://doi.org/10.1186/s13617-015-0028-8

Marzocchi W, Newhall C, Woo G (2012) The scientific management of volcanic crises. J Volcanol Geoth Res. https://doi.org/10.1016/j.volgeorgres.2012.08.016

McGuire WJ, Kilburn CRJ (1997) Forecasting volcanic events: some contemporary issues Geo. Rundsch 86:439–445

MIAVITA (Mitigate and Assess risk from Volcanic Impact on Terrain and human Activities) (2012) Handbook for volcanic risk management: prevention, crisis management, Resilience, MIAVITA team, Orleans. http://miavita.brgm.fr/default.aspx

Mothes PA, Yepes HA, Hall ML, Ramón PA, Steel AL, Ruiz MC (2015) The scientific-community interface over the fifteen-year eruptive episode of Tungurahua Volcano, Ecuador. J Appl Volcanol 4:9. https://doi.org/10.1186/s13617-015-0025-y

Newhall CG (2010) A checklist for volcanic risk mitigation. Cities on volcanoes 6 presentation, earth observatory of Singapore

Newhall C, Aramaki S, Barberi F, Blong R, Calavache M, Cheminee J-L, Punongbayan R, Siebe C, Simkin T, Sparks RSJ, Tjetjep W (1999) International association of volcanology and chemistry (IAVCEI) subcommittee for crisis protocols - professional conduct of scientists during volcanic crises. Bull Volcanol 60:323–334

NZ/MCDEM (2008) Exercise Ruaumoko 2008: final exercise report. Ministry of Civil Defence & Civil Management, Wellington, NZ

OECD (2015), Scientific advice for policy making: the role and responsibility of expert bodies and individual scientists. OECD science, technology and industry policy papers, no. 21, OECD Publishing, Paris. http://dx.doi.org/10.1787/5js33l1jcpwb-en

Paton D, Jackson D (2002) Developing disaster management capability: an assessment centre approach. Disaster Prev Manage 11(2):115–122

Renn O (2008) Risk governance – coping with uncertainty in a complex world. Earthscan, London, New York

Ricci T, Nave R, Barberi F (2013) Vesuvio civil protection exercise MESIMEX: survey on volcanic risk perception. Ann Geophys 56(4):s0452. https://doi.org/10.4401/ag-6458

Ronan KR, Johnston DM, Daly M, Fairley R (2008) School children's risk perception and preparedness: a hazard education survey. Aust J Disaster Trauma Stud 1. ISSN: 1174-4707. Available at: http://www.massey.ac.uk.nz/~trauma/issues/201-1/ronan.htm

Rothstein H (2002) Neglected risk regulation: the institutional attenuation problem, centre for analysis of risk and regulation. London School of Economics and Political Science, London

Salas E, Stout RJ, Cannon-Bowers JA (1994) The role of shared mental models in developing shared situational awareness. In: Gilson RD, Garland DJ, Koonce JM (eds) Situational awareness in complex systems: proceedings of a cahfa conference. Embry-Riddle Aeronautical University Press, Daytona Beach, FL, pp 298–304

Sandri L, Guidoboni E, Marzocchi W, Selva J (2009) Bayesian event tree for eruption forecasting (BET_EF) at Vesuvius, Italy: a retrospective forward application to the 1631 eruption Bull. Volcanol 71:729–745

Sobradelo R, Bartolini S (2014) Marti J (2014) HASSET: a probability tool to evaluate future volcanic scenarios using Bayesian inference. Bull Volc 76:770

Solana MC, Kilburn CRJ, Rolandi G (2008) Communicating eruption and hazard forecasts on Vesuvius, Southern Italy. J Volcanol Geoth Res 189(2010):308–314

Sparks RSJ, Aspinall WP, Auker M, Crosweller S, Hincks T, Mahony S, Nadim F, Polley J, Syre E (2012a) Mapping and characterising volcanic risk, magmatic rifting and active volcanism conference, Addis Ababa, 12 Jan 2012

Sparks RSJ, Aspinall WP, Auker M, Crosweller S, Hincks T, Mahony S, Nadim F, Polley J, Syre E (2012b) Volcano hazard and exposure in global facility for disaster reduction and recovery priority countries and risk management measures, GVM 1st workshop 30 Apr 2012, University of Bristol and NGI Norway for the World Bank

Treadwell JR, Lucas S, Tsou AY (2014) Surgical checklists: a systematic review of impacts and implementation. BMJ Qual Saf 23:299–318. https://doi.org/10.1136/bmjqs-2012-001797

UN/ISDR (2015) Sendai framework for disaster risk reduction 2015–30

van Ments M (1999) The effective use of role-play: practical techniques for improving learning. Konan Page Ltd, London

Woo G (2008) Probabilistic criteria for volcano evacuation decisions. Nat Hazards 45:87–97. https://doi.org/10.1007/s11069-007-9171-9

Further reading sources on VHUB website

VUELCO resources

Colima Exercise Plan

Colima Debriefing Report

Ciolli S, De la Cruz-Reyna S (2013) Colima volcano exercise debriefing report. https://vhub.org/resources/2477

Campi Flegrei Simulation Plan

Campi Flegrei Simulation Debriefing Report

Ciolli S, Cristiani C, Golisciani GA (2014) VUELCO—Campi Flegrei Caldera unrest scientific simulation—debriefing report. https://vhub.org/resources/3640

Cotopaxi Debriefing Report

Dominica Exercise Plan

Dominica Exercise Debriefing Report

Ciolli S, Robertson REA (2015) VUELCO—Dominica exercise debriefing report. https://vhub.org/resources/3822

Appendix
Volcanic Unrest: Terminology and Definitions

Servando De la Cruz-Reyna, Ana Teresa Mendoza Rosas, Joachim Gottsmann

Motivation and Background

This document provides a set of definitions of scientific and legal terminologies used in the context of volcanic unrest. The underpinning research behind this document was performed at the Universidad Nacional Autónoma de México—Geophysics Institute (Instituto de Geofísica, IGEF-UNAM) in collaboration with the National Center for Disaster Prevention (Centro Nacional de Prevención de Desastres, CENAPRED, Secretaría de Gobernación) and Colima University (Universidad de Colima, UCOL). Therefore, many definitions of the relevant legal terminology are based on Mexican law and may not be applicable *verbatim* to other jurisdictions. However, this document has been designed carefully to provide a common set of terms that are fundamental for volcanic unrest crisis management. This should provide a common platform for the interaction amongst different actors and stakeholders involved in the management of volcanic crises. The proposed definitions inherently follow on from investigations and interactions between scientists and decision-makers within the VUELCO project. It is not our intention to impose these definitions on volcanic unrest risk communication world-wide. However, we expect this compilation to instigate a dialogue and a shared platform of relevant terminologies between the different actors and stakeholders of volcanic risk governance [see also (Bretton et al. 2017)].

Analysis of past volcanic disasters shows that one of their causes were communication failures amongst stakeholders (Fiske 1984; Hall 1990; Tilling 2009; Fearnley et al. 2017). Some failures resulted from different perception of the meaning of terms used to describe or define the level of volcanic hazard or risk. This is in fact a problem related to semiotics. Definitions of scientific, social and legal concepts that are fundamental for volcanic crisis management having semantic contents that can be translated into other languages are a necessary first step to homogenize terminologies and improve risk communications. It is thus necessary to develop efficient communication protocols that univocally use the same meaning of the terminology (semantic content), integrate that meaning into a communication code that can be understood by all stakeholders (syntactic content), and translate the effects and consequences that volcanic activity may cause (pragmatic content). The following sections provide some context for a shared terminology.

S. De la Cruz-Reyna
Universidad Nacional Autónoma de México (UNAM), Mexico City, Mexico

A. T. Mendoza Rosas
Universidad Nacional Autónoma de México (UNAM), Mexico City, Mexico

J. Gottsmann
School of Earth Sciences, University of Bristol, Bristol, UK

Advs in Volcanology (2019) 299–313
DOI 10.1007/978-3-319-58412-6

Scientific Terminology

Term (English)	Término (Español)	Meaning	Significado	Further reading
Unrest (volcanic)	Activación; reactivación; agitación (volcánica)	A deviation from a background or baseline level behavior of a volcano which might be a prelude to an eruption, or to another hazardous event	Incremento del nivel de actividad volcánica respecto a un nivel de fondo o nivel base, el cual puede ser preludio de una erupción o de alguna otra manifestación peligrosa	Phillipson et al. (2013)
Crisis (volcanic)	Crisis (volcánica)	Situation during which a volcano shows signs of unrest, that can be interpreted as indication for potential impending eruptive activity and associated hazard. A crisis may or may not culminate in a dangerous eruption, but it may cause anxiety and socioeconomic unrest among the affected population	Situación en la que un volcán muestra signos de agitación que pueden interpretarse como precursora de actividad eruptiva inminente y de los peligros que conlleva. Una crisis volcánica puede o no culminar en una erupción peligrosa, pero puede producir ansiedad e inquietud socioeconómica en la población afectada	Tilling (1989)
Volcanic hazard (H)	Peligro volcánico (H)	Potentially damaging pre-eruptive, syn-eruptive or post- eruptive phenomena such as pyroclastic flows, windborne ash, lava flows, volcanic gases, lahars, ground deformation, major volcano-tectonic earthquakes, and landslides. In probabilistic assessments, "hazard" is sometimes referred to the probability that a specific volcanic manifestation or phenomenon will occur in a given area, within a given time frame (i.e., a threat)	Fenómenos potencialmente dañinos pre- eruptivos, co-eruptivos o post-eruptivos, como los flujos piroclásticos, ceniza transportada por el viento, flujos de lava, gases volcánicos, deformación del suelo, lahares, sismos volcano-tectónicos de magnitud considerable y deslizamientos. Para la evaluación probabilística, a veces "peligro" es la probabilidad de que una manifestación volcánica específica ocurra en una zona determinada, en un intervalo de tiempo dado	De la Cruz-Reyna and Tilling (2008)
Volcanic Risk (R)	Riesgo volcánico (R)	Probable loss (assets such as lives, property, productive capacity, etc.) in the event of the manifestation of a volcanic hazard. It is a conditional probability resulting from the product of the probability (P) that a specific volcanic hazard will occur (i.e., a threat) times the value (L) of the exposed assets times their vulnerability (V): R=P*L*V	Pérdida probable (de activos como vidas, propiedades, capacidad productiva, infraestructura, etc.) en un área sujeta a un peligro volcánico. Es la probabilidad condicional que resulta de multiplicar la probabilidad (P) de que una manifestación o fenómeno volcánico específico ocurra en un área y en un tiempo determinados (peligro), por el valor (L) de los activos por: R=P*L*V	Tilling (1989) De la Cruz-Reyna and Tilling (2008)

(continued)

Term (English)	Término (Español)	Meaning	Significado	Further reading
Acceptable risk	Riesgo aceptable	Risk that an individual or community is willing to accept, or that a public official is prepared to allow persons in his or her charge to accept. Acceptable risk is a function of the benefits of risk mitigation (safety and avoided losses) and its costs (losses of jobs and business, community disruption, the costs of transportation, housing, and food for evacuees, and costs of any structures to divert hazards). Tolerance for risk varies considerably, mainly because the benefits and costs of mitigating risks vary greatly from individual to individual and community to community	Riesgo que un individuo o comunidad están dispuestos a aceptar, o que un funcionario público puede permitir que las personas bajo su responsabilidad acepten. La aceptación del riesgo depende de los beneficios de su mitigación (seguridad y pérdidas evitadas) y su costo (pérdida de trabajos y negocios, disfunción de la comunidad, gastos de transporte, alojamiento y alimento para evacuados, y costos de estructuras para reducir peligros). La tolerancia al riesgo es muy variable, ya que los beneficios y costos de mitigación de riesgos pueden ser muy diferentes entre individuos y entre comunidades	Peterson (1988)
Vulnerability	Vulnerabilidad	The susceptibility of physical and human systems to be affected by a natural hazard. It can be expressed as a probability of damage, or as the proportion of the total exposed assets expected to be affected by a given manifestation. In probabilistic assessment "vulnerability" is the expected loss of the exposed value should a hazardous manifestation occur (i.e., the probability of loss)	La susceptibilidad de sistemas físicos y humanos que pueden ser afectados por fenómenos naturales peligrosos. Se puede expresar como una probabilidad de daño, o como la proporción del total de activos expuestos a daño por una manifestación determinada. Para evaluación probabilística, "vulnerabilidad" es el porcentaje esperado de perdida de los valores expuestos dado que una manifestación o evento volcánico ocurra (es decir, probabilidad de perdida)	Jolly and De la Cruz-Reyna (2015) De la Cruz-Reyna and Tilling (2008)
Preparedness	Preparación	A state of readiness or preparation for use or action in the case of volcanic activity or any other threat. It involves a clear understanding by the population and authorities of the natural phenomena and their destructive effects. Series of measures to reduce vulnerability	Estado de preparación y conciencia para prevenir y actuar en el caso de alguna actividad volcánica o cualquier otra amenaza. Involucra una clara comprensión de los fenómenos naturales y de sus efectos destructivos por parte de la población y las autoridades. Serie de medidas para reducir la vulnerabilidad	Jolly and De la Cruz-Reyna (2015) De la Cruz-Reyna and Tilling (2008)
Volcanic disaster	Desastre volcánico	An event or a series of events marked by a significant loss of life and/or other assets that exceed(s) the local response capacity and results from one or more volcanic hazards	Evento o serie de eventos caracterizados por una pérdida significativa de activos y del valor, que excede la capacidad de respuesta local y que resultan de uno o más peligros volcánicos	Jolly and De la Cruz-Reyna (2015)
Alert	Alerta	A condition of heightened watchfulness or preparation for	Condición de atención, vigilancia y preparación para la acción. Un	

(continued)

Term (English)	Término (Español)	Meaning	Significado	Further reading
		action. A warning method or system to make people aware of impending danger	método o sistema de aviso para que la gente perciba un peligro inminente	Jolly and De la Cruz-Reyna (2015)
Alert code	Código de Alerta	A set of rules or a system used for transmitting alert messages requiring brevity and clarity. The symbolic arrangement of instructions in a warning system	Un conjunto de reglas o un sistema utilizado para la transmisión de mensajes de alerta breves y claros. Estructura simbólica de instrucciones en un sistema de alerta.	Jolly and De la Cruz-Reyna (2015)
Consensus	Consenso	General agreement in the diagnostics and the prognosis of the majority of scientists involved in the study of an episode of volcanic activity. Consensus of the involved scientists is important for the decision making of authorities	Acuerdo general en el diagnóstico y el pronóstico de la mayoría de los científicos involucrados en el estudio de un episodio de actividad volcánica. El consenso de los científicos involucrados es muy importante para la toma de decisiones de las autoridades	Jolly and De la Cruz-Reyna (2015)
Crisis	Crisis	An unstable situation of increased danger. A crucial stage in the course of volcanic unrest, when the threat of an eruption, or the possibility of an eruption, exceeds some threshold or reference level	Una situación inestable de amenaza en aumento. Etapa crucial en el curso de una reactiviación/agitación volcánica en que la amenaza de una erupción, o la posibilidad de una erupción exceden un nivel de referencia	Jolly and De la Cruz-Reyna (2015)
Pre-event	Pre-evento	Period of time when there is significant evidence that a hazard could occur	Período de tiempo en que se tiene evidencia significativa de que un fenómeno potencialmente causante de daños puede ocurrir	Jolly and De la Cruz-Reyna (2015)
Volcanic Traffic Light Alert System (VTLAS)	Semáforo de Alerta Volcánica	A basic communications protocol that translates volcanic threat into seven levels of preparedness for the emergency-management authorities, but only three unambiguous levels of alert for the public (color coded green–yellow–red). The changing status of the volcano threat is represented as the most likely scenarios according to the opinions of an official scientific committee that analyzes all available data	Protocolo de comunicación básica que traduce las amenzas volcanicas en siete niveles de preparación para las autoridades de gestión de emergencias, pero en solo tres niveles de alerta sin ambigüedades para el público (código de colores verde-amarillo-rojo). Los cambios en el estado del volcán se representan como los escenarios de riesgo más probables de acuerdo al consenso de un Comité Científico oficial que analiza los datos disponibles	Fearnley (2013)
Seismic swarm	Enjambre sísmico	A group of many earthquakes of similar size occurring closely clustered in space and time with no dominant main shock	Grupo de múltiples sismos de tamaño similar estrechamente agrupados en el espacio y en el tiempo sin evento principal dominante	McNutt and Roman (2015)
Forecast	Pronóstico	A general description of future events, including rough estimates of time, location, and likely activity	Una descripción general de eventos futuros incluyendo estimaciones de tiempo, localización y actividad probable	Connor et al. (2015)

(continued)

Term (English)	Término (Español)	Meaning	Significado	Further reading
Precursor	Precursor	A geological event that occurs prior to an eruption and is related to the preparation processes of the forthcoming eruption	Un evento geológico que se produce antes de una erupción y se relaciona con los procesos de preparación de la próxima erupción	National Academies of Sciences, Engineering Medicine (2017)
Prediction	Predicción	A specific description of future events, including time, size, type, location, and formal errors for each	Una descripción específica de los acontecimientos futuros, incluyendo el tiempo, el tamaño, el tipo, la ubicación y los errores formales para cada uno	National Academies of Sciences, Engineering Medicine (2017)
Probabilistic	Probabilístico	A statement of the relative likelihood of an event based on study of a population of similar events that have occurred in the past	Una declaración de la probabilidad relativa de un evento basado en el estudio de una población de eventos similares que han ocurrido en el pasado	Connor et al. (2015)

Legal Terminology

Term (English)	Término (Español)	Meaning	Significado	Further reading
Accident (legal definition, Mexico)	Siniestro (definición legal, México)	Critical and harmful situation caused by the incidence of one or more hazards in a building or facility affecting the occupants and installations, with possible effects on surrounding facilities	Situación crítica y dañina generada por la incidencia de uno o más fenómenos perturbadores en un inmueble o instalación afectando a su población y equipo, con posible afectación a instalaciones circundantes	Ley General Protección Civil, 6 junio 2012, Capitulo l, Artículo 2, Fracción LV (Mexican Civil Protection General Law, 6 June 2012, Chapter 1, 2nd Article, Section LV)
Alert (legal definition, Mexico)	Alerta (definición legal, México)	Warning of the proximity of an anthropogenic or natural hazard, or of the increase of risk associated to it	El aviso de la proximidad de un Fenómeno Antropogénico o Natural Perturbador, o el incremento del Riesgo asociado al mismo	Reglamento de la Ley General Protección Civil, 6 junio 2012, Artículo 2, Fracción I (Bylaws of the General Civil Protection Law, June 6, 2012, 2nd Article, Fraction 1)
Brigade (legal definition, Mexico)	Brigada (definición legal, México)	Basic Civil Protection Unit. A group of people organized within a facility trained in basic emergency response functions such as: first aid, fighting fires, evacuations, search and rescue; designated in the Internal Civil Protection Unit as responsible for the development and implementation of prevention, rescue and	Grupo de personas que se organizan dentro de un inmueble, capacitadas y adiestradas en funciones básicas de respuesta a emergencias tales como: primeros auxilios, combate a conatos de incendio, evacuación, búsqueda y rescate; designados en la Unidad Interna de Protección Civil como encargados del	Ley General Protección Civil, 6 junio 2012, Capitulo l, Artículo 2, Fracción VI (Mexican Civil Protection General Law, 6 June 2012, Chapter 1, 2nd Article, Section VI)

(continued)

Term (English)	Término (Español)	Meaning	Significado	Further reading
		recovery based on the provisions of the internal civil protection program of the facility	desarrollo y ejecución de acciones de prevención, auxilio y recuperación, con base en lo estipulado en el Programa Interno de Protección Civil del inmueble	
Disaster (legal definition, Mexico)	Desastre (definición legal, México)	A consequence of the occurrence of one or more severe (and/or extreme) hazards, associated or not, with a natural origin, or caused by human activity, or from outer space that cause an amount of damage in a given time and in a determined area that exceeds the response capacity of the affected community	Resultado de la ocurrencia de uno o más agentes perturbadores severos y/o extremos, concatenados o no, de origen natural, de la actividad humana o aquellos provenientes del espacio exterior, que cuando acontecen en un tiempo y en una zona determinada, causan daños que por su magnitud exceden la capacidad de respuesta de la comunidad afectada	Ley General Protección Civil, 6 junio 2012, Capitulo l, Artículo 2, Fracción XVI. (Mexican Civil Protection General Law, 6 June 2012, Chapter 1, 2nd Article, Section XVI)
Disaster area (legal definition, Mexico)	Zona de desastre (definición legal, México)	Specified area upon which a competent authority formally declares a temporal condition of disrupted social structure, which precludes the normal community activities. Such declaration may involve funding from FONDEN (A special federal fund that can be only used to finance the recovery of disaster areas.)	Espacio territorial determinado en el tiempo por la declaración formal de la autoridad competente, en virtud del desajuste que sufre en su estructura social, impidiéndose el cumplimiento normal de las actividades de la comunidad. Puede involucrar el ejercicio de recursos públicos a través del Fondo de Desastres	Ley General Protección Civil, 6 junio 2012, Capitulo l, Artículo 2, Fracción LIX (Mexican Civil Protection General Law, 6 June 2012, Chapter 1, 2nd Article, Section LIX)
Disruptive natural phenomenon (legal definition, Mexico)	Fenómeno natural perturbador (definición legal, México)	Hazard of natural origin	Agente perturbador producido por la naturaleza	Ley General Protección Civil, 6 junio 2012, Capitulo l, Artículo 2, Fracción XXII (Mexican Civil Protection General Law, 6 June 2012, Chapter 1, 2nd Article, Section XXII)
Early warning systems (legal definition, Mexico)	Sistemas de alerta temprana (definición legal, México)	A set of elements providing timely and efficient information, allowing individuals exposed to a threat to take avoidance actions aimed to prevent or reduce risk, and to be prepared for an effective response. Early Warning Systems include	El conjunto de elementos para la provisión de información oportuna y eficaz, que permiten a individuos expuestos a una amenaza tomar acciones para evitar o reducir su riesgo, así como prepararse para una respuesta efectiva. Los	Reglamento de la Ley General Protección Civil, 6 junio 2012, Artículo 2, Fracción XIII (Bylaws of the General Civil Protection Law, June 6, 2012, 2nd Article, Fraction XIII)

(continued)

Term (English)	Término (Español)	Meaning	Significado	Further reading
		hazard recognition and mapping; monitoring and forecasting of impending events; processing and dissemination of warnings or alerts understandable to authorities and population and adopting appropriate and timely response actions to such alerts	Sistemas de Alerta Temprana incluyen conocimiento y mapeo de amenazas; monitoreo y pronóstico de eventos inminentes; proceso y difusión de Alertas comprensibles a las autoridades y población; así como adopción de medidas apropiadas y oportunas en respuesta a tales Alertas	
Emergency (legal definition, Mexico)	Emergencia	Abnormal situation in which damage to society is likely, posing an excessive risk to the safety and integrity of the general population, generated or associated with the presence, imminence, or high probability of occurrence of a hazard	Situación anormal que puede causar un daño a la sociedad y propiciar un riesgo excesivo para la seguridad e integridad de la población en general, generada o asociada con la inminencia, alta probabilidad o presencia de un agente perturbador	Ley General Protección Civil, 6 junio 2012, Capitulo 1, Artículo 2, Fracción XVII (Mexican Civil Protection General Law, 6 June 2012, Chapter 1, 2nd Article, Section XVII)
Evacuee (legal definition, Mexico)	Evacuado (definición legal, México)	Person who withdraws or is removed from a place of residence in a situation of emergency or likelihood of disaster, as a preventive and temporal measure to assure his or her safety and survival	Persona que, con carácter preventivo y provisional ante la posibilidad o certeza de una emergencia o desastre, se retira o es retirado de su lugar de alojamiento usual, para garantizar su seguridad y supervivencia	Ley General Protección Civil, 6 junio 2012, Capitulo 1, Artículo 2, Fracción XIX (Mexican Civil Protection General Law, 6 June 2012, Chapter 1, 2nd Article, Section XIX)
Geological phenomenon (legal definition, Mexico)	Fenómeno geológico (definición legal, México)	Hazard caused by actions and movements of the earth's crust. This category includes earthquakes, volcanic eruptions, tsunamis, unstable slopes, flows, landslides, terrain sinking and subsidence, and ground cracks	Agente perturbador que tiene como causa directa las acciones y movimientos de la corteza terrestre. A esta categoría pertenecen los sismos, las erupciones volcánicas, los tsunamis, la inestabilidad de laderas, los flujos, los caídos o derrumbes, los hundimientos, la subsidencia y los agrietamientos	Ley General Protección Civil, 6 junio 2012, Capitulo 1, Artículo 2, Fracción XXIII (Mexican Civil Protection General Law, 6 June 2012, Chapter 1, 2nd Article, Section XXIII)
Hazard (legal definition, Mexico)	Peligro (definición legal, México)	Probability of occurrence of a potentially damaging and disruptive phenomenon with a certain intensity, in a specific region and in a given time interval	Probabilidad de ocurrencia de un agente perturbador potencialmente dañino de cierta intensidad, durante un cierto periodo y en un sitio determinado	Ley General Protección Civil, 6 junio 2012, Capitulo 1, Artículo 2, Fracción XXXVII (Mexican Civil Protection General Law, 6 June 2012, Chapter 1, 2nd Article, Section XXXVII)

(continued)

Term (English)	Término (Español)	Meaning	Significado	Further reading
Help (legal definition, Mexico)	Auxilio (definición legal, México)	Support to people at risk or victims of a mishap, emergency or disaster, as a response by official or private specialized groups, or by internal civil protection units, as well as actions to protect other exposed people and valued non-human assets	Respuesta de ayuda a las personas en riesgo o las víctimas de un siniestro, emergencia o desastre, por parte de grupos especializados públicos o privados, o por las unidades internas de protección civil, así como las acciones para salvaguardar los demás agentes afectables	Ley General Protección Civil, 6 junio 2012, Capítulo 1 Artículo 2, Fracción V (Mexican Civil Protection General Law, 6 June 2012, Chapter 1, 2nd Article, Section V)
High-risk zone (legal definition, Mexico)	Zona de riesgo grave (definición legal, México)	Human settlement located within an area of high risk caused by a hazard	Asentamiento humano que se encuentra dentro de una zona de grave riesgo, originado por un posible fenómeno perturbador	Ley General Protección Civil, 6 junio 2012, Capitulo l, Artículo 2, Fracción LXI (Mexican Civil Protection General Law, 6 June 2012, Chapter 1, 2nd Article, Section LXI)
Imminent (impending) risk (legal definition, Mexico)	Riesgo Inminente (definición legal, México)	A risk situation that, according to a specialized technical body, requires immediate actions as there are conditions or increased probabilities of adverse effects on an endangered person or valued non-human asset	Aquel riesgo que según la opinión de una instancia técnica especializada, debe considerar la realización de acciones inmediatas en virtud de existir condiciones o altas probabilidades de que se produzcan los efectos adversos sobre un agente afectable	Ley General Protección Civil, 6 junio 2012, Capitulo l, Artículo 2, Fracción L (Mexican Civil Protection General Law, 6 June 2012, Chapter 1, 2nd Article, Section L)
Incident (legal definition, Mexico)	Incidente (definición legal, México)	A minor event that may be considered normal and not necessarily caused by hazards, but that may generate conditions for the occurrence of a mishap, an emergency or a disaster	El suceso que sin constituir una situación anormal ni haber sido provocado por fenómenos perturbadores severos, puede crear condiciones precursoras de Siniestros, Emergencias o Desastres	Reglamento de la Ley General Protección Civil, 6 junio 2012, Artículo 2, Fracción XI (Mexican Civil Protection General Law, 6 June 2012, Chapter 1, 2nd Article, Section XXVIII)
Injured party (legal definition, Mexico)	Damnificado (definición legal, México)	Person affected by a hazard who suffered damage to their person or to their property in such a degree that requires external assistance for his subsistence, remaining is such condition as long as the emergency persists, or the pre-disaster normality is restored	Persona afectada por un agente perturbador, ya sea que haya sufrido daños en su integridad física o un perjuicio en sus bienes de tal manera que requiere asistencia externa para su subsistencia; considerándose con esa condición en tanto no se concluya la emergencia o se restablezca la situación de normalidad previa al desastre	Ley General Protección Civil, 6 junio 2012, Capitulo l, Artículo 2, Fracción XIV (Mexican Civil Protection General Law, 6 June 2012, Chapter 1, 2nd Article, Section XIV)

(continued)

Term (English)	Término (Español)	Meaning	Significado	Further reading
Integrated management of risks (legal definition, Mexico)	Gestión integral de riesgos (definición legal, México)	Set of actions aimed to the identification, analysis, evaluation, control and reduction of risk, considering that it has a multifactorial origin, and that it is a process in continuous construction. Such actions involve all of the three levels of government, and all other sectors of society promoting the design and implementation of public policies, strategies and procedures that, within a sustained development scheme, reduce the disaster-prone structural vulnerabilities, and increase the resilience or strength of society. It involves identification of risks and their sources, and development of prevision, prevention, mitigation, preparedness, relief, recovery and reconstruction schemes	El conjunto de acciones encaminadas a la identificación, análisis, evaluación, control y reducción de los riesgos, considerándolos por su origen multifactorial y en un proceso permanente de construcción, que involucra a los tres niveles de gobierno, así como a los sectores de la sociedad, lo que facilita la realización de acciones dirigidas a la creación e implementación de políticas públicas, estrategias y procedimientos integrados al logro de pautas de desarrollo sostenible, que combatan las causas estructurales de los desastres y fortalezcan las capacidades de resiliencia o resistencia de la sociedad. Involucra las etapas de: identificación de los riesgos y/o su proceso de formación, previsión, prevención, mitigación, preparación, auxilio, recuperación y reconstrucción	Ley General Protección Civil, 6 junio 2012, Capitulo l, Artículo 2, Fracción XXVIII (Mexican Civil Protection General Law, 6 June 2012, Chapter 1, 2nd Article, Section XXVIII)
Internal Civil Protection program (legal definition, Mexico)	Programa interno de Protección Civil (definición legal, México)	Planning and operational tool circumscribed to an agency, or organization public or private and social sector, which is composed by the operating plan for the Internal Civil Protection Unit, the plan for continuity of operations and contingency plan, and aims to mitigate the previously identified risks and define preventive and response to be able to meet the event of an emergency or disaster	Es un instrumento de planeación y operación, circunscrito al ámbito de una dependencia, entidad, institución u organismo del sector público, privado o social; que se compone por el plan operativo para la Unidad Interna de Protección Civil, el plan para la continuidad de operaciones y el plan de contingencias, y tiene como propósito mitigar los riesgos previamente identificados y definir acciones preventivas y de respuesta para estar en condiciones de atender la eventualidad de alguna emergencia o desastre	Ley General Protección Civil, 6 junio 2012, Capitulo l, Artículo 2, Fracción XLI (Mexican Civil Protection General Law, 6 June 2012, Chapter 1, 2nd Article, Section XLI)

(continued)

Term (English)	Término (Español)	Meaning	Significado	Further reading
Mitigation (legal definition, Mexico)	Mitigación (definición legal, México)	Any action aimed to reduce the impact or damage by a hazard on a person or valued non-human asset	Es toda acción orientada a disminuir el impacto o daños ante la presencia de un agente perturbador sobre un agente afectable	Ley General Protección Civil, 6 junio 2012, Capitulo l, Artículo 2, Fracción XXXVI (Mexican Civil Protection General Law, 6 June 2012, Chapter 1, 2nd Article, Section XXXVI)
Monitoring systems (legal definition, Mexico)	Sistemas de monitoreo (definición legal, México)	A set of elements allowing detection, measurement, processing, prediction and study of the behavior of hazards, in order to assess them and their risks	El conjunto de elementos que permiten detectar, medir, procesar, pronosticar y estudiar el comportamiento de los agentes perturbadores, con la finalidad de evaluar Peligros y Riesgos	Reglamento de la Ley General Protección Civil, 6 junio 2012, Artículo 2, Fracción XIV (Bylaws of the General Civil Protection Law, June 6, 2012, 2nd Article, Fraction XIV)
National Risk Atlas (legal definition, Mexico)	Atlas Nacional de Riesgos (definición legal, México)	Integral geographic information system of threats and expected damages, resulting from a spatial and temporal analysis of the interaction between hazards, vulnerability and exposure of affected entities	Sistema integral de información sobre los agentes perturbadores y daños esperados, resultado de un análisis espacial y temporal sobre la interacción entre los peligros, la vulnerabilidad y el grado de exposición de los agentes afectables	Ley General Protección Civil, 6 junio 2012, Capitulo l, Artículo 2, Fracción IV (Mexican Civil Protection General Law, 6 June 2012, Chapter 1, 2nd Article, Section IV)
Prevention (legal definition, Mexico)	Prevención (definición legal, México)	Group of actions implemented in advance of the occurrence of hazards aimed to foresee and reduce the related risks, identifying, reducing or eliminating them, and to avert or reduce their destructive impact on populace, property, infrastructure, as well as prevent the social construction of risks	Conjunto de acciones y mecanismos implementados con antelación a la ocurrencia de los agentes perturbadores, con la finalidad de conocer los peligros o los riesgos, identificarlos, eliminarlos o reducirlos; evitar o mitigar su impacto destructivo sobre las personas, bienes, infraestructura, así como anticiparse a los procesos sociales de construcción de los mismos	Ley General Protección Civil, 6 junio 2012, Capitulo l, Artículo 2, Fracción XXXIX (Mexican Civil Protection General Law, 6 June 2012, Chapter 1, 2nd Article, Section XXXIX)
Prevision (legal definition, Mexico)	Previsión (definición legal, México)	To acquire awareness of the risks, and of what is needed to face them through defined steps of risk identification, prevention, mitigation, preparedness, emergency response, recovery and reconstruction	Tomar conciencia de los riesgos que pueden causarse y las necesidades para enfrentarlos a través de las etapas de identificación de riesgos, prevención, mitigación, preparación, atención de emergencias, recuperación y reconstrucción	Ley General Protección Civil, 6 junio 2012, Capitulo l, Artículo 2, Fracción XL (Mexican Civil Protection General Law, 6 June 2012, Chapter 1, 2nd Article, Section XL)

(continued)

Term (English)	Término (Español)	Meaning	Significado	Further reading
Reconstruction (legal definition, Mexico)	Reconstrucción (definición legal, México)	Transient actions aimed to recover the social and economic environment that prevailed among people before suffering the effects of a hazard in a given area or jurisdiction. This process should be aimed as much as possible to reduce existing risks, ensuring that no new risks would be created, so improving the pre-existing conditions	La acción transitoria orientada a alcanzar el entorno de normalidad social y económica que prevalecía entre la población antes de sufrir los efectos producidos por un agente perturbador en un determinado espacio o jurisdicción. Este proceso debe buscar en la medida de lo posible la reducción de los riesgos existentes, asegurando la no generación de nuevos riesgos y mejorando para ello las condiciones preexistentes	Ley General Protección Civil, 6 junio 2012, Capitulo l, Artículo 2, Fracción XLIV (Mexican Civil Protection General Law, 6 June 2012, Chapter 1, 2nd Article, Section XLIV)
Recovery (legal definition, Mexico)	Recuperación (definición legal, México)	Process starting during an emergency, consisting of actions aimed to return to the normality of the affected community	Proceso que inicia durante la emergencia, consistente en acciones encaminadas al retorno a la normalidad de la comunidad afectada	Ley General Protección Civil, 6 junio 2012, Capitulo l, Artículo 2, Fracción XLV (Mexican Civil Protection General Law, 6 June 2012, Chapter 1, 2nd Article, Section XLV)
Regulation (legal definition, Mexico, as opposite to disruption)	Agente regulador (definición legal, México)	Set of actions, tools, standards, works and more generally everything intended to protect people, property, strategic and productive infrastructure and the environment, and to reduce risk and control and prevent adverse effects from a hazard	Lo constituyen las acciones, instrumentos, normas, obras y en general todo aquello destinado a proteger a las personas, bienes, infraestructura estratégica, planta productiva y el medio ambiente, a reducir los riesgos y a controlar y prevenir los efectos adversos de un agente perturbador	Ley General Protección Civil, 6 junio 2012, Capitulo l, Artículo 2, Fracción l. (Mexican Civil Protection General Law, 6 June 2012, Chapter 1, 2nd Article, Section I)
Resilience (legal definition, Mexico)	Resiliencia (definición legal, México)	It is the ability of a hazard-exposed system, community or society to resist, assimilate, withstand and quickly and efficiently recover from damaging effects, preserving and recovering its basic and functional structures, and achieving a better protection and improved risk reduction for the future	Es la capacidad de un sistema, comunidad o sociedad potencialmente expuesta a un peligro para resistir, asimilar, adaptarse y recuperarse de sus efectos en un corto plazo y de manera eficiente, a través de la preservación y restauración de sus estructuras básicas y funcionales, logrando una mejor protección futura y mejorando las medidas de reducción de riesgos	Ley General Protección Civil, 6 junio 2012, Capitulo l, Artículo 2, Fracción XLVIII (Mexican Civil Protection General Law, 6 June 2012, Chapter 1, 2nd Article, Section XLVIII)

(continued)

Term (English)	Término (Español)	Meaning	Significado	Further reading
Risk (legal definition, Mexico)	Riesgo (definición legal, México)	Damage or probable loss of an asset, resulting from the interaction between its vulnerability and the presence of a hazard	Daños o pérdidas probables sobre un agente afectable, resultado de la interacción entre su vulnerabilidad y la presencia de un agente perturbador	Ley General Protección Civil, 6 junio 2012, Capitulo l, Artículo 2, Fracción XLIV. (Mexican Civil Protection General Law, 6 June 2012, Chapter 1, 2nd Article, Section XLIV)
Risk identification (legal definition, Mexico)	Identificación de riesgos (definición legal, México)	Recognizing and assessing potential damage or probable loss from hazards in terms of their geographical distribution through the analysis of hazard and vulnerability	Reconocer y valorar las pérdidas o daños probables sobre los agentes afectables y su distribución geográfica, a través del análisis de los peligros y la vulnerabilidad	Ley General Protección Civil, 6 junio 2012, Capitulo l, Artículo 2, Fracción XXXI (Mexican Civil Protection General Law, 6 June 2012, Chapter 1, 2nd Article, Section XXXI)
Risk reduction (legal definition, Mexico)	Reducción de riesgos (definición legal, México)	Preventive intervention of individuals, institutions and communities aimed to remove or reduce the adverse impact of disasters through preparedness and mitigation actions. It involves the identification of risks, and the analyses of vulnerabilities, resilience and response capabilities. Also involves developing a culture of civil protection, the public engagement and the development of an institutional framework, as well as the implementation of environmental protection measures, use of land and urban planning, protection of critical infrastructure, building partnerships for the implementation of financial and risk transfer instruments, and the development of early warning systems	Intervención preventiva de individuos, instituciones y comunidades que nos permite eliminar o reducir, mediante acciones de preparación y mitigación, el impacto adverso de los desastres. Contempla la identificación de riesgos y el análisis de vulnerabilidades, resiliencia y capacidades de respuesta, el desarrollo de una cultura de la protección civil, el compromiso público y el desarrollo de un marco institucional, la implementación de medidas de protección del medio ambiente, uso del suelo y planeación urbana, protección de la infraestructura crítica, generación de alianzas y desarrollo de instrumentos financieros y transferencia de riesgos, y el desarrollo de sistemas de alertamiento	Ley General Protección Civil, 6 junio 2012, Capitulo l, Artículo 2, Fracción XLVI (Mexican Civil Protection General Law, 6 June 2012, Chapter 1, 2nd Article, Section XLVI)
Risk zone (legal definition, Mexico)	Zona de riesgo (definición legal, México)	Defined territorial space upon which there is a likelihood of harm caused by a hazard	Espacio territorial determinado en el que existe la probabilidad de que se produzca un daño, originado por un fenómeno perturbador	Ley General Protección Civil, 6 junio 2012, Capitulo l, Artículo 2, Fracción LX (Mexican Civil Protection General Law, 6 June 2012, Chapter 1, 2nd Article, Section LX)

(continued)

Term (English)	Término (Español)	Meaning	Significado	Further reading
Shelter (legal definition, Mexico)	Albergue (definición legal, México)	Facility settled to provide shelter to people whose habitations have been affected by a hazard. Such people may dwell in the shelter until the recovery or reconstruction of their habitations	Instalación que se establece para brindar resguardo a las personas que se han visto afectadas en sus viviendas por los efectos de fenómenos perturbadores y en donde permanecen hasta que se da la recuperación o reconstrucción de sus viviendas	Ley General Protección Civil, 6 junio 2012, Capítulo 1 Artículo 2, Fracción III (Mexican Civil Protection General Law, 6 June 2012, Chapter 1, 2nd Article, Section III)
Sheltered (legal definition, Mexico)	Albergado (definición legal, México)	Person who is temporarily granted help, accommodation and protection against a threat, imminence or occurrence of a hazard	Persona que en forma temporal recibe asilo, amparo, alojamiento y resguardo ante la amenaza, inminencia u ocurrencia de un agente perturbador	Ley General Protección Civil, 6 junio 2012, Capítulo 1 Artículo 2, Fracción II (Mexican Civil Protection General Law, 6 June 2012, Chapter 1, 2nd Article, Section II)
Temporary shelter (legal definition, Mexico)	Refugio temporal (definición legal, México)	Physical installation enabled to temporarily provide protection and welfare to people who do not have immediate accessibility to a safe dwelling in case of impending hazard, an emergency, a major accident or a disaster	La instalación física habilitada para brindar temporalmente protección y bienestar a las personas que no tienen posibilidades inmediatas de acceso a una habitación segura en caso de un riesgo inminente, una emergencia, siniestro o desastre	Ley General Protección Civil, 6 junio 2012, Capitulo l, Artículo 2, Fracción XLVII (Mexican Civil Protection General Law, 6 June 2012, Chapter 1, 2nd Article, Section XLVII)
Vulnerability (legal definition, Mexico)	Vulnerabilidad (definición legal, México)	Susceptibility or propensity of an asset to suffer damage or loss in the presence of a disrupting phenomenon determined by physical, social, economic and environmental factor	Susceptibilidad o propensión de un agente afectable a sufrir daños o pérdidas ante la presencia de un agente perturbador, determinado por factores físicos, sociales, económicos y ambientales	Ley General Protección Civil, 6 junio 2012, Capitulo l, Artículo 2, Fracción LVIII. (Mexican Civil Protection General Law, 6 June 2012, Chapter 1, 2nd Article, Section LVIII)

Bibliography

Bretton RJ, Gottsmann J, Christie R (2017) The role of laws within the governance of volcanic risks advances in volcanology. Springer, Berlin, Heidelberg

Connor C, Bebbington M, Marzocchi W (2015) Chapter 51—probabilistic volcanic hazard assessment A2. In: Sigurdsson H (ed) The encyclopedia of volcanoes, 2nd edn. Academic Press, Amsterdam, pp 897–910

De la Cruz-Reyna S, Tilling RI (2008) Scientific and public responses to the ongoing volcanic crisis at Popocatépetl Volcano, Mexico: importance of an effective hazards-warning system. J Volcanol Geoth Res 170:121–134

Fearnley C, Winson AEG, Pallister J, Tilling R (2017) Volcano crisis communication: challenges and solutions in the 21st Century: Berlin. Springer, Berlin Heidelberg, Heidelberg, pp 1–19

Fearnley CJ (2013) Assigning a volcano alert level: negotiating uncertainty, risk, and complexity in decision-making processes. Environ Plan A 45:1891–1911.

Fiske RS (1984) Volcanologists, journalists, and the concerned local public: a tale of two crises in the eastern Caribbean, 110–121 p

Hall ML (1990) Chronology of the principal scientific and governmental actions leading up to the November 13, 1985 eruption of Nevado del Ruiz, Colombia. J Volcanol Geoth Res 42:101–115

Jolly G, De la Cruz-Reyna S (2015) Chapter 68—volcanic crisis management A2. In: Sigurdsson H (ed) The encyclopedia of volcanoes, 2nd edn. Academic Press, Amsterdam, pp 1187–1202

McNutt SR, Roman DC (2015) Chapter 59—volcanic seismicity A2. In: Sigurdsson H (ed) The encyclopedia of volcanoes, 2nd edn. Academic Press, Amsterdam, pp 1011–1034

National Academies of Sciences, Engineering Medicine (2017) Volcanic eruptions and their repose, unrest, precursors, and timing. The National Academies Press, Washington, DC.

Peterson DW (1988) Volcanic hazards and public response. J Geophy Res: Solid Earth 93(B5):4161–4170

Phillipson G, Sobradelo R, Gottsmann J (2013) Global volcanic unrest in the 21st century: an analysis of the first decade. J Volcanol Geoth Res 264:183–196

Tilling RI (1989) Volcanic hazards and their mitigation—progress and problems. Rev Geophy 27(2):237–269

Tilling RI (2009) El Chichón's "surprise" eruption in 1982: lessons for reducing volcano risk. Geofísica Internacional 48:3–19

Advs in Volcanology (2019) 313–313
DOI 10.1007/978-3-319-58412-6

www.ingramcontent.com/pod-product-compliance
Lightning Source LLC
LaVergne TN
LVHW082117200926
848484LV00008B/1970
9781013273810